Logical Analysis of Hybrid Systems

André Platzer

Logical Analysis of Hybrid Systems

Proving Theorems for Complex Dynamics

Springer

Dr. André Platzer
Carnegie Mellon University
School of Computer Science
5000 Forbes Ave.
Pittsburgh PA 15213
USA
aplatzer@cs.cmu.edu

ISBN 978-3-642-44474-6 ISBN 978-3-642-14509-4 (eBook)
DOI 10.1007/978-3-642-14509-4
Springer Heidelberg Dordrecht London New York

ACM Computing Classification (1998): F.4.1, F.3, D.2.4, I.2.3, G.1.7, I.2.8

Softcover re-print of the Hardcover 1st edition 2010

Cover design: KuenkelLopka GmbH, Heidelberg

Printed on acid-free paper

Springer is part of Springer Science+Business Media (www.springer.com)

Foreword

Hybrid Systems are notoriously hard to analyze and verify. So far, techniques based on either explicit or implicit state reachability have failed to scale with the size of such systems. Statistical Model Checking may prove useful, but sacrifices absolute certainty about the correctness of the answer obtained. In both cases, numerical error may change the truth-value of the result from False to True or vice versa. An alternative is to use a combination of decision procedures for real arithmetic and interactive theorem proving. Andre Platzer's Ph.D. thesis explores this alternative approach in great depth. He proposes a logic called Differential Dynamic Logic for specifying properties of Hybrid Systems, investigates the meta-theory of the logic, and gives inference rules for it. He has developed an extremely impressive graphical interface for the resulting tool KeYmaera, which is based on the KeY Prover for verifying Java programs, developed at the University of Karlsruhe, Chalmers, and Koblenz. Particularly noteworthy are the use of Differential Invariants for reasoning about complex Hybrid Systems and the examples that he is able to do: "The European Train Control System" and a curved flight roundabout maneuver for aircraft collision avoidance. Both examples are beyond the scope of current Hybrid System Model Checking tools. I believe that his verification tool is unique – there is no other one like it. I heartily recommend his book and theorem prover for those who need to verify complex cyber-physical systems.

Pittsburgh, February 2010 *Edmund M. Clarke*

Preface

The design of complex systems is essential to much of engineering and science. Equally essential is the effort to fully understand these systems and to develop tools and techniques that can steer us away from unsafe or incorrect designs. In civil engineering, for example, well-understood principles like statics can be used to analyse buildings before they are built, and refined architectural models can be used to predict whether a building will be safe or whether it might collapse during an earthquake. Similarly, in auto body design, wind tunnels and corresponding computer models based on computational fluid dynamics help engineers to gain an understanding of aerodynamic forces and wind resistance for energy efficiency before constructing the actual car and its chassis. Models and their analysis also play an important role in chip design and are used extensively in the semiconductor industry to prevent expensive bugs in hardware. Modelling and model analysis is thus an integral part of science and engineering and is used very effectively in many areas to ensure high-quality system designs, saving replacement cost and preventing dangerous side effects of malfunctioning designs.

Hybrid systems is an emergent area of growing importance, emphasising a systematic understanding of systems that combine discrete (e.g., digital) and continuous (e.g., analog or physical) effects. In fact, it is foreseeable that hybrid systems and the closely related notion of cyber-physical systems will soon play a ubiquitous role in engineering. Combinations of computation and control can lead to very powerful system designs, and computational aspects are being integrated into classical physical, mechanical, and chemical process controls on a routine basis today. The number of systems where both computational and physical aspects are important for really understanding them grows exponentially with modern technological advances. Hybrid systems occur frequently in automotive industries, aviation, railway applications, factory automation, process control, medical devices, mobile robotics, and mixed analog–digital chip design.

Despite the growing relevance in complex system designs, hybrid systems is an area where analytic approaches are still in their infancy. Hybrid systems occur ubiquitously and their analysis faces inherent complexity challenges. Hence, there is probably no other area where the gap is more noticeable between the tremendous complexity of the systems we can build and the modest size of systems that we can analyse. Mankind can build systems that are significantly more complicated than people can understand analytically. This book presents an approach with logical analysis techniques that are intended to help overcome these difficulties and bridge the gap between design demand and analysis power.

In light of this growing interest in the field, the purpose of this book is to provide an introduction to hybrid systems analysis and, in particular, to present a coherent logical analysis approach for hybrid systems. One of the highly successful techniques used for analysing finite-state models in chip designs today is *model checking*, which was pioneered in 1981 by the 2007 ACM Turing Award Laureates Edmund M. Clarke, Allen Emerson, and Joseph Sifakis. Nowadays, model checking is used routinely in the semiconductor industry. Model checking is one of the inspirations for this work. Another area that is strongly related is interactive and *automated theorem proving*, which is also used in advanced industrial settings. Model checking and automated theorem proving complement each other to tackle various aspects of formal system verification. While both areas are ultimately rooted in logic, the basic operating principles are somewhat different. Model checking is based on systematically exploring the state space of a system in a clever way. Model checking searches for counterexamples, i.e., traces of a system that lead to a bug and that serve as a falsification of a correctness property. Impressive results have been demonstrated for finite-state systems where model checking is decidable. In theorem proving, in contrast, the notion of a proof is fundamental and represents a verification of a correctness property. In particular, a proof is a reason and explanation for *why* a system works. Automated theorem proving techniques that construct proofs automatically are another deep source of inspiration for the work presented here. In fact, several of the proof procedures presented in this book are inspired by theorem proving principles that have been used successfully for conventional object-oriented programs.

One important new aspect in hybrid systems is the cardinality and structure of the state space. In (sufficiently small) finite state spaces, for instance, exhaustive state exploration is still feasible, but becomes inherently impossible for the uncountable continuous state spaces of hybrid systems, especially with respect to their complicated interacting discrete and continuous dynamics. Most notably, the continuous dynamics of hybrid systems that is commonly described by differential equations poses significant new challenges compared to classical settings. Thus, verification techniques for differential equations are one very important part of hybrid systems analysis.

Outline

This book is intended as an introduction to hybrid systems and advanced analysis techniques for their dynamics. It covers basic and advanced notions of hybrid systems, specification languages for hybrid systems, verification approaches for hybrid systems, and application scenarios for hybrid systems verification. Starting from a basic background in mathematics and computer science, this book develops all notions required for understanding and analysing hybrid systems. It also provides background material about logic and differential equations in the appendix.

This book presents a coherent logical foundation for hybrid systems analysis that will help the reader understand how behavioural properties of hybrid systems can be analysed successfully. The foundation developed here serves as a basis for advanced hybrid system analysis techniques. The hybrid systems analysis approach has also been implemented in the verification tool KeYmaera for hybrid systems, which is available for download at the book's Web page.

Part I describes specification and verification logics for hybrid systems that are the basis of hybrid systems analysis. It also covers constructive proof calculi that can be used to analyse and verify hybrid systems, including approaches for handling real arithmetic and differential equations. The chapters in Part I show a series of logical systems that are each presented in terms of their syntax, semantics, axiomatics, proof theory, and pragmatics. The syntax defines what can be said about hybrid system behaviour. The semantics gives meaning to the symbolic formulas and shows what we are ultimately interested in: truth, or, more precisely, what is true about the behaviour of the particular dynamics of a hybrid system. In the axiomatic parts, this book develops formal proof techniques that can be used by a human or machine to establish truth by proof. After all, truth that we do not know about is less helpful than truth that we can justify by giving a proof. The development of proof theory connects the semantic notion of truth in the real world with the syntactic device of formal proofs and shows that, in a sense of relative completeness, we can prove all true facts about hybrid systems from elementary properties of differential equations. Finally, this book shows the pragmatics of using verification procedures for analysing hybrid system scenarios. This includes both practical, algorithmic considerations of developing system analysis tools and various examples, application scenarios, and case studies that can be proven with the logics developed in Part I.

Part II focuses on the practical and algorithmic questions of how to turn the theoretical foundation from Part I into automated theorem proving procedures. This part also shows techniques for generating invariants and differential invariants of hybrid systems that are crucial for proving correctness, and shows how to overcome complexity challenges in real arithmetic verification. Part III shows how safety-critical properties of more advanced applications of hybrid systems in railway and aircraft control can be proven with the approach presented in Parts I and II. This part includes a study of collision avoidance in the European Train Control System (ETCS) and roundabout collision avoidance manoeuvres in air traffic control. Numerous examples, illustrations, and proofs throughout the text will also help the reader develop an intuition about hybrid systems behaviour and master the intricacies of the more subtle aspects in hybrid systems analysis.

How to Read This Book

The basic suggested reading sequence is linear (with additional consultation of the appendices for background information as needed). Except for the foundation of this work that is laid out in Part I, however, the chapters are mostly kept self-contained so that they can also be studied independently. The following figure shows the reading order dependencies among the chapters (solid lines) and the partial dependencies of suggested reading sequences that hold for the advanced material of the respective chapters (dashed lines).

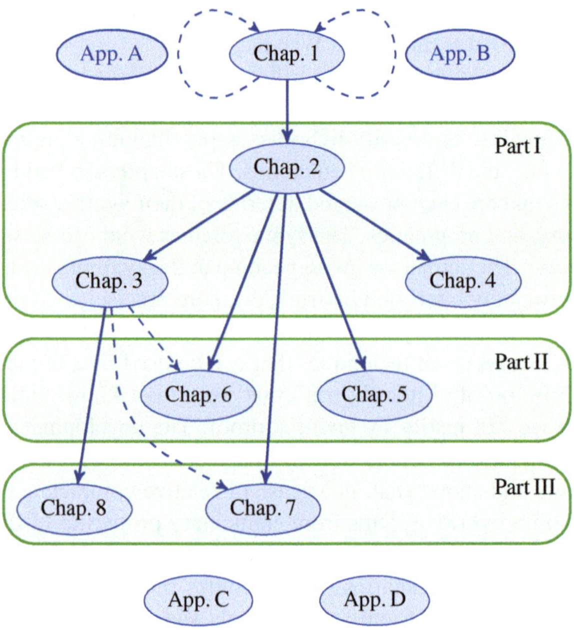

For background on classical first-order logic, we recommend you review App. A as needed. Depending on your interest, field of study, and preference, we recommend you either study the background information in App. A on first-order logic before reading Part I or use the material in App. A as a background reference book on demand while reading the main part of this book. Similarly, we recommend you review the background on ordinary differential equations in App. B either before or during the study of the main part. An intuitive approach to understanding differential equations and formal definitions of their semantics will be given throughout the text. Logic itself is also explained and illustrated intuitively during the main chapters, but some readers may also find it helpful to refresh, update, or learn about the basics of first-order logic from App. A before proceeding to the main part.

While there is a lot of flexibility in the reading sequence of the chapters, we strongly recommend you study the logical foundations of hybrid systems analysis in Chap. 2 of Part I before reading any other chapter of Parts I–III. Some more advanced sections in the applications in Part III also depend on the theory of differential invariants that is developed together with other extensions in Chap. 3.

Appendix C shows a formal relation of hybrid automata with hybrid programs. Appendix D gives more detail on the implementation of the approach put forth in this book in the verification tool KeYmaera. It also presents a survey of computational techniques for handling real arithmetic. Both App. C and D can be read as needed, after studying the introductory material and notions in Chap. 2. The most important formation rules for the logic and proof rules for the calculi are summarised at the end of the book.

Online Material for This Book

The Web page for this book provides online material, including the verification tool KeYmaera that implements our logical analysis approach for hybrid systems. We also provide slide material for parts of this book, an online tutorial for KeYmaera, and several KeYmaera problem files for examples from this book, including train and air traffic control studies. The book Web page is at the following URL:

`http://symbolaris.com/lahs/`

Acknowledgements

This book is based on my Ph.D. thesis and would not have been possible without the support of the PIs and collaborators on the projects that I have been working on. My sincere thanks go to Prof. Ernst-Rüdiger Olderog for his excellent advice and support, and for giving me the opportunity to work in one of the most fascinating areas of science in a group with a friendly and productive atmosphere. My advisor, Prof. Olderog, and the Director of AVACS, Prof. Werner Damm, both deserve my highest gratitude, not only for their continuous support and for their faith, but also for allowing me the freedom to pursue my own research ambitions in the stimulating context of the AVACS project ("Automatic Verification and Analysis of Complex Systems"). Ultimately, this made it possible for me to develop the logic and verification approach presented in this book.

I want to thank the external referees of my Ph.D. thesis, Prof. Tobias Nipkow from the Technical University of Munich and Prof. George J. Pappas from the University of Pennsylvania. It is an honour for me that they were willing to invest their valuable time and effort in the careful reviewing of my thesis. In fact, I am thankful to all members of my Ph.D. committee, Werner Damm, Ernst-Rüdiger Olderog, George J. Pappas, Tobias Nipkow, and Hardi Hungar for fruitful discussions and for the highest support they offered for my work.

I am especially grateful to Prof. Edmund M. Clarke, who invited me to Carnegie Mellon University several times, for his support, interest, and collaboration, and for sharing with me parts of his huge knowledge in all areas of formal methods. I further want to acknowledge the help by Prof. Peter H. Schmitt from the University of

Karlsruhe (TH), Profs. Bernhard Beckert and Ulrich Furbach from the University of Koblenz-Landau, Prof. Reiner Hähnle from the Chalmers University of Technology, Gothenburg, Sweden, Profs. Edmund M. Clarke and Frank Pfenning from Carnegie Mellon University, and Prof. Rajeev Goré from the Australian National University, Canberra, at various stages of my career.

I want to thank the program committee of the TABLEAUX 2007 conference for selecting my first paper on differential dynamic logic for the Best Paper Award, the first award at any TABLEAUX conference. This recognition has encouraged me to continue pursuing my research direction, which ultimately led to the results described in this book. I also thank the program committee of the FM 2009 conference for selecting my paper on formal verification of curved flight collision avoidance maneuvers for the Best Paper Award. I am very grateful to the ACM Doctoral Dissertation Award committee for honoring my Ph.D. thesis with the 2009 ACM Doctoral Dissertation Honorable Mention Award.

I am truly thankful to my colleagues at Carnegie Mellon University for their encouraging feedback about my work and for the friendly and constructive atmosphere at CMU. For many fruitful discussions I thank my colleagues and friends from Oldenburg, Ingo Brückner, Henning Dierks, Johannes Faber, Sibylle Fröschle, Jochen Hoenicke, Stephanie Kemper, Roland Meyer, Michael Möller, Jan-David Quesel, Tim Strazny, and especially my office mate Andreas Schäfer. Ernst-Rüdiger Olderog, Johannes Faber, Ingo Brückner, Roland Meyer, Henning Dierks, Silke Wagner, Nicole Betz, Alex Donzé, and especially Andreas Schäfer also deserve credit for proofreading some of my earlier papers, which formed the basis for this book. I also acknowledge Andreas Schäfer's helpful feedback from proofreading parts of this book. I appreciate the feedback of my students on this book.

Furthermore, I thank Jan-David Quesel for writing a Master's thesis under my supervision and for his invaluable support with the implementation of the verification tool KeYmaera based on the techniques that I present in this book and in prior publications. I also thank him for help with the experiments and ETCS. I am also thankful for indispensable and reliable help from Richard Bubel and Philipp Rümmer with the implementation internals of the KeY basis. I thank the whole KeY team for providing the impressive Java verification tool KeY as a basis for our implementation of KeYmaera.

For help with the book process, I thank Ronan Nugent from Springer.

Especially, I thank my parents, Rudolf and Brigitte Platzer, and my sister, Julia, for their continuous support and encouragement, and I thank my wife, Nicole, for her true faith in me. She also deserves credit for her invaluable help with some of the illustrations in this book.

Funding

This research was partly supported by the German Research Council (DFG) under grant SFB/TR 14 AVACS ("Automatic Verification and Analysis of Complex Systems", see `http://www.avacs.org`); a Transregional Collaborative Research

Center of the Max Planck Institute and the Universities of Oldenburg, Saarbrücken, and Freiburg in Germany, with associated cooperations with the University of Pennsylvania, ETH Zürich, and the Academy of Sciences of the Czech Republic. It was further supported partly by a research fellowship of the German Academic Exchange Service (DAAD) and by a research award of the Floyd und Lili Biava Stiftung. Some part of this work was also supported by the National Science Foundation under grant nos. CNS-0931985 and CNS-0926181, including the NSF Expedition on Computational Modeling and Analysis of Complex Systems (CMACS); see `http://cmacs.cs.cmu.edu` for more information.

The views and conclusions contained in this book are those of the author and should not be interpreted as representing the official policies, either expressed or implied, of any sponsoring institution or government.

Further Sources

This book is based on several sources, most notably the author's Ph.D. thesis [236]. Chapter 2 is an extended version of an article in the Journal of Automated Reasoning [235] and also covers some material from previous work at TABLEAUX [231] and HSCC [232]. Chapter 3 is an extended version of an article in the Journal of Logic and Computation [237], to which we now add a relative completeness argument and prove that DAL is a conservative extension of the sublogic $\mathsf{d}\mathcal{L}$. We further combine the solution-based techniques from Chap. 2 with differential induction-based techniques from Chap. 3 by introducing the new extension of differential monotonicity relaxations. Chapter 4 is a substantially extended version of a previous paper at LFCS [233], to which we now add a complete and more elegant calculus and provide a modular relative completeness proof.

In Chap. 5, we extend a previous paper at VERIFY [230] with more details on iterative background closure strategies, including experimental evaluation, and complement this proof technique with a new iterative inflation strategy. Chapter 6 is based on joint work with Edmund M. Clarke at CAV [239] and in Formal Methods in System Design [240].

Chapter 7 is a substantially revised and improved version of joint work with Jan-David Quesel at HSCC [243] with extensions from follow-up work [244]. Chapter 8 is a significantly improved and detailed case study developed on the basis of joint work with Edmund M. Clarke at HSCC [238] and CAV [239] with subsequent extensions at FM [241].

Appendix B summarises classical results from the theory of differential equations from the literature [297]. Finally, App. D uses a few excerpts from joint work with Jan-David Quesel at IJCAR [242], adding an overall discussion of the KeYmaera verification tool that implements the approach presented in this book. Appendix D also adds a thorough description of computational back-ends for real arithmetic, with extensions from joint work with Philipp Rümmer and Jan-David Quesel [246].

Pittsburgh, February 2010 *André Platzer*

Contents

1 Introduction **1**
1.1 Technical Context 4
1.1.1 Hybrid Systems 4
1.1.2 Model Checking 12
1.1.3 Deductive Verification 14
1.1.4 Compositional Verification 16
1.1.5 Lifting Quantifier Elimination 19
1.1.6 Differential Induction and Differential Strengthening 20
1.2 Related Work 21
1.3 Contributions 25
1.4 Structure of This Book 25

Part I Logics and Proof Calculi for Hybrid Systems 31

2 Differential Dynamic Logic $\mathsf{d}\mathcal{L}$ **33**
2.1 Introduction 34
2.1.1 Structure of This Chapter 35
2.2 Syntax 35
2.2.1 Terms 37
2.2.2 Hybrid Programs 41
2.2.3 Formulas 47
2.3 Semantics 49
2.3.1 Valuation of Terms 50
2.3.2 Valuation of Formulas 51
2.3.3 Transition Semantics of Hybrid Programs 54
2.4 Collision Avoidance in Train Control 61
2.5 Proof Calculus 64
2.5.1 Substitution 65
2.5.2 Proof Rules 76

2.5.3 Deduction Modulo with Invertible Quantifiers and Real Quantifier Elimination . . . 88
2.5.3.1 Lifting Quantifier Elimination by Invertible Quantifier Rules . . . 88
2.5.3.2 Admissibility in Invertible Quantifier Rules . . . 91
2.5.3.3 Quantifier Elimination and Modalities . . . 93
2.5.3.4 Global Invertible Quantifier Rules . . . 93
2.5.4 Verification Example . . . 94
2.6 Soundness . . . 97
2.7 Completeness . . . 101
2.7.1 Incompleteness . . . 102
2.7.2 Relative Completeness . . . 103
2.7.3 Characterising Real Gödel Encodings . . . 105
2.7.4 Expressibility and Rendition of Hybrid Program Semantics . 106
2.7.5 Relative Completeness of First-Order Assertions . . . 109
2.7.6 Relative Completeness of the Differential Logic Calculus . . 113
2.8 Relatively Semidecidable Fragments . . . 114
2.9 Train Control Verification . . . 118
2.9.1 Finding Inductive Candidates . . . 118
2.9.2 Inductive Verification . . . 119
2.9.3 Parameter Constraint Discovery . . . 120
2.10 Summary . . . 122

3 Differential-Algebraic Dynamic Logic DAL 123
3.1 Introduction . . . 124
3.1.1 Related Work . . . 128
3.1.2 Structure of This Chapter . . . 130
3.2 Syntax . . . 130
3.2.1 Terms . . . 132
3.2.2 Differential-Algebraic Programs . . . 132
3.2.3 Formulas . . . 139
3.3 Semantics . . . 141
3.3.1 Transition Semantics of Differential-Algebraic Programs . . 141
3.3.2 Valuation of Formulas . . . 145
3.3.3 Time Anomalies . . . 145
3.3.4 Conservative Extension . . . 147
3.4 Collision Avoidance in Air Traffic Control . . . 148
3.4.1 Flight Dynamics . . . 148
3.4.2 Differential Axiomatisation . . . 149
3.4.3 Aircraft Collision Avoidance Manoeuvres . . . 150
3.4.4 Tangential Roundabout Manoeuvre . . . 151
3.5 Proof Calculus . . . 152
3.5.1 Motivation . . . 153
3.5.2 Derivations and Differentiation . . . 154
3.5.3 Differential Reduction and Differential Elimination . . . 160

3.5.4 Proof Rules 162
3.5.5 Deduction Modulo by Side Deduction 168
3.5.6 Differential Induction with Differential Invariants 170
3.5.7 Differential Induction with Differential Variants 181
3.6 Soundness 185
3.7 Restricting Differential Invariants 188
3.8 Differential Monotonicity Relaxations 189
3.9 Relative Completeness 193
3.10 Deductive Strength of Differential Induction 194
3.11 Air Traffic Control Verification 197
3.11.1 Characterisation of Safe Roundabout Dynamics 197
3.11.2 Tangential Entry Procedures 200
3.11.3 Discussion 201
3.12 Summary 201

4 Differential Temporal Dynamic Logic dTL 203
4.1 Introduction 204
4.1.1 Related Work 205
4.1.2 Structure of This Chapter 206
4.2 Syntax 206
4.2.1 Hybrid Programs 207
4.2.2 State and Trace Formulas 207
4.3 Semantics 210
4.3.1 Trace Semantics of Hybrid Programs 210
4.3.2 Valuation of State and Trace Formulas 213
4.3.3 Conservative Temporal Extension 215
4.4 Safety Invariants in Train Control 216
4.5 Proof Calculus 217
4.5.1 Proof Rules 218
4.5.2 Verification Example 221
4.6 Soundness 221
4.7 Completeness 223
4.7.1 Incompleteness 223
4.7.2 Relative Completeness 224
4.7.3 Expressibility and Rendition of Hybrid Trace Semantics . . 225
4.7.4 Modular Relative Completeness Proof 226
4.8 Verification of Train Control Safety Invariants 227
4.9 Liveness by Quantifier Alternation 228
4.10 Summary 230

Part II Automated Theorem Proving for Hybrid Systems 231

5 Deduction Modulo Real Algebra and Computer Algebra 233
5.1 Introduction 234

5.1.1 Related Work 234
5.1.2 Structure of This Chapter 235
5.2 Tableau Procedures Modulo 235
5.3 Nondeterminisms in Tableau Modulo 238
5.3.1 Nondeterminisms in Branch Selection 238
5.3.2 Nondeterminisms in Formula Selection 239
5.3.3 Nondeterminisms in Mode Selection 240
5.4 Iterative Background Closure 243
5.5 Iterative Inflation 246
5.6 Experimental Results 248
5.7 Summary 251

6 Computing Differential Invariants as Fixed Points 253
6.1 Introduction 254
6.1.1 Related Work 255
6.1.2 Structure of This Chapter 256
6.2 Inductive Verification by Combining Local Fixed Points 256
6.2.1 Verification by Symbolic Decomposition 257
6.2.2 Discrete and Differential Induction, Differential Invariants . 258
6.2.3 Flight Dynamics in Air Traffic Control 260
6.2.4 Local Fixed-Point Computation for Differential Invariants . 262
6.2.5 Dependency-Directed Induction Candidates 263
6.2.6 Global Fixed-Point Computation for Loop Invariants 265
6.2.7 Interplay of Local and Global Fixed-Point Loops 268
6.3 Soundness 269
6.4 Optimisations 271
6.4.1 Sound Interleaving with Numerical Simulation 271
6.4.2 Optimisations for the Verification Algorithm 272
6.5 Experimental Results 272
6.6 Summary 273

Part III Case Studies and Applications in Hybrid Systems Verification 275

7 European Train Control System 277
7.1 Introduction 278
7.1.1 Related Work 280
7.1.2 Structure of This Chapter 281
7.2 Parametric European Train Control System 281
7.2.1 Overview of the ETCS Cooperation Protocol 281
7.2.2 Formal Model of Fully Parametric ETCS 284
7.3 Parametric Verification of Train Control 286
7.3.1 Controllability Discovery 287
7.3.2 Iterative Control Refinement 288

7.3.3 Safety Verification . . . 291
7.3.4 Liveness Verification . . . 293
7.3.5 Full Correctness of ETCS . . . 294
7.4 Disturbance and the European Train Control System . . . 295
7.4.1 Controllability Discovery . . . 296
7.4.2 Iterative Control Refinement . . . 298
7.4.3 Safety Verification . . . 298
7.5 Experimental Results . . . 299
7.6 Summary . . . 301

8 Air Traffic Collision Avoidance 303
8.1 Introduction . . . 304
8.1.1 Related Work . . . 307
8.1.2 Structure of This Chapter . . . 308
8.2 Curved Flight in Roundabout Manoeuvres . . . 309
8.2.1 Flight Dynamics . . . 309
8.2.2 Roundabout Manoeuvre Overview . . . 310
8.2.3 Compositional Verification Plan . . . 311
8.2.4 Tangential Roundabout Manoeuvre Cycles . . . 312
8.2.5 Bounded Control Choices . . . 315
8.2.6 Flyable Entry Procedures . . . 315
8.2.7 Bounded Entry Duration . . . 318
8.2.8 Safe Entry Separation . . . 319
8.3 Synchronisation of Roundabout Manoeuvres . . . 322
8.3.1 Successful Negotiation . . . 322
8.3.2 Safe Exit Separation . . . 326
8.4 Compositional Verification . . . 328
8.5 Flyable Tangential Roundabout Manoeuvre . . . 329
8.6 Experimental Results . . . 331
8.7 Summary . . . 333

9 Conclusion 335

Part IV Appendix 339

A First-Order Logic and Theorem Proving 341
A.1 Overview . . . 341
A.2 Syntax . . . 346
A.2.1 Terms . . . 346
A.2.2 Formulas . . . 347
A.3 Semantics . . . 348
A.3.1 Valuation of Terms . . . 349
A.3.2 Valuation of Formulas . . . 349
A.4 Proof Calculus . . . 350
A.4.1 Proof Rules . . . 351

A.4.2 Proof Example: Ground Proving Versus Free-Variable Proving 354
A.5 Soundness 356
A.6 Completeness 356
A.7 Computability Theory and Decidability 357

B Differential Equations **359**
B.1 Ordinary Differential Equations 359
B.2 Existence Theorems 363
B.3 Existence and Uniqueness Theorems 364
B.4 Linear Differential Equations with Constant Coefficients 365

C Hybrid Automata **369**
C.1 Syntax and Traces of Hybrid Automata 369
C.2 Embedding Hybrid Automata into Hybrid Programs 371

D KeYmaera Implementation **377**
D.1 KeYmaera: A Hybrid Theorem Prover for Hybrid Systems 377
D.1.1 Structure of This Appendix 379
D.2 Computational Back-ends for Real Arithmetic 380
D.2.1 Real-Closed Fields 381
D.2.2 Semialgebraic Geometry and Cylindrical Algebraic Decomposition 383
D.2.3 Nullstellensatz and Gröbner Bases 386
D.2.4 Real Nullstellensatz 392
D.2.5 Positivstellensatz and Semidefinite Programming 394
D.3 Discussion 396
D.4 Performance Measurements 399

References **401**

Index **415**

Operators and Proof Rules **423**

List of Figures

1.1 European Train Control System 2
1.2 ETCS: discrete evolution of acceleration a, continuous evolution of velocity v and of position z over time t 3
1.3 Collision avoidance manoeuvres in air traffic control 3
1.4 Hybrid automaton for an (overly) simplified train control system . 5
1.5 Hybrid automaton and hybrid program of a simple bouncing ball . 7
1.6 Switching between two damped oscillators 8
1.7 Hybrid automaton for switching damped oscillators 9
1.8 Stable trajectory switching between two damped oscillators 9
1.9 Instable trajectory switching between two damped oscillators . . . 10
1.10 Simple water tank system 11
1.11 Successive state space exploration in finite-state model checking . 12
1.12 Failed hybrid automaton decomposition attempt 17
1.13 Successful hybrid program decomposition 18
1.14 Dependencies and suggested reading sequence of chapters and appendices 28

2.1 Hybrid program rendition of hybrid automaton for (overly) simplified train control 36
2.2 Parametric bouncing ball 45
2.3 Parametric bouncing ball (with abbreviations resolved) 46
2.4 Transition semantics of modalities in $\mathsf{d}\mathcal{L}$ formulas 52
2.5 **Transition semantics and example dynamics of hybrid programs** 56
2.6 Continuous flow along differential equation $x' = \theta$ over time 57
2.7 Transition structure and transition example in (overly) simple train control 59
2.8 ETCS train coordination protocol using dynamic movement authorities 61
2.9 ETCS transition structure and various choices of speed regulation for train speed control 63
2.10 Application of simultaneous substitutions 65

2.11 **Rule schemata of the free-variable calculus for differential dynamic logic** . . . 79
2.12 **Correspondence of dynamic proof rules and transition semantics** 83
2.13 Simple propositional example proof . . . 87
2.14 Deduction modulo for analysis of MA violation in braking mode . 89
2.15 Controllable region of ETCS dynamics . . . 90
2.16 Deduction modulo for analysis of MA-safety in braking mode . . . 90
2.17a Wrong rearrangement with deduction modulo by invertible quantifiers . . . 91
2.17b Correct reintroduction order with deduction modulo by invertible quantifiers . . . 91
2.18 Bouncing ball proof (no evolution domain) . . . 95
2.19a Unsound attempt of induction without universal closure $\forall^\alpha$ 95
2.19b Correct use of induction with universal closure $\forall^\alpha$, i.e., $\forall x$ 95
2.20 Bouncing ball proof (with evolution domain) . . . 97
2.21 Characterisation of $\mathbb{N}$ as zeros of solutions of differential equations 103
2.22 Fractional encoding principle of $\mathbb{R}$-Gödel encoding by bit interleaving . . . 105
2.23 FOD definition characterising Gödel encoding of $\mathbb{R}$-sequences in one real number . . . 106
2.24 Explicit rendition of hybrid program transition semantics in FOD . 107
2.25 Evolution domain checks along backwards flow over time 108

3.1 Controllability violated in the presence of disturbance 138
3.2 Differential state flow . . . 143
3.3 Zeno system run . . . 146
3.4 Aircraft dynamics . . . 148
3.5 Reparametrise for differential axiomatisation . . . 149
3.6 Flight manoeuvres for collision avoidance in air traffic control . . . 151
3.7 Flight control with tangential roundabout collision avoidance manoeuvres . . . 152
3.8 Vector field and a solution of a differential equation . . . 153
3.9 **Rule schemata of the proof calculus for differential-algebraic dynamic logic** . . . 164
3.10 **Side deduction for quantifier elimination rules** . . . 164
3.11 Nested side deductions and differential variants for progress property 169
3.12 Differential invariants . . . 171
3.13a Cubic dynamics proof . . . 172
3.13b Cubic dynamics . . . 172
3.14 Unsound restriction of differential invariance . . . 173
3.15a Restricting differential invariance . . . 174
3.15b Linear dynamics . . . 174
3.16 Proof of MA-safety in braking mode with disturbance 176
3.17 Trajectory and evolution of a damped oscillator . . . 177
3.18 Trajectory switching between two damped oscillators 178

3.19 Parametric switched damped oscillator system 178
3.20 Instable trajectory switching between two damped oscillators . . . 179
3.21 Parametric switched damped oscillator proof 180
3.22 Differential variants 182
3.23a Monotonically decreasing convergent counterexample 184
3.23b Convergent descent dynamics 184
3.23c Non-inductive property in convergent descent 184
3.24a Counterexample of unbounded dynamics without Lipschitz continuity 184
3.24b Explosive dynamics with limited duration of solutions 184
3.25 Differential induction splitting over disjunctions for negative equations 189
3.26a Counterexample for disjunctive monotonicity 193
3.26b Interrupted dynamics 193
3.27 Quadrant sign selection regions of differential invariant 196
3.28 Circular dependencies for differential strengthening 196
3.29 Tangential construction for characteristics of roundabout dynamics 198

4.1 Trace semantics of dTL formulas 214
4.2 ETCS train coordination protocol phases 216
4.3 **Rule schemata of the proof calculus for temporal differential dynamic logic** 218
4.4 **Correspondence of temporal proof rules and trace semantics** . 219
4.5 Explicit rendition of hybrid program trace semantics in FOD 225
4.6 Transformation rules for alternating temporal path and trace quantifiers 229

5.1 Deductive, real algebraic, computer algebraic prover combination . 236
5.2 Tableau procedure for differential dynamic logics 237
5.3 Nondeterminisms in the tableau procedure for differential dynamic logics 237
5.4 Computational distraction in quantifier elimination 240
5.5 Eager and lazy quantifier elimination in proof search space 241
5.6 A large subgoal of first-order real arithmetic during ETCS verification 242
5.7a Proof strategy priorities 244
5.7b Iterative background closure (IBC) proof strategy 244
5.8 Iterative background closure (IBC) algorithm schema 245
5.9 General and/or-branching in proof strategies for differential dynamic logics 245
5.10 Iterative inflation order (IIO) algorithm schema 247

6.1 $d\mathcal{L}$-based verification by symbolic decomposition 257
6.2 Aircraft dynamics 261
6.3 Fixed-point algorithm for differential invariants (*Differential Saturation*) 262

6.4 Differential dependencies and variable clusters of flight dynamics . 264
6.5 Fixed-point algorithm for discrete loop invariants (loop saturation) 266
6.6 Hybrid program rendition of hybrid automaton for simple water tank 267
6.7 Interplay of local and global fixed-point verification loops during symbolic decomposition . . . 268
6.8 Robustness in counterexamples . . . 271
6.9 Flyable aircraft roundabout . . . 272

7.1 ETCS train cooperation protocol phases and dynamic movement authorities . . . 282
7.2 ETCS track profile . . . 283
7.3 Formal model of parametric ETCS cooperation protocol (skeleton) 284
7.4 Transition structure of ETCS skeleton . . . 286
7.5 Controllable region of ETCS . . . 288
7.6 ETCS cooperation protocol refined with parameter constraints . . . 291
7.7 Proof sketch for ETCS safety . . . 292
7.8 Controllability region changes in the presence of disturbance . . . 295
7.9 Proof of ETCS controllability despite disturbance . . . 297
7.10 Parametric ETCS cooperation protocol with disturbances . . . 299
7.11 Parametric ETCS cooperation protocol with disturbances (full instantiation) . . . 300

8.1 Evolution of collision avoidance manoeuvres in air traffic control . 304
8.2 Non-flyable straight-line manoeuvre with instant turns . . . 305
8.3 Flyable aircraft roundabout . . . 309
8.4 Flight dynamics . . . 309
8.5 Protocol cycle and construction of flyable roundabout manoeuvre . 310
8.6 Non-flyable tangential roundabout collision avoidance manoeuvre NTRM . . . 312
8.7 Tangential configuration $\mathscr{T}$ . . . 313
8.8 Flyable aircraft roundabout (multiple aircraft) . . . 314
8.9 Tangential roundabout collision avoidance manoeuvre (four aircraft) 314
8.10 Flyable entry characteristics . . . 316
8.11 Entry separation by bounded nondeterministic overapproximation . 320
8.12 Some mutually agreeable negotiation choices for aircraft . . . 323
8.13 Far separation for mutually agreeable negotiation choices . . . 325
8.14a Exit ray separation . . . 327
8.14b Incompatible exit rays . . . 327
8.15 Flight control with flyable tangential roundabout collision avoidance 329
8.16 Verification loop for flyable tangential roundabout manoeuvres . . 330
8.17 Flight control with FTRM (synchronous instantiation) . . . 332

9.1 Topics contributing to the logical analysis of hybrid systems . . . 336

A.1 **Rule schemata of the sequent calculus for first-order logic** . . . 352
A.2a Ground proof example . . . 354

A.2b Free-variable proof example 354
A.3 Wrong proof attempt in first-order logic 355

B.1 Vector field and a solution of a differential equation 360

C.1 Hybrid automaton and corresponding hybrid program 370
C.2a Hybrid automaton for water tank 373
C.2b Hybrid program for water tank 373
C.3 Parametric bouncing ball 374

D.1 Architecture and plug-in structure of the KeYmaera prover 378
D.2 Screenshot of the KeYmaera user interface 379
D.3 KeYmaera proof strategy options 380
D.4 Projection of semialgebraic sets and quantifier elimination 384
D.5 Rule schemata of Gröbner calculus rules 389
D.6 Some algebraic varieties generated by one polynomial equation in two variables 391
D.7 Example proof using the real Nullstellensatz 393
D.8 Rule schema of Positivstellensatz calculus rule 395
D.9 Example proof using the Positivstellensatz 395

List of Tables

2.1 **Statements and effects of hybrid programs (HPs)** 42
2.2 Statements and control structures definable with hybrid programs . . 44
2.3 **Operators and meaning in differential dynamic logic ($d\mathcal{L}$)** . . . 47

3.1 Comparison of DAL with DA-programs versus $d\mathcal{L}$ with hybrid programs 127
3.2 **Statements and effects of differential-algebraic programs** 137
3.3 Classification of differential-algebraic programs and correspondence to dynamical systems 139
3.4 **Operators and meaning in differential-algebraic dynamic logic (DAL)** 140
3.5 Embedding hybrid programs as DA-programs 147

4.1 **Operators and meaning in differential temporal dynamic logic (dTL)** 208

5.1 Experimental results for proof strategies (with standalone QE) I . . 249
5.2 Experimental results for proof strategies (with standalone QE) II . . 249
5.3 Experimental results for proof strategies (no standalone QE) I . . . 250
5.4 Experimental results for proof strategies (no standalone QE) II . . . 250

6.1 Experimental results for differential invariants as fixed points 273

7.1 Experimental results for the European Train Control System 300

8.1 Verification loop properties for flyable tangential roundabout manoeuvres 330
8.2 Experimental results for air traffic control (initial timeout = 10s) . . 331
8.3 Experimental results for air traffic control (initial timeout = 4s) . . . 333

A.1 Intuitive meaning of logical operators in first-order logic 343

List of Theorems

L 2.1	Uniqueness	57
L 2.2	Substitution Lemma	70
L 2.3	Substitution property	75
L 2.4	Substitutions preserve validity	76
L 2.5	Quantifier elimination lifting	92
L 2.6	Coincidence lemma	93
T 2.1	Soundness of $d\mathcal{L}$	98
T 2.2	Incompleteness of $d\mathcal{L}$	102
T 2.3	Relative completeness of $d\mathcal{L}$	104
L 2.7	$\mathbb{R}$-Gödel encoding	105
L 2.8	Hybrid program rendition	106
L 2.9	$d\mathcal{L}$ Expressibility	108
L 2.10	Derivability of sequents	109
L 2.11	Generalisation	110
P 2.1	Relative completeness of first-order safety	111
P 2.2	Relative completeness of first-order liveness	112
T 2.4	Relatively semidecidable fragment	114
L 2.12	Uniform Skolem symbols	115
P 3.1	Conservative extension	147
L 3.1	Derivation lemma	156
L 3.2	Differential substitution property	158
L 3.3	Differential transformation principle	158
L 3.4	Differential inequality elimination	161
L 3.5	Differential equation normalisation	161
L 3.6	Differential weakening	175
L 3.7	Closure properties of differential invariants	181
T 3.1	Soundness of DAL	185
P 3.2	Open differential induction	188
P 3.3	Differential monotonicity	191
T 3.2	Relative completeness of DAL	193
P 3.4	Equational deductive power	194

T 3.3	Deductive power	194
T 3.4	Safety of tangential roundabout manoeuvre	199
P 3.5	External separation of roundabout manoeuvres	200
P 4.1	Conservative temporal extension	215
L 4.1	Trace relation	215
T 4.1	Soundness of dTL	221
T 4.2	Incompleteness of dTL	223
T 4.3	Relative completeness of dTL	224
L 4.2	Hybrid program trace rendition	225
L 4.3	dTL Expressibility	225
P 4.2	Local soundness for temporal quantifier alternation	229
P 6.1	Principle of differential induction	260
P 6.2	Differential saturation	262
P 6.3	Loop saturation	265
T 6.1	Soundness of fixed-point verification algorithm	269
L 7.1	Principle of separation by movement authorities	282
P 7.1	Controllability	288
P 7.2	RBC preserves train controllability	289
P 7.3	Reactivity of ETCS	290
P 7.4	Reactivity constraint	290
P 7.5	Safety of ETCS	291
P 7.6	Liveness of ETCS	293
T 7.1	Correctness of ETCS cooperation protocol	294
P 7.7	Controllability despite disturbance	296
P 7.8	Reactivity constraint despite disturbance	298
P 7.9	Safety despite disturbance	298
T 8.1	Safety property of flyable tangential roundabouts	331
T A.1	Soundness of FOL	356
T A.2	Completeness of FOL	356
T B.1	Existence theorem of Peano	363
T B.2	Uniqueness theorem of Picard-Lindelöf	364
P B.1	Continuation of solutions	365
P B.2	Linear systems with constant coefficients	365
P C.1	Hybrid automata embedding	371
T D.1	Tarski-Seidenberg principle	383
T D.2	Semialgebraic sets	383
T D.3	Hilbert's basis theorem	388
P D.1	Soundness of Gröbner basis rules	390
T D.4	Hilbert's Nullstellensatz	391
T D.5	Real Nullstellensatz for real-closed fields	392
T D.6	Positivstellensatz for real-closed fields	394

Chapter 1
Introduction

Time is defined so that motion looks simple [209, p. 23]

Ensuring correct functioning of complex physical systems is among the most challenging and most important problems in computer science, mathematics, and engineering. In addition to nontrivial underlying physical system dynamics, the behaviour of complex systems is determined increasingly by computerised control and automatic analog or digital decision-making, e.g., in aviation, railway, and automotive applications. At the same time, correct decisions and control of these systems are becoming increasingly important, because more and more safety-critical processes are regulated using automatic or semiautomatic controllers, including the European Train Control System [117], collision avoidance manoeuvres in air traffic control [293, 196, 104, 238, 129, 171], car platooning technology for highways following the California PATH project [166], recent driverless vehicle technology [64], and biomedical applications like automatic glucose regulation for diabetes patients [223]. As a more general phenomenon of complex physical systems that are exemplified in these scenarios, correct system behaviour depends on correct functioning of the *interaction* of control with physical system dynamics and is not just an isolated property of only the control logic or only the physical system dynamics. These interactions of computation and control are more difficult to understand and get right than isolated systems. Even if the control software does not crash, the system may still malfunction, because the control software could issue unsafe control actions to the physical process. And even if the pure physics of the process is well understood, an attempt to control the process may still become unsafe, e.g., when the controller reacts to situation changes too slowly because computations take too long, or when sensor values are already outdated once the control actions finally take effect. It is the *interaction* of computation and control that must be taken into account. Systems with such an interaction of discrete dynamics and continuous dynamics are called *hybrid dynamical systems*, or just *hybrid systems* for short.

To illustrate typical aspects and effects in application areas, we take a look at two examples in more detail, which will serve as running examples and case studies throughout this book. In high-speed trains like the ICE (InterCityExpress), TGV (train à grande vitesse), Shinkansen, and the upcoming California High-Speed Rail, whose high mass (1,000–3,000 tons) and high speed (320km/h) cause them to re-

A. Platzer, Logical Analysis of Hybrid Systems,
DOI 10.1007/978-3-642-14509-4_1,

quire fairly long braking distances (more than 3.8 km), safe driving is impossible just based on sight without automatic technical means that enforce a safe minimum distance between trains. The *European Train Control System* (ETCS), which is currently being developed and installed in Europe [117], regulates and protects train movement according to movement authorities (MAs). A movement authority represents permission for the train to move up to a certain point on the track (the end of the movement authority) and is negotiated dynamically in rapid succession by wireless communication with decentralised radio block controllers (RBCs); see Fig. 1.1.

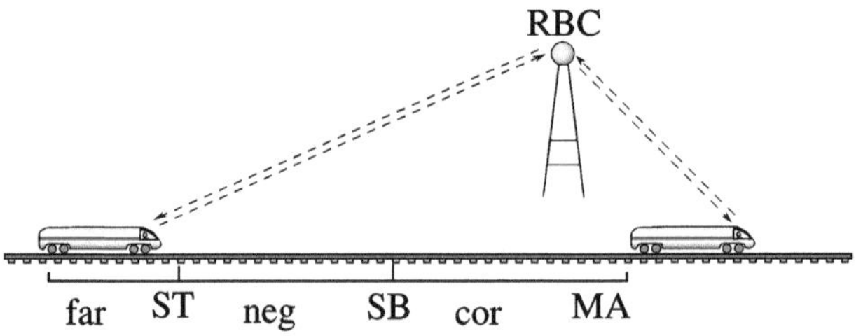

Fig. 1.1 European Train Control System

With the next generation development of ETCS, Level 3, all classical fixed trackside signalling and fixed track segment partitioning with physical separation of the train segments will become obsolete, thereby advancing to a fully autonomous operation of ETCS in order to achieve its performance goals of maximum speed and density on the track. Yet, a safe operation of ETCS requires that—while achieving these performance goals—the train controllers still always respect their local movement authorities and that the radio block controllers only grant compatible movement authorities to each of the trains. Even in emergency situations, the overall train control system must always ensure that the trains cannot crash into one another. To determine correct functioning of these controllers it has to be shown that the train positions, which evolve dynamically over time, are always safely separated. For this, however, we need to be able to analyse the interaction of the train control logic and the wireless ETCS cooperation protocol with a model of the actual physical train dynamics, because collision freedom is not an isolated property of only the discrete cooperation-layer control protocol, only the local train control decision process, or only the continuous train dynamics, but a joint property of their superposition or combination.

Train dynamics is a typical example of what is known as a *hybrid system*: it combines instantaneous discrete jump dynamics with continuous evolution. See Fig. 1.2 for a typical evolution of the acceleration a, which changes instantaneously by discrete control decisions at some points in time, and of the train velocity v and position z, which evolve continuously over time. The continuous dynamics can be described by a differential equation system like $z' = v, v' = a$, saying that the time derivat-

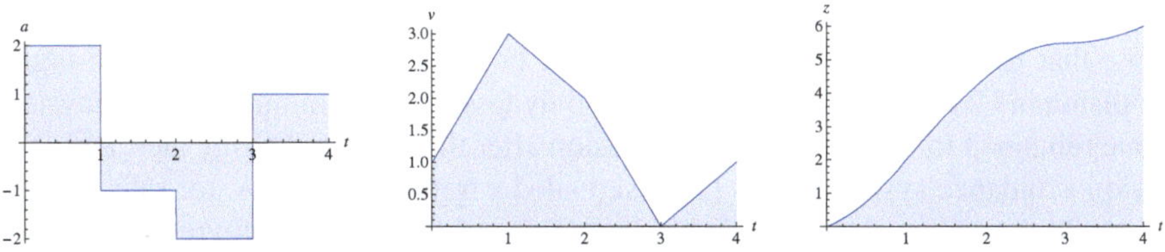

Fig. 1.2 ETCS: discrete evolution of acceleration a, continuous evolution of velocity v and of position z over time t

ive z' of the position equals the velocity v and the time derivative v' of the velocity equals the chosen acceleration a. The discrete dynamics can be described by an assignment such as $a := -2$ together with the conditions of when and under which circumstances this discrete jump will be executed. For example, the discrete dynamics can be described by the operations of the switching control logic, possibly as a small control program that describes under which circumstances the acceleration a switches to braking (negative acceleration), and under which circumstances it switches to positive acceleration. By combining the continuous differential equations and discrete control logic appropriately, we obtain a description of the interacting hybrid dynamics of the hybrid (dynamical) system. Yet the effects of such a combination also need to be well-defined, including a faithful model defining unambiguously how discrete and continuous dynamics work together. Finally, note that full formal verification and proof that the system operates correctly is indeed quite important and of particular practical relevance for safety-critical complex physical systems like ETCS. Despite careful development and testing, safety violations have recently been reported in ETCS [143] even at its moderate currently deployed level.

In air traffic control, collision avoidance manoeuvres [293, 196, 104, 238, 129, 171] are used to help pilots resolve conflicting flight paths that arise during arbitrary free flight of the aircraft. See Fig. 1.3 for an overview of different collision avoidance approaches. These manoeuvres are last-resort means for resolving air traffic

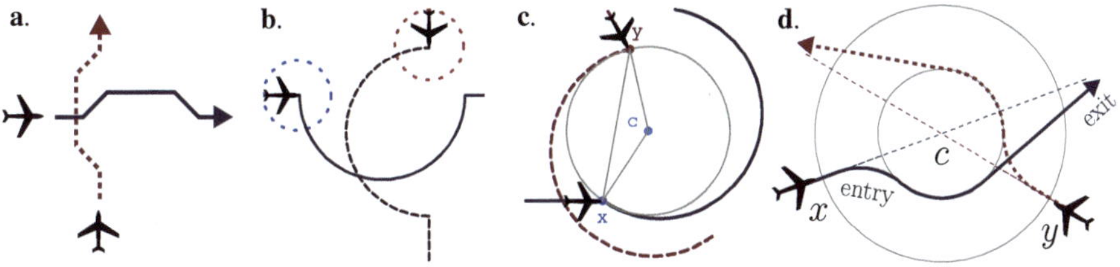

Fig. 1.3 Collision avoidance manoeuvres in air traffic control

conflicts that could otherwise lead to collisions and have not been detected before by the pilots during free flight or by the flight directors of the Air Route Traffic Control Centres. Consequently, complicated online trajectory prediction, trajectory evaluation, or lengthy manoeuvre negotiation may no longer be feasible in the short time that remains for resolving the conflict. For example, in the tragic mid-flight collision in Überlingen [43] in July 2002, only less than one minute of manoeuvring time remained to try to prevent the collision after the on-board traffic alert and collision avoidance system TCAS [196] signalled a traffic alert. Thus, for safe aircraft control we need particularly reliable instant reactions with manoeuvres whose correctness has been established previously by a thorough offline analysis. To ensure correct functioning of aircraft collision avoidance manoeuvres under all circumstances, the temporal evolution of the aircraft in space must be analysed carefully together with the effects that manoeuvring control decisions have on their dynamics, giving again a superposition or combination of physical system dynamics with control, and thus a hybrid system.

The continuous dynamics in air traffic control can again be described by differential equations for the flight dynamics. The relevant system variables for which differential equations need to be found include the two- or three-dimensional position in space and orientation in space for each of the aircraft at positions $x = (x_1, x_2, x_3) \in \mathbb{R}^3$ and $y = (y_1, y_2, y_3) \in \mathbb{R}^3$, for instance. Because rails are essentially a one-dimensional space (except for the track topology) but airspace is three-dimensional, the resulting differential equations for flight are more involved than for trains. The discrete dynamics in air traffic control comes, e.g., from the decisions of when and how to initiate a collision avoidance manoeuvre, and from the discrete change in direction or the angular velocity of the aircraft.

These examples are prototypical and similar phenomena occur in many other application scenarios of hybrid systems. The continuous dynamics often comes from physical movement or physical processes. Discrete dynamics often (but not always) comes from control decisions or digital control implementations.

1.1 Technical Context

In this section we survey the technical context of this book, briefly summarise some important technical concepts in the domain, and preview some technical highlights of this book informally. We also give a short overview of related work.

1.1.1 Hybrid Systems

As a common mathematical model for complex physical systems, *hybrid dynamical systems*, or just *hybrid systems* for short [289, 9, 218, 8, 56, 156, 11, 58, 189, 97, 228, 90, 194], are dynamical systems [227, 177, 279] where the system state evolves over time according to interacting laws of discrete and continuous dynamics, the idea being to capture the superposition or combination of physical system dynamics with control at a natural modelling level. For discrete transitions, the hybrid system changes state instantaneously and possibly discontinuously. During continuous transitions, the system state is a continuous function of continuous time and varies according to a differential equation, which is possibly subject to domain restrictions or algebraic relations resulting from physical circumstances or the interaction of continuous dynamics with discrete control. Continuous dynamics results, for example, from the continuous movement of a train along the track (train position z evolves with velocity v along the differential equation $z' = v$ where z' is the time derivative of z) or from the continuous variation of its velocity over time ($v' = a$ with acceleration a); see evolutions in Fig. 1.2. Other behaviour can be modelled more naturally by discrete dynamics, for example, the instantaneous change of control variables like the acceleration (e.g., the changing of a by setting $a := -b$ with braking force $b > 0$) or the change of status information in discrete controllers; see Fig. 1.2 again. Both kinds of dynamics interact, e.g., when measurements of the continuous state affect decisions of discrete controllers (the train switches to braking mode when v is too high). Likewise, they interact when the resulting control choices take effect by changing the control variables of the continuous dynamics (e.g., changing control variable a in $z'' = a$). The superposition of continuous dynamics with analog or discrete control causes complex system behaviour, which can be analysed neither by purely continuous reasoning (because of the discontinuities caused by discrete transitions) nor by considering discrete change in isolation (because safety depends on continuous states).

Among several other models for hybrid systems [69, 100, 58, 270, 272, 40, 183], the model of *hybrid automata* [156, 8] is one of the more widely used notations. Hybrid automata specify discrete and continuous dynamics in a graph. Even though hybrid automata are a fairly common notation for hybrid systems, there are numerous slightly different notions of hybrid automata or automata-based models for hybrid systems [289, 9, 218, 8, 56, 156, 11, 58, 189, 97, 228, 90]. We consider a number of examples to introduce hybrid automata and illustrate simple hybrid systems informally.

Example 1.1 (Simplistic train control). See Fig. 1.4 for a (much too) simple train control example written as a hybrid automaton. The nodes of the hybrid automaton

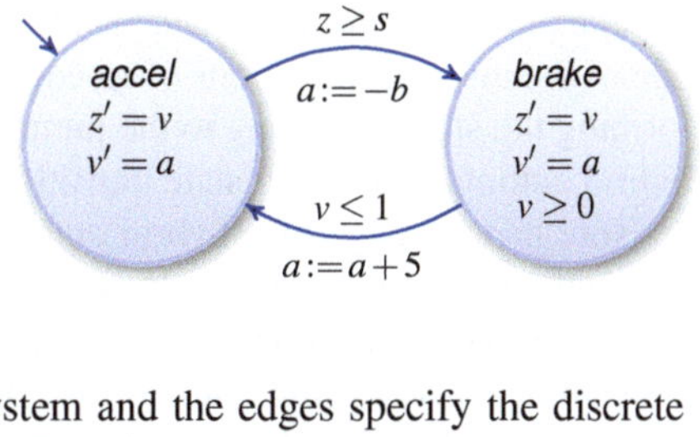

Fig. 1.4 Hybrid automaton for an (overly) simplified train control system

specify the continuous dynamics of the system and the edges specify the discrete switching behaviour between the various modes of continuous dynamics.

Each node of the graph structure corresponds to a continuous dynamical system and is annotated with its differential equations and an evolution domain specifying the maximum domain of evolution. The basic differential equation in both nodes of the train automaton here is $z' = v, v' = a$, because the time derivative of the position z is the velocity v, and the time derivative of the velocity v is the acceleration a. Yet both nodes have different choices for the acceleration a. By specifying a maximum domain of evolution in addition to the differential equations in a node, we restrict the circumstances under which the system is allowed to stay in a node, and enforce when it has to transition to another node instead. The system cannot stay in a node outside this maximum domain of evolution, and is then forced to switch to another node. The system is not required to stay as long as possible in a node until it reaches the border of its maximum domain of evolution, however. Instead, the system is allowed to leave a node earlier if one of the outgoing edges can be used at any time. For example, in the node *brake* of Fig. 1.4, the differential equations $z' = v, v' = a$ only apply within the evolution domain $v \geq 0$ (because the train does not move backwards when braking). Thus, the system is allowed to stay in node *brake* arbitrarily long, but it has to leave before the evolution domain $v \geq 0$ is violated. Edges specify the discrete switching behaviour between the respective modes of continuous evolution. They can be annotated with conditions (*guards*) that need to hold for the state when the system follows the edge, and with discrete state transformations (*jumps*) that take instantaneous effect and transform the continuous state when the system follows the edge. For example, the automaton in Fig. 1.4 can take an edge to leave node *accel* when train position z passes point s (i.e., when the current state satisfies $z \geq s$), which sets the acceleration to braking by changing the state variables according to $a := -b$, and then enter node *brake*. The edge from node *brake* to node *accel* can be taken only when the guard $v \leq 1$ is true for the current continuous state; the transition will then increase the value of the acceleration by $a := a + 5$ and the system will enter node *accel*. The short edge pointing to *accel* from nowhere in the top-left corner indicates that the system will initially start in the *accel* node. In addition, we also need to know what values the continuous variables (z, v, a, b, s) can have in the initial state.

To identify in which state a hybrid automaton is at some point in time, we need to know real number values for all the continuous state variables (z, v, a and the parameters b and s) and we need to know in which node the automaton currently is (either *accel* or *brake*). Over time, the state evolves according to the discrete or continuous dynamics. The state either evolves continuously according to the differential equation (if the state stays inside the evolution domain restriction), or follows one of the discrete transitions of an edge (when the guard is satisfied, thereby transforming the state according to the jump relation). For reference, we formally define hybrid automata and their state transitions in App. C.

Note that an edge of a hybrid automaton can only be taken if the guard holds true and, after updating the state according to the jump, the state is in the evolution domain of the target node. For instance, the automaton starts in *accel*, and may stay some time in *accel* while following the differential equations $z' = v, v' = a$. Then the system could switch to *brake* at some point where the guard $z \geq s$ is satisfied for the current state, thereby setting the acceleration to $a := -b$. Next the system can stay in node *brake*, following the differential equation $z' = v, v' = a$ (now with a different value for a than before), but only while staying inside the permitted evolution domain $v \geq 0$. In particular, the system has to leave node *brake* and switch back to *accel* before the velocity is negative. Because the only outgoing edge from *brake* has a guard $v \leq 1$, the system can switch back to *accel* if and only if the current velocity satisfies $0 \leq v \leq 1$. The system can switch at any point in time where this is satisfied. Suppose the system waits in *brake* until the speed is, say, 0.5. Then the current speed satisfies $v \leq 1$ and the automaton can switch to *accel* again by updating $a := a+5$.

Note, however, that the resulting acceleration does not have to be positive! If $b > 5$, then the increase $a := a+5$ for switching to *accel* is less than the braking reset $a := -b$; hence, the value of a after switching from *accel* to *brake* and back to *accel* again will still be negative. Suppose the position still satisfies the guard $z \geq s$ of the outgoing edge back to *brake*. Then, the system might not be allowed to follow this edge, even though $z \geq s$ holds, because the maximum evolution domain of *brake* requires $v \geq 0$ to be true all the time, including when entering the state. That is, in such a run, where the resulting velocity is negative, the automaton is stuck and can never do any other transitions again. In summary, we can see that the operational effect of an edge in an automaton depends not only on the guard and jump of the edge, but also on the evolution domain restrictions in the nodes.

How can we find out if the train model always stays within the safe region on the track? How do we find out for which choices of the parameters s and b the model works as expected? How can we determine if the train is always able to make progress and is never stuck? We will revisit these questions in Chaps. 2 and 7. □

Example 1.2 (Bouncing ball). Another very intuitive example of a hybrid system is the bouncing ball [110]; see Fig. 1.5. The bouncing ball is let loose in the air and is

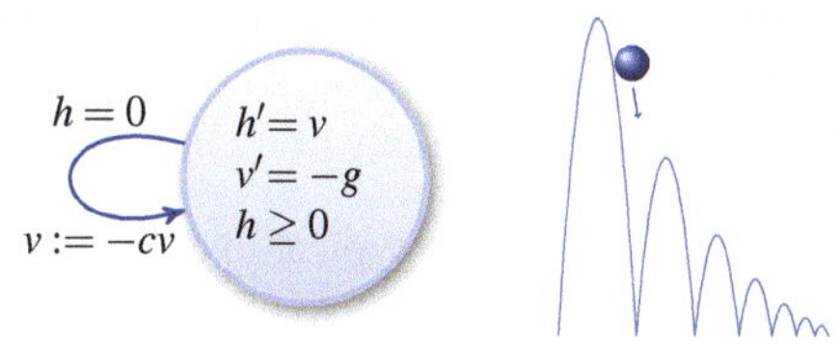

$$\begin{array}{l}(h' = v, v' = -g \,\&\, h \geq 0;\\ \quad \text{if } (h = 0) \text{ then}\\ \qquad v := -cv\\ \quad \text{fi})^*\end{array}$$

Fig. 1.5 Hybrid automaton and hybrid program of a simple bouncing ball

falling towards the ground. When it hits the ground, the ball bounces back up and climbs until gravity wins and it starts to fall again. The bouncing ball follows the continuous dynamics of physical movement by gravity. It can be understood naturally as a hybrid system, because its continuous movement switches from falling to climbing by reversing its velocity whenever the ball hits the ground and bounces back. Let us denote the height of the ball by h and the current velocity of the ball by v. The bouncing ball is affected by gravity of force $g > 0$, so its height follows the differential equation $h'' = -g$, i.e., the second time derivative of height equals the negative gravity force. The ball bounces back from the ground (which is at height $h = 0$) after an elastic deformation. At every bounce, the ball loses energy according to a damping factor $0 \le c < 1$. Figure 1.5 depicts a hybrid automaton, an illustration of the system dynamics, and a representation of the same system in a program notation for hybrid systems. The hybrid automaton has only one node: falling along the differential equation system $h' = v, v' = -g$ (which is equivalent to $h'' = -g$) restricted to the evolution domain $h \ge 0$, above the floor. In particular, the bouncing ball can never fall through the floor. The hybrid automaton also has only one jump edge: on the ground (when $h = 0$), it can reset the velocity v to $-cv$ and continue in the same node. This jump will change the direction from falling (the velocity v was negative before) to climbing (the velocity $-cv$ is nonnegative again) after dampening the velocity v by c.

The program on the right of Fig. 1.5 represents the same bouncing ball system textually (we will call these textual representations hybrid programs and examine them in detail in Chap. 2). The first line of the program describes the continuous dynamics along the differential equation $h' = v, v' = -g$ restricted to (written &) the maximum evolution domain region $h \ge 0$. After the sequential composition (;), an if-then statement is executed, which resets velocity v to $-cv$ by assignment $v := -cv$ if $h = 0$ holds at the current state. Finally, the sequence of continuous and discrete statements can be repeated arbitrarily often, as indicated by the regular-expression-style repetition operator (*) at the end.

Note one strange phenomenon in the bouncing ball automaton and program. It seems like the bouncing ball will bounce over and over again, switching its direction in shorter and shorter periods of time as indicated in Fig. 1.5 (unless $c = 0$, which means that the ball will just lie flat right away). Even worse, the ball will end up switching directions infinitely often in a short amount of time. This controversial phenomenon is called Zeno behaviour, which we discuss in more detail in later parts of this book.

How can we see that the bouncing ball never bounces higher than its initial height? How can we ensure that the bouncing ball fits closer to the physical reality that it stops after some time, instead of bouncing infinitely often in finite time? □

Example 1.3 (Switching damped oscillators). Another typical example of a hybrid system is a switching damped oscillator with continuous variables x, y; see Fig. 1.6. This system consists of two continuous modes of damped oscillators with different

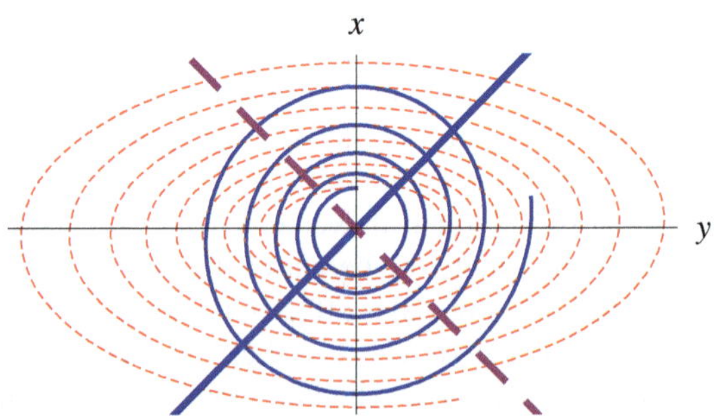

Fig. 1.6 Switching between two damped oscillators

settings. The solid spiral in Fig. 1.6 illustrates the one continuous mode, and the dashed spiral the other continuous mode. The overall system is a hybrid system that switches between those two modes at the switching surfaces depicted by the diagonal lines in Fig. 1.6. Also see a hybrid automaton model of this system in Fig. 1.7.

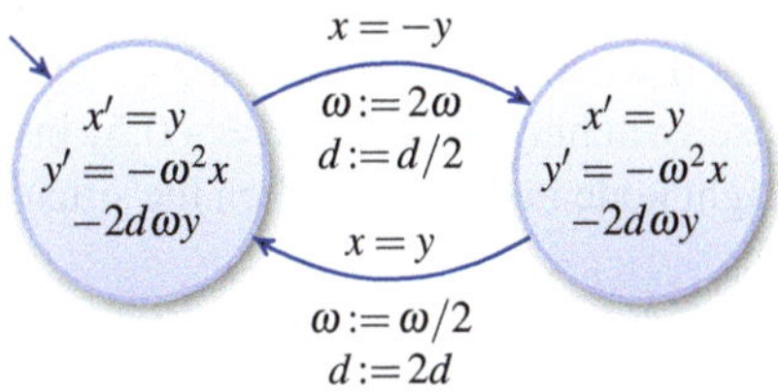

Fig. 1.7 Hybrid automaton for switching damped oscillators

The system starts with the solid dynamics in the node on the left. When passing the dashed secondary diagonal ($x = -y$), it switches to following the dashed continuous mode (right node in Fig. 1.7). When passing the solid main diagonal ($x = y$), the hybrid system switches to the solid continuous dynamics (left node). An example of

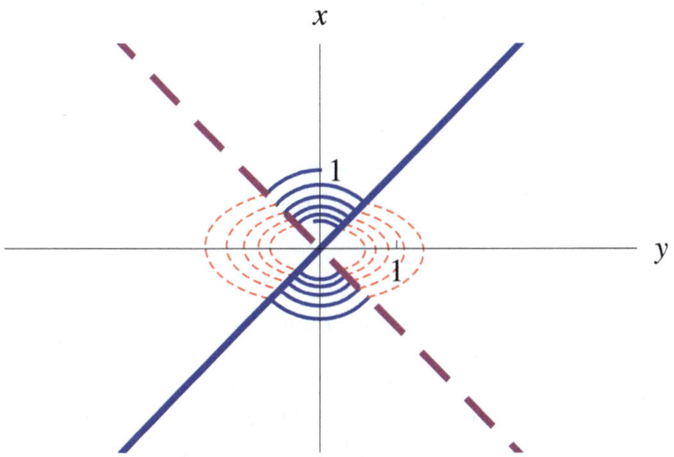

Fig. 1.8 Stable trajectory switching between two damped oscillators

an evolution of this hybrid system that switches between both continuous modes according to these switching conditions is shown in Fig. 1.8. The differential equations for the damped oscillators are of the form

$$x' = y, y' = -\omega^2 x - 2d\omega y$$

with the parameters ω (for the undamped angular frequency) and d (for the damping ratio). These parameters will be adjusted whenever the hybrid system follows an edge in the automaton to a different continuous evolution mode. The solid parts of the curves in Fig. 1.8 correspond to the parts where the system is in the left node of the automaton in Fig. 1.7, and thus follow the solid spiral in Fig. 1.6 The dashed parts of the curves in Fig. 1.8 correspond to the parts where the system has switched to the right node of the automaton in Fig. 1.7, and thus follow the dashed spiral in Fig. 1.6. The switching happens at the diagonals.

The evolution in Fig. 1.8 is *stable*, i.e., it converges to the origin $x = 0, y = 0$. Yet if we tilt the switching axes only slightly and switch a little later —for instance use switching condition $x = -0.5y$ instead of $x = -y$ on the edge from the left to the right node in Fig. 1.7— then the situation is quite different; see Fig. 1.9. Under these

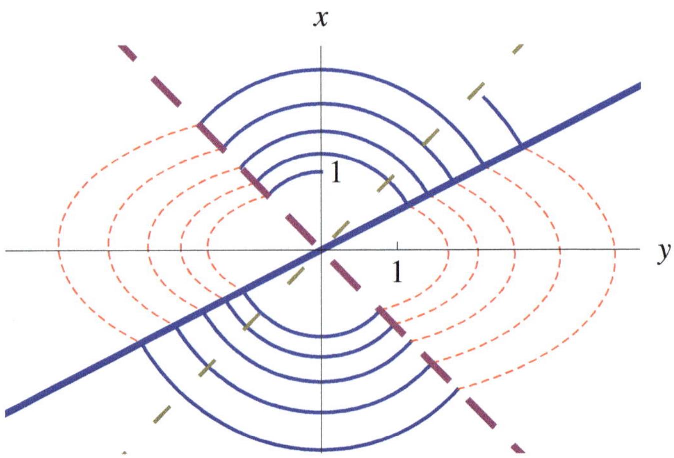

Fig. 1.9 Instable trajectory switching between two damped oscillators

circumstances, the evolution from the same initial point $x = 1,\ y = 0$ is suddenly *instable*, i.e., it does not converge to the origin $x = 0,\ y = 0$. Instead, the evolution diverges and x, y grow infinitely large. This is quite a remarkable difference caused by a seemingly minor variation in the switching conditions of the hybrid systems. In fact, this instability is caused only by the hybrid system switching, because the continuous dynamics alone is stable in both evolution modes (i.e., if both spirals in Fig. 1.6 are considered separately). How do we find out which parameters for the switching surfaces make the system stable, and which parameter choices render it instable? How can we establish whether safety and/or stability depend on the parameters ω and d?

Finally note that the evolutions in Figs. 1.8 and 1.9 both switch nodes in the automaton eagerly, i.e., as soon as the guards of outgoing transitions are true. This is certainly one choice, but hybrid systems do not restrict the evolution to follow permitted transitions as early as possible. Instead, the hybrid automaton in Fig. 1.7 would allow the system to skip a switching possibility and wait for the next one to come along. Nondeterminism like this occurs quite frequently when modelling natural phenomena in complex physical systems, especially in the presence of external control decisions or input. How does this flexibility in switching affect the behaviour of the system? How do we establish for which choices of switching diagonals the overall system behaviour is well behaved. We will come back to these questions about switching damped oscillators in Sect. 3.5.6. □

Example 1.4 (Simple water tank). As a further illustration of a hybrid system, consider a simple water tank model, where a controller tries to keep the water level x in a tank between a lower bound of 1 and an upper bound of 10; see Fig. 1.10. The

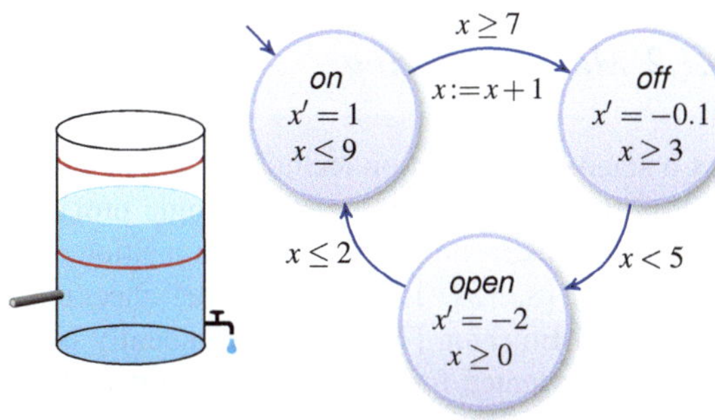

Fig. 1.10 Simple water tank system

water tank can be filled with a pump (node *on*), or the pump can be turned off (node *off*). For a twist, let us assume that, before activating the pump again, the valve has to be opened completely, so that water pours out quickly (node *open*) until the pump is turned *on* again. While the pump is *on*, water enters the tank, increasing the water level continuously by the differential equation $x' = 1$. The tank has a water level sensor that shuts off the pump before the water level exceeds 9; hence, the continuous evolution in mode *on* is restricted to the maximum evolution domain $x \leq 9$. When the water level is $x \geq 7$, the controller is allowed to shut the pump *off* by taking the edge from node *on* to node *off*, but during the shutdown of the valve, more water pours in. That is, the water level increases from x to $x + 1$ by the discrete change $x := x + 1$ on the edge. Since the evolution in node *on* is restricted to the evolution domain $x \leq 9$, but the transition from node *on* to node *off* requires guard $x \geq 7$, the controller can switch from *on* to *off* no earlier than when $x \geq 7$, but no later than when the water level still is $x \leq 9$. Any point between the water levels 7 and 9 (including 7 and 9) is a permitted switching point.

In node *off*, the pump is off and the valve closed, but, nevertheless, water leaks slowly by the continuous dynamics $x' = -0.1$. The controller knows about this water leakage, so it stays in *off* at most as long as $x \geq 3$. Whenever $x < 5$, the controller can switch along the edge from *off* to *open* for opening the valve. Again, because of the evolution domain restriction $x \geq 3$ in *off* and the guard condition $x < 5$ on the edge, the switch may happen at any water level between 3 (inclusive) and 5 (exclusive). In node *open*, where the valve is open, water drains quickly following the differential equation $x' = -2$, but restricted to the evolution domain $x \geq 0$, because water stops pouring out when there is no water left. Finally, when the water level satisfies guard $x \leq 2$, the system can take the edge from node *open* back to node *on*. It is easy to see that the controller in Fig. 1.10 does not guarantee keeping the water level within the target range of 1 to 10 always, because it can stay in node *open* too long until $x = 0$.

How can we detect this problem and how can we find more subtle problems as automatically as possible? If we found problems, how can we find suggestions for fixes? And, once there are no more bugs, how do we prove that the system always works as intended under all circumstances? □

1.1.2 Model Checking

As a standard verification technique, model checking [75, 114, 78] has been used successfully for verifying temporal logic properties [247, 114, 115, 6, 7, 284] of finite-state abstractions of automata transition structures by exhaustive state space exploration [8, 159, 156, 217]. Model checking was pioneered [76] in 1981 by Clarke and Emerson [75] and independently by Queille and Sifakis [255]. It was originally invented for finite-state systems and has since been extended to handle finite-state abstractions of infinite-state systems. The working principle of model checking is to successively explore all state transitions of a transition system from the set of initial states until an unsafe state is found; see Fig. 1.11 for an illustration. One of the biggest advantages of model checking is this search for concrete

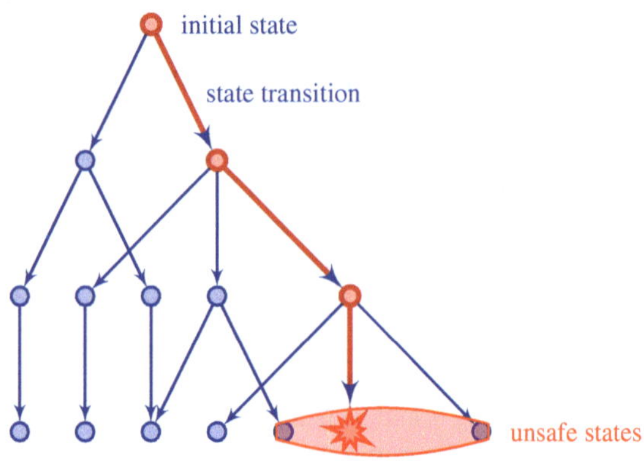

Fig. 1.11 Successive state space exploration in finite-state model checking

counterexamples, because such a sequence of transitions to an unsafe state can be very informative for debugging the system. If no unsafe state can be found this way, however, then model checkers can stop only if they have made sure that they have explored all possibilities. In finite-state systems, this exploration terminates trivially, because the search space is finite. Still, numerous successful optimisations and refinements have been developed to ensure faster termination and avoid exhaustive memory usage, including symbolic model checking with BDDs [66, 65], bounded model checking with propositional SAT solvers [47, 74], and model checking based on Craig interpolants [173, 174]. See the 2007 ACM Turing Award lecture by Edmund M. Clarke, Allen Emerson, and Joseph Sifakis [76] for a detailed survey of model checking, its successes, and its history.

For infinite-state systems the situation is more difficult than for finite automata because no model checker can actually explore an infinite set of states, so various abstractions have been developed that group sets of states into equivalence classes. If a model checker finds an unsafe reachable state in an infinite-state system, this counterexample is still extremely useful. Yet in case it does not find a counterexample after finitely many state explorations, it becomes more difficult to make an argument why no unsafe state could ever be reached by further exploration of the—still infinite—remaining state space. After all, finite exploration of an infinite state space still leaves infinitely many possibilities to consider. For example, timed automata (finite automata with clocks that measure the progress of time) can be abstracted with only finitely many equivalence classes that need to be considered, so that model checking is still guaranteed to terminate by considering one representative of each of these finitely many equivalent behaviours [6, 7]. These timed automata [10, 5, 86] can be understood as hybrid automata in which all differential equations are of the form $x' = 1$, all guards are of a form such as $x \leq c$ or $x - y \leq c$ for numbers $c \in \mathbb{Q}$, and all jumps (if any) are of the form $x := 0$. But even seemingly small extensions like stopwatch automata, which are timed automata with clocks that can be stopped (differential equation $x' = 0$) and resumed ($x' = 1$), have no equivalent finite-state abstractions anymore, so that model checking becomes undecidable [67].

The more general continuous state spaces of hybrid automata are inherently infinite, and do not admit equivalent finite-state abstractions either [156]. Because of this, model checkers for hybrid automata use various approximations [159, 8, 156, 70, 125, 15, 77, 21, 291, 217] and are still more successful in falsification than in verification. Furthermore, for hybrid systems with symbolic parameters in the dynamics, correctness crucially depends on the free parameters (e.g., b and s in Fig. 1.4). It is, however, quite difficult to determine corresponding symbolic parameter constraints from concrete values of a counterexample trace produced by a model checker, especially if they rely on nonstructural state splitting [70, 77, 21, 126]. Finally, in hybrid systems with nontrivial interaction of discrete and continuous dynamics, parameters also have a nontrivial impact on the system behaviour, leading to nonlinear parameter constraints and nonlinearities in the discrete and continuous dynamics. For instance, nonlinear constraints of the form $s \geq \frac{v^2}{2b} + \ldots$ will turn out to be important for the safety of train systems such as the one in Fig. 1.4. Thus, the standard model checking approaches [156, 8, 158, 70, 126, 127] cannot be used, because they require linear dynamics in the discrete transitions and various forms of linear approximations in the continuous dynamics. Even in simpler timed automata, in which clocks are the only source of continuous dynamics, parameters immediately make the model checking problem undecidable [13, 61]. Still, parameters occur frequently in practical applications.

1.1.3 Deductive Verification

Deductive approaches [35, 37, 170, 149, 148, 306, 95, 97] have been used for verifying systems by proofs instead of by state space exploration and, thus, do not require finite-state abstractions. Deduction and theorem proving have already been used very successfully for program verification [35, 23, 170], in which proofs are used to show that a conventional program such as a Java program operates as specified. Davoren and Nerode [97] have also argued that deductive methods support formulas with free parameters better than, e.g., the standard model checking problem does. First-order logic, for instance, has widely proven its power and flexibility in handling symbolic parameters as free or quantified logical variables. However, first-order logic has no built-in means for referring to state transitions, which are crucial for verifying dynamical systems where states change over time.

In *temporal logics* [247, 114, 115, 6, 284], state transitions can be referred to using modal operators. For instance, the temporal formula $\Box\phi$ would hold for a system if formula ϕ is true always in all runs of the system. A temporal formula $\Box\phi$ that holds for all systems would be called valid. In deductive approaches, temporal logics have been used to prove the validity of formulas in calculi [97, 306]. Valid formulas of temporal logic, however, only express generic facts that are true for all systems, regardless of their actual behaviour. But these are not exactly the formulas we are interested in. We want to know the special properties of the actual system at hand, not those generic properties shared by all possible systems at once (including the broken implementations).

Hence, the behaviour of a specific hybrid system would need to be characterised declaratively with temporal formulas to obtain meaningful results. It is quite difficult, however, (and generally impossible for reasons of lack of expressive power), to characterise the behaviour of a hybrid system just with a set of temporal formulas such as $\Box x \geq 0$, because there are often myriads of systems satisfying the same formulas. Most temporal logics are not expressive enough to characterise the dynamics of a single system this way. Furthermore, the equivalence of declarative temporal representations and actual system operations would still need to be proven separately using other techniques. Finally even for finite-state systems, direct temporal characterisations can become computationally infeasible. In discrete finite-state systems, for example, direct temporal characterisations of the operational behaviour transform the linear-time model checking problem of the temporal logic CTL [75] into an EXPTIME-complete satisfiability problem [113].

Dynamic logic (DL) [253, 148, 149] is a successful approach for verifying infinite-state discrete systems deductively [35, 37, 170, 149, 148]. Like model checking, DL does not need declarative characterisations of system behaviour but can express and analyse the transition behaviour of actual operational system models directly. Yet, operational models are fully *internalised* within DL-formulas, and DL is closed under logical operators. That is, correctness statements about systems can be combined into bigger formulas with arbitrary propositional operators or quantifiers, and even into nestings of formulas. Within a single specification and verification language, DL combines operational system models with means to talk about the states that are reachable by system transitions. DL provides parametrised modal operators $[\alpha]$ and $\langle\alpha\rangle$ that refer to the states reachable by system α and can be placed in front of any formula. The formula $[\alpha]\phi$ expresses that all states reachable by system α satisfy formula ϕ. Likewise, $\langle\alpha\rangle\phi$ expresses that there is at least one state reachable by α for which ϕ holds. These modalities can be used to express necessary or possible properties of the transition behaviour of α in a natural way. They can be nested or combined propositionally. In first-order dynamic logic [253], where also quantifiers are allowed, $\exists p\,[\alpha]\langle\beta\rangle\phi$ says that there is a choice of parameter p (expressed by $\exists p$) such that for all possible behaviours of system α (expressed by $[\alpha]$) there is a reaction of system β (i.e., $\langle\beta\rangle$) that ensures ϕ. Likewise, $\exists p\,([\alpha]\phi \wedge [\beta]\psi)$ says that there is a choice of parameter p that makes both $[\alpha]\phi$ and $[\beta]\psi$ true, simultaneously, i.e., that makes the conjunction $[\alpha]\phi \wedge [\beta]\psi$ true, saying that formula ϕ holds for all states reachable by α executions and, independently, ψ holds after all β executions. The logical operator $\wedge$ is for conjunction ("and").

On the basis of first-order logic over the reals [288], which we use to describe safe regions of hybrid systems and to quantify over parameter choices, we introduce a first-order dynamic logic over the reals with modalities that directly quantify over the possible transition behaviour of hybrid systems. Since hybrid systems are subject to both continuous evolution and discrete state change, we generalise dynamic logic so that operational models α of hybrid systems can be used in modal formulas such as $[\alpha]\phi$.

With our extension of dynamic logic we can specify desirable properties of hybrid systems. Suppose *train* denotes the hybrid system in Fig. 1.4; then the formula $[train](z \leq s+10)$ expresses that the train always ($[train]$) stays in the region $z \leq s+10$, i.e., it never passes the position $s+10$. This formula cannot be true for fast trains, however, because the model in Fig. 1.4 starts braking at position $z = s$ at the earliest, but there may not be sufficient braking distance to $s+10$ for high velocities v. For some other values of the variables and parameters, instead, the formula may be true, but how can we find out if it is? And how do we determine corresponding parameter constraints that characterise under which circumstances such a formula is true? If we knew these parameter constraints, we may be able to choose s wisely in practise, so that the train model always works as intended.

As a further example, let *ball* be the hybrid system for the bouncing ball from Fig. 1.5. Then the formula $[ball](0 \leq h \leq H)$ expresses that the bouncing ball always ($[ball]$) stays between the ground at height 0 and maximum height H. The answer to the question about whether this formula is true or not depends on the value of H and on the initial height and velocity of the ball (the initial values of h, v) as well as the value of the gravity constant g. For $H = -5$, the formula is certainly false, because the ball can never be lower than the ground ($h < 0$). But for $h = v = 0$ and $H > 1$ it is definitely true, because the ball lies still without moving then. The most interesting cases, however, are those that are neither trivially true nor trivially false, but those where the ball actually bounces for a certain period of time and only the dynamics and relations of parameters and state variables determine the truth of the formula $[ball](0 \leq h \leq H)$.

Finally, suppose *wctrl* is the hybrid system for water control from Fig. 1.10. Then the formula $[wctrl](1 \leq x \leq 10)$ expresses that the water level always ($[wctrl]$) stays between 1 and 10. This formula may be true or false. In fact, it is not true under all circumstances for the model in Fig. 1.10, because the system could still stay in node *open* too long so that $1 \leq x$ no longer holds true. Instead, the dual formula $\langle wctrl \rangle x < 1$ is true, because there is a run of the water tank model ($\langle wctrl \rangle$) that reaches a state where the water level is less than 1 ($x < 1$), which, in fact, holds for any initial water level.

1.1.4 Compositional Verification

As a verification technology for our dynamic logic for hybrid systems, we devise a compositional proof calculus for verifying properties of a hybrid system by proving properties of its parts. Compositional verification is a very powerful proof technique with interesting scalability properties compared to monolithic approaches. It is, in essence, a divide-and-conquer technique for verification. Compositional proof calculi prove properties of a complex system by reducing them to subproperties of its subsystems recursively. Compositional approaches can have the advantage that properties for subsystems are often easier to prove, because they are simpler. For that purpose, our calculus recursively decomposes formula $[\alpha]\phi$ symbolically into an equivalent formula, for instance, $[\alpha_1]\phi_1 \wedge [\alpha_2]\phi_2$ about subsystems α_i of α and subproperties ϕ_i of ϕ. With this, formula $[\alpha]\phi$ can simply be verified by proving the formulas $[\alpha_i]\phi_i$ separately—using the same symbolic decomposition principles recursively—and combining the results conjunctively. In particular, synthesised parameter constraints that are required for the desired property to hold carry over from the latter formula to the former formula just by conjunction. Indeed, our compositional reasoning techniques turn out to be be quite successful in reducing the complexity of more complicated hybrid systems by decomposing them into their submodules.

Unfortunately, hybrid automata are not suitably compositional for this purpose. Their graph structures cannot be decomposed into subgraphs α_i such that the formula $[\alpha_1]\phi_1 \wedge [\alpha_2]\phi_2$ is equivalent to $[\alpha]\phi$, because of the dangling edges between the subgraphs α_i. For instance, the automaton in Fig. 1.4 cannot simply be verified by proving $[accel]\phi \wedge [brake]\phi$, because the effects of edges between the nodes need to be taken into account. Figure 1.12 shows an unsuccessful decomposition attempt of the hybrid automaton for train control from Fig. 1.4. Suppose we had a logic for hybrid automata[1] in which we could formulate logical formulas like $[\alpha](v \leq 8)$, where α is the hybrid automaton from Fig. 1.4. This graphical "formula" is depicted at the bottom of Fig. 1.12. The problem is that this formula at the bottom cannot

[1] The syntax and semantics of such a dynamic logic for hybrid automata can actually be derived from the presentation in Part I quite easily. It is the development of a compositional proof calculus for such a logic for hybrid automata that is technically more involved than the constructions for the logic we present in this book.

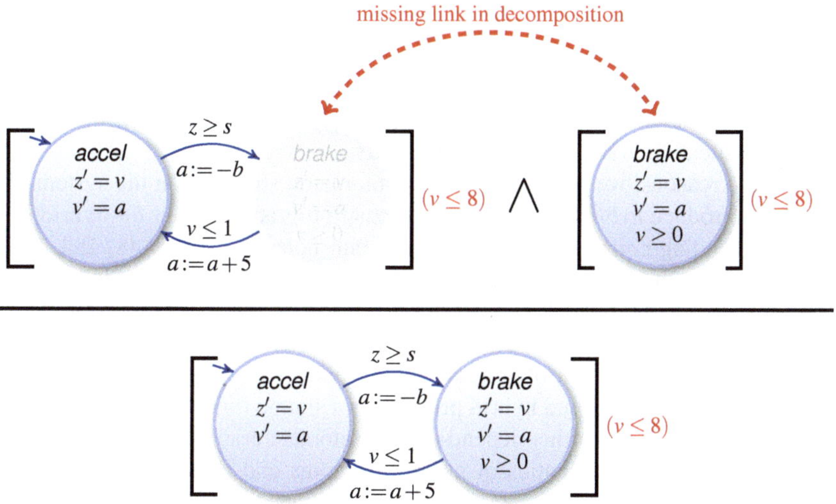

Fig. 1.12 Failed hybrid automaton decomposition attempt

just be proven by decomposing it into a proof of $[accel](v \leq 8) \wedge [brake](v \leq 8)$, because the discrete transition edges are important for the behaviour of the automaton and cannot be ignored. For several reasons, the formula at the bottom of Fig. 1.12 cannot be proven by decomposing it into the conjunction of the two graphical formulas at the top of Fig. 1.12 (the top-left formula results from the bottom formula by isolating node *brake* into the top-right formula, indicated by the grey colour).

One reason why this decomposition fails is that the automaton in the top left formula, which results from the automaton in the bottom by splitting off the *brake* phase, is no longer a hybrid automaton: it now has dangling edges pointing into nowhere. But what is the meaning and behaviour of such a broken, incomplete automaton fragment supposed to be? Remember that node *brake* is not present in the top-left formula, because we wanted to decompose the system in the first place. What should the dynamics look like when the automaton makes a transition into the undefined void? Another part that is missing in the failed decomposition attempt in Fig. 1.12 is the missing (dashed) link from the transitions leaving the automaton in the left formula to the automaton in the right formula and a link that says "if the operations of the automaton in the right formula stop, then the system jumps back to the automaton in the left formula, resuming action with a transition to *accel*." This missing link is pretty strange. It would be a link between the operational behaviour of parts of multiple different formulas. What is the meaning of the left formula supposed to be then if we want to consider it and reason about it alone, without the context of the right formula?

In other words, these decomposition attempts of hybrid automata do not lead to a compositional approach where we can reason separately about the fragments of the system and give a compositional and denotational [277] meaning to all formulas and system models, i.e., a semantics of the whole, solely defined in terms of the semantics of its parts, and not in terms of exterior formulas. The semantics of hybrid automata is just not sufficiently compositional for our purposes. It is not impossible to follow such a graphical approach, but it is technically quite involved and makes decompositions of the hybrid system unnecessarily hard, which is an obstacle to scalable verification, traceability, and tangibility.

For these reasons, we do not impose an automaton structure on the system. Instead, we introduce *hybrid programs* as a textual program notation for hybrid systems by extending conventional discrete program notations [253, 148, 182, 149]. The program on the right of Fig. 1.5 on p. 7, for instance, would be a hybrid program notation for the hybrid automaton on the left of Fig. 1.5. Hybrid programs allow for flexible programmatic combinations of elementary discrete or continuous transitions by structured control programs with a perfectly compositional semantics: The semantics of a compound hybrid program is a simple function of the semantics of its parts and does not further depend on any automata graph structures.

This compositional semantics of hybrid programs and a logic for hybrid programs has significant advantages. Figure 1.13, for instance, shows how the ana-

Fig. 1.13 Successful hybrid program decomposition

$$\frac{[accel](v \leq 8) \land [brake](v \leq 8)}{[accel \cup brake](v \leq 8)}$$

lysis of a property of the hybrid program $accel \cup brake$ that can choose between the operations of $accel$ and of $brake$ by a nondeterministic choice ($\cup$) can be decomposed directly into the formula $[accel](v \leq 8) \land [brake](v \leq 8)$ on the top. By decomposing the property at the bottom about a more complicated hybrid program into independent conjunctions—$[accel](v \leq 8)$ and $[brake](v \leq 8)$—of properties of simpler hybrid programs, we can recursively make the hybrid program simpler by decomposing it until we can prove the remaining properties of subprograms. As a special case, these decompositions also include cases that define switching conditions between acceleration and braking, but no special non-compositional handling of these is required.

In this book we study systematic analysis principles that can be used to decompose properties of complicated hybrid programs into simpler properties as in Fig. 1.13 and that handle the remaining discrete dynamics (discrete assignments), continuous dynamics (differential equations), and their hybrid interactions. We introduce a first-order dynamic logic for hybrid programs that makes these informal ideas rigorous. The resulting logic, which we call *differential dynamic logic* ($\mathsf{d}\mathcal{L}$), constitutes a natural specification and verification logic for hybrid systems. With the goal of developing a solid theoretical, practical, and applicable foundation for deductive verification of hybrid systems by automated theorem proving, the focus of this book is a thorough analysis of the logic $\mathsf{d}\mathcal{L}$, its calculus, extensions, verification procedures, and applications.

1.1.5 Lifting Quantifier Elimination

When proving formulas of differential dynamic logic ($\mathsf{d}\mathcal{L}$), interacting hybrid dynamics cause interactions of arithmetic quantifiers and dynamic modalities, which both affect the values of symbols. For continuous evolutions, we have to prove formulas like $\forall t\,[\alpha]x{\geq}0$, expressing that, for all durations t of some continuous evolution in α, the property $x \geq 0$ holds after all runs ($[\alpha]$) of system α. The standard techniques for handling quantifiers in first-order logic [122, 147, 123] are incomplete for handling these situations, because they are based on instantiation or unification (see App. A), which is already insufficient for proving the tautology $\forall z\,(z^2 \geq 0)$ of real arithmetic. Unfortunately, decision procedures for real arithmetic, such as real *quantifier elimination* [288, 81], cannot handle the real quantifier $\forall t$ in $\forall t\,[\alpha]x{\geq}0$ either, because of the modality $[\alpha]$. The dynamics of the hybrid program α may depend on the value of t, but, at the same time, the constraints on t depend on the effect of α. Quantifier elimination in real-closed fields [288, 81], however, is only defined for first-order logic, not for first-order differential dynamic logic with modalities for hybrid programs. Indeed, the actual algebraic constraints on the quantified real variable t still depend on how the system variables evolve along the dynamics of α. This effect inherently results from the interacting dynamics of hybrid systems, where the duration t of a continuous evolution determines the resulting state and, hence, affects all subsequent discrete or continuous evolutions in α. Thus, the effect of α first needs to be analysed with respect to the arithmetical constraints it imposes on t for $x \geq 0$ to hold, before the quantifier $\forall t$ can be handled.

In this book, we present a calculus that is suitable for automation and combines deductive and arithmetical quantifier reasoning within a single compositional proof calculus. It introduces *real-valued free variables* and *real Skolem terms* to postpone quantifier elimination and continue reasoning beyond the occurrence of a real quantifier in front of a modality. Quite unlike classical proof techniques, however, our calculus later reintroduces a corresponding quantifier into the proof when its algebraic constraints have been discovered completely. For $\forall t\,[\alpha]x \geq 0$, our calculus will, for instance, continue with the unquantified kernel $[\alpha]x \geq 0$ after replacing variable t with a Skolem term $s(x)$. Once all arithmetical constraints on $s(x)$ are known, a quantifier for $s(x)$ will be reintroduced and handled by real quantifier elimination [288, 81]. The Skolem term $s(x)$ can be thought of as a new name for the universally quantified variable t. In a similar manner, our calculus combines quantifier elimination with deduction for handling existential real quantifiers using real-valued free variables.

In Chap. 2, we introduce a calculus that makes this intuition formally precise. Crucially, we exploit the relationship of Skolem terms and free variables in order to keep track of the lost quantifier nesting to prohibit unsound rearrangements of quantifiers when they are reintroduced. After all, it would be disastrous if we could start out proving a property $\exists x \forall y\,[\alpha]\phi$, turn the quantifiers for x and y into free variables and Skolem terms, and accidentally reintroduce quantifiers in a swapped order to prove the weaker statement $\forall y \exists x \ldots$ only. Clearly, it would not be sufficient to find a different x for each value of y, if, what we started looking for in the first place is one common x for all the y such that $[\alpha]\phi$ holds. The corresponding calculus rules that we introduce, which comply with all soundness requirements, are perfectly natural and coincide nicely with the prerequisites for quantifier elimination over the reals. Further, the d$\mathcal{L}$ semantics and calculus are fully compositional so that properties of a hybrid program can be proven by the reduction to properties of its parts following a structural symbolic decomposition within the compositional d$\mathcal{L}$ proof calculus.

1.1.6 Differential Induction and Differential Strengthening

In Chap. 3, we extend our logic d$\mathcal{L}$ to the *differential-algebraic dynamic logic* DAL, which is the logic of general hybrid change. DAL provides *differential-algebraic programs* (DA-programs) as general models for hybrid systems by allowing for propositional operators and quantifiers in discrete and continuous transitions. DA-programs have a very natural semantics of simultaneous change using just conjunctions ($\wedge$) as logical operators. For instance, the conjunctive *differential-algebraic constraint* $x' = y \wedge y' = -x$, in which each conjunct needs to hold during continuous evolutions, gives a natural semantics to differential equation systems $x' = y, y' = -x$, such that $x' = y, y' = -x$ is a notational variant of $x' = y \wedge y' = -x$. Disjunctions, in contrast, are a natural way to express switching conditions and finite nondeterminism in the dynamics. The disjunctive differential-algebraic constraint $x' = y \vee x' = -y$, for instance, allows x to evolve with velocity y or with velocity $-y$ and even switch between both cases, because only one of the two disjuncts needs to hold at any time. Quantifiers in the dynamics of DA-programs are a very expressive extension with which differential-algebraic equations and differential inequalities can be represented directly, and with which quantified disturbance in the dynamics can be characterised in an elegant and uniform way. The quantified differential-algebraic constraint $\exists y\,(x' = y \wedge y^2 < 5)$, for example, allows *all* evolutions of x with *some* velocity y (i.e., $\exists y$) whose square is less than 5. This quantified nondeterminism makes DA-programs quite expressive for characterising more involved system dynamics.

The standard approach to dealing with continuous dynamics for hybrid systems is to use symbolic or numerical solutions of their respective differential equations. Unfortunately, the range of systems that is amenable to these techniques is fairly limited, because even solutions of simple linear differential equations quickly fall into undecidable classes of arithmetic. For instance, the solutions of the linear differential equation system $s' = c, c' = -s$ are trigonometric functions like sin and cos. But first-order arithmetic with trigonometric functions is undecidable by a simple corollary to Gödel's famous incompleteness theorem [137]. As a means for verifying hybrid systems with challenging continuous dynamics without having to solve their differential equations, we complement discrete induction for loops and repetitions with a new form of *differential induction* for differential equations. Differential induction is a natural induction technique for differential equations. It is based on the local dynamics of the (right-hand side of the) differential equations themselves and does not need closed-form solutions for the differential equations. Because differential equations are simpler than their solutions (which is part of the representational power of differential equations), differential induction techniques working with the differential equations themselves are more scalable than techniques that need solutions of differential equations. Our differential induction techniques even generalise to differential-algebraic constraints with differential inequalities or quantifiers in the dynamics.

To further increase the verification power, we add differential strengthening or differential cuts as a powerful proof technique for refining the system dynamics with auxiliary invariants that can simplify the proof of the original property significantly. The basic insight is that auxiliary properties that are provable invariants of the dynamics can help prove the original property even if it was not provable before. This phenomenon is unique to differential equations and does not happen in classical discrete systems. Overall, our combination of compositional proof calculi with differential induction and differential strengthening turns out to be very powerful for the analysis of advanced hybrid systems, including air traffic control applications.

1.2 Related Work

In this section, we briefly discuss related approaches to verification.

Model Checking of Hybrid Automata

Model checking approaches work by state space exploration and—due to their undecidable reachability problem—require [156] various abstractions or approximations [159, 8, 156, 125, 15, 77, 291, 217] for hybrid automata, including numerical approximations [70, 21].

Beyond standard approaches [8, 156, 126] for linear automata with constant dynamics, the seminal work of Lafferriere et al. [189, 188, 190] presented a decision procedure for o-minimal hybrid automata and classes of linear dynamics with a homogeneous eigenstructure. They have analysed the discrete and continuous dynamics independently, which requires completely decoupled dynamics with forgetful jumps, i.e., where the outcome of a jump is completely independent of the continuous state. Unfortunately, actual systems rarely forget about states completely. For modelling train dynamics accurately, for example, it is important that the train position z and velocity v stays the same when switching from acceleration to braking mode in Fig. 1.4, and these variables do not just change arbitrarily and independently of their prior values.

Chutinan and Krogh [70] presented polyhedral approximations of hybrid automata with polyhedral discrete dynamics, invariants, and initial state sets. The relevant discrete dynamics, initial state regions, and invariants for our train and aircraft applications are nonlinear and thus cannot be described accurately by polyhedra. Fränzle [125] showed that reachability is decidable for specific classes of robust polynomial hybrid automata, where the safe and unsafe states are sufficiently separate and the safe region is bounded. Asarin et al. [21] used piecewise linear numerical approximations in an approximate reachability algorithm for continuous systems with known Lipschitz bounds. Mysore et al. [217] showed decidability of bounded-time and bounded switching reachability prefixes of semialgebraic hybrid automata. Because there seldom is a known bound on the number of transitions that a system can make, nor a (small) bound on the lifetime of a system, we are mostly interested in unbounded horizon properties, which cannot be obtained with bounded model checking.

Model checking tools like HyTech [157], PHAVer [126, 127], d/dt [22], and CheckMate [70] are already quite successful. Still these or other approaches [156, 125, 228] cannot handle our train and aircraft applications with nonlinear switching, nonlinear discrete and continuous dynamics, and high-dimensional state spaces (more than 30 dimensions).

Because hybrid systems do not admit equivalent finite-state abstractions [156] and due to general limits of numerical approximation [238], model checkers are still more successful in falsification than in verification. To obtain a sound verification approach and for improved handling of free parameters [97], we follow a symbolic logic-based approach and support $d\mathcal{L}$ as a significantly more expressive specification language. Finally, we introduce hybrid programs as a more uniform model for hybrid systems that is amenable to compositional symbolic verification.

Logics for Real-time Systems

Logics for real-time systems [159, 275] are not expressive enough to capture the dynamics of hybrid systems, particularly their differential equations, which are the main focus of this book. For instance, Schobbens et al. [275] give complete axiomatisations of two decidable dense time propositional linear temporal logics. Unfortunately, in these propositional logics one cannot express that relevant separation properties hold always during the flight of aircraft guided by specific flight controllers. A crucial property, for instance, is that the system never leaves region $(x_1-y_1)^2+(x_2-y_2)^2 \geq p^2$, which expresses that the distance of the aircraft at (x_1,x_2) and the aircraft at (y_1,y_2) is at least the protected zone p.

Successful model checking approaches for timed automata, which are a real-time restriction of hybrid automata, have been developed before, including the tools KRONOS [98] and UPPAAL [193]. Unlike for hybrid systems, model checking for timed automata is decidable [10, 5, 86]. But the expressive power of timed automata is fairly limited. The only continuous variables are clocks t following the differential equation $t' = 1$. Positions or velocities evolving according to other laws of continuous dynamics are not allowed. In fact, even very minor extensions of timed automata make the verification problem undecidable, including stopwatches [67], which are clocks that can be stopped (dynamics $t' = 0$) and resumed (dynamics $t' = 0$). For a general survey of real-time verification approaches, we refer the reader to the book by Olderog and Dierks [221].

Logics for Hybrid Systems

Zhou et al. [306] extended the duration calculus [305] with mathematical expressions in derivatives of state variables. They use a multitude of calculus rules and a non-constructive oracle that requires external mathematical reasoning about the notions of derivatives and continuity. This undirected reasoning about mathematics, especially derivatives and continuity, is not suitable for automatic verification.

Davoren and Nerode [95, 97] presented a semantics of modal μ-calculus [181] in hybrid systems and examined topological aspects. They provided Hilbert-style calculi to prove formulas that are valid for all hybrid systems simultaneously. With this, however, only limited information can be obtained about a particular system: In propositional modal logics, system behaviour needs to be axiomatised declaratively in terms of abstract actions a, b, c of unknown effect. With unknown effects, however, we cannot characterise, say, a train control system or the flight dynamics for air traffic control, and not even the dynamics of a bouncing ball.

The strength of our logic primarily is that it is an expressive *first-order* dynamic logic: It handles actual operational models of hybrid systems such as $z'' = a$ instead of abstract propositional actions of unknown effect. The advantage of our calculus in comparison to others [306, 95, 97] is that it provides a constructive modular combination of arithmetic reasoning with reasoning about hybrid transitions and works by structural symbolic decomposition. With this, our calculus can be used easily for verifying actual operational hybrid system models, including railway and air traffic control systems, which are of considerable practical interest [156, 58, 70, 77, 217, 90, 238, 91]. Our proof calculus supports free parameters and first-order definable flows, which are well suited for verifying the coordination of train dynamics. First-order approximations of more general flows can be used according to [15, 238, 227]. More general dynamics can also be verified with the technique of differential induction that we introduce in Chap. 3.

Specification Languages for Hybrid Systems

Inspired by a proposal by He [175], Zhou et al. [69] presented a hybrid variant of CSP [162] as a language for describing hybrid systems. They gave a semantics in the extended duration calculus [306]. Zhou et al. did not yet give an actual verification approach for hybrid CSP.

Rönkkö et al. [270] extended guarded command programs with differential relations and gave a weakest-precondition semantics in higher-order logic with built-in derivatives. Without providing a means for verification of this higher-order logic, this approach is limited to providing a notational variant of standard mathematics.

Rounds and Song [272] have developed a hybrid version of the π-calculus [208] as a modelling language for embedded systems. Later, Rounds [271] gave a semantics in a rich set theory for an abstract spatial logic for the hybrid π-calculus, which is also called Φ-calculus. In the hybrid π-calculus, processes interact with a continuously changing environment, but cannot themselves evolve continuously, which would be crucial to capture the physical movement of traffic agents. From the semantics alone, however, no verification is possible in these approaches, except perhaps by manual semantic reasoning.

Other process-algebraic approaches, such as χ [40], have been developed for modelling and simulation purposes [39]. At present, verification is still limited to small fragments that can be translated directly to other verification tools such as PHAVer [127, 126] when converted to hybrid automata or to UPPAAL [193] when converted to timed automata.

Modelling languages for hybrid systems further include SHIFT [100] for networks of hybrid automata, and R-Charon for reconfigurable systems [183]. These approaches focus more on simulation and compilation [100] or the development of a semantics [183], so no verification or formal analysis is possible yet.

Uses of Deduction for Hybrid Systems

Manna et al. [201, 178] and Ábrahám et al. [1] used theorem provers for checking invariants of hybrid automata in STeP [201] and PVS [1], respectively. Their working principle is, however, quite different from ours. Given a hybrid automaton and a global system invariant, they compile, in a single step, a verification condition expressing that the invariant is preserved under all transitions of the hybrid automaton. Hence, hybrid aspects and transition structure vanish completely before the proof starts. All that remains is a flat quantified mathematical formula. Which hybrid systems can be verified with this approach in practise strongly depends on the general mathematical proving capabilities of STeP and PVS, which, because of their general-purpose focus, typically requires user interaction.

In contrast, we follow a fully symbolic approach using a genuine specification and verification logic for hybrid systems. Our dynamic logic for hybrid systems works deductively by symbolic decomposition and preserves the transition structure during the proof, which simplifies the traceability of results considerably. Further, the structure in this symbolic decomposition can be exploited for deriving invariants or parametric constraints. Consequently, invariants do not necessarily need to be given beforehand in our approach. Moreover, in practise, guiding quantifier elimination procedures along natural splitting possibilities of the structural decomposition performed by the our proof calculus turns out to be important for successful automatic proof strategies (Chaps. 5 and 6).

1.3 Contributions

Our main conceptual contribution is a series of differential dynamic logics for hybrid systems (these logics are called $d\mathcal{L}$, DAL, and dTL), which capture the logical quintessence of the dynamics of hybrid systems succinctly. Our logics provide a uniform semantics and concise language for specifying and verifying correctness properties of general hybrid systems with sophisticated dynamics. Our main practical contribution is a concise *free-variable calculus* that axiomatises the transition behaviour of hybrid systems relative to differential equation solving. With our generalisation of free-variable calculi to dynamic logic over the reals, the calculus is suitable for automated theorem proving and for verifying hybrid superpositions of interacting discrete and continuous dynamics compositionally.

Our main theoretical contribution is that we prove our calculi to be sound and complete relative to the handling of differential equations. To the best of our knowledge, this is the first relative completeness proof for a logic of hybrid systems, and even the first formal notion of hybrid completeness. Our results fully align hybrid and continuous reasoning proof-theoretically and show that hybrid systems with interacting repetitive discrete and continuous evolutions can be verified whenever differential equations can.

We further contribute a verification calculus that includes uniform proof rules for differential induction along differential equations or more general differential-algebraic constraints, using a combination of differential invariants, differential variants, and differential strengthening for verifying hybrid systems without having to solve their differential constraints. Based on these calculi, we develop a fixed-point verification algorithm that computes the required invariants and differential invariants for a formula and refines the underlying system dynamics as needed during the proof.

As applied contributions, we demonstrate the capabilities of our logics, calculi, and algorithms by verifying collision avoidance in realistic train control applications and challenging air traffic control manoeuvres. Overall, our logical analysis approach for hybrid systems can successfully verify realistic applications that were out of the scope of other approaches, for both theoretical and scalability reasons.

1.4 Structure of This Book

This book consists of three parts that basically correspond to the theory, practise, and applications, respectively, of the logical analysis of hybrid systems. You are now reading the introduction.

Logics and Calculi

In Part I, which is the core of this book, we introduce novel logics and proof calculi that form the new conceptual, formal, and technical basis for the logical analysis of hybrid systems. In Chap. 2, we introduce the differential dynamic logic $\mathsf{d}\mathcal{L}$ as a variant of dynamic logic that is suitable for specifying and verifying properties of hybrid systems. It generalises dynamic logic to dynamic logic over the reals in the presence of hybrid dynamics with discrete state transitions and with continuous state evolutions along differential equations. As a verification technique, we present a new compositional sequent calculus for $\mathsf{d}\mathcal{L}$ that is suitable for automation and integrates handling of real quantifiers by generalising Skolemisation and free variables to the reals. In Chap. 2, we also prove completeness relative to differential equations as the most fundamental theoretical result in this book.

In Chap. 3, we introduce the differential-algebraic logic DAL that extends the class of hybrid system models by allowing more general differential-algebraic equations, differential inequalities, and quantified nondeterminism. Further, we present a uniform theory of differential induction, differential invariants, differential variants, and differential strengthening as central symbolic verification techniques for handling challenging continuous dynamics in hybrid systems without having to solve their differential equations.

In Chap. 4, we address the handling of temporal properties and introduce the differential temporal dynamic logic dTL along with a calculus that reduces temporal properties to $\mathsf{d}\mathcal{L}$ properties. The extensions of $\mathsf{d}\mathcal{L}$ that we present in Chaps. 3 and 4 are complementary and compatible. Their direct modular combination immediately defines the differential-algebraic temporal dynamic logic DATL.

Automated Theorem Proving

In Part II, we focus on the practical aspects of implementing the verification calculi from Part I. The calculi in Part I have already been designed for the practical needs of automated theorem proving, most notably for the free-variable and Skolemisation techniques from Chap. 2 and the compositional proof calculi from Part I. Immediate implementations of the proof calculi from Part I in automated theorem provers can prove examples of medium complexity directly. Yet, more complex case studies require additional algorithmic techniques for achieving high degree automation and good scalability properties. In Chap. 5, we refine the calculi from Part I to tableau procedures that are suitable for automated theorem proving (ATP) and present proof strategies that navigate through their nondeterminisms to help overcome the complexity issues of integrating real quantifier elimination as a decision procedure for real arithmetic.

In Chap. 6, which is based on joint work with Edmund M. Clarke [239], we introduce the "differential invariants as fixed points" (DIFP) paradigm. We refine the differential induction techniques from Chap. 3 to a fully automatic verification algorithm for computing the required discrete and differential invariants of a hybrid system locally in a logic-based fixed-point loop.

Applications

In Part III, we shift our attention to application scenarios for our logical analysis approach for hybrid systems. Extending smaller hybrid systems which have served as running examples throughout this book, we show full verification case studies of the European Train Control System (ETCS) in Chap. 7, which is based on joint work with Jan-David Quesel [243]. We also extend and show verification case studies for aircraft collision avoidance manoeuvres in air traffic control (ATC) in Chap. 8, which is based on joint work with Edmund M. Clarke [238, 239].

Finally, Chap. 9 concludes this work with a discussion of the results and perspectives for future research.

Appendices

In Part IV, we provide the background in logic and differential equations that we need for the course of this book. In App. A, we give an introduction to basic first-order logic (FOL), its syntax, semantics, and proof techniques. For reference, App. B summarises some classical results about ordinary differential equations (ODEs) that we need as background for this book. In App. C, we formally investigate the relationship between hybrid automata [156] and hybrid programs by embedding hybrid automata into hybrid programs. In App. D, we briefly characterise the verification tool KeYmaera that implements the logics and automated theorem proving techniques presented in this book and that has been implemented in joint work with Jan-David Quesel [242]. We also survey various techniques that can be used to verify real arithmetic.

Online Material

At the Website for this book, we provide the KeYmaera verification tool for download and webstart. KeYmaera has been developed in joint work with Jan-David Quesel [242] and implements the logical analysis approach presented in this book. Slide material, an online tutorial, and KeYmaera problem files for several examples, including the case studies of train and air traffic control, can also be found on the Web.

The Web page for this book is at the following URL:

`http://symbolaris.com/lahs/`

Suggested Reading Sequence

The basic suggested reading sequence in this book is linear (with additional consultation of the appendices for background). Except for the foundation of this work that is laid out in Part I, however, the chapters in this book are mostly self-contained so that they can also be studied independently. Figure 1.14 shows the reading order dependencies among the chapters (solid lines) and the partial dependencies of suggested reading sequences that holds for the advanced material of the respective chapters (dashed lines).

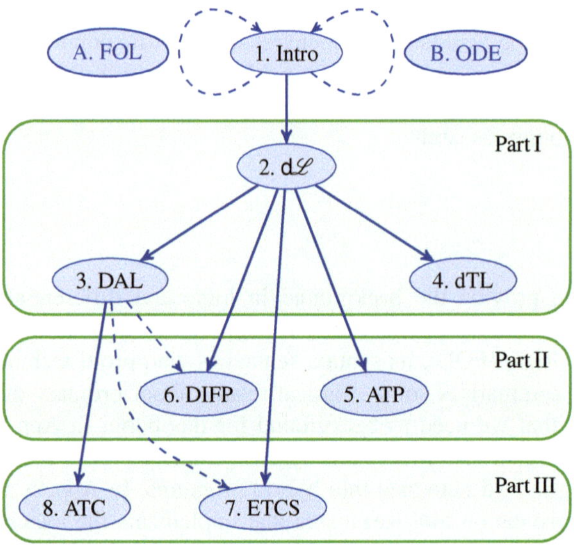

Fig. 1.14 Dependencies and suggested reading sequence of chapters and appendices

For a background in classical first-order logic (FOL), we recommend you review App. A. Depending on the interest, field of study, and preference of the reader, we recommend he either study the background information in App. A on first-order logic before proceeding to Part I or use the material in App. A as a background reference book while reading the main part of this book. Similarly, we recommend he review the background on ordinary differential equations (ODEs) in App. B either before or during the study of this book. An intuitive approach to understanding differential equations and formal definitions of their semantics will be given in the main parts of this book. Logic itself is also explained and illustrated intuitively during the main part of this book, but some readers may also find it helpful to refresh or learn about the basics of first-order logic from App. A before proceeding to the main part.

While there is a lot of flexibility in the reading sequence of the chapters, we strongly recommend you study the logical foundations of hybrid systems analysis and differential dynamic logic (d$\mathcal{L}$) in Chap. 2 of Part I before reading any other chapter of Parts I–III. Chapter 2 develops the logical foundations, the system model of hybrid programs, the differential dynamic logic ($d\mathcal{L}$) for expressing correctness properties, and the proof calculus for verifying these properties that will be needed in the remainder of this book.

Some more advanced sections in the applications in Part III also depend on the theory of differential invariants and the differential algebraic dynamic logic (DAL) that is developed together with other extensions in Chap. 3. We recommend you read Chap. 3 on DAL before studying the air traffic control (ATC) verification in Chap. 8. While most of Chap. 7 on the European Train Control System (ETCS) verification can be read with the foundation in $d\mathcal{L}$ from Chap. 2, some advanced parts also use results from Chap. 3 on DAL. We also recommend you study the theoretical foundations on DAL in Chap. 3 before reading the automation approach "differential invariants as fixed points" (DIFP) in Chap. 6. Still, some level of understanding of the DIFP automation approach in Chap. 6 can also be gained without your having read the full theoretical background on DAL in Chap. 3.

Part I
Logics and Proof Calculi for Hybrid Systems

Overview In this part, which is the core part of this book, we introduce novel logics and proof calculi that form the new conceptual, formal, and technical basis for the logical analysis of hybrid systems. In Chap. 2, we introduce the differential dynamic logic $d\mathcal{L}$ as a variant of dynamic logic that is suitable for specifying and verifying properties of hybrid systems. It generalises classical dynamic logic to dynamic logic over the reals in the presence of hybrid dynamics with interacting discrete state transitions and continuous state evolutions along differential equations. As a verification technique, we present a new compositional proof calculus for $d\mathcal{L}$ that is suitable for automation and integrates handling of real quantifiers by generalising Skolemisation and free variables to the reals. In Chap. 2, we also prove completeness relative to differential equations as the most fundamental theoretical result in this book.

In Chap. 3, we introduce the differential-algebraic logic DAL that extends the class of hybrid system models by allowing more general differential-algebraic equations, differential inequalities, and quantified nondeterminism. Further, we present a uniform theory of differential induction, differential invariants, differential variants, and differential strengthening as central symbolic verification techniques for handling challenging continuous dynamics in hybrid systems without having to solve their differential equations.

In Chap. 4, we address the handling of temporal properties and introduce the differential temporal dynamic logic dTL along with a calculus that reduces temporal properties to $d\mathcal{L}$ properties. The extensions of $d\mathcal{L}$ that we present in Chap. 3 and Chap. 4 are complementary and compatible. Their direct modular combination immediately defines the differential-algebraic temporal dynamic logic DATL.

The logics and proof techniques developed in this part will form the basis for the automation techniques developed in Part II. They also form the foundation for the formal verification tool KeYmaera. We will also use the differential dynamic logics to formalise safety-critical properties of the train and aircraft control studies in Part III and prove them with the proof techniques we develop in Part I.

Chapter 2
Differential Dynamic Logic d$\mathcal{L}$

Contents

2.1 Introduction **34**
  2.1.1 Structure of This Chapter 35
2.2 Syntax **35**
  2.2.1 Terms 37
  2.2.2 Hybrid Programs 41
  2.2.3 Formulas 47
2.3 Semantics **49**
  2.3.1 Valuation of Terms 50
  2.3.2 Valuation of Formulas 51
  2.3.3 Transition Semantics of Hybrid Programs 54
2.4 Collision Avoidance in Train Control **61**
2.5 Proof Calculus **64**
  2.5.1 Substitution 65
  2.5.2 Proof Rules 76
  2.5.3 Deduction Modulo with Invertible Quantifiers and Real Quantifier Elimination 88
  2.5.4 Verification Example 94
2.6 Soundness **97**
2.7 Completeness **101**
  2.7.1 Incompleteness 102
  2.7.2 Relative Completeness 103
  2.7.3 Characterising Real Gödel Encodings 105
  2.7.4 Expressibility and Rendition of Hybrid Program Semantics 106
  2.7.5 Relative Completeness of First-Order Assertions 109
  2.7.6 Relative Completeness of the Differential Logic Calculus 113
2.8 Relatively Semidecidable Fragments **114**
2.9 Train Control Verification **118**
  2.9.1 Finding Inductive Candidates 118
  2.9.2 Inductive Verification 119
  2.9.3 Parameter Constraint Discovery 120
2.10 Summary **122**

A. Platzer, Logical Analysis of Hybrid Systems,
DOI 10.1007/978-3-642-14509-4_2,

Synopsis Hybrid systems are models for complex physical systems and are defined as dynamical systems with interacting discrete transitions and continuous evolutions along differential equations. With the goal of developing a theoretical and practical foundation for deductive verification of hybrid systems, we introduce a dynamic logic for hybrid programs, which is a program notation for hybrid systems. As a verification technique that is suitable for automation, we introduce a free-variable proof calculus with a novel combination of real-valued free variables and Skolemisation for lifting quantifier elimination for real arithmetic to dynamic logic. The calculus is compositional, i.e., it reduces properties of hybrid programs to properties of their parts. Our main result proves that this calculus axiomatises the transition behaviour of hybrid systems completely relative to differential equations. In a study with cooperating traffic agents of the European Train Control System, we further show that our calculus is well suited for verifying realistic hybrid systems with parametric system dynamics.

2.1 Introduction

In this chapter, we introduce the differential dynamic logic $d\mathcal{L}$, its syntax, semantics, and proof calculus. It forms the core of this book and is the basis for the extensions, algorithmic refinements, and applications in subsequent chapters of this book.

Contributions

Our main conceptual contribution in this chapter is the differential dynamic logic $d\mathcal{L}$ for hybrid programs, which captures the logical quintessence of the dynamics of hybrid systems succinctly. Our main practical contribution is a concise free-variable calculus for $d\mathcal{L}$ that axiomatises the transition behaviour of hybrid systems relative to differential equation solving. It is suitable for automated theorem proving and for verifying hybrid interacting discrete and continuous dynamics compositionally. Our main theoretical contribution is that we prove the $d\mathcal{L}$ calculus to be sound and complete relative to the handling of differential equations. To the best of our knowledge, this is the first relative completeness proof for a logic of hybrid systems, and even the first formal notion of hybrid completeness. Our results fully align hybrid and continuous reasoning proof-theoretically and show that hybrid systems with interacting repetitive discrete and continuous evolutions can be verified whenever differential equations can. As an applied contribution, we further demonstrate that our logic and calculus can be used successfully for verifying collision avoidance in realistic train control applications.

2.1.1 Structure of This Chapter

After introducing syntax and semantics of the differential dynamic logic $d\mathscr{L}$ in Sects. 2.2 and 2.3, we introduce a free-variable sequent calculus for $d\mathscr{L}$ in Sect. 2.5 and prove soundness and relative completeness in Sects. 2.6 and 2.7, respectively. We present relatively semidecidable fragments of $d\mathscr{L}$ in Sect. 2.8. In Sect. 2.9, we use our calculus to prove an inductive safety property of the train control system that we present in Sect. 2.4. We draw conclusions and discuss future work in Sect. 2.10.

2.2 Syntax of Differential Dynamic Logic

In this section, we introduce the *differential dynamic logic* $d\mathscr{L}$ in which operational models of hybrid systems are internalised as first-class citizens, so that correctness statements about the transition behaviour of hybrid systems can be expressed as formulas. As a basis, $d\mathscr{L}$ includes (nonlinear) real arithmetic for describing concepts like safe regions of the state space. Further, $d\mathscr{L}$ supports real-valued quantifiers for quantifying over the possible values of system parameters or durations of continuous evolutions. For talking about the transition behaviour of hybrid systems, $d\mathscr{L}$ provides modal operators such as $[\alpha]$ or $\langle\alpha\rangle$ that refer to the states reachable by following the transitions of hybrid system α.

The logic $d\mathscr{L}$ is a first-order dynamic logic over the reals for hybrid programs, which is a compositional program notation for hybrid systems. Hybrid programs provide the following constructs.

Discrete jump sets. Discrete transitions are represented as instantaneous assignments of values to state variables, which are, essentially, difference equations. They can express resets like $a := -b$ or adjustments of control variables like $a := a+5$, as occurring in the discrete transformations attached to edges in hybrid automata; see Fig. 2.1. Likewise, implicit discrete state changes such as the changing of evolution modes from one node of an automaton to the other can be expressed uniformly as, e.g., $q := brake$, where variable q remembers the current node. To handle simultaneous changes of multiple variables, discrete jumps can be combined to sets of jumps with simultaneous effect following corresponding techniques in the discrete case [37]. For instance, the discrete jump set $a := a+5, A := 2a^2$ expresses that a is increased by 5 and, simultaneously, variable A is set to $2a^2$, which is evaluated *before* a receives its new value $a+5$.

Differential equation systems. Continuous variation in system dynamics is represented using differential equation systems as evolution constraints. For example the (second-order) differential equation $z'' = -b$ describes deceleration with braking force b and $z' = v, v' = -b \,\&\, v \geq 0$ expresses that the evolution only applies as long as the speed is $v \geq 0$, which represents mode *brake* of Fig. 2.1. This is an evolution along the differential equation system $z' = v, v' = -b$ that is restricted (written &) to remain within the evolution domain region $v \geq 0$, i.e., to stop braking before $v < 0$. Such an evolution can stop at any time within $v \geq 0$, it could even continue with transient grazing along the border $v = 0$, but it is never allowed to enter $v < 0$. The second-order differential equation $z'' = -b$ itself is equivalent to the first-order differential equation system $z' = v, v' = -b$, in which the velocity v is explicit. In this chapter, we separate the respective differential equations in a differential equation system by a comma (,) and separate the evolution domain region (if any) from the differential equations by an ampersand (&). We choose this notation for this chapter to make it easier to identify the evolution domain region. In Chap. 3, we will see that both (,) and (&) can be understood more uniformly as conjunctions.

Control structure. Discrete and continuous transitions—represented as difference or differential equations, respectively—can be combined to form a hybrid program with interacting hybrid dynamics using regular expression operators $(\cup, {}^*, ;)$ of regular programs [149] as control structure. For example, the hybrid program $q:=accel \cup z'' = -b$ describes a train controller that can choose to either switch to acceleration mode ($q:=accel$) or brake by the differential equation $z'' = -b$, by a nondeterministic choice ($\cup$). The nondeterministic choice $q:=accel \cup z'' = -b$ expresses that either $q:=accel$ or $z'' = -b$ happens. The sequential composition $q:=accel; z'' = -b$, instead, expresses that first $q:=accel$, and then $z'' = -b$ happens. In conjunction with other regular combinations, control constraints can be expressed using tests like $?z \geq s$ as guards for the system state. This test will succeed if, indeed, the current state of the system satisfies $z \geq s$; otherwise the test will fail and execution cannot proceed. In that respect, a test is like an assert statement in conventional programs and cuts the system run if the test is not successful.

Example 2.1 (Embedding hybrid automata). With these operations, hybrid systems can be represented naturally as hybrid programs. For example, the right of Fig. 2.1 depicts a hybrid program rendition of the hybrid automaton on the left, which repeats the automaton from Fig. 1.4 on p. 5. Line 1 represents that, in the beginning,

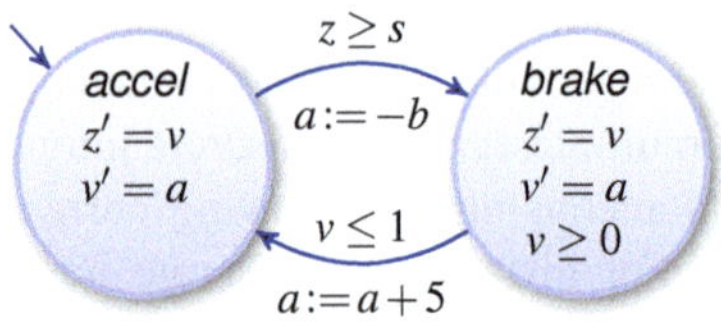

$$
\begin{array}{l}
q:=accel; \quad \text{/* initial mode is node accel */} \\
(\ \ (?q=accel;\ z'=v, v'=a) \\
\cup\ (?q=accel \land z \geq s;\ a:=-b;\ q:=brake;\ ?v \geq 0) \\
\cup\ (?q=brake;\ z'=v, v'=a \,\&\, v \geq 0) \\
\cup\ (?q=brake \land v \leq 1;\ a:=a+5;\ q:=accel))^*
\end{array}
$$

Fig. 2.1 Hybrid program rendition of hybrid automaton for (overly) simplified train control

the current node q of the system is the initial node *accel*. We represent each discrete and continuous transition of the automaton as a sequence of statements with a nondeterministic choice ($\cup$) between these transitions. Line 4 represents a continuous transition of the automaton. It tests if the current node q is *brake*, and then (i.e., if the test was successful) follows the differential equation system $z' = v, v' = a$ restricted to the evolution domain $v \geq 0$. Line 3 characterises a discrete transition of the automaton. It tests the guard $z \geq s$ when in node *accel*, and, if successful, resets $a := -b$ and then switches q to node *brake*. By the semantics of hybrid automata [8, 156], an automaton in node *accel* is only allowed to make a transition to node *brake* if the evolution domain restriction of *brake* is true when entering the node, which is expressed by the additional test $?v \geq 0$ at the end of line 3. Observe that this test of the evolution domain region generally needs to be checked as the last operation after the guard and reset, because a reset like $v := v - 1$ could affect the outcome of the evolution domain region test. In order to obtain a fully compositional model, hybrid programs make all these implicit side conditions explicit. Line 2 represents the continuous transition when staying in node *accel* and following the differential equation system $z' = v, v' = a$. Line 5 represents the discrete transition from node *brake* of the automaton to node *accel*.

Lines 2–5 cannot be executed unless their tests succeed. In particular, at any state, the nondeterministic choice ($\cup$) among lines 2–5 reduces de facto to a nondeterministic choice between either lines 2–3 or between lines 4–5. At any state, q can have value either *accel* or *brake* (assuming these are different constants), not both. Consequently, when $q = brake$, a nondeterministic choice of lines 2–3 would immediately fail the tests in the beginning and not execute any further. The only remaining choices that have a chance to succeed are lines 4–5 then. In fact, only the single successful choice of line 4 would remain if the second conjunct $v \leq 1$ of the test in line 5 does not hold for the current state. Note that, still, all four choices in lines 2–5 are available, but at least two of these nondeterministic choices will always be unsuccessful. Finally, the repetition operator (*) at the end of Fig. 2.1 expresses that the transitions of a hybrid automaton, as represented by lines 2–5, can repeat indefinitely, possibly taking different nondeterministic choices between lines 2–5 at every repetition. □

2.2.1 Terms

The construction of the logic d$\mathcal{L}$ starts with a set V of logical variables and a *signature* Σ, which is the set of names (called *symbols*) of all entities nameable in a certain context. The signature Σ and set V form the vocabulary or alphabet of signs from which well-formed formulas can be built. For d$\mathcal{L}$ we assume all variables in V are interpreted over the reals and that Σ is a (finite) set of real-valued function and predicate symbols, with the usual function and predicate symbols for real arithmetic, such as $0, 1, +, -, \cdot, /, =, \leq, <, \geq, >$, where $+$ is addition, $\cdot$ is multiplication, $/$ is division and so on. For each function and predicate symbol, we are given the number of arguments that it expects, which is called *arity*, and is a natural number. The arity can be zero, in which case the function or predicate symbol does not have any arguments. The function symbols for the numbers 0,1 have arity zero, because they do not need arguments. The binary arithmetic operators $+, -, \cdot, /$ have arity

2, because they expect two arguments. The binary predicate symbols $=, \leq, <, \geq, >$ also have arity 2, because they need two arguments to compare.

The difference between function and predicate symbols is that function symbols stand for functions that take the values of arguments and give back a function value. Predicate symbols, on the other hand, are interpreted either as true for a vector of arguments or as false. That is, they take the values of arguments and give either the truth-value "*true*" or the truth-value "*false*". No other result is permitted for predicate symbols. For instance, the predicate symbol $\geq$ will be understood such that $\geq(x,y)$ is true if and only if the value of x is greater than or equal to the value of y. For real arithmetic, we use standard notation and standard semantics. In particular, we write $x \geq y$ instead of $\geq(x,y)$. We fix the semantics of $\cdot$ to be multiplication, i.e., the value of $\cdot(x,y)$ is always meant to be the product of the value of x and the value of y. Again, we use the standard notation $x \cdot y$, or just xy if no confusion arises, instead of $\cdot(x,y)$. The denotation of a function symbol could also be, e.g., the function that takes an argument and gives back its cube. A predicate, in contrast, cannot give back any value other than "*true*" or "*false*", but could hold, say, for all real numbers larger than 5. Or it could be the relation that holds for all pairs where the second element of the pair is larger than the square of the first element of the pair. Function symbols are often written as f, g, h, a, b, c and predicate symbols are often written as p, q, r.

State variables of hybrid systems, such as positions, velocities, and accelerations, are represented as real-valued function symbols of Σ of arity 0. Unlike fixed symbols like the number 1, state variables are *flexible*, i.e., their interpretation can change from state to state during the execution of a hybrid program. Flexibility of symbols will be used to represent the progression of system values along states over time during a hybrid evolution. Symbols like 1, on the other hand, are *rigid*, i.e., they have the same value at all states. The symbols of real arithmetic like 1 and $+, \cdot$ are rigid, because we do not want them to change their meaning at any time. State variables like velocity v, in contrast, are flexible, because they can change their value depending on the state. While the velocity v may have been 0 in the beginning, the train could increase its velocity to 10 and then decrease it again later when approaching another train.

Note that there is no need to distinguish between discrete and continuous variables in d$\mathcal{L}$. The distinction between logical variables in V, which can be quantified universally or existentially, and state variables in Σ, which can change their value by discrete jumps and differential equations of hybrid programs in modalities, is not strictly required either. For instance, universal and existential quantification of state variables is definable using auxiliary logical variables. The distinction makes the semantics and soundness proof less subtle, though. Our calculus assumes that V contains sufficiently many variables and Σ contains additional Skolem function symbols, which are reserved for use by the calculus.

Terms

Well-formed arguments to function symbols and predicate symbols are called terms. Logical variables are well-formed terms, and functions applied to the appropriate number of terms as arguments are well-formed terms. The set $\mathrm{Trm}(\Sigma, V)$ of *terms* is defined as in classical first-order logic, yielding polynomial (or rational) expressions over V and over additional Skolem terms $s(t_1, \ldots, t_n)$ with terms t_i. Our calculus actually only uses Skolem terms $s(X_1, \ldots, X_n)$ with logical variables $X_i \in V$ as arguments.

Definition 2.1 (Terms). $\mathrm{Trm}(\Sigma, V)$ is the set of all *terms*, which is the smallest set such that:

- If $x \in V$, then $x \in \mathrm{Trm}(\Sigma, V)$.
- If $f \in \Sigma$ is a function symbol of arity $n \geq 0$ and, for $1 \leq i \leq n$, $\theta_i \in \mathrm{Trm}(\Sigma, V)$, then $f(\theta_1, \ldots, \theta_n) \in \mathrm{Trm}(\Sigma, V)$. The case $n = 0$ is permitted (e.g., for state variables).

More succinctly, we also say that the terms of $\mathsf{d}\mathscr{L}$ are defined by the following grammar (where $\theta_1, \ldots, \theta_n$ are terms, f a function symbol of arity n, and $x \in V$ is a logical variable):

$$\theta ::= x \mid f(\theta_1, \ldots, \theta_n).$$

Example 2.2. (Well-formed) terms of $\mathsf{d}\mathscr{L}$ include:

- Logical variables $X \in V$
- State variables $x \in \Sigma$ that may change their value during system evolution
- Expressions of nonlinear polynomial real arithmetic like $x + 5y \cdot (x - 3y + za)$, which we consider as a short notation for $x + 5 \cdot y \cdot (x - 3 \cdot y + z \cdot a)$. Here we assume that $x, y, z, a \in \Sigma$ are state variables. In principle, we also have to mention that the number symbols $5, 3 \in \Sigma$ are (rigid) function symbols without arguments. Yet these number symbols are what we assume as given throughout this book. Note that we could just as well have assumed that $x, y \in \Sigma$ are state variables, $a \in \Sigma$ is a rigid function symbol of arity 0, and $z \in V$ is a logical variable. Then $x + 5y \cdot (x - 3y + za)$ is still a term for this different signature and variables set. For terms, all ways of declaring symbols as state variables, rigid function symbols of arity 0, or logical variables are essentially equivalent. There are many ways to say the same thing. The differences only play a role later for quantification and state change.
- Expressions with Skolem function terms like $x + 5s(X_1, X_2) \cdot (x - 3y + z \cdot t(X_2))$. Here we assume that $x, y \in \Sigma$ are state variables, that $s, t \in \Sigma$ are rigid function symbols of arity 2 and 1, respectively, and that $X_1, X_2 \in V$ are logical variables.
- Real arithmetic expressions with integer powers like $8x^2 + 2x^3(y - a^2bc)$ that can easily be rewritten as $8 \cdot x \cdot x + 2 \cdot x \cdot x \cdot x \cdot (y - a \cdot a \cdot b \cdot c)$. Again, we assume that x, y, a, b, c are symbols in Σ or V.

The following, however, are *no terms* with respect to Σ and V:

- $1+x^y$, because the exponential function x^y cannot be rewritten as a finite product, quite unlike $x^3 = x\cdot x\cdot x$ or $x^4 = x\cdot x\cdot x\cdot x$. In fact, the logical properties of the exponential function are a very exciting and a challenging object of study in recent model theory [107, 206, 44, 108, 45, 2].
- $y^2-\pi$, unless the transcendental number $\pi = 3.1415926\ldots$ is explicitly added to Σ, because unlike rational numbers, the transcendental number cannot be characterised exactly with a finite combination of sums, products, and 0, 1. Arbitrarily precise approximations of π, instead, can be defined; see Example 2.3. □

First-Order Formulas

Meaningful propositions that are either true or false in a context are called (well-formed) formulas. The well-formed formulas of a logic form a formal language over the alphabet $\Sigma \cup V$ of symbols. The formulas consist of all words that can be built by recursively combining symbols of the signature with logical operator symbols appropriately. We first define only the fragment of first-order logic, then the syntax of hybrid programs, and define the actual formulas of differential dynamic logic afterwards.

The set of formulas of *first-order logic* is defined as usual (cf. App. A), giving first-order real arithmetic [288] augmented with Skolem terms. We will show the precise relationship to standard first-order real arithmetic without Skolem terms in Lemma 2.5 of Sect. 2.5.3.2.

Definition 2.2 (First-order formulas). The set $\mathrm{Fml_{FOL}}(\Sigma,V)$ of *formulas of first-order logic* is the smallest set with:

- If $p\in\Sigma$ is a predicate symbol of arity $n\geq 0$ and $\theta_i\in \mathrm{Trm}(\Sigma,V)$ for $1\leq i\leq n$, then $p(\theta_1,\ldots,\theta_n)\in \mathrm{Fml_{FOL}}(\Sigma,V)$.
- If $\phi,\psi\in \mathrm{Fml_{FOL}}(\Sigma,V)$, then $\neg\phi,(\phi\wedge\psi),(\phi\vee\psi),(\phi\to\psi)\in \mathrm{Fml_{FOL}}(\Sigma,V)$.
- If $\phi\in \mathrm{Fml_{FOL}}(\Sigma,V)$ and $x\in V$, then $(\forall x\,\phi),(\exists x\,\phi)\in \mathrm{Fml_{FOL}}(\Sigma,V)$.

More succinctly, we also say that first-order formulas are defined by the following grammar (where ϕ,ψ are first-order formulas, θ_i are terms, p is a predicate symbol of arity n, and $x\in V$ is a logical variable):

$$\phi,\psi ::= p(\theta_1,\ldots,\theta_n) \mid \neg\phi \mid \phi\wedge\psi \mid \phi\vee\psi \mid \phi\to\psi \mid \forall x\,\phi \mid \exists x\,\phi.$$

Example 2.3. (Well-formed) first-order formulas in our context include:

- $v\cdot v\leq 2b\cdot(m-z)$. Again we assume that v,b,m,z are symbols in the vocabulary Σ or V. For instance, we could assume that $z,v\in\Sigma$ are (flexible) state variables and $b,m\in\Sigma$ are rigid function symbols of arity 0. The rationale for this classification would be that z and v are meant to represent the position and velocity of a train, which of course can change over time (flexible). The symbols b and m are meant to represent the braking force and movement authority of a train, which we assume not to change in this formula (rigid). We could just as well assume that $z,v,b,m \in \Sigma$ are (flexible) state variables.

- $v > 0 \rightarrow v \cdot v \leq 2b \cdot (m-z) \vee b = 0$
- $\forall x \forall y \, (x > y \leftrightarrow x - y > 0)$. Here we assume $x,y \in V$ are logical variables; otherwise the syntax would not allow us to quantify over x,y.
- $x > 0 \wedge \forall y \exists z \, (x > z^2 + y \cdot z - 5)$. Here we assume $y,z \in V$ are logical variables and we could either assume $x \in \Sigma$ to be a state variable, or a rigid function symbol of arity zero, or a logical variable $x \in V$. All those choices are reasonable for this formula, and, in fact, it does not make a real difference here, because they will essentially have the same meaning. These distinctions become somewhat more important for $\mathsf{d}\mathcal{L}$ formulas later.
- Formulas with divisions like $b < x/y$, which can easily be defined in terms of multiplication $(b \cdot y < x \wedge y > 0) \vee (b \cdot y > x \wedge y < 0)$.
- Formulas with rational constants like $a > \frac{2}{3}x^2 + 3.1415x \cdot y^4$, which can easily be defined in terms of successive addition and inverses, say,

$$a > (1+1)/(1+1+1) \cdot x^2 + 31415/10000 \cdot x \cdot y^4.$$

- Arithmetic expressions with roots like $x^4 - y\sqrt{2z} > 0$, which can easily be defined in terms of their characteristic polynomials as $\exists r \, (r^2 = 2z \wedge r \geq 0 \wedge x^4 - y \cdot r > 0)$.

□

2.2.2 *Hybrid Programs*

As uniform compositional models for hybrid systems, hybrid programs can combine discrete and continuous transitions to structured control programs using the regular-expression-style operators of Kleene algebras [182].

Definition 2.3 (Hybrid programs). The set $\mathrm{HP}(\Sigma,V)$ of *hybrid programs*, with typical elements α,β, is defined inductively as the smallest set such that

1. If $x_i \in \Sigma$ is a state variable and $\theta_i \in \mathrm{Trm}(\Sigma,V)$ for $1 \leq i \leq n$, then the *discrete jump set* $(x_1 := \theta_1, \ldots, x_n := \theta_n) \in \mathrm{HP}(\Sigma,V)$ is a hybrid program. We assume that the $x_1,\ldots,x_n$ are pairwise different state variables.
2. If $x_i \in \Sigma$ is a state variable and $\theta_i \in \mathrm{Trm}(\Sigma,V)$ for $1 \leq i \leq n$, then $x_i' = \theta_i$ is a *differential equation* in which x_i' represents the time derivative of variable x_i. If χ is a first-order formula, then $(x_1' = \theta_1, \ldots, x_n' = \theta_n \,\&\, \chi) \in \mathrm{HP}(\Sigma,V)$. We assume that the $x_1,\ldots,x_n$ are pairwise different state variables.
3. If χ is a first-order formula, then $(?\chi) \in \mathrm{HP}(\Sigma,V)$.
4. If $\alpha,\beta \in \mathrm{HP}(\Sigma,V)$, then $(\alpha \cup \beta) \in \mathrm{HP}(\Sigma,V)$.
5. If $\alpha,\beta \in \mathrm{HP}(\Sigma,V)$, then $(\alpha;\beta) \in \mathrm{HP}(\Sigma,V)$.
6. If $\alpha \in \mathrm{HP}(\Sigma,V)$, then $(\alpha^*) \in \mathrm{HP}(\Sigma,V)$.

Table 2.1 summarises the statements and (informal) effects of hybrid programs. More succinctly, hybrid programs are defined by the following grammar (α,β are

Table 2.1 Statements and (informal) effects of hybrid programs (HPs)

HP Notation	Operation	Effect
$x_1 := \theta_1, \dots, x_n := \theta_n$	discrete jump set	simultaneously assigns terms θ_i to variables x_i
$x_1' = \theta_1, \dots, x_n' = \theta_n \,\&\, \chi$	continuous evolution	differential equations for x_i with terms θ_i within first-order constraint χ (evolution domain)
$?\chi$	state test / check	test first-order formula χ at current state
$\alpha; \beta$	seq. composition	HP β starts after HP α finishes
$\alpha \cup \beta$	nondet. choice	choice between alternatives HP α or HP β
α^*	nondet. repetition	repeats HP α n-times for any $n \in \mathbb{N}$

hybrid programs, θ_i are terms, $x_i \in \Sigma$ are state variables, and χ is a formula of first-order logic):

$$\alpha, \beta ::= x_1 := \theta_1, \dots, x_n := \theta_n \mid x_1' = \theta_1, \dots, x_n' = \theta_n \,\&\, \chi \mid ?\chi \mid \alpha \cup \beta \mid \alpha; \beta \mid \alpha^*.$$

The effect of the discrete jump set $x_1 := \theta_1, \dots, x_n := \theta_n$ is to simultaneously change the interpretations of the x_i to the respective θ_i by performing a discrete jump in the state space. In particular, the new values θ_i are evaluated before changing the value of any variable x_j. The effect of $x_1' = \theta_1, \dots, x_n' = \theta_n \,\&\, \chi$ is an ongoing continuous evolution respecting the differential equation system $x_1' = \theta_1, \dots, x_n' = \theta_n$ that is restricted to remain within the evolution domain region χ. The evolution is allowed to stop at any point in χ. It is, however, required to stop before it leaves χ. For unconstrained evolutions, we write $x' = \theta$ in place of $x' = \theta \,\&\, \mathit{true}$. For structural reasons, we expect both difference equations (discrete jump sets) and differential equations to be given in explicit form, i.e., with the affected variable on the left (we allow more general implicit forms in Chap. 3). The d$\mathcal{L}$ semantics allows arbitrary differential equations. To retain feasible arithmetic, some of our calculus rules in this chapter assume that, as in [8, 125, 217, 156], the differential equations have first-order definable flows or approximations. We assume that standard techniques are used to determine corresponding solutions or approximations, e.g., [15, 189, 238, 227, 297]. We consider verification techniques for more advanced differential equations in Chap. 3.

The test action or state check $?\chi$ is used to define conditions. Its semantics is that of a no-op if the formula χ is true in the current state; otherwise, like *abort*, it allows no transitions. That is, if the test succeeds because formula χ holds in the current state, then the state does not change, and the system execution continues normally. If the test fails because formula χ does not hold in the current state, then the system execution cannot even continue. Thus, the effect of a test action is similar to an *assert* statement in Java. Note that, according to Definition 2.3, we have only allowed first-order formulas as tests. Instead, we could actually allow *rich tests*, i.e., arbitrary d$\mathcal{L}$ formulas χ with nested modalities as tests $?\chi$ inside hybrid programs (and even in evolution domains χ of differential equations). The calculus and our meta-results, including soundness and relative completeness, directly carry over to this rich test version of d$\mathcal{L}$. To simplify the presentation, however, we refrain from allowing arbitrary $d\mathcal{L}$ formulas as tests, because that requires simultaneous inductive handling of hybrid programs and $d\mathcal{L}$ formulas in syntax, semantics, and completeness proofs, because $d\mathcal{L}$ formulas would then be allowed to occur in hybrid programs, and vice versa.

The nondeterministic choice $\alpha \cup \beta$, sequential composition $\alpha;\beta$, and nondeterministic repetition α^* of programs are as in regular expressions but generalised to a semantics in hybrid systems. Choices $\alpha \cup \beta$ are used to express behavioural alternatives between the transitions of α and β. That is, the hybrid program $\alpha \cup \beta$ can choose nondeterministically to follow the transitions of the hybrid program α, or, instead, to follow the transitions of the hybrid program β. The sequential composition $\alpha;\beta$ says that the hybrid program β starts executing after α has finished (β never starts if α does not terminate). In $\alpha;\beta$, the transitions of α take effect first, until α terminates (if it does), and then β continues. Observe that, like repetitions, continuous evolutions within α can take more or less time, which causes uncountable nondeterminism. This nondeterminism is inherent in hybrid systems, because they can operate in so many different ways, and as such reflected in hybrid programs. Repetition α^* is used to express that the hybrid process α repeats any number of times, including zero times. When following α^*, the transitions of hybrid program α can be repeated over and over again, any nondeterministic number of times (≥ 0). Hybrid programs form a regular-expression-style Kleene algebra with tests [182].

Example 2.4 (Simplistic train). The differential equation $z' = v, v' = a$ expresses continuous movement of position z with velocity v and acceleration a. A very simple (in fact much too simplistic) train controller could be the following hybrid program:

$$((a := -b) \cup (?v < 8; a := A));\ z' = v, v' = a.$$

By a nondeterministic choice ($\cup$), the system either chooses to set the acceleration a to the braking force $-b$ by executing $a := -b$, or the system tries to pass the test $?v < 8$ instead. If the system tries the second choice and the latter test succeeds, i.e., the current velocity is indeed less than 8, then the system sets the acceleration a to A. Otherwise, if it tries the second choice but the test fails, then nothing happens as this execution is blocked and cannot continue. Afterwards (after executing the first part of the sequential composition, which is the nondeterministic choice), the system follows the second part of the sequential composition, which is the differential equation $z' = v, v' = a$ with the previously chosen acceleration. The system then follows this differential equation for a certain (unspecified) period of time.

This controller leaves open too many aspects to be useful, but already illustrates a very simple hybrid program. One of the problems is that the controller can only take a control action for choosing the acceleration a once, at the beginning of the system evolution, and then follows the differential equation for an arbitrarily long time. But the above controller can never react to situation changes and change its mind with a different choice of a when necessary. To improve this issue, the following hybrid program allows repetitive choices by the repetition operator *:

$$(((a:=-b)\cup(?v<8;a:=A));\ z'=v,v'=a)^*. \tag{2.1}$$

Unlike the previous hybrid program, the hybrid program in (2.1) contains a repetition, which can change the acceleration repeatedly over and over again after following the continuous evolution for a certain period of time. While already an improvement over the last controller, this hybrid program has shortcomings. For one thing, the differential equation does not say when it stops. It has no evolution domain restriction and would thus be allowed to evolve as long or as short as it pleases. This may be unsafe if the differential equation would continue indefinitely without giving the controller for the acceleration a chance to react to situation changes. Furthermore, a velocity of 8 may not be a safe choice for the switching condition between acceleration and braking. We will see in Sect. 2.4 how a reasonable train controller can be designed as a hybrid program from first principles and elaborate on train control further in Chap. 7 in full detail. □

Definable Operations

The control flow operations of choice, sequential composition, and repetition in hybrid programs can be combined with $?\chi$ to form all other control structures [149]. All classical discrete control structures can be defined in terms of the basic hybrid program operators (it is easy to see that hybrid programs are Turing-complete). See Table 2.2 for a selection of control structures and statements that are definable as a hybrid program. For instance, $(?\chi;\alpha)^*;?\neg\chi$ corresponds to a while loop that re-

Table 2.2 Statements and control structures definable with hybrid programs

HP Notation	Operation	Effect
$x:=*$	nondet. assignment	assigns any real value to x equivalently definable, see Chap. 3
if χ then α else β	if-then-else	executes HP α if χ holds, otherwise HP β equivalent to $(?\chi;\alpha)\cup(?\neg\chi;\beta)$
if χ then α	if-then	executes HP α if χ holds, otherwise no effect equivalent to $(?\chi;\alpha)\cup(?\neg\chi)$
while χ do α	while loop	repeats α if χ holds, only stops if $\neg\chi$ holds at end equivalent to $(?\chi;\alpha)^*;?\neg\chi$
repeat α until χ	repeat until	repeats HP α until χ holds at end (at least once) equivalent to $\alpha;(?\neg\chi;\alpha)^*;?\chi$
skip	do nothing	has no effect and does not change the state space equivalent to $?true$
abort	aborts execution	blocks current execution and allows no transition equivalent to $?false$

peats α while χ holds and only stops when χ ceases to hold after α. Because the *-operator can repeat arbitrarily often, the subprogram $(?\chi;\alpha)^*$ can repeat α any number of times, but a repetition can only be successful if the test $?\chi$ succeeds. Hence the repetitions have to stop, at the latest, when the test $?\chi$ fails. Now the subprogram $(?\chi;\alpha)^*$ can repeat any number of times and is allowed to stop even if the test $?\chi$ is successful and the loop could be repeated again. But the subsequent sequential composition with the test $?\neg\chi$ makes sure that $(?\chi;\alpha)^*$ can only stop repeating when χ actually ceases to hold. Overall, the hybrid program $(?\chi;\alpha)^*;?\neg\chi$ executes α if χ holds and repeats α again exactly as often as χ still holds after executing α.

If-then-else can be defined with nondeterministic choices and tests. The corresponding hybrid program $(?\chi;\alpha)\cup(?\neg\chi;\beta)$ in Table 2.2 makes a nondeterministic choice between $?\chi;\alpha$ and $?\neg\chi;\beta$. While this choice is nondeterministic, at any state only one of the subsequent tests in the two cases can succeed, because they are complementary. Consequently, hybrid program α will be executed if and only if the test $?\chi$ succeeds because χ is true at the current state. Likewise, hybrid program β will be executed if and only if the dual test $?\neg\chi$ succeeds because $\neg\chi$ is true, i.e., χ is false at the current state. The nondeterministic assignment $x := *$ that assigns an arbitrary real number to state variable x is definable also. While it is possible define nondeterministic assignments in hybrid programs already, we will come back to this in Chap. 3, where the definition is easier to see.

Example 2.5 (Parametric bouncing ball). As a classical example from the hybrid systems literature [110], consider the bouncing ball. We will describe the bouncing ball as a hybrid program, using the definable hybrid program operations from Table 2.2. Figure 2.2 depicts a hybrid automaton, an illustration of the bouncing ball dynamics, and a representation of the system as a hybrid program.

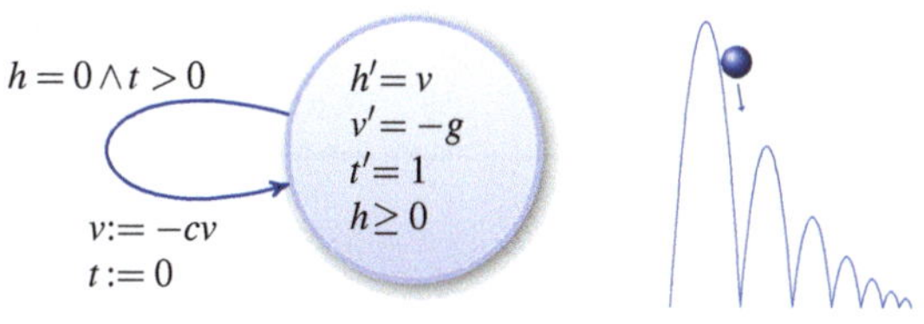

$ball \equiv ($
$\quad h'=v, v'=-g, t'=1 \,\&\, h\geq 0;$
$\quad$ if $(h=0 \land t>0)$ then
$\quad\quad c:=*;\ ?(0\leq c<1);$ *// extra*
$\quad\quad v:=-cv;\ t:=0$
$\quad$ fi$)^*$

Fig. 2.2 Parametric bouncing ball

The bouncing ball is let loose and falls from height h, but bounces back from the ground (which corresponds to height $h=0$) after an elastic deformation. The current speed of the ball is denoted by v, and t is a clock measuring the falling time. The bouncing ball follows the continuous dynamics of physical movement by gravity. The ball is affected by gravity of force g, so its height follows the differential equation $h''=-g$. This second-order differential equation is equivalent to the first-order differential equation system $h'=v, v'=-g$, with an explicit velocity v. Simultaneously, clock t evolves according to the differential equation $t'=1$. Finally, the ball always stays above the ground and cannot fall through, thus its evolution domain is restricted to $h\geq 0$. Altogether, this gives the continuous evolution $h'=v, v'=-g, t'=1 \,\&\, h\geq 0$ in the beginning of the hybrid program in Fig. 2.2.

At the ground (which is at height $h=0$), the ball bounces back after losing energy in an elastic deformation according to a damping factor $0\leq c<1$. That is, if the ball is on the ground ($h = 0$) and it actually fell (so time has passed, $t > 0$), then the ball changes its direction and bounces back into the air by reflecting its current velocity v by a discrete jump $v := -cv$ and resetting the falling time by $t := 0$.

Now for illustration purposes we have added an extra twist to the hybrid program in Fig. 2.2 that is not in the hybrid automaton. The automaton still enforces infinite bouncing so that the ball can never stop (unless $c = 0$, where it stops immediately). In reality, the ball bounces a couple of times and can then come to a standstill when its remaining kinetic energy is insufficient. To model this phenomenon without the need to have a precise physical model for all physical forces and frictions, we allow for the damping factor c to change at each bounce. Line 4 of the hybrid program in Fig. 2.2 represents a corresponding uncountably infinite nondeterministic choice for c as a nondeterministic assignment. The subsequent test $?(0 \leq c < 1)$ restricts the arbitrary choices for c to choices in the half-open interval $[0,1)$ and discards all other choices.

For comparison, Fig. 2.3 shows an equivalent hybrid program for the same bouncing, now with all abbreviations for extended statements resolved according to Table 2.2. Note that it is fairly easy to see that height h and clock t always stay nonnegative if they start nonnegative. For that reason, the last test $?(h \neq 0 \vee t \leq 0)$ in Fig. 2.3 could even be replaced equivalently by $?(h > 0 \vee t = 0)$. □

Fig. 2.3 Parametric bouncing ball (with abbreviations resolved)

$$
\begin{aligned}
&\mathit{ball} \equiv (\\
&\quad h' = v, v' = -g, t' = 1 \,\&\, h \geq 0;\\
&\quad (?(h = 0 \wedge t > 0);\\
&\qquad\quad (c' = 1 \cup c' = -1);\\
&\qquad\quad ?(0 \leq c < 1);\\
&\qquad\quad v := -cv;\ t := 0\\
&\quad) \cup ?(h \neq 0 \vee t \leq 0)\\
&\quad)^*
\end{aligned}
$$

Classification of Hybrid Programs

Hybrid programs are designed as a minimal extension of conventional discrete programs. They characterise hybrid systems succinctly by adding continuous evolution along differential equations as the only additional primitive operation to a regular basis of conventional discrete programs. To yield hybrid systems, their operations are interpreted over the domain of real numbers. This gives rise to an elegant syntactic hierarchy of discrete, continuous, and hybrid systems. Hybrid automata [156] can be represented as hybrid programs using a straightforward generalisation of standard program encodings of automata; see App. C for formal details. The fragment of hybrid programs without differential equations corresponds to conventional discrete programs generalised over the reals or to discrete-time dynamical systems [56]. The fragment without discrete jumps corresponds to switched continuous systems [56, 58], whereas the fragment of differential equations gives purely con-

Table 2.3 Operators and (informal) meaning in differential dynamic logic ($d\mathcal{L}$)

$d\mathcal{L}$ Notation	Operator	Meaning
$p(\theta_1,\dots,\theta_n)$	atomic formula	true iff predicate p holds for $(\theta_1,\dots,\theta_n)$
$\neg\phi$	negation / not	true if ϕ is false
$\phi \wedge \psi$	conjunction / and	true if both ϕ and ψ are true
$\phi \vee \psi$	disjunction / or	true if ϕ is true or if ψ is true
$\phi \rightarrow \psi$	implication / implies	true if ϕ is false or ψ is true
$\phi \leftrightarrow \psi$	bi-implication / equivalent	true if ϕ and ψ are both true or both false
$\forall x\, \phi$	universal quantifier / for all	true if ϕ is true for all values of variable x
$\exists x\, \phi$	existential quantifier / exists	true if ϕ is true for some values of variable x
$[\alpha]\phi$	$[\cdot]$ modality / box	true if ϕ is true after all runs of HP α
$\langle\alpha\rangle\phi$	$\langle\cdot\rangle$ modality / diamond	true if ϕ is true after at least one run of HP α

tinuous dynamical systems [279]. Only the composition of mixed discrete jumps and continuous evolutions gives rise to truly hybrid behaviour.

2.2.3 Formulas of Differential Dynamic Logic

The formulas of the differential dynamic logic $d\mathcal{L}$ are defined as in first-order dynamic logic [253, 148, 149] but with real arithmetic as a semantic domain and with hybrid programs as system models. That is, they are built using propositional connectives $\neg, \wedge, \vee, \rightarrow, \leftrightarrow$ and quantifiers $\forall, \exists$ over the reals (first-order part). In addition, if ϕ is a $d\mathcal{L}$ formula and α a hybrid program, then $[\alpha]\phi, \langle\alpha\rangle\phi$ are formulas (dynamic part).

Definition 2.4 ($d\mathcal{L}$ formulas). The set $\mathrm{Fml}(\Sigma, V)$ of *formulas* of $d\mathcal{L}$, with typical elements ϕ, ψ, is the smallest set such that

1. If p is a predicate symbol of arity $n \geq 0$ and $\theta_i \in \mathrm{Trm}(\Sigma, V)$ for $1 \leq i \leq n$, then $p(\theta_1,\dots,\theta_n) \in \mathrm{Fml}(\Sigma, V)$.
2. If $\phi, \psi \in \mathrm{Fml}(\Sigma, V)$, then $\neg\phi, (\phi \wedge \psi), (\phi \vee \psi), (\phi \rightarrow \psi) \in \mathrm{Fml}(\Sigma, V)$.
3. If $\phi \in \mathrm{Fml}(\Sigma, V)$ and $x \in V$, then $\forall x\, \phi, \exists x\, \phi \in \mathrm{Fml}(\Sigma, V)$.
4. If $\phi \in \mathrm{Fml}(\Sigma, V)$ and $\alpha \in \mathrm{HP}(\Sigma, V)$, then $[\alpha]\phi, \langle\alpha\rangle\phi \in \mathrm{Fml}(\Sigma, V)$.

For reference, the logical operators of differential dynamic logic are summarised in Table 2.3. More succinctly, we also say that the formulas of $d\mathcal{L}$ are defined by the following grammar (where ϕ, ψ are $d\mathcal{L}$ formulas, θ_i are terms, p a predicate symbol of arity n, $x \in V$ is a logical variable, and α is a hybrid program):

$$\phi, \psi ::= p(\theta_1,\dots,\theta_n) \mid \neg\phi \mid \phi \wedge \psi \mid \phi \vee \psi \mid \phi \rightarrow \psi \mid \forall x\, \phi \mid \exists x\, \phi \mid [\alpha]\phi \mid \langle\alpha\rangle\phi.$$

We consider the bi-implication or equivalence $\phi \leftrightarrow \psi$ as an abbreviation for $(\phi \rightarrow \psi) \wedge (\psi \rightarrow \phi)$ to simplify the calculus. We often leave out superfluous brackets and use binding priorities instead in order to improve readability. Quantifiers and modalities bind strongly, i.e., their scope only extends to the formula immediately after. Unary operators (negation $\neg$), quantifiers ($\forall, \exists$), and modalities ($[\alpha], \langle\alpha\rangle$) bind stronger than binary operators. Further, conjunction $\wedge$ and disjunction $\vee$ bind stronger than implication $\rightarrow$ and bi-implication $\leftrightarrow$. Thus

$$\phi_0 \wedge \langle\alpha\rangle\phi_1 \wedge \forall x\, \phi_2 \wedge \phi_3 \rightarrow \neg\phi_4 \vee [\alpha]\phi_5 \vee \phi_6$$

is taken to mean

$$(\phi_0 \wedge (\langle\alpha\rangle\phi_1) \wedge (\forall x\, \phi_2) \wedge \phi_3) \rightarrow ((\neg\phi_4) \vee ([\alpha]\phi_5) \vee \phi_6)$$

and does not mean

$$\phi_0 \wedge \Big(\langle\alpha\rangle(\phi_1 \wedge \forall x (\phi_2 \wedge \phi_3)) \rightarrow \neg(\phi_4 \vee [\alpha](\phi_5 \vee \phi_6))\Big).$$

Example 2.6 (Train control). When *train* denotes the hybrid program in Fig. 2.1 or the hybrid program in Example 2.4, or, in fact, any other hybrid program model for a train system, then the following $d\mathcal{L}$ formula expresses that this train is able ($\langle train\rangle$) to enter region $z \geq m$, thereby leaving region $z < m$ when it starts in region $z < m$ with nonnegative initial velocity $v \geq 0$:

$$v \geq 0 \wedge z < m \rightarrow \langle train\rangle\, z \geq m. \tag{2.2}$$

Dually, the following $d\mathcal{L}$ formula expresses that the train will always ($[train]$) stay inside the region $z < m$ when it starts inside it with an initial nonnegative velocity less than 5:

$$v \geq 0 \wedge v < 5 \wedge z < m \rightarrow [train]\, z < m.$$

For most train models *train*, the latter safety property will only be true for additional constraints on the initial state and on the internal parameter choices, including, e.g., braking forces, reaction times, and start braking points. □

Example 2.7 (Parametric bouncing ball). Let *ball* denote the hybrid program for the bouncing ball from Example 2.5. The ball loses energy at every bounce, thus the ball never bounces higher than the initial height. This can be expressed by the property $0 \leq h \leq H$, where H denotes the initial energy level (which corresponds to the initial height if $v = 0$ initially). Then, for instance, the following $d\mathcal{L}$ formula expresses that (under a list of assumptions on the free variables h, v, t and H, g, c) the ball always stays in the region $0 \leq h \leq H$:

$$(v^2 \leq 2g(H-h) \wedge h \geq 0 \wedge g > 0 \wedge H \geq 0 \wedge 1 > c \geq 0) \rightarrow [ball](0 \leq h \leq H). \tag{2.3}$$

This $d\mathcal{L}$ formula follows the pattern of Hoare triples [161]. It expresses that the bouncing ball, when started in an initial state satisfying the precondition on the left of the implication ($\rightarrow$), always respects the postcondition $0 \leq h \leq H$ of the dynamic modality $[ball]$, i.e., all runs of the bouncing ball stay in the region $0 \leq h \leq H$. □

A $\mathsf{d}\mathcal{L}$ formulas of the form $\psi \to [\alpha]\phi$ corresponds to Hoare triples [161], generalised for hybrid systems. They occur quite frequently, because they specify that system α, when starting in a state satisfying the *precondition* ψ, always respects the *postcondition* ϕ. That is, when started in a state satisfying ψ, all states reachable by α satisfy ϕ. There are several other relevant shapes of $\mathsf{d}\mathcal{L}$ formulas in practical systems verification; see Part III.

Note that, according to Definition 2.4, hybrid programs are not additional external objects but fully internalised [48] as first-class citizens within the logic $\mathsf{d}\mathcal{L}$ itself, and the logic is closed. That is, modalities can be combined propositionally, by quantifiers, or nested. For instance, $[\alpha]\langle\beta\rangle x \leq c$ says that, whatever hybrid program α is doing, hybrid program β can react in some way to reach a controlled state where x is less than some critical value c. That is, for all α actions, there is a β (re)action such that $x \leq c$ holds. Dually, $\langle\beta\rangle[\alpha]x \leq c$ expresses that hybrid program β can stabilise $x \leq c$, i.e., behave in such a way that $x \leq c$ remains true no matter how hybrid program α reacts. That is, there is a β action such that all α actions maintain $x \leq c$. Accordingly, $\exists p\,[\alpha]x \leq c$ says that there is a choice of parameter p such that α remains in $x \leq c$. Nesting modalities and quantifiers in this way can be quite useful for describing interactions of a hybrid program α with an environment β, or for describing the impact of parameter choices on properties of the system behaviour.

During our analysis, we assume differential equations and discrete transitions to be well-defined. In particular, we assume that all divisions p/q are guarded by conditions that ensure $q \neq 0$ as, otherwise, the system behaviour is not well-defined due to an undefined value at a singularity. It is simple but tedious to augment the semantics and the calculus with corresponding side conditions to show that this is respected. For instance, we assume that $x := p/q$ is guarded by $?q \neq 0$ and that continuous evolutions are restricted such that the differential equations are well-defined as $x' = p/q \,\&\, q \neq 0$. Also see our joint work with Beckert [37] for techniques of how such exceptional behaviour can be handled by program transformation while avoiding partial valuations of undefinedness in the semantics. In logical formulas, partiality can be avoided altogether by writing $p = c \cdot q \wedge q \neq 0$ rather than $p/q = c$, and writing $(p > c \cdot q \wedge q > 0) \vee (p < c \cdot q \wedge q < 0)$ rather than $p/q > c$.

2.3 Semantics of Differential Dynamic Logic

We define the semantics of $\mathsf{d}\mathcal{L}$ as a possible world Kripke semantics [185] with worlds representing the possible system states and with reachability along the hybrid transitions of the system representing accessibility relations between worlds. The interpretations of $\mathsf{d}\mathcal{L}$ consist of states (worlds) that are essentially first-order structures over the reals. In particular, real values are assigned to state variables, possibly different values in each state. A potential behaviour of a hybrid system corresponds to a succession of states that contain the observable values of system variables during its hybrid evolution.

2.3.1 Valuation of Terms

Symbols in the logic d$\mathcal{L}$ come from three different syntactic categories that we decided to distinguish in Sect. 2.2.1:

1. rigid symbols in Σ that cannot change their value, e.g., $0, 1, +, \cdot$;
2. flexible symbols in Σ, which are the state variables, whose value can change depending on the current state of the system;
3. logical variables in V that cannot change their value over time by running hybrid programs, but which can be quantified over universally or existentially.

All of those symbols need to be interpreted to give meaning to terms in which they occur. We associate values with rigid symbols by what we call an interpretation I, associate values with flexible symbols by a state ν, and associate values with logical variables by an assignment η. Recall that there is some leeway in declaring symbols as either rigid or flexible symbols or as logical variables. The semantics is unambiguous for each choice, though.

An *interpretation* I assigns functions and relations over the reals to the respective rigid symbols in Σ. The function and predicate symbols of real arithmetic are interpreted as usual by I. Especially, the interpretation $I(+)$ is addition and $I(\cdot)$ is multiplication of real numbers. A *state* is a map $\nu : \Sigma_{fl} \to \mathbb{R}$; the set of all states is denoted by $\mathrm{Sta}(\Sigma)$. Here, Σ_{fl} denotes the set of (flexible) state variables in Σ (they have arity 0, thus take no arguments). Finally, an *assignment for logical variables* is a map $\eta : V \to \mathbb{R}$. It contains the values for logical variables, which are not subject to change by modalities but only by quantification. Observe that flexible symbols (which represent state variables) are allowed to assume different interpretations in different states. Logical variable symbols, however, are rigid in the sense that their value is determined by η alone and does not depend on the state ν.

The *valuation* $val_{I,\eta}(\nu,\cdot)$ of terms is defined as usual [123, 149] with an extra distinction of rigid and flexible functions [37]. It is defined inductively by recursion on the structure of the term, based on the interpretation that assignment η assigns to logical variables, that interpretation I assigns to rigid function symbols, and that state ν assigns to flexible state variables. The semantics of terms is compositional and denotational [277], that is, the semantics of a complex term is defined as a combination of the semantics of its subterms.

Definition 2.5 (Valuation of terms). The *valuation of terms* with respect to interpretation I, assignment η, and state ν is defined by

1. $val_{I,\eta}(\nu,x) = \eta(x)$ if $x \in V$ is a logical variable.
2. $val_{I,\eta}(\nu,a) = \nu(a)$ if $a \in \Sigma$ is a state variable (flexible function symbol of arity 0).
3. $val_{I,\eta}(\nu, f(\theta_1,\dots,\theta_n)) = I(f)\big(val_{I,\eta}(\nu,\theta_1),\dots,val_{I,\eta}(\nu,\theta_n)\big)$ when $f \in \Sigma$ is a rigid function symbol of arity $n \geq 0$.

Example 2.8. Let interpretation I interpret the constant function symbol $b \in \Sigma$ as $I(b) = 2.14$, and interpret the unary function symbol $c \in \Sigma$ as the cubic function

$d \mapsto d^3$, i.e., $\big(I(c)\big)(d) = d^3$. Let the assignment η interpret the logical variable $X \in V$ as $\eta(X) = 4.2$. Finally let the state ν interpret the state variables $x, y, z \in \Sigma$ as $\nu(x) = 3$, $\nu(y) = -5.01$, $\nu(z) = 0$. Throughout this book we assume that the interpretation of $0, 1, +, -, \cdot$ always is as usual in real arithmetic, that is:

$$
\begin{aligned}
I(0) &= 0 \\
I(1) &= 1 \\
\big(I(+)\big)(d,e) &= d+e && \text{(addition)} \\
\big(I(-)\big)(d,e) &= d-e && \text{(subtraction)} \\
\big(I(\cdot)\big)(d,e) &= d \cdot e && \text{(multiplication)}
\end{aligned}
$$

With this we can valuate terms recursively with respect to I, η, ν as follows:

$$
\begin{aligned}
val_{I,\eta}(\nu, x+y) &= val_{I,\eta}(\nu, x) + val_{I,\eta}(\nu, y) = \nu(x) + \nu(y) \\
&= 3 + (-5.01) = -2.01, \\
val_{I,\eta}(\nu, y + 2 \cdot X) &= val_{I,\eta}(\nu, y) + val_{I,\eta}(\nu, 2) \cdot val_{I,\eta}(\nu, X) \\
&= \nu(y) + I(2) \cdot \eta(X) = -5.01 + 2 * 4.2 = 3.39, \\
val_{I,\eta}(\nu, X + b \cdot (x + y \cdot X)) &= \eta(X) + I(b) \cdot (\nu(x) + \nu(z) \cdot \eta(X)) \\
&= 4.2 + 2.14 \cdot (3 + 0 \cdot 4.2) = 10.62, \\
val_{I,\eta}(\nu, c(x+X) - x) &= val_{I,\eta}(\nu, c(x+X)) - val_{I,\eta}(\nu, x) \\
&= I(c)\big(val_{I,\eta}(\nu, x+X)\big) - \nu(x) \\
&= I(c)\big(\nu(x) + \eta(X)\big) \; - \; \nu(x) \\
&= (3+4.2)^3 - 3 = 370.248.
\end{aligned}
$$

Note here, that the decision about which symbols we consider as rigid function symbols, which ones we consider as flexible function symbols (state variables), and which ones we consider as logical variables is somewhat arbitrary in this example. This decision only becomes relevant when we add quantifiers (for only logical variables can be quantified over) or hybrid programs (for only state variables can be assigned to in hybrid programs). Overall, the syntactic category of symbols is not crucial, as there are often many equivalent ways to assign symbols to syntactic categories. But if we fix a choice of symbols, the semantics becomes less subtle, so we assume a choice has been made for every formula. □

2.3.2 Valuation of Formulas

The *valuation* $val_{I,\eta}(\nu, \cdot)$ of formulas is defined as usual for first-order modal logic [123, 149] with a distinction of rigid and flexible functions [37]. Modalities parametrised by a hybrid program α follow the accessibility relation spanned by the respective hybrid state transition relation $\rho_{I,\eta}(\alpha)$, which is simultaneously inductively defined in Definition 2.7.

The valuation of formulas is defined inductively by recursion on the structure of formulas, based on the interpretation of the terms occurring in it. The semantics of formulas is compositional and denotational, that is, the semantics of a complex formula is defined as a simple function of the semantics of its subformulas. We will use $\eta[x \mapsto d]$ to denote the *modification* of an assignment η that agrees with η except for the interpretation of the logical variable $x \in V$, which is assigned $d \in \mathbb{R}$ in $\eta[x \mapsto d]$.

Definition 2.6 (Valuation of d$\mathcal{L}$ formulas). The *valuation* $val_{I,\eta}(\nu, \cdot)$ *of formulas* with respect to interpretation I, assignment η, and state ν is defined as

1. $val_{I,\eta}(\nu, p(\theta_1, \dots, \theta_n)) = I(p)\big(val_{I,\eta}(\nu, \theta_1), \dots, val_{I,\eta}(\nu, \theta_n)\big)$.
2. $val_{I,\eta}(\nu, \phi \land \psi) = \mathit{true}$ iff $val_{I,\eta}(\nu, \phi) = \mathit{true}$ and $val_{I,\eta}(\nu, \psi) = \mathit{true}$.
3. $val_{I,\eta}(\nu, \phi \lor \psi) = \mathit{true}$ iff $val_{I,\eta}(\nu, \phi) = \mathit{true}$ or $val_{I,\eta}(\nu, \psi) = \mathit{true}$.
4. $val_{I,\eta}(\nu, \neg\phi) = \mathit{true}$ iff $val_{I,\eta}(\nu, \phi) \neq \mathit{true}$.
5. $val_{I,\eta}(\nu, \phi \to \psi) = \mathit{true}$ iff $val_{I,\eta}(\nu, \phi) \neq \mathit{true}$ or $val_{I,\eta}(\nu, \psi) = \mathit{true}$.
6. $val_{I,\eta}(\nu, \forall x\, \phi) = \mathit{true}$ iff $val_{I,\eta[x\mapsto d]}(\nu, \phi) = \mathit{true}$ for all $d \in \mathbb{R}$.
7. $val_{I,\eta}(\nu, \exists x\, \phi) = \mathit{true}$ iff $val_{I,\eta[x\mapsto d]}(\nu, \phi) = \mathit{true}$ for some $d \in \mathbb{R}$.
8. $val_{I,\eta}(\nu, [\alpha]\phi) = \mathit{true}$ iff $val_{I,\eta}(\omega, \phi) = \mathit{true}$ for all states ω for which the transition relation satisfies $(\nu, \omega) \in \rho_{I,\eta}(\alpha)$.
9. $val_{I,\eta}(\nu, \langle\alpha\rangle\phi) = \mathit{true}$ iff $val_{I,\eta}(\omega, \phi) = \mathit{true}$ for some state ω for which the transition relation satisfies $(\nu, \omega) \in \rho_{I,\eta}(\alpha)$.

Following the usual notation, we also write $I, \eta, \nu \models \phi$ iff $val_{I,\eta}(\nu, \phi) = \mathit{true}$. We then say that ϕ is *satisfied* in I, η, ν or holds in I, η, ν. We also say that I, η, ν is a *model* of ϕ. Dually, we write $I, \eta, \nu \not\models \phi$ iff $val_{I,\eta}(\nu, \phi) \neq \mathit{true}$. If ϕ is satisfied for at least one I, η, ν, then ϕ is called *satisfiable*. Occasionally, we write just $\models \phi$ iff $I, \eta, \nu \models \phi$ holds for all I, η, ν. Formula ϕ is then called *valid*, i.e., true in all I, η, ν.

The semantics of modal formulas $[\alpha]\phi$ and $\langle\alpha\rangle\phi$ in d$\mathcal{L}$ is illustrated in Fig. 2.4, showing how the truth of ϕ at (all or some) states ω_i reachable by α relates to the truth of $[\alpha]\phi$ or $\langle\alpha\rangle\phi$ at state ν.

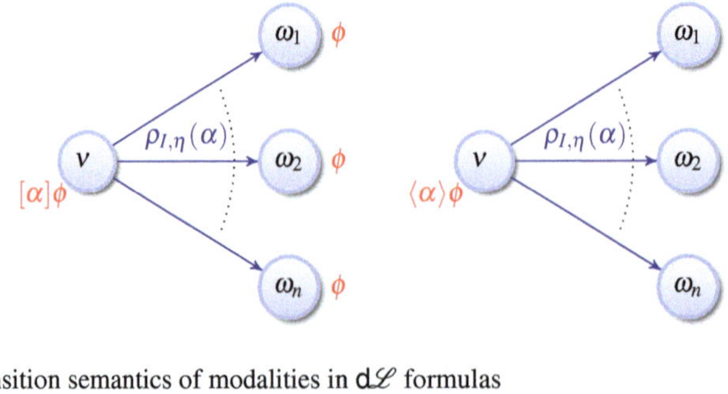

Fig. 2.4 Transition semantics of modalities in d$\mathcal{L}$ formulas

Example 2.9. Consider the following formula that we want to evaluate:

$$v > 0 \rightarrow v \cdot v \leq 2b \cdot (m - z) \vee b = 0. \tag{2.4}$$

First we have to declare which syntactic category the symbols are meant to come from. Suppose $z, v \in \Sigma$ are state variables (flexible function symbols of arity 0), because they represent position and velocity of the train, which are intended to be able to change over time from state to state. Further suppose that $m, b \in \Sigma$ are rigid function symbols, because, for the moment, movement authority and braking force are not allowed to change over time. Note that we could just as well have chosen all symbols z, v, m, b to be flexible state variables. The only notable difference is that if b is a flexible symbol, we would have to prove that a particular hybrid program never changes the value of b to know that it denotes the same value in every part of the program. Otherwise, if b is a rigid symbol, we already know that it cannot possibly change its value by running a hybrid program, because b then is a constant, and only flexible symbols are syntactically allowed to be assigned to or have differential equations in Definition 2.3. While it is certainly not crucial to make this distinction, it can make some things easier to see syntactically.

Now let interpretation I interpret rigid symbol $m \in \Sigma$ as $I(m) = 20$ and interpret $b \in \Sigma$ as $I(b) = 2.2$. Let state ω interpret state variables $v, z \in \Sigma$ as $\omega(v) = 10$, $\omega(z) = 0$. In formula (2.4), suppose we do not have any free logical variables, so that the assignment η of logical variables does not matter. Then we can evaluate formula (2.4) with respect to I, η, ω:

$$\begin{aligned}
&val_{I,\eta}(\omega, v > 0 \rightarrow v \cdot v \leq 2b \cdot (m - z) \vee b = 0) = \textit{true} \text{ iff} \\
&val_{I,\eta}(\omega, v > 0) \neq \textit{true}, \text{ or} \\
&val_{I,\eta}(\omega, v \cdot v \leq 2b \cdot (m - z)) = \textit{true}, \text{ or} \\
&val_{I,\eta}(\omega, b = 0) = \textit{true}.
\end{aligned}$$

Let us evaluate the terms to determine if the subformulas are true or not:

$$\begin{aligned}
val_{I,\eta}(\omega, v > 0) &= (val_{I,\eta}(\omega, v) \stackrel{?}{>} val_{I,\eta}(\omega, 0)) = (\omega(v) \stackrel{?}{>} I(0)) \\
&= (10 \stackrel{?}{>} 0) = \textit{true}, \\
val_{I,\eta}(\omega, v \cdot v \leq 2b \cdot (m - z))&(val_{I,\eta}(\omega, v \cdot v) \stackrel{?}{\leq} val_{I,\eta}(\omega, 2b \cdot (m - z))) \\
&= \big(\omega(v) \cdot \omega(v) \stackrel{?}{\leq} 2I(b) \cdot (I(m) - \omega(z))\big) \\
&= \big(10 \cdot 10 \stackrel{?}{\leq} 2 \cdot 2.2 \cdot (20 - 0)\big) = (100 \stackrel{?}{\leq} 88) = \textit{false}, \\
val_{I,\eta}(\omega, b = 0) &= (I(b) \stackrel{?}{=} 0) = (2.2 \stackrel{?}{=} 0) = \textit{false}.
\end{aligned}$$

Consequently the formula (2.4) evaluates to *false*. For a different state ν with slower speed $\nu(v) = 8$ and the same position $\nu(z) = 0$, we instead evaluate (2.4) to *true*:

$$val_{I,\eta}(\nu, v > 0 \rightarrow v \cdot v \leq 2b \cdot (m - z) \vee b = 0) = true.$$

Also for a different interpretation J with $J(m) = 20$ and another braking force $J(b) = 4$, but the same original state ω, we evaluate (2.4) to *true*:

$$val_{J,\eta}(\omega, v > 0 \rightarrow v \cdot v \leq 2b \cdot (m - z) \vee b = 0) = true.$$

So we see that the truth-value of formula (2.4) depends on I, η, ω. For some choices of I, η, ω, it evaluates to *true*, for others it evaluates to *false*. Thus, formula (2.4) is not valid, because

$$I, \eta, \omega \models v > 0 \rightarrow v \cdot v \leq 2b \cdot (m - z) \vee b = 0$$

does not hold for all I, η, ω. Still, the formula (2.4) is at least satisfiable, because it holds for some I, η, ω. □

Example 2.10. Consider the assignment η with $\eta(Z) = -2$ and the state ν with $\nu(x) = -4$. Then we can evaluate

$$val_{I,\eta}(\nu, x > -5 \wedge \forall y\, (y^2 + Z > x)) = true$$

because $\nu(x) > -5$ and all squares are greater than or equal zero, so that for all $d \in \mathbb{R}$:

$$\begin{aligned} val_{I,\eta[y \mapsto d]}(\nu, y^2 + Z > x) &= ((\eta[y \mapsto d](y))^2 + \eta[y \mapsto d](Z) \overset{?}{>} \nu(x) \\ &= \left(d^2 + (-2) \overset{?}{>} -4\right) = true. \end{aligned}$$

□

Note, that we have not yet explained how to evaluate formulas with modalities like $x > 0 \rightarrow [ctrl; drive^*]\, z \leq m$ in any I, η, ν, because we first have to define the transition semantics $\rho_{I,\eta}(\alpha)$ of hybrid programs, which we will do next.

2.3.3 Transition Semantics of Hybrid Programs

Now we define the transition semantics, $\rho_{I,\eta}(\alpha)$, of hybrid program α. The semantics of a hybrid program is captured by its hybrid state transition relation. For discrete jumps this transition relation holds for pairs of states that respect the discrete jump set. For continuous evolutions, the transition relation holds for pairs of states that can be interconnected by a continuous flow respecting the differential equations and evolution domain restriction throughout the evolution.

The transition semantics of hybrid programs is defined by induction based on the structure of the programs. The semantics of hybrid programs is compositional, that is, the semantics of a complex program is defined as a simple function of the transition semantics of its parts. We will use $\nu[x \mapsto d]$ to denote the *modification* of a state ν that agrees with ν except for the interpretation of the symbol $x \in \Sigma_{fl}$, which is changed to $d \in \mathbb{R}$ in $\nu[x \mapsto d]$.

Definition 2.7 (Transition semantics of hybrid programs). The *valuation of a hybrid program* α, denoted by $\rho_{I,\eta}(\alpha)$, is a *transition relation* on states. It specifies which state ω is reachable from a state ν by operations of the hybrid program α and is defined as follows

1. $(\nu,\omega) \in \rho_{I,\eta}(x_1 := \theta_1, \ldots, x_n := \theta_n)$ iff the state ω equals the state obtained by semantic modification of state ν as $\nu[x_1 \mapsto val_{I,\eta}(\nu,\theta_1)]\ldots[x_n \mapsto val_{I,\eta}(\nu,\theta_n)]$. Particularly, the values of other variables $z \notin \{x_1,\ldots,x_n\}$ remain constant, i.e., $val_{I,\eta}(\omega,z) = val_{I,\eta}(\nu,z)$, and the x_i receive their new values simultaneously, i.e., $val_{I,\eta}(\omega,x_i) = val_{I,\eta}(\nu,\theta_i)$.
2. $(\nu,\omega) \in \rho_{I,\eta}(x_1' = \theta_1, \ldots, x_n' = \theta_n \,\&\, \chi)$ iff there is a *flow* f of some duration $r \geq 0$ from state ν to state ω along $x_1' = \theta_1, \ldots, x_n' = \theta_n \,\&\, \chi$, i.e., a function $f : [0,r] \to \text{Sta}(\Sigma)$ such that:
   - $f(0) = \nu, f(r) = \omega$;
   - f respects the differential equations: For each variable x_i, the valuation $val_{I,\eta}(f(\zeta),x_i) = f(\zeta)(x_i)$ of x_i at state $f(\zeta)$ is continuous in ζ on $[0,r]$ and has a derivative of value $val_{I,\eta}(f(\zeta),\theta_i)$ at each time $\zeta \in (0,r)$;
   - the value of other variables $z \notin \{x_1,\ldots,x_n\}$ remains constant, that is, we have $val_{I,\eta}(f(\zeta),z) = val_{I,\eta}(\nu,z)$ for all $\zeta \in [0,r]$;
   - and f respects the invariant: $val_{I,\eta}(f(\zeta),\chi) = true$ for each $\zeta \in [0,r]$.
3. $\rho_{I,\eta}(?\chi) = \{(\nu,\nu) \,:\, val_{I,\eta}(\nu,\chi) = true\}$
4. $\rho_{I,\eta}(\alpha \cup \beta) = \rho_{I,\eta}(\alpha) \cup \rho_{I,\eta}(\beta)$
5. $\rho_{I,\eta}(\alpha;\beta) = \{(\nu,\omega) : (\nu,\mu) \in \rho_{I,\eta}(\alpha), (\mu,\omega) \in \rho_{I,\eta}(\beta)$ for a state $\mu\}$
6. $(\nu,\omega) \in \rho_{I,\eta}(\alpha^*)$ iff there is an $n \in \mathbb{N}$ and states $\nu = \nu_0, \ldots, \nu_n = \omega$ such that $(\nu_i,\nu_{i+1}) \in \rho_{I,\eta}(\alpha)$ for all $0 \leq i < n$.

For graphical illustrations of the transition semantics of hybrid programs and example dynamics, see Fig. 2.5. On the left of Fig. 2.5, we illustrate the generic shape of the transition structure $\rho_{I,\eta}(\alpha)$ for transitions along various cases of hybrid programs α from state ν to state ω. On the right of Fig. 2.5, we show examples of how the value of a variable x may evolve over time t when following the dynamics of the respective hybrid program α. The shape of the transition structure of a discrete jump $x := \theta$ (row 1) and of a differential equation $x' = \theta \,\&\, \chi$ (row 2) is an elementary one-step transition from ν to ω. For discrete jumps, however, the transition is an instant jump in the state space (row 1 on the right), while the transition for a differential equation is a continuous evolution in the state space (row 2 on the right). Note that the modifications of a discrete jump set $x_1 := \theta_1, \ldots, x_n := \theta_n$ are executed simultaneously in Definition 2.7 in the sense that all terms θ_i are evaluated in the initial state ν. For simplicity, we assume the x_i to be different, and refer to previous work [37] for a compatible semantics and calculus handling concurrent modifications of the same x_i.

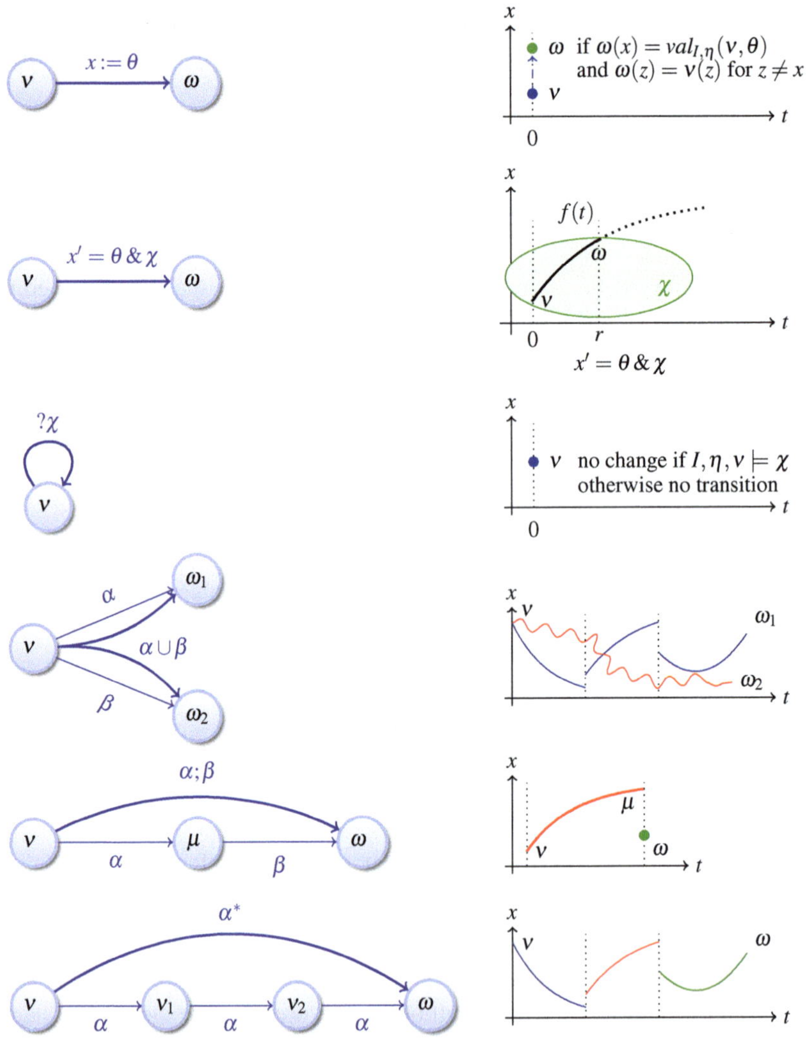

Fig. 2.5 Transition semantics (left) and example dynamics (right) of hybrid programs

For test $?\chi$ (row 3), the only possible transitions in the transition structure are self-loops that do not change the state ν, but even those transitions are only possible if the test succeeds, i.e., $I,\eta,\nu \models \chi$; see Fig. 2.5. The transition structure for choice $\alpha \cup \beta$ (row 4) is a choice between any transition of α and any transition of β. Thus, in the example on the right of row 4, the system can choose between either an evolution like the hybrid evolution (consisting, in this example, of 3 continuous flows and 2 intermediate jumps) leading to ω_1 or the squiggly evolution from ν to ω_2. The transition structure for sequential composition $\alpha;\beta$ (row 5) is that of any α transition to an intermediate state μ, followed by any β transition to ω. In the example evolution on the right of row 5, the system first follows a continuous evolution (which would come from α in this example) and then a discrete jump (which would come from β). The transition structure of a repetition α^* (row 6) repeats any number of α transitions to go from ν to ω via some number of intermediate states ν_i. In the example on the right, the system follows a sequence of various continuous evolutions and discrete jumps, giving truly hybrid behaviour.

For differential equations like $x' = \theta$, Definition 2.7 characterises transitions along a continuous evolution respecting the differential equation; see Fig. 2.6a. A continuous transition along $x' = \theta$ is possible from state ν to state ω whenever there is a continuous flow f of some duration $r \geq 0$ connecting state ν with ω such that f gives a solution of the differential equation $x' = \theta$. That is, its value is continuous on the closed interval $[0,r]$ and differentiable with the value of θ as derivative on the open interval $(0,r)$. Further, only variables subject to a differential equation change during such a continuous transition. Similarly, the continuous transitions of $x' = \theta \,\&\, \chi$ with evolution domain χ are those where f always resides within χ during the whole evolution; see Fig. 2.6b. The evolutions of $x' = \theta \,\&\, \chi$ may still stop at any point in time, but they are no longer allowed to leave χ and have to stop at an arbitrary point in time before that happens; see Fig. 2.6c.

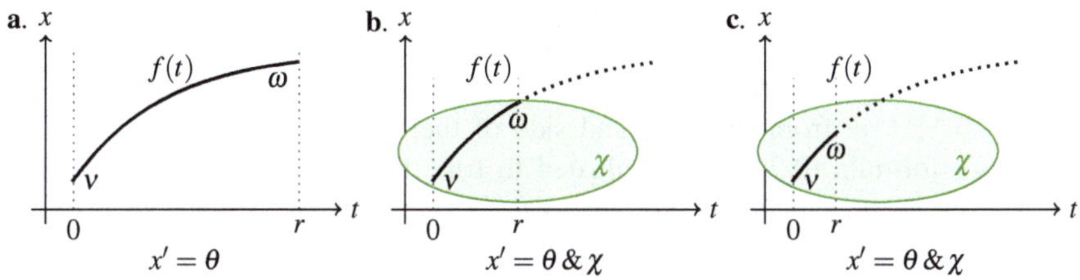

Fig. 2.6 Continuous flow along differential equation $x' = \theta$ over time t

For the semantics of differential equations, derivatives are well defined on the open interval $(0,r)$, because the set $\mathrm{Sta}(\Sigma)$ of states is isomorphic to some finite-dimensional metric real vector space spanned by the variables of the differential equations (derivatives are not defined on the closed interval $[0,r]$ if $r = 0$). For the purpose of a differential equation system, states are fully determined by an assignment of a real value to each occurring variable, which are finitely many. Furthermore, the terms of $\mathsf{d}\mathscr{L}$ are continuously differentiable on the open domain where divisors are nonzero, because the zero set of divisors is closed. Hence, solutions in $\mathsf{d}\mathscr{L}$ are unique:

Lemma 2.1 (Uniqueness). *Differential equations of $\mathsf{d}\mathscr{L}$ have unique solutions, i.e., for each differential equation system, each state ν, and each duration $r \geq 0$, there is at most one flow $f\colon [0,r] \to \mathrm{Sta}(\Sigma)$ satisfying the conditions of Case 2 of Definition 2.7.*

Proof. Let $x_1' = \theta_1, \ldots, x_n' = \theta_n \,\&\, \chi$ be a differential equation system with evolution domain χ. Using simple computations in the field of rational fractions, we can assume the right-hand sides θ_i of the differential equations to be of the form p_i/q_i for polynomials p_i, q_i. The set of points in real space where $q_i = 0$ holds is closed. As a finite union of closed sets, the set where $q_1 = 0 \vee \cdots \vee q_n = 0$ holds is closed. Hence, the valuations of the θ_i are continuously differentiable on the complement of the latter set, which is open. Thus, as a consequence of Picard-Lindelöf's theorem, a.k.a. the Cauchy-Lipschitz theorem (Theorem B.2), the solutions are unique on each connected component of this open domain. Consequently, solutions are unique when restricted to χ, which, by assumption, entails $q_1 \neq 0 \wedge \cdots \wedge q_n \neq 0$. □

Example 2.11 (Evaluation of formula and transition semantics). Recall the following hybrid program from Example 2.4 that models an (overly) simplistic train controller:

$$train \equiv (((a:=-b) \cup (?v < 8; a:=A)); \; z' = v, v' = a)^* \tag{2.1*}$$

Recall the $d\mathcal{L}$ formula (2.2) from p. 48 that claims that this simplistic train model can leave the movement authority region m:

$$v \geq 0 \wedge z < m \rightarrow \langle train \rangle z \geq m \tag{2.2*}$$

Let us evaluate this $d\mathcal{L}$ formula. Consider the interpretation I that interprets rigid symbol $m \in \Sigma$ as $I(m) = 20$ and interprets $b \in \Sigma$ as $I(b) = 2$ and $I(A) = 1$. Let state ν interpret state variables $v, z \in \Sigma$ as $\nu(v) = 9$, $\nu(z) = 0$. Then the assumptions $v \geq 0 \wedge z < m$ from the left-hand side of the implication of (2.2) are satisfied, so for formula (2.2) to be evaluated to true, the right-hand side of the implication needs to evaluate to $val_{I,\eta}(\nu, \langle train \rangle z \geq m) = true$. To find out if this is the case, the semantics of $\langle train \rangle$ in Definition 2.6 requires us to find a transition $(\nu, \omega) \in \rho_{I,\eta}(train)$ of the hybrid program *train* from ν to some state ω, according to Definition 2.7, after which $val_{I,\eta}(\omega, z \geq m)$ holds true. Let us try to find such a state ω by following the transition structure $\rho_{I,\eta}(train)$ depicted in Fig. 2.7a. Essentially, we obtain the transition structure in Fig. 2.7a by gluing the elementary transition patterns from Fig. 2.5 together according to the structure of hybrid program (2.1). We will find a path in the transition structure Fig. 2.7a from ν to ω along the transitions illustrated in Fig. 2.7b, as we explain in the following.

The top-level statement in *train* is a repetition (corresponding to the outer loop in Fig. 2.7a). We are allowed to execute the repetition twice as illustrated in the double unrolling in Fig. 2.7b (we could also repeat it any other number of times, but two times is sufficient). Thus, we hope to find an intermediate state σ_2 such that both

$$(\nu, \sigma_2) \in \rho_{I,\eta}(((a:=-b) \cup (?v < 8; a:=A)); \; z' = v, v' = a) \tag{2.5}$$

and

$$(\sigma_2, \omega) \in \rho_{I,\eta}(((a:=-b) \cup (?v < 8; a:=A)); \; z' = v, v' = a) \tag{2.6}$$

In the first transition (2.5), the top-level statement is a sequential composition (;) with a nondeterministic choice ($\cup$) as its first action. This nondeterministic choice can choose either side, indicated as upper and lower choices on the left of Fig. 2.7b.

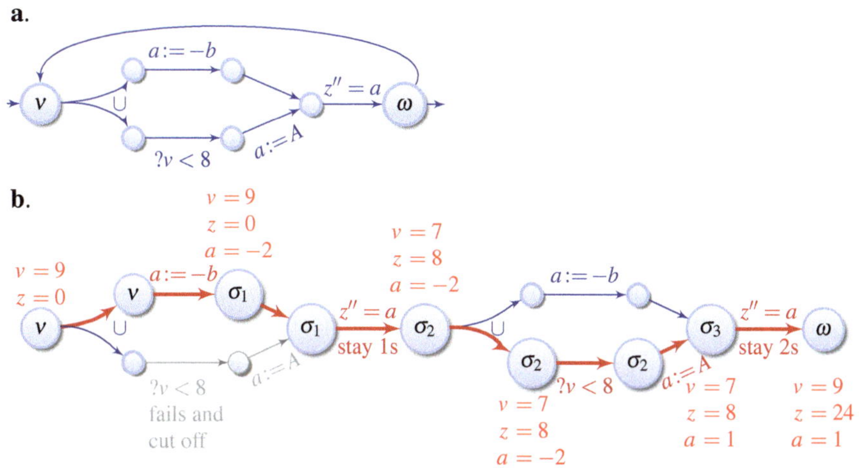

Fig. 2.7 Transition structure and transition example in (overly) simple train control

If it chooses to try to run the second (lower) choice $(?v<8;a:=A)$, however, the hybrid program cannot run successfully, because the test $?v<8$ will fail and abort the transition as a dead end, since this test evaluates to $val_{I,\eta}(\nu, v<8) = \mathit{false}$ at state ν. Hence, the hybrid program can only choose the first (upper) option $(a:=-b)$ to a state σ_1 whose only difference with ν is that $\sigma_1(a) = -2 = val_{I,\eta}(\nu,-b)$. For this state, we have $(\nu,\sigma_1) \in \rho_{I,\eta}(a:=-b)$. Next, the differential equation will run from σ_1 as the second part of the sequential composition. Because $\sigma_1(a)<0$, it will brake and the velocity v will decrease over time. If we just follow this differential equation long enough, say for one second, then the velocity at the end of it will be less than 8. Indeed, after staying in the differential equation for one second, we reach a state σ_2 with $(\sigma_1,\sigma_2) \in \rho_{I,\eta}(z'=v,v'=a)$ and $\sigma_2(v)=7$ and $\sigma_2(z)=8$, because $z(t) := \frac{-2}{2}t^2+9t+0$ and $v(t) := -2t+9$ is the solution of the differential equation when starting in state σ_1 with the interpretation I and staying for t time units. Thus, by the semantics of sequential composition and nondeterministic choice, Definition 2.7, we have that relation (2.5) holds for this state σ_2, and all we need to do is make sure that relation (2.6) holds as well.

For the transition (2.6), we can choose the second (lower) part of the nondeterministic choice, because, unlike before, the test $?v<8$ succeeds in the new state now: $val_{I,\eta}(\sigma_2, v<8) = \mathit{true}$. Hence, we follow $?v<8;a:=A$ to the state σ_3 that is like σ_2 except that we now have $\sigma_3(a) = I(A) = 1$. From state σ_3, we can stay with and follow the subsequent differential equation as long as we want to, because there is no evolution domain restriction on it. From this particular initial state σ_3, the solution of the differential equation is $z(t) := \frac{1}{2}t^2+7t+8$ and $v(t) := 1t+7$ when staying for t time units. If now we just follow the continuous evolution along this differential equation for long enough, we will eventually reach a state ω with $(\sigma_3,\omega) \in \rho_{I,\eta}(z'=v,v'=a)$ such that $val_{I,\eta}(\omega, z \geq m) = \mathit{true}$. In fact, the minimum time for this to happen is 1.544 time units, after which z has a value greater or equal $I(m) = 20$. But any longer period of time will do too. For instance, after 2 time units, we would have $\omega(z) = 24 \geq 20$ and $\omega(v) = 9$. Thus we have shown $I, \eta, \omega \models z \geq m$ and

$$(\nu, \omega) \in \rho_{I,\eta}((((a:=-b) \cup (?v < 8; a:=A)); z' = v, v' = a)^*) ,$$

which implies that formula (2.2) holds for I, η, ν.

So far, we have shown by semantic reasoning that formula (2.2) is true for I, η, ν. Yet formula (2.2) does not evaluate to *true* under all interpretations and states. For a different interpretation J with $J(m) = 1{,}000$, braking force $J(b) = 4$, and (now negative) acceleration $J(A) = -2$, but the same original state ν, we evaluate (2.4) to *false*. The reason is that, no matter which choice the hybrid program uses, the train always brakes, either with braking acceleration $-J(b) = -4$ or with negative acceleration $J(A) = -2$. Either way, the initial velocity $\nu(v) = 9$ is not high enough to reach $I(m) = 1{,}000$ from the initial position $\nu(z) = 0$. Eventually, the train velocity will be 0 and it cannot move forward anymore.

Another trivial example to show that formula (2.2) can evaluate to *false* for some J, η, ω is the interpretation J with $J(m) = 20$, $J(b) = 2$, and $I(A) = 0$ along with the state ω that interprets $\omega(v) = 0$, $\omega(z) = 0$. Then, the train stands still in the beginning and cannot move forward to $I(m) = 20$ at all. In particular, formula (2.2) is not valid, because it does not evaluate to *true* for all I, η, ν. □

What we notice in this example is that it is quite difficult and cumbersome to reason about d$\mathcal{L}$ formulas and the dynamics of hybrid systems on the level of semantics. Especially, we have only analysed particular behaviours starting at specific initial values for all the variables. For validity, we are interested in analysing all possible initial values, all numbers of repetitions, and arbitrary durations of staying in the continuous evolution modes. To do this in an elegant and coherent way, we introduce a proof calculus for d$\mathcal{L}$ in Sect. 2.5. After all, the semantics gives meaning to formulas and captures the intended meaning and behaviour in real systems. The semantics is intended to be intuitive to relate to the real world, not necessarily for being easy to use in meta-reasoning. Supporting simple analysis and proofs is the task of the proof calculus for d$\mathcal{L}$ that we develop in Sect. 2.5. Still, a good semantics like the one we chose is compositional, which makes the proof calculus simpler.

Further note that, for control-feedback loops α with a discrete controller regulating a continuous plant, transition structures involve all safety-critical states; hence, $\psi \to [\alpha]\phi$ is a natural rendition of the safety property that ϕ holds at all states reachable by α from initial states that satisfy ψ. Otherwise, d$\mathcal{L}$ can be augmented with temporal operators to refer to intermediate states or nonterminating traces. The corresponding calculus is compatible and reduces temporal properties to nontemporal properties at intermediate states of the hybrid program, as we illustrate in Chap. 4.

2.4 Collision Avoidance in Train Control

As a case study to illustrate how d$\mathcal{L}$ can be used for specifying and verifying hybrid systems, we examine a scenario of cooperating traffic agents in the European Train Control System (ETCS) [91]. The purpose of ETCS is to ensure that trains cannot crash into other trains or pass open gates. Its secondary objective is to maximise throughput and velocity without endangering safety. To achieve these objectives, ETCS discards the static partitioning of the track into fixed segments of mutually exclusive and physically separated access by trains, which has been used traditionally. Instead, permission to move is granted dynamically by decentralised Radio Block Controllers (RBCs) depending on the current track situation and movement of other traffic agents within the region of responsibility of the RBC; see Fig. 2.8.

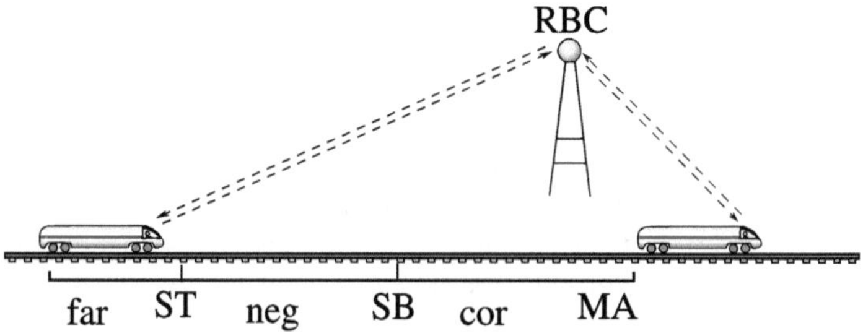

Fig. 2.8 ETCS train coordination protocol using dynamic movement authorities

Movement Authorities

This moving block principle is achieved by dynamically giving a *movement authority* (MA) to each traffic agent, within which it is obliged to remain. Before a train moves into a part of the track for which it does not have MA, it asks the RBC for an MA extension (during the negotiation phase indicated *neg* in Fig. 2.8). Depending on the MA that the RBC has currently given to other traffic agents or gates, the RBC will grant this extension and the train can move on. If the requested MA extension is still in the possession of another train that could possibly occupy the same part of the track, or if the MA is still consumed by an open gate, the RBC will deny the MA extension such that the requesting train needs to reduce speed or start braking in order to safely remain within its old MA. This is the correction phase *cor* in Fig. 2.8, which has to happen at the point SB (for start braking) at the latest. As the negotiation process with the RBC can take time because of possibly unreliable wireless communication and negotiation of the RBC with other agents, the train initiates negotiation well before reaching the end of its MA. This negotiation phase *neg* starts at the start talking point ST at the latest. Only if the train has a very large distance to the end of its MA (phase *far* in Fig. 2.8) is it safe to drive freely and not yet necessary to request MA extensions. When the rear end of a train has safely left a part of a track, the train can give that part of its MA back to RBC control such that it can be used by other traffic agents, including trains or gates.

In addition to increased flexibility and throughput of this moving block principle, the underlying technical concept of movement authorities can be exploited for verifying ETCS. It can be shown that a system of arbitrarily many trains, gates, and RBCs, which communicate in the aforementioned manner, safely avoids collisions if each traffic agent always resides within its MA under all circumstances, provided that the RBCs grant MAs mutually exclusively so that the MAs dynamically partition the track (Chap. 7). This way, verification of a system of unboundedly many traffic agents can be reduced to an analysis of individual agents with respect to their specific MA.

Train Control Model

In trains, speed supervision and automatic train protection are responsible for locally controlling the movement of a train such that it always respects its MA [90]. Depending on the current driving situation, the train controller determines a point SB (for start braking) up to which driving is safe, and adjusts its acceleration a in accordance with SB. Before SB, speed can be regulated freely (to keep the desired speed and throughput of a track profile). Beyond SB (correcting phase *cor* in Fig. 2.8), the train starts braking in order to make sure it remains within its MA if the RBC does not grant an extension in time.

We assume that an MA has been granted up to some track position, which we call m, and the train is located at position z, heading with current speed v towards m. We represent the point SB as the safety distance s relative to the end m of the MA (i.e., $m - s = \text{SB}$). In this situation, $d\mathcal{L}$ can analyse the following crucial safety property of ETCS, which we state as a $d\mathcal{L}$ formula:

$$\begin{aligned} \psi &\rightarrow [(ctrl;drive)^*]\, z \leq m \qquad (2.7)\\ \text{where } ctrl &\equiv (?m - z \leq s; a := -b) \cup (?m - z \geq s; a := A),\\ drive &\equiv \tau := 0; (z' = v, v' = a, \tau' = 1 \,\&\, v \geq 0 \land \tau \leq \varepsilon). \end{aligned}$$

It expresses that a train always ($[(ctrl;drive)^*]$) remains within its MA ($z \leq m$), assuming some constraint ψ for its parameters. The operational system model is a control-feedback loop of the digital controller *ctrl* and the plant *drive*. In *ctrl*, the train controller corrects its acceleration or brakes on the basis of the remaining distance ($m - z$). As a fail-safe recovery manoeuvre [90], it applies brakes with force b if the remaining MA is less than or equal to s. Otherwise, speed is regulated freely. The controller *ctrl* has a nondeterministic choice ($\cup$) where the left option starts with test $?m - z \leq s$ and the right option starts with test $?m - z \geq s$. The controller can try both options, but the left test will only succeed if $m - z \leq s$ holds for the current state, and the right test will only succeed if $m - z \geq s$ holds. In particular, if

$m-z<s$ holds the controller can only choose the left option, leading to braking by the assignment $a:=-b$. If $m-z>s$ holds the controller can only choose the right option, leading to acceleration by the assignment $a:=A$. If both tests could succeed, i.e., $m-z=s$, then either choice can be taken, nondeterministically. For simplicity, we assume the train uses a fixed acceleration A before passing s and does not choose any other accelerations than full braking b and full acceleration A (bang-bang control). The verification is quite similar when the controller can dynamically choose any acceleration $a \leq A$ instead, as we illustrate in Chap. 7.

After acceleration a has been set in *ctrl*, the second half of the sequential composition *ctrl*;*drive* executes, and the train continues moving in *drive*. There, the position z of the train evolves according to the differential equation system $z'=v, v'=a$ (i.e., $z''=a$). The evolution in *drive* stops when the speed v drops below zero (or earlier), because the train would not drive backwards just by braking. Thus, $v \geq 0$ is in the maximum evolution domain of *drive*. Simultaneously, clock τ measures the duration of the current *drive* phase before the controllers react to situation changes again. Clock τ is reset to zero by $\tau:=0$ when entering *drive*, constantly evolves along $\tau'=1$ together with the differential equations $z'=v, v'=a$, and is restricted by the evolution domain $\tau \leq \varepsilon$. Hence, the system can only follow *drive* for up to ε time units and at most as long as $v \geq 0$. The effect is that a *drive* phase is interrupted for reassessing the driving situation after at most ε seconds, and the *ctrl*;*drive* feedback loop repeats by the repetition operator (*). In particular, the continuous evolution cannot just be followed indefinitely without giving the controller *ctrl* a chance to react to situation changes. The corresponding transition structure $\rho_{I,\eta}((ctrl;drive)^*)$ is depicted in Fig. 2.9a. Essentially, we obtain the transition structure in Fig. 2.9a by gluing the elementary transition patterns from Fig. 2.5 together according to the structure of the hybrid program in (2.7).

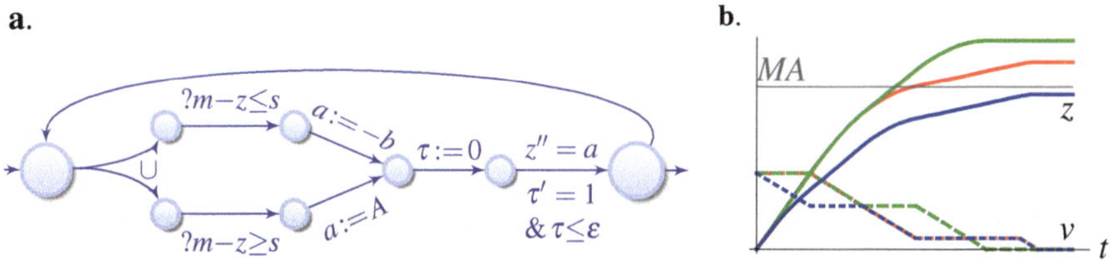

Fig. 2.9 ETCS transition structure and various choices of speed regulation for train speed control

Figure 2.9b shows possible runs of the train where speed regulation successively decreases velocity v because its MA has not been extended in time. Figure 2.9b shows three different runs (three upper position curves and three lower, partially overlapping velocity curves) which correspond to different choices of parameter s, where only the lowest velocity choice is safe. Finally, observe that the evolution domain $v \geq 0 \land \tau \leq \varepsilon$ needs to be true at *all times* during continuous evolutions of *drive*; otherwise there is no corresponding transition in $\rho_{I,\eta}(drive)$. This not only restricts the maximum duration of *drive*, but also imposes a constraint on permitted initial states: The arithmetic constraint $v \geq 0$ expresses that the differential equation only applies for nonnegative speed. Hence, as in a test $?v \geq 0$, program *drive* allows no transitions at all when v is initially less than 0. In that case, $\rho_{I,\eta}((ctrl;drive)^*)$ collapses to the trivial identity transition where only zero repetitions are possible.

Discussion

Here, we explicitly take into account possibly delayed controller reactions to bridge the gap of continuous-time models and discrete-time control design. To get meaningful results, we need to assume a maximum reaction delay ε, because safety cannot otherwise be guaranteed (the system would not be safe if the controllers can never execute). Polling cycles of sensors and digital controllers as well as latencies of actuators such as brakes contribute to ε. Instead of using specific estimates for ε for a particular train, we accept ε as a fully symbolic parameter. Further, instead of manually choosing specific values for the free parameters of (2.7) as in model checking approaches [91], we will use our calculus to synthesise constraints on the relationship of parameters that are required for safe operation of train control. We do not model weather conditions, slope of track, wheel friction, or train mass, because these are less relevant for the cooperation layer of train control [90].

Because of its nonlinear behaviour and nontrivial reset relations, system (2.7) is beyond the modelling capabilities of linear hybrid automata [8, 156, 126] and beyond o-minimal automata [189]. Previous approaches need linear flows [8, 156], do not support the coupled dynamics caused by nontrivial resets [189], require polyhedral initial sets and discrete dynamics [70], only handle robust systems with bounded regions [125] although parametric systems are not robust uniformly for all parameter choices, or handle only bounded-time safety for systems with bounded switching [217]. Finally, in addition to general numerical limits [238], numerical approaches [70, 21] quickly become intractable due to the exponential impact of the number of variables (curse of dimensionality).

2.5 Free-Variable Proof Calculus for Differential Dynamic Logic

In this section, we introduce a sequent calculus for formally verifying hybrid systems by proving validity of corresponding $\mathsf{d}\mathcal{L}$ formulas. The basic idea is to symbolically compute the effects of hybrid programs and successively transform them into logical formulas describing these effects by structural symbolic decomposition. The calculus consists of standard propositional rules, rules for dynamic modalities that are generalised to hybrid programs, and novel quantifier rules that integrate real quantifier elimination (or, in fact, any other quantifier elimination procedure) into the modal calculus using free variables and Skolemisation.

2.5.1 Substitution

The d$\mathcal{L}$ calculus uses substitutions that take effect within formulas and programs. The result of applying to a d$\mathcal{L}$ formula ϕ the *substitution* that simultaneously replaces variable y_i by term θ_i (for $1 \leq i \leq m$) is defined as usual. Figure 2.10 shows

$$
\begin{array}{rll}
\sigma(y_i) &= \theta_i & \text{for } 1 \leq i \leq n\\
\sigma(z) &= y & \text{if } z \notin \{y_1,\dots,y_m\} \text{ is a variable}\\
\sigma(f(\theta_1,\dots,\theta_n)) &= f(\sigma(\theta_1),\dots,\sigma(\theta_n)) & \text{if } f \text{ is a function symbol}\\
\\
\sigma(p(\theta_1,\dots,\theta_n)) &= p(\sigma(\theta_1),\dots,\sigma(\theta_n)) & \text{if } p \text{ is a predicate symbol}\\
\sigma(\neg\phi) &= \neg\sigma(\phi) &\\
\sigma(\phi\wedge\psi) &= \sigma(\phi)\wedge\sigma(\psi) &\\
\sigma(\phi\vee\psi) &= \sigma(\phi)\vee\sigma(\psi) &\\
\sigma(\phi\rightarrow\psi) &= \sigma(\phi)\rightarrow\sigma(\psi) &\\
\sigma(\forall x\,\phi) &= \forall x\,\sigma(\phi) & \text{if admissible}\\
\sigma(\exists x\,\phi) &= \exists x\,\sigma(\phi) & \text{if admissible}\\
\sigma([\alpha]\phi) &= [\sigma(\alpha)]\sigma(\phi) & \text{if admissible}\\
\sigma(\langle\alpha\rangle\phi) &= \langle\sigma(\alpha)\rangle\sigma(\phi) & \text{if admissible}\\
\\
\sigma(x_1:=\theta_1,..,x_n:=\theta_n) &= x_1:=\sigma(\theta_1),..,x_n:=\sigma(\theta_n) & \text{if admissible}\\
\sigma(x_1'=\theta_1,..,x_n'=\theta_n\,\&\,\chi) &= x_1'=\sigma(\theta_1),..,x_n'=\sigma(\theta_n)\,\&\,\sigma(\chi) & \text{if admissible}\\
\sigma(?\chi) &= ?\sigma(\chi) &\\
\sigma(\alpha;\beta) &= \sigma(\alpha);\sigma(\beta) &\\
\sigma(\alpha\cup\beta) &= \sigma(\alpha)\cup\sigma(\beta) &\\
\sigma(\alpha^*) &= (\sigma(\alpha))^* &
\end{array}
$$

Fig. 2.10 Application of substitution σ that simultaneously replaces variable y_i by term θ_i (for $1 \leq i \leq m$)

how the substitution σ that replaces variable y_i by term θ_i (for each $1 \leq i \leq m$) can be applied to a term, d$\mathcal{L}$ formula, or hybrid program, respectively. The first line in Fig. 2.10 represents that the substitution σ matches on the replaced (logical or state) variables y_i and replaces them by θ_i, respectively. The second line represents that no logical or state variable z other than $y_1,\dots,y_n$ are affected by σ. The third line maps the substitution σ homomorphically over function applications by applying σ recursively to all argument terms. Similarly, the next block of cases in Fig. 2.10 maps substitutions homomorphically over all subformulas. Yet for quantifiers ($\forall,\exists$) and modalities ($[\alpha],\langle\alpha\rangle$), the substitution is only applicable if admissible (as defined below) so that the bound variable x of the quantifier does not interfere with the substitution. We assume *bound variable renaming* (also known as α conversion) for renaming as needed: bound variables can be renamed to resolve conflicts, e.g., $\forall x\,\phi(x) \equiv \forall z\,\phi(z)$. Likewise, for applying the substitution homomorphically to hybrid programs (last block in Fig. 2.10) admissibility of the substitution is crucial in all cases. Admissibility implies, for instance, that the variables y_i replaced by the substitution are different from the changed variables x_j on the left-hand sides of the assignments or differential equations of the hybrid program.

Definition 2.8 (Admissible substitution). An application of a substitution σ is *admissible* if no variable x that σ replaces by $\sigma(x)$ occurs in the scope of a quantifier or modality binding x or a (logical or state) variable of the replacement $\sigma(x)$. A modality *binds* a state variable x iff it contains a discrete jump set assigning to x (such as $x:=\theta$) or a differential equation containing x' (such as $x'=\theta$).

In this book, only admissible substitutions are applicable, which is crucial for soundness. Admissible substitutions are denotation-preserving: They ensure that symbols still denote the same values after a substitution when they did so before.

Example 2.12 (Non-admissible substitution). It is important that only admissible substitutions are applicable. For the following formula, ϕ,

$$x=z \rightarrow \langle z:=z+1\rangle(z \geq x+1),$$

the substitution σ that replaces all occurrences of x by z is not admissible. This is due to the fact that for when we try to apply σ to ϕ forming

$$z=z \rightarrow \langle z:=z+1\rangle(z \geq z+1),$$

the substitution replaces x in postcondition $z \geq x+1$ by z, which is bound by modality $\langle z:=z+1\rangle$. Hence, within the scope of the modality, symbol z denotes a different value than outside the modality, thereby destroying the property of the occurrences of x—or, after the substitution, those of z—to share the same value throughout the formula. Instead, a substitution σ_2 of x by $y+1$ in ϕ to form $\sigma_2(\phi)$ is admissible for other symbols y, giving the formula $\sigma_2(\phi)$:

$$y+1=z \rightarrow \langle z:=z+1\rangle(z \geq y+1+1).$$

□

More succinctly, we abbreviate the result of applying to ϕ the substitution σ that replaces variable y_i with term θ_i (for $1 \leq i \leq m$) by $\phi_{y_1}^{\theta_1} \dots {}_{y_m}^{\theta_m}$. Thus $\phi_{y_1}^{\theta_1} \dots {}_{y_m}^{\theta_m}$ is an abbreviation for $\sigma(\phi)$ defined according to Fig. 2.10. When no confusion arises, we also use implicit notation for substitutions to improve readability. Let $\phi(z)$ be a formula with a free variable z. Then for any term θ, we use $\phi(\theta)$ as an abbreviation for the formula $\phi(z)_z^{\theta}$ that results from $\phi(z)$ by substituting θ for z.

Example 2.13 (Admissible versus non-admissible substitutions). Consider the (valid) d$\mathcal{L}$ formula ϕ defined as

$$\phi \equiv x>0 \land y>1 \land z \geq x \rightarrow [z:=z+xy]z>x.$$

Now the substitution that replaces x by $5a+x^2-y$ is admissible for ϕ, giving the result $\phi_x^{5a+x^2-y}$:

$$5a+x^2-y>0 \land y>1 \land z \geq 5a+x^2-y \rightarrow [z:=z+(5a+x^2-y)y]\,z>5a+x^2-y.$$

This formula $\phi_x^{5a+x^2-y}$, which results by an admissible substitution from ϕ, is valid, just like ϕ.

However, the substitution that replaces x with the term az is not admissible for ϕ, because variable z occurs in the replacement az but is bound in ϕ, and could thus have a different value at its various occurrences. So we cannot apply this substitution to ϕ. Yet if we choose a fresh variable u and use bound variable renaming to rename all occurrences of bound variable z to u, we obtain the formula

$$\tilde{\phi} \equiv x>0 \land y>1 \land z \geq x \rightarrow [u:=z+xy]\,u>x.$$

This variant $\tilde{\phi}$ is equivalent to ϕ, because only bound variables have been renamed. After this bound variable renaming, the substitution replacing x by az becomes admissible and we obtain

$$\tilde{\phi}_x^{az} \equiv az>0 \land y>1 \land z \geq az \rightarrow [u:=z+(az)y]\,u>az.$$

This formula is valid (just like ϕ and $\tilde{\phi}$). But it is quite different from the formula we would obtain if we had just naïvely replaced every occurrence of x (admissible or not) by az, instead of using more careful admissible substitutions:

$$az>0 \land y>1 \land z \geq az \rightarrow [z:=z+(az)y]\,z>az.$$

The latter formula is clearly false for all I, η, ν with $val_{I,\eta}(\nu, a) = 1$, because z cannot possibly be greater than az then. Contrast this with the validity of the original formula ϕ and its (admissible) substitution instance $\tilde{\phi}_x^{az}$.

Similarly, the substitution that replaces z with ax is not admissible for ϕ, because the replaced variable z is bound in ϕ, and could thus have a different value at its various occurrences. So we cannot apply this substitution. Yet if we again choose a fresh variable u and use bound variable renaming to rename all occurrences of bound variable z to u, we obtain the formula $\tilde{\phi}$ above. After this bound variable renaming, the substitution replacing z with ax becomes admissible and we obtain

$$\tilde{\phi}_z^{ax} \equiv x>0 \land y>1 \land ax \geq x \rightarrow [u:=ax+xy]\,u>x.$$

Again, this formula is valid and quite different from the formula we would obtain if we had just naïvely replaced every occurrence of z (admissible or not) by ax:

$$x>0 \land y>1 \land ax \geq x \rightarrow [z:=ax+xy]\,ax>x.$$

The latter formula is again false for all I, η, ν with $val_{I,\eta}(\nu, a) = 1$, because x cannot possibly be greater than ax then. Contrast this with the validity of the original formula ϕ and the admissible substitution instance $\tilde{\phi}_x^{ax}$.

Thus, there is a close connection between the formula ϕ and its various substitution instances (if admissible!), which we will identify in the next lemma. As part of that, we will show that, since ϕ is valid, all of its (admissible) substitution instances are valid. This close connection (and every other similarity) breaks when we naïvely replace terms in ϕ instead of obeying the requirements of admissible substitutions.

□

Example 2.14 (Non-admissibility in repetitions). The last example is prototypical for several d$\mathcal{L}$ formulas and works similarly for all d$\mathcal{L}$ formulas without repetitions or differential equations. Yet repetitions and differential equations themselves are more involved. Consider a (valid) d$\mathcal{L}$ formula with a repetition:

$$\psi \equiv x>0 \land y>1 \land z>x \rightarrow [(z:=z+xy)^*]z>x. \tag{2.8}$$

As with the formula ϕ from the last example, the substitution that replaces x by the term $5a+x^2-y$ is admissible for ψ, giving the result $\psi_x^{5a+x^2-y}$:

$$5a+x^2-y>0 \land y>1 \land z>5a+x^2-y \rightarrow [(z:=z+(5a+x^2-y)y)^*]z>5a+x^2-y$$

This formula $\psi_x^{5a+x^2-y}$, which results by an admissible substitution from ψ is valid, just like ψ.

Again, the substitution that replaces x by the term az is not admissible for ψ, because variable z occurs in the replacement az but is bound in ψ, and could have different values at its occurrences. Hence, we cannot apply this substitution. However, for repetitions, it is not so easy to do bound variable renaming to get around this! We cannot simply replace all bound occurrences of z by one fresh variable u, which would give

$$x>0 \land y>1 \land z>x \rightarrow [(u:=\check{u}+xy)^*]u>x.$$

But here the connection of u with the input z has been lost and the formula is no longer valid. The reason is that the occurrence of z on the right-hand side $z+xy$ of the jump (which corresponds to the occurrence of u we marked ˇ in the last formula) is neither just free nor just bound. During the first iteration of the repetition, it would be free (because it receives its value from outside); during subsequent iterations, however, it would be bound (because it receives its value from the last assignment). The formula we would obtain if we had just naïvely replaced every occurrence of x (admissible or not) by az is also quite different and not valid:

$$x>0 \land y>1 \land z>x \rightarrow [(z:=z+(a\check{z})y)^*]z>az.$$

The reason is that the occurrence marked with ˇ is neither just free nor just bound, because it depends on the number of iterations of the loop.

Likewise, the substitution that replaces z by ax is not admissible for ψ and cannot be applied, because the replaced variable z is bound in ψ. We thus cannot apply this substitution. Once more, it is not so easy to do bound variable renaming to get around this and we cannot just rename z to a fresh variable u to resolve this issue. The formula we would obtain if we had just naïvely replaced every occurrence of x (admissible or not) by az is also quite different and not valid:

$$x > 0 \land y > 1 \land ax > x \rightarrow [(z := \check{a}x + xy)^*]\, ax > x.$$

The reason is again that the occurrence of z (prior to replacing) at the position marked ˇ is neither just free nor just bound. While it would be perfectly alright to replace the first dynamic occurrence of z (in the sequential execution order) by ax, subsequent occurrences (including those in repetitions) have a different operational value and cannot be replaced.

In these two cases, the substitutions are just not admissible for ψ and cannot be applied, because the modalities of ψ bind relevant replaced variables or variables in the replacements. Our proof calculus in Sect. 2.5.2 will use other ways that do not need substitution to prove formulas with repetitions like these. □

Example 2.15 (Non-admissibility in differential equations). The situation with differential equations is quite similar. In the d$\mathcal{L}$ formula

$$\psi \equiv x > 0 \land y > 1 \land z > x \rightarrow [z' = z + xy]\, z > x \tag{2.9}$$

the occurrences of z in the differential equation are neither just free nor just bound: The value z affects the initial value z of the differential equation, but the value of z also evolves over time when following the differential equation to a new value. Thus, z is both a free initial value and bounded or updated during the evolution. The substitution that replaces x with $5a^2 + x^2 - y$ is still admissible for ψ, giving $\psi_x^{5a^2+x^2-y}$:

$$5a^2 + x^2 - y > 0 \land y > 1 \land z > 5a^2 + x^2 - y \rightarrow [z' = z + (5a^2 + x^2 - y)y]z > 5a^2 + x^2 - y$$

But we cannot substitute x with az, because the substitution is not admissible for ψ as bound variable z occurs in the replacement az. Nor can we substitute z with ax, because this substitution is not admissible for ψ either, as the replaced variable z is bound in ψ. The formula we would obtain if we had just naïvely replaced every occurrence of x (admissible or not) with az, is different and not valid:

$$az > 0 \land y > 1 \land z > x \rightarrow [z' = z + (a\check{z})y]\, z > az.$$

The formula we would obtain, instead, if we had just naïvely replaced every occurrence of z (admissible or not) by ax, is also different and not valid:

$$x > 0 \land y > 1 \land ax > x \rightarrow [z' = \check{a}x + xy]\, ax > x.$$

In both cases, we marked the positions where the occurrences have been neither free nor bound with ˇ once again.

In these two cases, the substitutions are not admissible for ψ and cannot be applied, because the modalities of ψ bind relevant replaced variables or variables in the replacements. Our proof calculus in Sect. 2.5.2 will prove such properties of differential equations differently. □

Example 2.16 (Bound variable renaming for repetitions and differential equations). On a side note, it would not be impossible to define bound variable renaming for repetitions and differential equations. We decide not to use these extensions in our approach, because they are technically more involved and not necessary for our proof calculus. The purpose of this example is to show how bound variable renaming could be extended appropriately, nevertheless. When we add an extra discrete jump, we could define the following extended bound variable renaming variant of formula (2.8):

$$x > 0 \land y > 1 \land z > x \to [u := z;\ (u := u + xy)^*]\, u > x.$$

This formula separates the initial value assignment from the loop. Similarly when we add an extra discrete jump, we could define the following extended bound variable renaming variant of formula (2.9):

$$x > 0 \land y > 1 \land z > x \to [u := z;\ u' = u + xy]\, u > x.$$

Again, this formula separates the initial value assignment from the differential equation. For both variants, the substitution replacing x with az is admissible, and so is the substitution replacing z with ax. Essentially, the above two variants retain the initial value z explicitly before the repetition or differential equation. We have chosen not to use these extended bound variable renamings in this book and, instead, follow our choice that non-admissible substitutions are not applicable at all. □

There is a direct connection between a formula ϕ and its substitution instance $\sigma(\phi)$, provided that the substitution σ is admissible for ϕ. In fact, the valuation of ϕ and $\sigma(\phi)$ coincide if only we change the interpretation of the replaced symbols appropriately when evaluating ϕ. That is, semantically evaluating ϕ (after modifying the interpretation of the symbols replaced by σ in I, η, ν) is the same as semantically evaluating ϕ in the original I, η, ν after applying the substitution (resulting in $\sigma(\phi)$). Stated differently, we can show that, for admissible substitutions, syntactic substitution in the formula and semantic modification of I, η, ν have the same effect:

Lemma 2.2 (Substitution Lemma). *Let σ be an admissible substitution for the (term or) formula ϕ and let σ replace only logical variables; then*

$$\textit{for each } I, \eta, \nu : \ val_{I,\eta}(\nu, \sigma(\phi)) = val_{I,\sigma^*(\eta)}(\nu, \phi),$$

where the semantic modification $\sigma^*(\eta)$ of assignment η is adjoint *to σ, i.e., $\sigma^*(\eta)$ is identical to η, except that $\sigma^*(\eta)(x) = val_{I,\eta}(\nu, \sigma(x))$ for all logical variables $x \in V$.*

Proof. In essence, the proof of this lemma is a simple corollary to the fact that both substitution and valuation are homomorphisms defined inductively on formulas from their effect on atomic symbols. The application of an admissible substitution σ is a homomorphic continuation of its effect on atomic symbols to all d$\mathcal{L}$ formulas by way of Fig. 2.10. That is, the effect of an admissible(!) substitution on a compound formula is just defined by applying the substitution recursively to all subformulas. Likewise, the valuation is a homomorphic continuation of the interpretation I, state ν, and assignment η on atomic symbols to all $\mathsf{d}\mathscr{L}$ formulas by way of Definition 2.6. That is, the valuation of a compound formula is just defined by using the valuation on all subformulas.

First we prove the substitution lemma applied to terms θ:

$$\text{for each } I,\eta,\nu: \; val_{I,\eta}(\nu,\sigma(\theta)) = val_{I,\sigma^*(\eta)}(\nu,\theta).$$

The proof is by induction on the structure of the term θ.

1. If θ is a logical variable $x \in V$, then, by definition of $\sigma^*(\eta)$:

$$val_{I,\eta}(\nu,\sigma(x)) = \sigma^*(\eta)(x) = val_{I,\sigma^*(\eta)}(\nu,x).$$

2. If θ is a state variable $x \in \Sigma$, then it is different from replaced logical variables $u \in V$ and $\sigma(x) = x$. Hence

$$val_{I,\eta}(\nu,\sigma(x)) = val_{I,\eta}(\nu,x) = \nu(x) = val_{I,\sigma^*(\eta)}(\nu,x).$$

3. If θ is of the form $f(\theta_1,\dots,\theta_n)$ for a function symbol f of arity $n \geq 1$, then

$$\begin{aligned}
&val_{I,\eta}(\nu,\sigma(f(\theta_1,\dots,\theta_n)))\\
&= val_{I,\eta}(\nu,f(\sigma(\theta_1),\dots,\sigma(\theta_n)))\\
&= I(f)\big(val_{I,\eta}(\nu,\sigma(\theta_1)),\dots,val_{I,\eta}(\nu,\sigma(\theta_n))\big)\\
&= I(f)\big(val_{I,\sigma^*(\eta)}(\nu,\theta_1),\dots,val_{I,\sigma^*(\eta)}(\nu,\theta_n)\big)\\
&= val_{I,\sigma^*(\eta)}(\nu,f(\theta_1,\dots,\theta_n))
\end{aligned}$$

because the θ_i are simpler than $f(\theta_1,\dots,\theta_n)$ so that, by induction hypothesis, we have for each i:

$$val_{I,\eta}(\nu,\sigma(\theta_i)) = val_{I,\sigma^*(\eta)}(\nu,\theta_i).$$

Next we prove the substitution lemma applied to $\mathsf{d}\mathscr{L}$ formulas ϕ:

$$\text{for each } I,\eta,\nu: \; val_{I,\eta}(\nu,\sigma(\phi)) = val_{I,\sigma^*(\eta)}(\nu,\phi).$$

The proof is by induction on the structure of the formula ϕ.

1. If ϕ is of the form $p(\theta_1,\dots,\theta_n)$ for a predicate symbol p of arity $n \geq 1$, then the proof is almost identical to that for function symbols above.
2. If ϕ is of the form $\phi_1 \vee \phi_2$, then we use the induction hypothesis on ϕ_1 and ϕ_2 to conclude

$$\begin{aligned}
&val_{I,\eta}(\nu,\sigma(\phi_1 \vee \phi_2))\\
&= val_{I,\eta}(\nu,\sigma(\phi_1) \vee \sigma(\phi_2)) = \mathit{true}\\
&\text{iff } val_{I,\eta}(\nu,\sigma(\phi_1)) = \mathit{true} \text{ or } val_{I,\eta}(\nu,\sigma(\phi_2)) = \mathit{true}
\end{aligned}$$

$$\text{iff } val_{I,\sigma^*(\eta)}(\nu,\phi_1) = true \text{ or } val_{I,\sigma^*(\eta)}(\nu,\phi_2) = true$$
$$\text{iff } val_{I,\sigma^*(\eta)}(\nu,\phi_1 \vee \phi_2) = true$$

If ϕ is of the form $\phi_1 \wedge \phi_2$ or of the form $\phi_1 \to \phi_2$ or $\neg\phi_1$, then the proof is similar.

3. If ϕ is of the form $\exists x\,\psi$, then we use that σ was assumed to be admissible for ϕ. In particular (by bound variable renaming), x is not one of the replaced variables u and x does not occur in any of the replacements $\sigma(u)$. We use the induction hypothesis on ψ to conclude

$$val_{I,\eta}(\nu,\sigma(\exists x\,\psi)) = val_{I,\eta}(\nu,\exists x\,\sigma(\psi)) = true$$
$$\text{iff there is a } d \text{ such that } val_{I,\eta[x\mapsto d]}(\nu,\sigma(\psi)) = true$$
$$\text{iff there is a } d \text{ such that } val_{I,\sigma^*(\eta[x\mapsto d])}(\nu,\psi) = true$$
$$\text{iff there is a } d \text{ such that } val_{I,\sigma^*(\eta)[x\mapsto d]}(\nu,\psi) = true$$
$$\text{iff } val_{I,\sigma^*(\eta)}(\nu,\exists x\,\psi) = true.$$

Note that $\sigma^*(\eta[x\mapsto d]) = \sigma^*(\eta)[x\mapsto d]$, because x is not affected by the substitution σ (since admissible); hence x is not affected by adjoint assignments. If ϕ is of the form $\forall x\,\psi$, the proof is similar.

4. If ϕ is of the form $[\alpha]\psi$, then we use that the substitution σ is admissible by assumption. Hence, α does not bind any of the replaced variables nor any of the variables that occur in any of the replacements $\sigma(u)$. We use the induction hypothesis on ψ to conclude

$$val_{I,\eta}(\nu,\sigma([\alpha]\psi)) = val_{I,\eta}(\nu,[\sigma(\alpha)]\sigma(\psi)) = true$$
$$\text{iff for all } \omega \text{ with } (\nu,\omega) \in \rho_{I,\eta}(\sigma(\alpha)) : val_{I,\eta}(\omega,\sigma(\psi)) = true$$
$$\text{iff for all } \omega \text{ with } (\nu,\omega) \in \rho_{I,\eta}(\sigma(\alpha)) : val_{I,\sigma^*(\eta)}(\omega,\psi) = true$$
$$\text{iff}^\star \text{ for all } \omega \text{ with } (\nu,\omega) \in \rho_{I,\sigma^*(\eta)}(\alpha) : val_{I,\sigma^*(\eta)}(\omega,\psi) = true$$
$$\text{iff } val_{I,\sigma^*(\eta)}(\nu,[\alpha]\psi) = true.$$

For the middle step marked with $\star$, we still have to prove the substitution lemma for hybrid programs:

$$\rho_{I,\eta}(\sigma(\alpha)) = \rho_{I,\sigma^*(\eta)}(\alpha). \tag{2.10}$$

If α is of the form $\langle\alpha\rangle\psi$ then the proof is similar.

Finally we prove the substitution lemma for hybrid programs α as formulated in (2.10). The proof is by induction on the structure of hybrid program α.

1. If α is of the form $x_1 := \theta_1, \dots x_n := \theta_n$, then we use the substitution lemma on the terms θ_i to show

$$\begin{aligned}
&(\nu,\omega) \in \rho_{I,\eta}(\sigma(x_1 := \theta_1, \ldots x_n := \theta_n)) = \rho_{I,\eta}(x_1 := \sigma(\theta_1), \ldots x_n := \sigma(\theta_n)) \\
&\text{iff } \nu[x_1 \mapsto val_{I,\eta}(\nu, \sigma(\theta_1))] \ldots [x_n \mapsto val_{I,\eta}(\nu, \sigma(\theta_n))] = \omega \\
&\text{iff } \nu[x_1 \mapsto val_{I,\sigma^*(\eta)}(\nu, \theta_1)] \ldots [x_n \mapsto val_{I,\sigma^*(\eta)}(\nu, \theta_n)] = \omega \\
&\text{iff } (\nu,\omega) \in \rho_{I,\sigma^*(\eta)}(x_1 := \theta_1, \ldots x_n := \theta_n).
\end{aligned}$$

2. If α is of the form $?\chi$ for a (first-order) d$\mathscr{L}$ formula χ, then we use the substitution lemma on the (simpler and even first-order) d$\mathscr{L}$ formula χ:

$$\begin{aligned}
&(\nu,\omega) \in \rho_{I,\eta}(\sigma(?\chi)) = \rho_{I,\eta}(?\sigma(\chi)) \\
&\text{iff } \nu = \omega \text{ and } val_{I,\eta}(\nu, \sigma(\chi)) = true \\
&\text{iff } \nu = \omega \text{ and } val_{I,\sigma^*(\eta)}(\nu, \chi) = true \\
&\text{iff } (\nu,\omega) \in \rho_{I,\sigma^*(\eta)}(?\chi).
\end{aligned}$$

3. If α is of the form $x_1' = \theta_1, \ldots x_n' = \theta_n \,\&\, \chi$, then we use the substitution lemma on terms and on the (first-order) d$\mathscr{L}$ formula χ to conclude:

$$\begin{aligned}
&(\nu,\omega) \in \rho_{I,\eta}(\sigma(x_1' = \theta_1, \ldots x_n' = \theta_n \,\&\, \chi)) \\
&\quad = \rho_{I,\eta}(x_1' = \sigma(\theta_1), \ldots x_n' = \sigma(\theta_n) \,\&\, \sigma(\chi)),
\end{aligned}$$

which holds if and only if there is a continuous flow function $f : [0,r] \to \mathrm{Sta}(\Sigma)$ with $f(0) = \nu, f(r) = \omega$ and $val_{I,\eta}(f(\zeta), z) = val_{I,\eta}(\nu, z)$ for all $\zeta \in [0,r]$ and all $z \notin \{x_1, \ldots, x_n\}$ such that:

- for each x_i, $val_{I,\eta}(f(\zeta), x_i) = f(\zeta)(x_i)$ is continuous in ζ on $[0,r]$ and has a derivative of value $val_{I,\eta}(f(\zeta), \sigma(\theta_i))$ at each time $\zeta \in (0,r)$,
- and $val_{I,\eta}(f(\zeta), \sigma(\chi)) = true$ for each $\zeta \in [0,r]$.

By the substitution lemma for terms and formulas, respectively, these conditions are equivalent to

- for each x_i, $val_{I,\sigma^*(\eta)}(f(\zeta), x_i) = f(\zeta)(x_i)$ is continuous in ζ on $[0,r]$ and has a derivative of value $val_{I,\sigma^*(\eta)}(f(\zeta), \theta_i)$ at each time $\zeta \in (0,r)$,
- and $val_{I,\sigma^*(\eta)}(f(\zeta), \chi) = true$ for each $\zeta \in [0,r]$,

which hold if and only if

$$(\nu,\omega) \in \rho_{I,\sigma^*(\eta)}(x_1' = \theta_1, \ldots x_n' = \theta_n \,\&\, \chi).$$

4. If α is of the form $\beta \cup \gamma$, then we can use the induction hypothesis on β and γ to conclude

$$\begin{aligned}
\rho_{I,\eta}(\sigma(\beta \cup \gamma)) &= \rho_{I,\eta}(\sigma(\beta) \cup \sigma(\gamma)) = \rho_{I,\eta}(\sigma(\beta)) \cup \rho_{I,\eta}(\sigma(\gamma)) \\
&= \rho_{I,\sigma^*(\eta)}(\beta) \cup \rho_{I,\sigma^*(\eta)}(\gamma) = \rho_{I,\sigma^*(\eta)}(\beta \cup \gamma).
\end{aligned}$$

5. If α is of the form $\beta;\gamma$, then we use the induction hypothesis on β and on γ to conclude

$$\begin{aligned}&(\nu,\omega)\in\rho_{I,\eta}(\sigma(\beta;\gamma))=\rho_{I,\eta}(\sigma(\beta);\sigma(\gamma))\\ &\text{iff there is a }\mu\text{ with }(\nu,\mu)\in\rho_{I,\eta}(\sigma(\beta))\text{ and }(\mu,\omega)\in\rho_{I,\eta}(\sigma(\gamma))\\ &\text{iff there is a }\mu\text{ with }(\nu,\mu)\in\rho_{I,\sigma^*(\eta)}(\beta)\text{ and }(\mu,\omega)\in\rho_{I,\sigma^*(\eta)}(\gamma)\\ &\text{iff }(\nu,\omega)\in\rho_{I,\sigma^*(\eta)}(\beta;\gamma).\end{aligned}$$

6. The case where α is of the form β^* is similar, again using admissibility:

$$\begin{aligned}&(\nu,\omega)\in\rho_{I,\eta}(\sigma(\beta^*))=\rho_{I,\eta}(\sigma(\beta)^*)\\ &\text{iff there are }n\in\mathbb{N},\mu_0=\nu,\mu_1,\dots,\mu_n=\omega:(\mu_i,\mu_{i+1})\in\rho_{I,\eta}(\sigma(\beta))\\ &\text{iff there are }n\in\mathbb{N},\mu_0=\nu,\mu_1,\dots,\mu_n=\omega:(\mu_i,\mu_{i+1})\in\rho_{I,\sigma^*(\eta)}(\beta)\\ &\text{iff }(\nu,\omega)\in\rho_{I,\sigma^*(\eta)}(\beta^*).\end{aligned}$$

□

The substitution lemma implies a simple corollary for substituting program variables instead of (or in addition to) logical variables. The proof is an immediate consequence of a double application of the substitution lemma, so that, in the remainder of this book, we do not distinguish between Lemma 2.2 and the following corollary.

Corollary 2.1. *Let σ be an admissible substitution for the (term or) formula ϕ; then*

$$\textit{for each } I,\eta,\nu:\ val_{I,\eta}(\nu,\sigma(\phi))=val_{I,\sigma^*(\eta)}(\sigma^*(\nu),\phi),$$

where the semantic modification $\sigma^*(\nu)$ of state ν is adjoint *to σ. The adjoint $\sigma^*(\nu)$ is identical to ν, except that $\sigma^*(\nu)(x)=val_{I,\eta}(\nu,\sigma(x))$ for all state variables $x\in\Sigma$. The adjoint $\sigma^*(\eta)$ is defined as in Lemma 2.2.*

Proof. The proof is a simple corollary to Lemma 2.2, using fresh logical variables z_i to relate $\sigma(\phi)$ with ϕ for gluing two uses of Lemma 2.2 together. To simplify notation, assume that σ only replaces a single state variable x by θ and let us denote the result of applying this substitution to ϕ by ϕ_x^θ. Let z be a fresh logical variable. Since the substitution σ is admissible for ϕ, the replaced variable x and all variables in its replacement θ are not bound in ϕ. Thus, ϕ is of the form ψ_z^x for the formula ψ, which is like ϕ except that it has z in place of x everywhere. Now abbreviate $val_{I,\eta}(\nu,\theta)$ as e, and abbreviate $val_{I,\eta}(\nu[x\mapsto e],x)$ as d. Then, we use Lemma 2.2 at the positions indicated $\star$ to conclude:

$$\begin{aligned}val_{I,\eta}(\nu,\phi_x^\theta)&=val_{I,\eta}(\nu,\psi_{z\,x}^{x\,\theta})=val_{I,\eta}(\nu,\psi_z^\theta)\overset{\star}{=}val_{I,\eta[z\mapsto e]}(\nu,\psi)\\ &=val_{I,\eta[z\mapsto d]}(\nu[x\mapsto e],\psi)\overset{\star}{=}val_{I,\eta}(\nu[x\mapsto e],\psi_z^x)=val_{I,\eta}(\nu[x\mapsto e],\phi).\end{aligned}$$

Note that the two lines are equal because the value of state variable x in the state does not matter for ψ, where x does not occur, and because $d=val_{I,\eta}(\nu[x\mapsto e],x)=e$.

□

Example 2.17. Again consider the formula ϕ, and an instance $\phi_x^{5a^2+b}$ under an admissible substitution:

$$\phi \equiv x = z \to \langle z{:=}z+1 \rangle (z \geq x+1),$$
$$\phi_x^{5a^2+b} \equiv 5a^2+b = z \to \langle z{:=}z+1 \rangle (z \geq 5a^2+b+1).$$

Using the substitution lemma, we can conclude that with respect to any I,η,ν, the formula ϕ and its instance $\phi_x^{5a^2+b}$ evaluate to the same truth-value when adapting the value of x appropriately. That is, let σ be the substitution that replaces x with $5a^2+b$, i.e., $\sigma(\phi) \equiv \phi_x^{5a^2+b}$; then (the corollary to) Lemma 2.2 implies:

$$val_{I,\eta}(\nu, \phi_x^{5a^2+b}) = val_{I,\sigma^*(\eta)}(\sigma^*(\nu), \phi).$$

Let us abbreviate the value $val_{I,\eta}(\nu, 5a^2+b)$ of the replacement $5a^2+b$ of x by e. Then if $x \in V$ is a logical variable, then $\sigma^*(\nu) = \nu$ and $\sigma^*(\eta) = \eta[x \mapsto e]$; hence

$$val_{I,\eta}(\nu, \phi_x^{5a^2+b}) = val_{I,\eta[x\mapsto e]}(\nu, \phi).$$

If, instead, $x \in \Sigma$ is a state variable, then $\sigma^*(\eta) = \eta$ and $\sigma^*(\nu) = \nu[x \mapsto e]$; hence

$$val_{I,\eta}(\nu, \phi_x^{5a^2+b}) = val_{I,\eta}(\nu[x \mapsto e], \phi).$$

In either case (either $x \in V$ or $x \in \Sigma$), if the value of x and its replacement $5a^2+b$ agree in the original I,η,ν already, i.e., if $val_{I,\eta}(\nu, x) = val_{I,\eta}(\nu, 5a^2+b)$, then their valuations agree according to the substitution lemma:

$$val_{I,\eta}(\nu, \phi_x^{5a^2+b}) = val_{I,\eta}(\nu, \phi).$$

□

The substitution lemma is a very powerful tool, because, among other things, we can use it to replace equals for equals without changing the valuation (substitution property). If we know that x and θ have the same value in I,η,ν, then we can substitute θ for x in a formula ϕ (if admissible) without changing the truth-value of ϕ, that is:

Lemma 2.3 (Substitution property). *If $I,\eta,\nu \models x = \theta$, then $I,\eta,\nu \models \phi \leftrightarrow \phi_x^\theta$ for any formula ϕ for which the substitution replacing x with θ is admissible.*

Proof. Consider any I,η,ν with $I,\eta,\nu \models x = \theta$. First, note that this assumption is equivalent to $val_{I,\eta}(\nu, x) = val_{I,\eta}(\nu, \theta)$. We have to show $I,\eta,\nu \models \phi \leftrightarrow \phi_x^\theta$, or, equivalently, $val_{I,\eta}(\nu, \phi) = val_{I,\eta}(\nu, \phi_x^\theta)$. This follows from the Substitution Lemma 2.2 when we choose σ to be the substitution that replaces x by θ since

$$val_{I,\eta}(\nu, \phi_x^\theta) = val_{I,\sigma^*(\eta)}(\sigma^*(\nu), \phi) = val_{I,\eta}(\nu, \phi).$$

The last step follows from the fact that I, η, ν equals $I, \sigma^*(\eta), \sigma^*(\nu)$, respectively, because the substitution σ only replaces x by θ, which already have the same value to begin with, as we assumed $val_{I,\eta}(\nu, x) = val_{I,\eta}(\nu, \theta)$. □

In addition, whenever a formula ϕ is valid (ϕ is true in all I, η, ν), the substitution lemma implies that all of its (admissible) substitution instances $\sigma(\phi)$ are valid too for any substitution σ that is admissible for ϕ.

Lemma 2.4 (Substitutions preserve validity). *If* $\vDash \phi$*, i.e.,* ϕ *is valid, then* $\vDash \sigma(\phi)$ *for any substitution* σ *that is admissible for* ϕ*.*

Proof. Let ϕ be valid, i.e., $I, \eta, \nu \models \phi$ for all I, η, ν. Consider any I, η, ν and any substitution σ that is admissible for ϕ. Now the Substitution Lemma 2.2 implies

$$val_{I,\eta}(\nu, \sigma(\phi)) = val_{I,\sigma^*(\eta)}(\sigma^*(\nu), \phi) = true.$$

The last step holds because ϕ is valid and, in particular, holds for $I, \sigma^*(\eta), \sigma^*(\nu)$. □

Observe that, for soundness, the notion of bound variables in Definition 2.8 could in fact be *any* overapproximation of the set of variables that possibly change their value during a hybrid program. In vacuous identity changes like $x := x$ or $x' = 0$, variable x will not really change its value, but we still consider x as a bound variable for simplicity. For a hybrid program α, we denote by $\forall^\alpha \phi$ the *universal closure* of formula ϕ with respect to all state variables bound in α. Quantification over state variable x is definable as $\forall X\, [x := X]\Phi$ using an auxiliary logical variable X.

2.5.2 *Rules of the Calculus for Differential Dynamic Logic*

We present a proof calculus for d$\mathcal{L}$ as a Gentzen-style sequent calculus [133]. Sequents are essentially a standard form for logical formulas that is convenient for proving. A *sequent* is of the form $\Gamma \vdash \Delta$, where the *antecedent* Γ and *succedent* Δ are finite sets of formulas. The semantics of $\Gamma \vdash \Delta$ is that of the formula $\bigwedge_{\phi \in \Gamma} \phi \to \bigvee_{\psi \in \Delta} \psi$. For quantifier elimination rules, we make use of this fact by considering sequent $\Gamma \vdash \Delta$ as an abbreviation for the latter formula. The antecedent Γ can be thought of as the formulas we assume to be true, whereas the succedent Δ can be understood as formulas for which we want to show that at least one of them is true assuming all formulas of Γ are true. So for proving a sequent $\Gamma \vdash \Delta$, we assume all Γ and want to show that one of the Δ is true. For some simple sequents like $\Gamma, \phi \vdash \phi, \Delta$, we directly know that they are valid, because we can certainly show ϕ if we assume ϕ (in fact, we will use this as an axiom). For other sequents, it is more difficult to see whether they are valid (true under all circumstances) and it is the purpose of a proof calculus to provide a means to find out.

For handling quantifiers in the d$\mathcal{L}$ calculus, we cannot use the standard proof rules [147, 122, 123], because these are for uninterpreted first-order logic and (ultimately) work by instantiating quantifiers, either eagerly as in ground tableaux or lazily by unification as in free-variable tableaux [147, 122, 123]. Also, see App. A for an exposition of proving in uninterpreted first-order logic. The basis of $\mathsf{d}\mathcal{L}$, in contrast, is first-order logic interpreted over the reals or in the theory of real-closed fields [287, 288]. A formula like $\exists a \forall x (x^2 + a > 0)$ cannot be proven by instantiation-based quantifier rules but is valid in the theory of real-closed fields. Unfortunately, quantifier elimination (QE) over the reals [81, 288], which is the standard decision procedure for real arithmetic, cannot be applied to formulas with modalities either. Hence, we introduce new quantifier rules that integrate quantifier elimination in a way that is compatible with dynamic modalities (as we illustrate in Sect. 2.5.3).

Definition 2.9 (Quantifier elimination). A first-order theory admits *quantifier elimination* if, with each formula ϕ, a quantifier-free formula $\mathrm{QE}(\phi)$ can be associated effectively that is equivalent (i.e., $\phi \leftrightarrow \mathrm{QE}(\phi)$ is valid) and has no additional free variables or function symbols. The operation QE is further assumed to evaluate ground formulas (i.e., without variables), yielding a decision procedure for closed formulas of this theory (i.e., formulas without free variables).

Example 2.18. Quantifier elimination uses the special structure of real arithmetic to express quantified arithmetic formulas equivalently without quantifiers and without using more free variables. For instance, QE yields the following equivalence:

$$\mathrm{QE}(\exists x (ax^2 + bx + c = 0)) \equiv (a \neq 0 \wedge b^2 - 4ac \geq 0) \vee (a = 0 \wedge (b = 0 \rightarrow c = 0)).$$

In this particular case, the equivalence can be found by using the generic condition for solvability of quadratic equations over the reals plus special cases when coefficients are zero. For details on quantifier elimination in real-closed fields and an overview of decision procedures for real arithmetic, also see App. D.2. □

As usual in sequent calculus rules—although the direction of entailment in the proof rules is from *premises* (above rule bar) to *conclusion* (below)—the order of reasoning is *goal-directed*: Rules are applied backwards, i.e., starting from the desired conclusion at the bottom (*goal*) to the resulting premises (*subgoals*). To highlight the logical essence of the $\mathsf{d}\mathcal{L}$ calculus, Fig. 2.11 provides *rule schemata* with which the following definition associates the calculus rules that are applicable in $\mathsf{d}\mathcal{L}$ proofs. The calculus consists of propositional rules ($\neg$r–*cut*), first-order quantifier rules ($\forall$r–i$\exists$), rules for dynamic modalities ($\langle;\rangle$–$[']$), and global rules ($[]$*gen*–*con*). All substitutions in the rules in Fig. 2.11 need to be admissible for the rules to be applicable, including the substitution that inserts $s(X_1,..,X_n)$ into $\phi(s(X_1,..,X_n))$. Proof schemata come in three kinds with which the following definition associates proof rules: 1) sequent proof schemata that mention the sequent symbol $\vdash$, 2) symmetric proof schemata that do not mention the sequent symbol $\vdash$ and can be applied on either side of the sequent, 3) the special proof schema i$\exists$ that merges multiple branches.

Definition 2.10 (Rules). The *rule schemata* in Fig. 2.11—in which *all* substitutions need to be admissible for the rules to be applicable, including the substitution that inserts $s(X_1,..,X_n)$ into $\phi(s(X_1,..,X_n))$—induce *calculus rules* by:

1. If
$$\frac{\Phi_1 \vdash \Psi_1 \quad \dots \quad \Phi_n \vdash \Psi_n}{\Phi_0 \vdash \Psi_0} \tag{2.11}$$
is an instance of a rule schema in Fig. 2.11 (rules ∀r–∀l, i∀, and the propositional and global rule schemata have this form), then
$$\frac{\Gamma, \langle \mathcal{J} \rangle \Phi_1 \vdash \langle \mathcal{J} \rangle \Psi_1, \Delta \quad \dots \quad \Gamma, \langle \mathcal{J} \rangle \Phi_n \vdash \langle \mathcal{J} \rangle \Psi_n, \Delta}{\Gamma, \langle \mathcal{J} \rangle \Phi_0 \vdash \langle \mathcal{J} \rangle \Psi_0, \Delta}$$
can be applied as a proof rule of the d$\mathcal{L}$ calculus, where Γ, Δ are arbitrary finite sets of additional context formulas (including empty sets) and $\mathcal{J}$ is a discrete jump set (including the empty set). Hence, the rule context Γ, Δ and prefix $\langle \mathcal{J} \rangle$ remain unchanged during rule applications; only the formulas mentioned in (2.11) are affected.
2. Symmetric schemata can be applied on either side of the sequent: If
$$\frac{\phi_1}{\phi_0}$$
is an instance of one of the symmetric rule schemata (the dynamic rules) in Fig. 2.11, then
$$\frac{\Gamma \vdash \langle \mathcal{J} \rangle \phi_1, \Delta}{\Gamma \vdash \langle \mathcal{J} \rangle \phi_0, \Delta} \quad \text{and} \quad \frac{\Gamma, \langle \mathcal{J} \rangle \phi_1 \vdash \Delta}{\Gamma, \langle \mathcal{J} \rangle \phi_0 \vdash \Delta}$$
can both be applied as proof rules of the d$\mathcal{L}$ calculus, where Γ, Δ are arbitrary finite sets of context formulas (including the empty set) and $\mathcal{J}$ is a discrete jump set (including empty sets). In particular, symmetric schemata yield equivalence transformations, because the same rule applies in the antecedent as in the succedent.
3. Schema i∃ applies to *all* goals containing X at once: If $\Phi_1 \vdash \Psi_1, \dots, \Phi_n \vdash \Psi_n$ is the list of all open goals of the proof that contain free variable X, then an instance
$$\frac{\vdash \mathrm{QE}(\exists X \bigwedge_i (\Phi_i \vdash \Psi_i))}{\Phi_1 \vdash \Psi_1 \quad \dots \quad \Phi_n \vdash \Psi_n}$$
of rule schema i∃ can be applied as a proof rule of the d$\mathcal{L}$ calculus.

Propositional Rules

For propositional logic, standard propositional rules ¬r–*cut* with the cut rule are listed in the first block of Fig. 2.11. They decompose the propositional structure of formulas. Rules ¬r and ¬l use simple dualities caused by the implicative semantics of sequents. Essentially, instead of showing $\neg\phi$ in the succedent, we assume the contrary ϕ in the antecedent with rule ¬r. In rule ¬l, instead of assuming $\neg\phi$ in

$$(\neg r)\ \frac{\phi \vdash}{\vdash \neg\phi} \qquad (\vee r)\ \frac{\vdash \phi, \psi}{\vdash \phi \vee \psi} \qquad (\wedge r)\ \frac{\vdash \phi \quad \vdash \psi}{\vdash \phi \wedge \psi} \qquad (\rightarrow r)\ \frac{\phi \vdash \psi}{\vdash \phi \rightarrow \psi} \qquad (ax)\ \frac{}{\phi \vdash \phi}$$

$$(\neg l)\ \frac{\vdash \phi}{\neg\phi \vdash} \qquad (\vee l)\ \frac{\phi \vdash \quad \psi \vdash}{\phi \vee \psi \vdash} \qquad (\wedge l)\ \frac{\phi, \psi \vdash}{\phi \wedge \psi \vdash} \qquad (\rightarrow l)\ \frac{\vdash \phi \quad \psi \vdash}{\phi \rightarrow \psi \vdash} \qquad (cut)\ \frac{\vdash \phi \quad \phi \vdash}{\vdash}$$

$$(\langle ; \rangle)\ \frac{\langle \alpha \rangle \langle \beta \rangle \phi}{\langle \alpha ; \beta \rangle \phi} \qquad (\langle {}^{*n} \rangle)\ \frac{\phi \vee \langle \alpha \rangle \langle \alpha^* \rangle \phi}{\langle \alpha^* \rangle \phi} \qquad (\langle := \rangle)\ \frac{\phi_{x_1}^{\theta_1} \cdots {}_{x_n}^{\theta_n}}{\langle x_1 := \theta_1, .., x_n := \theta_n \rangle \phi}$$

$$([;])\ \frac{[\alpha][\beta]\phi}{[\alpha ; \beta]\phi} \qquad ([{}^{*n}])\ \frac{\phi \wedge [\alpha][\alpha^*]\phi}{[\alpha^*]\phi} \qquad ([:=])\ \frac{\langle x_1 := \theta_1, .., x_n := \theta_n \rangle \phi}{[x_1 := \theta_1, .., x_n := \theta_n]\phi}$$

$$(\langle \cup \rangle)\ \frac{\langle \alpha \rangle \phi \vee \langle \beta \rangle \phi}{\langle \alpha \cup \beta \rangle \phi} \qquad (\langle ? \rangle)\ \frac{\chi \wedge \psi}{\langle ?\chi \rangle \psi} \qquad (\langle ' \rangle)\ \frac{\exists t{\geq}0\, \big((\forall 0{\leq}\tilde{t}{\leq}t\, \langle \mathcal{S}_{\tilde{t}} \rangle \chi) \wedge \langle \mathcal{S}_t \rangle \phi\big)}{\langle x_1' = \theta_1, .., x_n' = \theta_n \,\&\, \chi \rangle \phi}\ {}^{1}$$

$$([\cup])\ \frac{[\alpha]\phi \wedge [\beta]\phi}{[\alpha \cup \beta]\phi} \qquad ([?])\ \frac{\chi \rightarrow \psi}{[?\chi]\psi} \qquad ([\,'\,])\ \frac{\forall t{\geq}0\, \big((\forall 0{\leq}\tilde{t}{\leq}t\, \langle \mathcal{S}_{\tilde{t}} \rangle \chi) \rightarrow \langle \mathcal{S}_t \rangle \phi\big)}{[x_1' = \theta_1, .., x_n' = \theta_n \,\&\, \chi]\phi}\ {}^{1}$$

$$(\forall r)\ \frac{\vdash \phi(s(X_1, .., X_n))}{\vdash \forall x\, \phi(x)}\ {}^{2} \qquad (\exists r)\ \frac{\vdash \phi(X)}{\vdash \exists x\, \phi(x)}\ {}^{4}$$

$$(\exists l)\ \frac{\phi(s(X_1, .., X_n)) \vdash}{\exists x\, \phi(x) \vdash}\ {}^{2} \qquad (\forall l)\ \frac{\phi(X) \vdash}{\forall x\, \phi(x) \vdash}\ {}^{4}$$

$$(i\forall)\ \frac{\vdash \mathrm{QE}(\forall X\, (\Phi(X) \vdash \Psi(X)))}{\Phi(s(X_1, .., X_n)) \vdash \Psi(s(X_1, .., X_n))}\ {}^{3} \qquad (i\exists)\ \frac{\vdash \mathrm{QE}(\exists X \bigwedge_i (\Phi_i \vdash \Psi_i))}{\Phi_1 \vdash \Psi_1 \quad \ldots \quad \Phi_n \vdash \Psi_n}\ {}^{5}$$

$$([]gen)\ \frac{\vdash \forall^\alpha (\phi \rightarrow \psi)}{[\alpha]\phi \vdash [\alpha]\psi} \qquad (\langle\rangle gen)\ \frac{\vdash \forall^\alpha (\phi \rightarrow \psi)}{\langle \alpha \rangle \phi \vdash \langle \alpha \rangle \psi}$$

$$(ind)\ \frac{\vdash \forall^\alpha (\phi \rightarrow [\alpha]\phi)}{\phi \vdash [\alpha^*]\phi} \qquad (con)\ \frac{\vdash \forall^\alpha \forall v{>}0\, (\varphi(v) \rightarrow \langle \alpha \rangle \varphi(v-1))}{\exists v\, \varphi(v) \vdash \langle \alpha^* \rangle \exists v{\leq}0\, \varphi(v)}\ {}^{6}$$

---

[1] t and $\tilde{t}$ are fresh logical variables and $\langle \mathcal{S}_t \rangle$ is the jump set $\langle x_1 := y_1(t), .., x_n := y_n(t) \rangle$ with simultaneous solutions $y_1, .., y_n$ of the respective differential equations with constant symbols x_i as symbolic initial values.

[2] s is a new (Skolem) function symbol and $X_1, .., X_n$ are all free logical variables of $\forall x\, \phi(x)$.

[3] X is a new logical variable. Further, QE needs to be defined for the formula in the premise.

[4] X is a new logical variable.

[5] Among all open branches, free logical variable X only occurs in the branches $\Phi_i \vdash \Psi_i$. Further, QE needs to be defined for the formula in the premise, especially, no Skolem dependencies on X can occur.

[6] Logical variable v does not occur in α.

Fig. 2.11 Rule schemata of the free-variable proof calculus for differential dynamic logic

the antecedent, we show the contrary ϕ in the succedent. Rule $\vee r$ uses the fact that formulas are combined disjunctively in succedents, rule $\wedge l$ that they are conjunctive in antecedents. The comma between formulas in an antecedent has the same effect as a conjunction, and the comma between formulas in the succedent has the same effect as a disjunction. Rules $\vee l$ and $\wedge r$ split the proof into two cases, because conjuncts in the succedent can be proven separately ($\wedge r$) and, dually, disjuncts of the antecedent can be assumed separately ($\vee l$). For $\wedge r$ we want to show conjunction $\phi \wedge \psi$, so in the left branch we proceed to show $\Gamma \vdash \phi, \Delta$ and, in addition, in the right branch we show $\Gamma \vdash \psi, \Delta$, which, together, entail $\Gamma \vdash \phi \wedge \psi, \Delta$. If, as in rule $\vee$l, we assume disjunction $\phi \vee \psi$ as part of the antecedent, then we do not know if we can assume ϕ to hold or if we can assume ψ to hold in the antecedent, but know only that one of them holds. Hence, as in a case distinction, $\vee$l considers both cases, the case where we assume ϕ in the antecedent, and the case where we assume ψ. If both subgoals can be proven, this entails $\Gamma, \phi \vee \psi \vdash \Delta$. Rules $\rightarrow$r and $\rightarrow$l can be derived from the equivalence of $\phi \rightarrow \psi$ and $\neg\phi \vee \psi$. Rule $\rightarrow$r uses the fact that implication $\rightarrow$ has the same meaning as the sequent arrow $\vdash$ of a sequent. Intuitively, to show implication $\phi \rightarrow \psi$, rule $\rightarrow$r assumes ϕ (in the antecedent) and shows ψ (in the succedent). Rule $\rightarrow$l assumes an implication $\phi \rightarrow \psi$ to hold in the antecedent, but we do not know if this implication holds because ϕ is false, or because ψ is true, so $\rightarrow$l splits into those two branches.

The axiom rule *ax* closes a goal (there are no further subgoals, which we sometimes mark $*$ explicitly), because assumption ϕ in the antecedent trivially entails ϕ in the succedent (sequent $\Gamma, \phi \vdash \phi, \Delta$ is a simple syntactic tautology). Rule *cut* is the *cut* rule that can be used for case distinctions: The right subgoal assumes any additional formula ϕ in the antecedent that the left subgoal shows in the succedent. Dually: regardless of whether ϕ is actually true or false, both cases are covered by proof branches. We only use cuts in an orderly fashion to derive simple rule dualities and to simplify meta-proofs. In practical applications, cuts are not usually needed and we conjecture that this is no coincidence.

According to the definition in Definition 2.10, all propositional rules can be applied with an additional context Γ, Δ. In particular, rules *ax* and *cut* can also be applied as:

$$\text{ax}\,\frac{}{\Gamma, \phi \vdash \phi, \Delta} \quad \text{and} \quad \text{cut}\,\frac{\Gamma \vdash \phi, \Delta \quad \Gamma, \phi \vdash \Delta}{\Gamma \vdash \Delta}$$

First-Order Quantifier Rules

The quantifier rules $\forall$r,$\exists$l,$\exists$r,$\forall$l,i$\forall$,i$\exists$ constitute a purely modular interface to arithmetic mathematical reasoning. They can use any theory that admits quantifier elimination and has a decidable ground theory (formulas without quantifiers or variables), including the theory of real arithmetic or real-closed fields [288, 81]. Rules $\forall$r,$\exists$l,$\exists$r,$\forall$l handle quantifiers and replace quantified variables by Skolem function terms ($\forall$r,$\exists$l) or free logical variables ($\exists$r,$\forall$l), respectively. Later in the proof, rules i$\forall$,i$\exists$ can reintroduce quantifiers for these previously quantified symbols and apply quantifier elimination in real-closed fields once the remaining formulas are first-order in the relevant symbols.

Rule $\exists$l, with which we want to show $\exists x\, \phi(x)$ in the succedent, introduces a new free logical variable X for an existentially quantified variable x. Essentially, free variable X can be thought of as a variable for which an appropriate value still needs to be found for the proof to close. This makes sense, because at the time of applying proof rule $\exists$l, it is mostly impossible to know which particular instance to choose for X that will help. But once we find such an X that proves the subgoal $\Gamma \vdash \phi(X), \Delta$

later, we have also proven the goal $\Gamma \vdash \exists x\, \phi(x), \Delta$, because X will be a witness for the existence. We have also proven the goal if, later during the proof, we prove the existence of an X satisfying the constraints indirectly, without directly instantiating a witness. This is what rule i$\exists$ is for.

The dual rule $\forall$l, which assumes $\forall x\, \phi(x)$ in the antecedent, introduces a new free logical variable X for the universally quantified variable x in the antecedent. If, later, we have found an instance of X that proves subgoal $\Gamma, \phi(X) \vdash \Delta$, then we have also proven goal $\Gamma, \forall x\, \phi(x) \vdash \Delta$, because if we can prove the subgoal just from assuming the particular $\phi(X)$ in the antecedent, then the goal also holds where we even assume $\phi(x)$ holds for all x. While this reasoning is perfectly good if it works, it is somewhat surprising why this should always work for all cases. Why should one instance be enough? Why should it not be necessary to assume two different instances $\phi(X)$ and $\phi(Y)$ during the proof? The fact that this is not necessary comes from proof rule i$\exists$, which can reintroduce quantifiers and eliminate them equivalently.

Rule $\forall$r, with which we want to show $\forall x\, \phi(x)$ in the succedent, introduces a new (Skolem) function symbol s for the previously quantified variable x and replaces x by a (Skolem) term $s(X_1, \dots, X_n)$ where $X_1, \dots, X_n$ are all the free logical variables of the original formula $\forall x\, \phi(x)$. This works like a proof in mathematics, where we want so show $\forall x\, \phi(x)$ in the succedent and do so by choosing a fresh symbol s for which we prove that $\phi(s(X_1, \dots, X_n))$ holds. Because s was arbitrary and we did not assume anything special about the value of s, this implies that $\forall x\, \phi(x)$ holds. The free variables $X_1, \dots, X_n$ of the Skolem terms keep track of the dependencies of the symbols for nested quantifiers. Having all free logical variables $X_1, \dots, X_n$ in the Skolem term is important for soundness in order to prevent unsound rearrangements of quantifiers, as we elaborate in Sect. 2.5.3.

The dual rule $\exists$l is similar. When we assume $\exists x\, \phi(x)$ in the antecedent, then we only know that such an x exists, not what value it has. Hence, $\exists$l introduces a new name for this object in the form of a new (Skolem) function symbol s and replaces x by a (Skolem) term $s(X_1, \dots, X_n)$ where $X_1, \dots, X_n$ are all the free logical variables of the original formula $\exists x\, \phi(x)$. If we can prove the subgoal, the subgoal entails the goal, because we did not assume anything special about s. Having all free logical variables $X_1, \dots, X_n$ in the Skolem term to track the dependencies of the symbols is again important for soundness to prevent unsound rearrangements of quantifiers. Intuitively, for a formula like $\forall x \exists y\, \phi(x,y)$ in the antecedent—which will yield $\exists y\, \phi(X,y)$ after applying $\forall$l—we need to track the dependency of y on X, which yields $\phi(s(X),X)$ when applying $\exists$l. We need to remember that the choice for s may depend on X, because the choice of y may depend on x.

With the rule i$\forall$, we can reintroduce a universal quantifier for a Skolem term $s(X_1, \dots, X_n)$, which corresponds to a previously universally quantified variable in the succedent or a previously existentially quantified variable in the antecedent. The point of reintroducing the quantifier is that this makes sense when the remaining formulas are first-order in the quantified variable so that they can be handled equivalently by quantifier elimination in real-closed fields. When we have proven the subgoal (with for all X) then this entails the goal for the particular $s(X_1, \dots,_n)$. In particular, when we remove a quantifier with $\forall$r,$\exists$l to obtain a Skolem term, we can continue with other proof rules to handle the dynamic modalities and then reintroduce the quantifier for the Skolem term with i$\forall$ once quantifier elimination for real arithmetic becomes applicable.

The dual rule i$\exists$ can reintroduce an existential quantifier for a free logical variable that was previously existentially quantified in the succedent or previously universally quantified in the antecedent. Again, this makes sense when the resulting formula in the premise is first-order in the quantified variable X so that quantifier elimination can eliminate the quantifier equivalently. When we remove a quantifier with $\exists$r,$\forall$l to obtain a free logical variable, we can continue using other proof rules to handle the dynamic modalities and then reintroduce the quantifier for the free logical variable with i$\exists$ once quantifier elimination is applicable.

The quantifier rules $\forall$r and $\exists$l correspond to the liberalised δ^+-rule of Hähnle and Schmitt [147]. Rules $\exists$r and $\forall$l resemble the usual γ-rule but, unlike in [122, 123, 147, 134], they cannot be applied twice because the original formula is removed ($\exists x\, \phi(x)$ in $\exists$r). The calculus still has a complete handling of quantifiers due to i$\forall$ and i$\exists$, which can reconstruct and eliminate quantifiers once QE is applicable as the remaining constraints are first-order in the respective variables. In the premise of i$\forall$ and i$\exists$, we again consider sequents $\Phi \vdash \Psi$ as abbreviations for formulas. For closed formulas, we do not need other arithmetic rules. We defer illustrations and further discussion of quantifier rules to Sect. 2.5.3. For comparison, App. A gives a summary of the standard γ-rules and δ^+-rules that are used for handling quantifiers in uninterpreted first-order logic. In Sect. 3.5.5, we show an alternative way of handling real arithmetic in a modular way using deduction modulo by side deductions.

Dynamic Rules

The dynamic modality rules transform a hybrid program into structurally simpler logical formulas by symbolic decomposition. Rules $\langle;\rangle$,$[;]$,$\langle\cup\rangle$,$[\cup]$,$\langle^{*n}\rangle$,$[^{*n}]$,$\langle?\rangle$,$[?]$ are as in discrete dynamic logic [149, 37]. Also, see Fig. 2.12 for an illustration of the correspondence of a representative set of proof rules for dynamic modalities to the transition semantics of hybrid programs (from Definition 2.7).

Nondeterministic choices split into their alternatives ($\langle\cup\rangle$,$[\cup]$). For rule $[\cup]$: If all α transitions lead to states satisfying ϕ (i.e., $[\alpha]\phi$ holds) and all β transitions lead to states satisfying ϕ (i.e., $[\beta]\phi$ holds), then, all transitions of program $\alpha \cup \beta$ that choose between following α and following β also lead to states satisfying ϕ (i.e., $[\alpha \cup \beta]\phi$ holds). Dually for rule $\langle\cup\rangle$, if there is an α transition to a ϕ state ($\langle\alpha\rangle\phi$) or a β-transition to a ϕ state ($\langle\beta\rangle\phi$), then, in either case, there is a transition of $\alpha \cup \beta$ to ϕ ($\langle\alpha \cup \beta\rangle\phi$ holds), because $\alpha \cup \beta$ can choose which of those transitions to follow. A general principle behind the $d\mathcal{L}$ proof rules that is most noticeable in $\langle\cup\rangle$,$[\cup]$ is that these proof rules symbolically decompose the reasoning into two separate parts and analyse the fragments α and β separately, which is good for scalability. For these symbolic structural decompositions, it is very helpful that $d\mathcal{L}$ is a full logic that is closed under all logical operators, including disjunction and conjunction, for then the premises in $[\cup]$,$\langle\cup\rangle$ are $d\mathcal{L}$ formulas again (unlike in Hoare logic [161]).

$$([:=])\ \frac{\phi_x^\theta}{[x:=\theta]\phi}$$

$$(\langle'\rangle)\ \frac{\exists t{\geq}0\,\langle x:=y_x(t)\rangle\phi}{\langle x'=\theta\rangle\phi}$$

$$(\langle'\rangle)\ \frac{\exists t{\geq}0\,(\hat{\chi}\wedge\langle x:=y_x(t)\rangle\phi)}{\langle x'=\theta\,\&\,\chi\rangle\phi}$$
$$\hat{\chi}\equiv\forall 0{\leq}s{\leq}t\,\langle x:=y_x(s)\rangle\chi$$

$$([\cup])\ \frac{[\alpha]\phi\wedge[\beta]\phi}{[\alpha\cup\beta]\phi}$$

$$([;])\ \frac{[\alpha][\beta]\phi}{[\alpha;\beta]\phi}$$

$$(ind)\ \frac{\vdash\forall^\alpha(\phi\rightarrow[\alpha]\phi)}{\phi\vdash[\alpha^*]\phi}$$

$$(con)\ \frac{\vdash\forall^\alpha\forall v{>}0\,(\varphi(v)\rightarrow\langle\alpha\rangle\varphi(v-1))}{\exists v\,\varphi(v)\vdash\langle\alpha^*\rangle\exists v{\leq}0\,\varphi(v)}$$

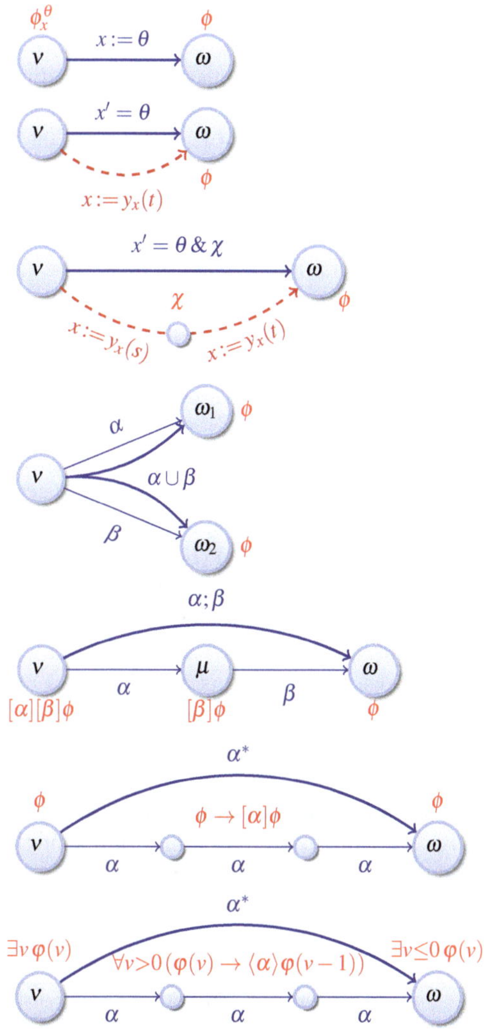

Fig. 2.12 Correspondence of dynamic proof rules and transition semantics

Sequential compositions are proven using nested modalities ($\langle;\rangle$,$[;]$). For rule $[;]$: If after all α-transitions, all β-transitions lead to states satisfying ϕ (i.e., $[\alpha][\beta]\phi$ holds), then also all transitions of the sequential composition $\alpha;\beta$ lead to states satisfying ϕ (i.e., $[\alpha;\beta]\phi$ holds). See, again, Fig. 2.12 for a graphical illustration of this proof principle. The dual rule $\langle;\rangle$ uses the fact that if there is an α-transition, after which there is a β-transition leading to ϕ (i.e., $\langle\alpha\rangle\langle\beta\rangle\phi$), then there is a transition of $\alpha;\beta$ leading to ϕ (that is, $\langle\alpha;\beta\rangle\phi$), because the transitions of $\alpha;\beta$ are just those that first do any α-transition, followed by any β-transition (Definition 2.7).

Rules $\langle *^n \rangle$,$[*^n]$ are the usual iteration rules, which partially unwind loops. Rule $\langle *^n \rangle$ uses the fact that ϕ holds after repeating α (i.e., $\langle \alpha^* \rangle \phi$), if ϕ holds at the beginning (for ϕ holds after zero repetitions then), or if, after one execution of α, ϕ holds after any number of repetitions of α, including zero repetitions (i.e., $\langle \alpha \rangle \langle \alpha^* \rangle \phi$). So rule $\langle *^n \rangle$ expresses that for $\langle \alpha^* \rangle \phi$ to hold, ϕ must hold either immediately or after one or more repetitions of α. Rule $[*^n]$ is the dual rule expressing that ϕ must hold after all of those combinations for $[\alpha^*]\phi$ to hold.

Tests are proven by showing (with a conjunction in rule $\langle ? \rangle$) or assuming (with an implication in rule $[?]$) that the test succeeds, because test $?\chi$ can only make a transition when condition χ actually holds true (Definition 2.7). Thus, for d$\mathcal{L}$ formula $\langle ?\chi \rangle \phi$ rule $\langle ? \rangle$ is used to prove that χ holds true (otherwise there is no transition and thus the reachability property is false) and that ϕ holds after the resulting no-op. Rule $[?]$ for d$\mathcal{L}$ formula $[?\chi]\phi$, in contrast, assumes that χ holds true (otherwise there is no transition and thus nothing to show) and that ϕ holds after the resulting no-op.

Rule $\langle := \rangle$ uses simultaneous substitutions from Fig. 2.10 for handling discrete jump sets. To show that ϕ is true after a discrete jump, $\langle := \rangle$ shows that ϕ has been true before, when replacing the affected variables x_i with their new values θ_i in ϕ by an admissible substitution (Definition 2.8). Alternatively, the discrete jump set can also remain an unchanged prefix ($\mathcal{J}$ in Definition 2.10) for other d$\mathcal{L}$ rules applied to ϕ, until the substitution for rule $\langle := \rangle$ becomes admissible. This is what our proof calculus uses instead of what we have shown in Example 2.16. Rule $[:=]$ uses the fact that discrete jump sets characterise a unique deterministic transition. Hence, its premise and conclusion are actually equivalent, because there is exactly one terminating transition for each discrete jump set. Assuming the presence of vacuous identity jumps $a := a$ for variables a that do not otherwise change (vacuous identity jumps can be added as they do not change state), we can further use rule $\langle := \rangle$ to *merge* subsequent discrete jumps into a single discrete jump set (see previous results [37] for a compatible calculus detailing jump set merging, which works without the need to add vacuous identity jumps $a := a$):

$$\langle := \rangle \frac{\vdash \langle z := -\frac{b}{2}t^2 + Vt, v := V+1, a := -b \rangle [\beta]\phi}{\vdash \langle a := -b, v := V \rangle \langle z := \frac{a}{2}t^2 + vt, v := v+1, a := a \rangle [\beta]\phi}$$
$$[:=] \frac{\vdash \langle a := -b, v := V \rangle \langle z := \frac{a}{2}t^2 + vt, v := v+1, a := a \rangle [\beta]\phi}{\vdash \langle a := -b, v := V \rangle [z := \frac{a}{2}t^2 + vt, v := v+1, a := a][\beta]\phi}$$
$$[;] \frac{\vdash \langle a := -b, v := V \rangle [z := \frac{a}{2}t^2 + vt, v := v+1, a := a][\beta]\phi}{\vdash \langle a := -b, v := V \rangle [z := \frac{a}{2}t^2 + vt, v := v+1, a := a; \beta]\phi}$$

More generally, $\langle x_1 := \theta_1, \ldots, x_n := \theta_n \rangle \langle x_1 := \vartheta_1, \ldots, x_n := \vartheta_n \rangle \phi$ can be merged by $\langle := \rangle$ to $\langle x_1 := \vartheta_1{}_{x_1}^{\theta_1} \ldots {}_{x_n}^{\theta_n}, \ldots, x_n := \vartheta_n{}_{x_1}^{\theta_1} \ldots {}_{x_n}^{\theta_n} \rangle \phi$. Also see previous work [37] for more advanced and optimised merging techniques for state changes.

Given first-order definable flows for their differential equations, proof rules $\langle ' \rangle$,$[']$ handle continuous evolutions (see [15, 189, 238] and App. B for flow approximation and solution techniques). These flows are combined in the discrete jump set $\mathcal{S}_t$. Given a solution $\mathcal{S}_t$ for the differential equation system with symbolic initial values $x_1, \ldots, x_n$, continuous evolution along differential equations can be replaced by a discrete jump $\langle \mathcal{S}_t \rangle$ with an additional quantifier for the evolution time t. The effect of the constraint on χ is to restrict the continuous evolution such that its solution $\mathscr{S}_{\tilde{t}}$ remains in the evolution domain χ at all intermediate times $\tilde{t} \leq t$. This constraint simplifies to *true* if the evolution domain restriction χ is *true*, which makes sense, because there are no special constraints on the evolution (other than the differential equations) if the evolution domain region is described by *true*, hence the full space $\mathbb{R}^n$. A notable special case of rules $[']$ and $\langle ' \rangle$ is when the evolution domain χ is *true*:

$$\frac{\forall t{\geq}0\,\langle \mathscr{S}_t \rangle \phi}{[x_1' = \theta_1,\ldots,x_n' = \theta_n]\phi} \qquad \frac{\exists t{\geq}0\,\langle \mathscr{S}_t \rangle \phi}{\langle x_1' = \theta_1,\ldots,x_n' = \theta_n\rangle\phi} \tag{2.12}$$

Similar simplifications can be made for convex invariant conditions (Sects. 2.9 and 3.8).

Global Rules

The last block of rules $[]gen,\langle\rangle gen,ind,con$ are *global rules*. They depend on the truth of their premises in all states reachable by hybrid program α, which is ensured by the universal closure $\forall^\alpha$ with respect to all bound state variables (Definition 2.8) of the respective hybrid program α. This universal closure overapproximates all possible change caused by α, because it comprises all bound variables. This universal closure is required for soundness in the presence of contexts Γ, Δ (Definition 2.10) or free variables. The global rules are given in a form that best displays their underlying logical principles. The general pattern for applying global rules to prove that the succedent of their conclusion holds is to prove that both their premise and the antecedent of their conclusion hold. In particular, the antecedent can be thought of as holding in the current state, whereas the premise can be thought of as holding in all reachable states because of the universal closure.

Rules $[]gen,\langle\rangle gen$ are generalisation rules and can be used to strengthen postconditions: antecedent $[\alpha]\phi$ is sufficient for proving succedent $[\alpha]\psi$ when postcondition ϕ entails ψ in all relevant states reachable by α, which are overapproximated by the universal closure $\forall^\alpha$ with respect to the bound variables of α. Clearly, for rule $[]gen$, if all states reachable by α satisfy ϕ ($[\alpha]\phi$) and ϕ implies ψ in all these states ($\forall^\alpha \phi \to \psi$), then ψ also holds in all states reachable by α ($[\alpha]\psi$). Similarly, for rule $\langle\rangle gen$, if some state reachable by α satisfies ϕ ($\langle\alpha\rangle\phi$) and ϕ implies ψ in all reachable states ($\forall^\alpha \phi \to \psi$), then ψ also holds in some state reachable by α ($\langle\alpha\rangle\psi$).

Rule *ind* is an induction schema with *inductive invariant* ϕ. Similarly, *con* is a generalisation of Harel's convergence rule [149] to the hybrid case with decreasing *variant* φ. Both rules are given in a form that best displays their underlying logical principles and similarity. Rule *ind* says that ϕ holds after any number of repetitions of α if it holds initially (antecedent) and, for all reachable states (as overapproximated by $\forall^\alpha$), invariant ϕ remains true after one iteration of α (premise). If ϕ is true after executing α whenever ϕ has been true before, then, if ϕ holds in the beginning, ϕ will continue to hold, no matter how often we repeat α in $[\alpha^*]\phi$; again, see Fig. 2.12 for an illustration. Rule *con* expresses that the variant $\varphi(v)$ holds for some real number $v \leq 0$ after repeating α sufficiently often if $\varphi(v)$ holds for some real number at all in the beginning (antecedent) and, by premise, $\varphi(v)$ decreases after every execution of α by 1 (or another positive real constant). This rule can be used to show positive progress (by 1) with respect to $\varphi(v)$ by executing α.

For practical verification, rules *ind* or *con* can be combined with generalisation ($[]gen$, $\langle\rangle gen$) to prove a postcondition ψ of a loop α^* by showing that (*a*) the antecedents of the respective goals of *ind* and *con*, which represent the induction start, holds initially (*b*) their subgoals, which represent the induction step, hold and (*c*) the postcondition of the succedentd in their goals entails ψ. The corresponding variants of *ind* and *con* are *derived rules*. That is, these rules are non-essential, because they can be derived easily by chaining the proof rules from Fig. 2.11 together in an appropriate way.

$$(ind')\ \frac{\vdash \phi \qquad \vdash \forall^{\alpha}(\phi \to [\alpha]\phi) \qquad \vdash \forall^{\alpha}(\phi \to \psi)}{\vdash [\alpha^*]\psi}$$

$$(con')\ \frac{\vdash \exists v\, \varphi(v) \qquad \vdash \forall^{\alpha}\forall v{>}0\,(\varphi(v) \to \langle\alpha\rangle\varphi(v-1)) \qquad \vdash \forall^{\alpha}(\exists v{\le}0\, \varphi(v) \to \psi)}{\vdash \langle\alpha^*\rangle\psi}$$

For example, using a cut with $\phi \to [\alpha^*]\phi$, rule *ind'* can be derived from *ind* and $[]gen$ as follows:

$$cut\ \dfrac{\to r\ \dfrac{ind\ \dfrac{\vdash \forall^{\alpha}(\phi \to [\alpha]\phi)}{\phi \vdash [\alpha^*]\phi}}{\vdash \phi \to [\alpha^*]\phi} \qquad \to l\ \dfrac{\vdash \phi \qquad []gen\ \dfrac{\vdash \forall^{\alpha}(\phi \to \psi)}{[\alpha^*]\phi \vdash [\alpha^*]\psi}}{\phi \to [\alpha^*]\phi \vdash [\alpha^*]\psi}}{\vdash [\alpha^*]\psi}$$

These derived rules are not necessary in theory, but still useful in practise.

Derivability and Proofs

We call any formula ϕ provable or derivable (in the d$\mathcal{L}$ calculus) if we can find a d$\mathcal{L}$ proof for it that starts with axioms (rule *ax*) at the leaves and ends with a sequent $\vdash \phi$ at the bottom. While constructing proofs, however, we would start with the desired goal $\vdash \phi$ at the bottom and work our way backwards to the subgoals until they can be proven to be valid as axioms (*ax*). Once all subgoals have been proven to be valid axioms, they entail their consequences, which, recursively, entail the original goal $\vdash \phi$. This property of preserving truth or preserving entailment, which we prove in Sect. 2.6, is called soundness. Thus, while constructing proofs, we work bottom-up from the goal. When we have found a proof, we justify formulas from the axioms top-down to the original goal.

The notions of derivations and proofs for the d$\mathcal{L}$ calculus are standard, except that $i\exists$ produces multiple conclusions. Hence, we define derivations as finite acyclic graphs instead of trees. We want proofs to be acyclic and not accept a formula that is used to prove itself.

Definition 2.11 (Provability). A *derivation* is a finite acyclic graph labelled with sequents such that, for every node, the (set of) labels of its children must be the (set of) premises of an instance of one of the calculus rules (Definition 2.10) and the (set of) labels of the parents of these children must be the (set of) conclusions of that rule instance. A formula ψ is *provable* from a set Φ of formulas, denoted by $\Phi \vdash_{\mathsf{d}\mathcal{L}} \psi$, iff there is a finite subset $\Phi_0 \subseteq \Phi$ for which the sequent $\Phi_0 \vdash \psi$ is derivable, i.e., there is a derivation with a single root (i.e., node without parents) labelled $\Phi_0 \vdash \psi$.

Example 2.19. A very simple (in fact essentially propositional) proof of the formula

$$v^2 \leq 10 \wedge b > 0 \rightarrow b > 0 \wedge (\neg(v \geq 0) \vee v^2 \leq 10) \tag{2.13}$$

is shown in Fig. 2.13. The proof starts with the proof goal as a sequent at the bottom:

$$\vdash v^2 \leq 10 \wedge b > 0 \rightarrow b > 0 \wedge (\neg(v \geq 0) \vee v^2 \leq 10).$$

$$\text{→r}\ \dfrac{\text{∧r}\ \dfrac{\text{∧l}\ \dfrac{\text{ax}\ \dfrac{*}{v^2 \leq 10, b > 0 \vdash b > 0}}{v^2 \leq 10 \wedge b > 0 \vdash b > 0} \qquad \text{∨r}\ \dfrac{\text{∧l}\ \dfrac{\text{ax}\ \dfrac{*}{v^2 \leq 10, b > 0 \vdash \neg(v \geq 0), v^2 \leq 10}}{v^2 \leq 10 \wedge b > 0 \vdash \neg(v \geq 0), v^2 \leq 10}}{v^2 \leq 10 \wedge b > 0 \vdash \neg(v \geq 0) \vee v^2 \leq 10}}{v^2 \leq 10 \wedge b > 0 \vdash b > 0 \wedge (\neg(v \geq 0) \vee v^2 \leq 10)}}{\vdash v^2 \leq 10 \wedge b > 0 \rightarrow b > 0 \wedge (\neg(v \geq 0) \vee v^2 \leq 10)}$$

Fig. 2.13 Simple propositional example proof

The first (i.e., bottom most) proof step applies proof rule →r to turn the implication (→) to the sequent level by moving the assumption into the antecedent. The next proof step applies rule ∧r to split the proof into the left branch for showing that conjunct $b > 0$ follows from the assumptions in the antecedent and into the right branch for showing that conjunct $\neg(v \geq 0) \vee v^2 \leq 10$ follows from the antecedent also. On the left branch, the proof closes with an axiom *ax* after splitting the conjunction ∧ on the antecedent with rule ∧l. We mark closed proof goals with $*$. The right branch closes with an axiom *ax* after splitting the disjunction (∨) in the succedent with rule ∨r and then splitting the conjunction (∧) in the antecedent with rule ∧l. Now that all branches of the proof have closed (with *ax*), we know that all leaves at the top are valid, and, hence, since the premises are valid, each application of a proof rule ensures that their respective conclusions are valid also. By recursively following this derivation from the leaves at the top to the original root at the bottom, we see that the original goal is valid and formula (2.13) is, indeed, true under all circumstances (valid).

While this proof does not show anything particularly exciting, because it only uses propositional rules, it shows how a proof can be build systematically in the $\mathsf{d}\mathcal{L}$ calculus and gives an intuition about how validity is inherited from the premises to the conclusions. □

2.5.3 *Deduction Modulo with Invertible Quantifiers and Real Quantifier Elimination*

The first-order quantifier rules in Fig. 2.11 lift quantifier elimination to d$\mathcal{L}$ by following a generalised deduction modulo approach. They integrate decision procedures, e.g., for real quantifier elimination as a background prover [32], into the deductive proof system. Yet, unlike in the approaches of Dowek et al. [103] and Tinelli [290], the information given to the background prover is not restricted to ground formulas [290] or atomic formulas [103]. Further, real quantifier elimination is different from uninterpreted logic [147, 122, 134] in that the resulting formulas are not obtained by instantiation but by intricate arithmetic recombination. The quantifier rules can use any theory that admits quantifier elimination (see Definition 2.9) and has a decidable ground theory, for instance, the first-order theory of real arithmetic (which is equivalent to the theory of real-closed fields [288, 81]). A *formula of real arithmetic* is a first-order formula with $+,-,\cdot,/,=,\leq,<,\geq,>$ as the only function or predicate symbols besides constant symbols of Σ and logical variables of V. Also see App. D.2.

Integrating quantifier elimination to deal with statements about real quantities is quite challenging in the presence of modalities that influence the values of flexible symbols. In principle, quantifier elimination can be used to handle quantified constraints such as those arising for continuous evolutions. In d$\mathcal{L}$, however, real quantifiers interact with modalities containing further discrete or continuous transitions, which is an effect nherent in the interacting nature of hybrid systems. A hybrid formula like $\exists z\,\langle z'' = -b; ?m - z \geq s; z'' = 0\rangle\, m - z < s$ is not first-order; hence quantifier elimination cannot be applied. Even more so, the effect of a modality depends on the solutions of the differential equations contained therein. The dynamics of a hybrid program depends on the values of its parameters (z, b, m, s in the above case), but, at the same time, the constraints on a quantified variable like z depend on the effect of the hybrid program. For instance, it is hard to know in advance, which first-order constraints need to be solved by QE for the above formula. To find out how z evolves from quantifier $\exists z$ to postcondition $m - z < s$, the system dynamics in the modality needs to be taken into account (as for repetitions). Hence, our calculus first unwraps the first-order structure before applying QE to the resulting arithmetic formulas.

2.5.3.1 Lifting Quantifier Elimination by Invertible Quantifier Rules

The purpose of the quantifier rules in Fig. 2.11 is to postpone QE until the actual arithmetic constraints become apparent. The idea is that $\forall$r,$\exists$l,$\exists$r, and $\forall$l temporarily remove quantifiers by introducing new auxiliary symbols for quantified variables such that the proof can be continued beyond the occurrence of the quantifier to further analyse the modalities contained therein. Later, when the actual first-order constraints for the auxiliary symbol have been discovered, the corresponding quantifier

$$
\begin{array}{ll}
 & v \geq 0, z < m \vdash v^2 > 2b(m-z) \\
\hline
{\to}\mathrm{r},\wedge\mathrm{l} & \vdash v \geq 0 \wedge z < m \to v^2 > 2b(m-z) \\
\hline
 & \qquad\qquad\qquad v \geq 0, z < m \vdash -\frac{b}{2}T^2 + vT + z > m \\
\mathrm{i}\exists & v \geq 0, z < m \vdash T \geq 0 \qquad {}^{\langle:=\rangle}\, v \geq 0, z < m \vdash \langle z := -\frac{b}{2}T^2 + vT + z \rangle z > m \\
\hline
\wedge\mathrm{r} & v \geq 0, z < m \vdash T \geq 0 \wedge \langle z := -\frac{b}{2}T^2 + vT + z \rangle z > m \\
\hline
\exists\mathrm{r} & v \geq 0, z < m \vdash \exists t{\geq}0\, \langle z := -\frac{b}{2}t^2 + vt + z \rangle z > m \\
\hline
\langle ' \rangle & v \geq 0, z < m \vdash \langle z' = v, v' = -b \rangle z > m \\
\hline
{\to}\mathrm{r},\wedge\mathrm{l} & \vdash v \geq 0 \wedge z < m \to \langle z' = v, v' = -b \rangle z > m
\end{array}
$$

Fig. 2.14 Deduction modulo for analysis of MA violation in braking mode

can be reintroduced (i$\forall$, i$\exists$) and quantifier elimination QE is applied to reduce the sequents equivalently to a simpler formula with less (distinct) symbols. In $\exists$r,$\forall$l,i$\exists$, the respective auxiliary symbols are free logical variables. In $\forall$r,$\exists$l,i$\forall$, Skolem function terms are used instead for reasons that are crucial for soundness and will be illustrated in the remainder of this section. In this context, we think of *free* logical variables as being introduced by γ-rules ($\exists$r and $\forall$l), and hence implicitly existentially quantified.

To illustrate how quantifier and dynamic rules of $\mathsf{d}\mathcal{L}$ interact to combine arithmetic with dynamic reasoning in hybrid systems, we analyse the braking behaviour in train control. The proof in Fig. 2.14 can be used to analyse whether a train can violate its MA although it is braking. That is, if the train position z can leave m ($z > m$) although it starts inside ($z < m$) and is braking will full braking force all the time:

$$v \geq 0 \wedge z < m \to \langle z' = v, v' = -b \rangle z > m.$$

As the proof reveals, the answer depends on the initial velocity v. The proof starts with the conjecture at the bottom and applies propositional transformation rules $\to$r,$\wedge$l to obtain a decomposed sequent form. Then it uses rule $\langle ' \rangle$ to replace the differential equation with a quantified formula about its solution. For notational convenience, we use the simplified $\langle ' \rangle$ rule from (2.12), as the differential equation is not restricted to an evolution domain. Now we have a quantified modal formula, $\exists t{\geq}0\, \langle z := -\frac{b}{2}t^2 + vt + z \rangle z > m$, which, unfortunately, cannot be handled by quantifier elimination in real-closed fields, because it is not first-order. Using rule $\exists$r, however, the proof can continue by introducing a new free variable T for the quantified variable t and postpone QE. After introducing T, the proof can continue by splitting a conjunction in the succedent into two branches (rule $\wedge$r) and applying the assignment with a substitution on the right branch (rule $\langle := \rangle$). Finally, the previously quantified free variable T only occurs in first-order formulas on all open goals. Then rule i$\exists$ can be applied in Fig. 2.14 to merge all open proof goals mentioning T, reintroduce the quantifier for T, and apply quantifier elimination. The conjunction of the two goals can be handled by QE and simplification, yielding the resulting subgoal:

$$\text{QE}\left(\exists T\,((v \geq 0 \wedge z < m \rightarrow T \geq 0) \wedge (v \geq 0 \wedge z < m \rightarrow -\frac{b}{2}T^2 + vT + z > m))\right)$$
$$\equiv v \geq 0 \wedge z < m \rightarrow v^2 > 2b(m-z).$$

After applying rules $\rightarrow$r,$\wedge$l for structural reasons again, the open branch with this formula reveals the speed limit and can be used to synthesise a corresponding parameter constraint. When $v^2 > 2b(m-z)$ holds initially, m can eventually be violated even in braking mode, as the velocity exceeds the braking force.

Similarly, the dual constraint $v^2 \leq 2b(m-z)$ guarantees that m can be respected by appropriate braking. The constraint so discovered thus forms a *controllability constraint* of ETCS, i.e., a constraint that characterises from which states control choices exist that guarantee safety. It is essentially equivalent to $[z'' = -b]z \leq m$ and $\exists a\,(-b \leq a \leq A \wedge [z'' = a]z \leq m)$. The resulting controllable region of the state space of ETCS is illustrated in Fig. 2.15.

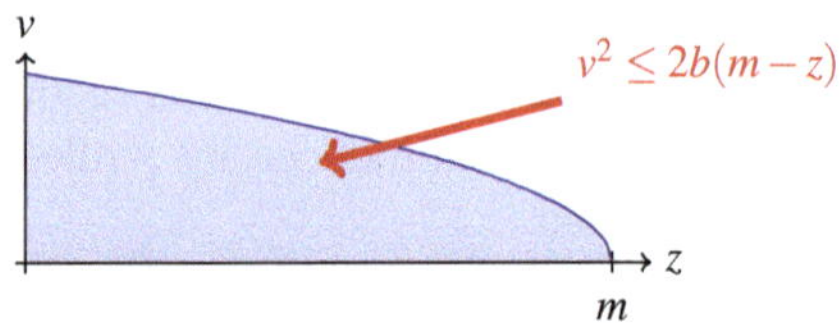

Fig. 2.15 Controllable region of ETCS dynamics

For comparison, the dual formula $v \geq 0 \wedge z < m \rightarrow [z' = v, v' = -b]\,z \leq m$ can be analysed as shown in Fig. 2.16 to study under which circumstances the MA is always respected ($[z'' = -b]z \leq m$) rather than under which it can fail ($\langle z'' = -b\rangle z > m$). The outcome again discovers the controllability constraint. The difference of the

$$\begin{array}{ll}
 & v \geq 0, z < m \vdash v^2 \leq 2b(m-z) \\ \hline
\rightarrow\text{r},\wedge\text{l} & \vdash v \geq 0 \wedge z < m \rightarrow v^2 \leq 2b(m-z) \\ \hline
\text{i}\forall & v \geq 0, z < m, s \geq 0 \vdash -\frac{b}{2}s^2 + vs + z \leq m \\ \hline
\langle:=\rangle & v \geq 0, z < m, s \geq 0 \vdash \langle z := -\frac{b}{2}s^2 + vs + z\rangle z \leq m \\ \hline
[:=] & v \geq 0, z < m, s \geq 0 \vdash [z := -\frac{b}{2}s^2 + vs + z]z \leq m \\ \hline
\rightarrow\text{r} & v \geq 0, z < m \vdash s \geq 0 \rightarrow [z := -\frac{b}{2}s^2 + vs + z]z \leq m \\ \hline
\forall\text{r} & v \geq 0, z < m \vdash \forall t{\geq}0\,[z := -\frac{b}{2}t^2 + vt + z]z \leq m \\ \hline
['] & v \geq 0, z < m \vdash [z' = v, v' = -b]z \leq m \\ \hline
\rightarrow\text{r},\wedge\text{l} & \vdash v \geq 0 \wedge z < m \rightarrow [z' = v, v' = -b]z \leq m
\end{array}$$

Fig. 2.16 Deduction modulo for analysis of MA-safety in braking mode

deduction in Fig. 2.16 compared to that in Fig. 2.14 is that we now use rule $[']$, which gives a universal quantifier for time t. With rule $\forall$r, the quantifier can be turned into a Skolem constant term s, which does not have any arguments, because no free logical variables occur. After applying the solution of the differential equation with

$[:=],\langle:=\rangle$, the resulting formula is first-order in the Skolem term s. Then rule i$\forall$ can be used to reintroduce a universal quantifier for the previously quantified variable, and to apply quantifier elimination:

$$\begin{aligned} &\mathrm{QE}\left(\forall s\,(v \geq 0 \land z < m \land s \geq 0 \rightarrow -\frac{b}{2}s^2 + vs + z \leq m)\right) \\ \equiv\; & v \geq 0 \land z < m \rightarrow v^2 \leq 2b(m-z). \end{aligned}$$

2.5.3.2 Admissibility in Invertible Quantifier Rules

The requirement that substitutions in i$\forall$ are admissible implies that no occurrence of $s(X_1,\dots,X_n)$ is within the scope of a quantifier for any of these X_i. Admissibility makes sense, because variables in $s(X_1,\dots,X_n)$ would otherwise be captured by quantifiers when substituting. The admissibility condition prevents i$\forall$ from rearranging the order of quantifiers from $\exists X_i \forall s$ to the weaker $\forall s \exists X_i$. Such a rearrangement would be unsound, because it is not sufficient to show the weak subgoal $\forall s \exists X_i$ (each s has an X_i) in order to prove the strong statement $\exists X_i \forall s$ saying that the same X_i works for all s. Because this is an important part of soundness, we illustrate in detail why unsound rearrangements are prevented.

$$\begin{array}{ll} & \text{i}\forall \text{ is not applicable} \\ \hline & \vdash \mathrm{QE}(\exists X\,(2X+1 < s(X))) \\ \hline \text{i}\exists & \vdash 2X+1 < s(X) \\ \hline \langle:=\rangle & \vdash \langle x:=2X+1\rangle(x < s(X)) \\ \hline \forall\text{r} & \vdash \forall y\,\langle x:=2X+1\rangle(x<y) \\ \hline \exists\text{r} & \vdash \exists x \forall y\,\langle x:=2x+1\rangle(x<y) \end{array}$$

Fig. 2.17a Wrong rearrangement with deduction modulo by invertible quantifiers

$$\begin{array}{ll} & \overbrace{\vdash \mathrm{QE}\,(\exists X\,\mathrm{QE}(\forall s\,(2X+1 < s)))}^{\text{false}} \\ \hline \text{i}\exists & \vdash \mathrm{QE}(\forall s\,(2X+1 < s)) \\ \hline \text{i}\forall & \vdash 2X+1 < s(X) \\ \hline \langle:=\rangle & \vdash \langle x:=2X+1\rangle(x < s(X)) \\ \hline \forall\text{r} & \vdash \forall y\,\langle x:=2X+1\rangle(x<y) \\ \hline \exists\text{r} & \vdash \exists x \forall y\,\langle x:=2x+1\rangle(x<y) \end{array}$$

Fig. 2.17b Correct reintroduction order with deduction modulo by invertible quantifiers

For the moment, suppose the rules did not contain QE. The requirement for admissible substitutions (Definition 2.8) ensures that the proof attempt of an invalid formula in Fig. 2.17a cannot close in the $\mathsf{d}\mathcal{L}$ calculus. At the indicated position at the top, i$\forall$, which would unsoundly invert the quantifier order to $\forall S \exists X$, cannot be applied: In i$\forall$, the substitution inserting $s(X)$ gives $\exists Y\,(2Y+1 < s(X))$ by bound variable renaming instead of $\exists X\,(2X+1 < s(X))$, because the substitution would not otherwise be admissible. Thus, i$\forall$ is not applicable, because the quantified formula is not of the form $\Psi(s(X))$.

Now, we consider what happens in the presence of QE. The purpose of QE is to (equivalently) remove quantifiers like $\exists X$. Thus it is no longer obvious that the admissibility argument applies, because the blocking variable X would have disappeared after successful quantifier elimination. However, quantifier elimination over the reals is defined in the first-order theory of real arithmetic [288, 81]. Yet, when eliminating X in Fig. 2.17a, the Skolem term $s(X)$ is no term of real arithmetic, as, unlike that of $+$, the interpretation of the Skolem function s is arbitrary. The truth-value of $\exists X\,(2X+1<s(X))$ depends on the interpretation of s. If $I(s)$ happens to be a constant function, the formula is true, if $I(s)(a)=2a$, however, it is false. In general, such cases cannot be distinguished without quantifiers, because two functions cannot be shown to be identical by evaluating them at finitely many points. Thus, in the presence of uninterpreted function terms, real arithmetic does not generally admit quantifier elimination. Consequently, i$\exists$ and i$\forall$ are only applicable if QE is defined. Yet, we show that QE can be lifted to formulas with Skolem functions, nevertheless, when these are instances of real arithmetic formulas:

Lemma 2.5 (Quantifier elimination lifting). *Quantifier elimination can be lifted to instances of formulas of first-order theories that admit quantifier elimination, i.e., to formulas that result from the base theory by substitution.*

Proof. Let formula ϕ be an instance of ψ, with ψ being a formula of the base theory, i.e., ϕ is $\psi_{z_1}^{\theta_1}\cdots_{z_n}^{\theta_n}$ for some variables z_i and arbitrary terms θ_i. As QE is defined for the base theory, let $\mathrm{QE}(\psi)$ be the quantifier-free formula belonging to ψ according to Definition 2.9. Then $\mathrm{QE}\,(\psi)_{z_1}^{\theta_1}\cdots_{z_n}^{\theta_n}$ satisfies the requirements of Definition 2.9 for ϕ, because $\models \psi_{z_1}^{\theta_1}\cdots_{z_n}^{\theta_n} \leftrightarrow \mathrm{QE}\,(\psi)_{z_1}^{\theta_1}\cdots_{z_n}^{\theta_n}$: For F defined as $\psi \leftrightarrow \mathrm{QE}(\psi)$, we have that $\models F$ implies $\models F_{z_1}^{\theta_1}\cdots_{z_n}^{\theta_n}$ by a standard consequence of the Substitution Lemma 2.2. And $\psi \leftrightarrow \mathrm{QE}(\psi)$ is indeed valid, by the properties of QE; see Definition 2.9. □

With this, consider again the example in Fig. 2.17a. By Lemma 2.5, QE is defined in the presence of Skolem terms that do not depend on quantified variables, e.g., for $\exists X\,(2X+1<t(Y,Z))$, which is an instance of the form $(\exists X\,(2X+1<z))_z^{t(Y,Z)}$. However, QE is *not defined* in the premise of i$\exists$ when Skolem dependencies on X occur. In Fig. 2.17a, $\exists X\,(2X+1<s(X))$ is no instance of first-order real arithmetic, because, by bound variable renaming $(\exists X\,(2X+1<z))_z^{s(X)}$ yields a different formula $\exists Y\,(2Y+1<s(X))$. An occurrence of $s(X)$, which corresponds to a quantifier nesting of $\exists X\,\forall s$, thus requires $s(X)$ to be eliminated by i$\forall$ before i$\exists$ can eliminate X; see Fig. 2.17b. Hence, inner universal quantifiers are enforced to be handled first and unsound quantifier rearrangements are prevented even in the presence of QE.

Finally, observe that i$\forall$ and i$\exists$ do not require quantifiers to be eliminated in the exact same order in which they occurred in the original formula. The elimination order within homogeneous quantifier blocks like $\forall x_1\,\forall x_2$ is not restricted as there are no Skolem dependencies among the corresponding auxiliary Skolem terms. Yet, eliminating such a quantifier block is sound in any order (accordingly for $\exists x_1\,\exists x_2$). Similarly, i$\exists$ and i$\forall$ could interchange the order of $\forall x\,\exists y$ to the stronger $\exists y\,\forall x$, because the resulting Skolem term s for x in the former formula does not depend on y. In this direction, however, the interchange is sound, as it amounts to proving a stronger statement. This quantifier rearrangement is not necessarily wise, because it requires proving a stronger formula, but it is at least sound.

2.5.3.3 Quantifier Elimination and Modalities

Quantifier elimination over the first-order theory of reals cannot handle modal formulas. Hence, the d$\mathcal{L}$ calculus first reduces modalities to first-order constraints before applying QE. Yet, this is not necessary for all modalities. The modal subformula in the following example does not impose any constraints on X, but its truth-value only determines which first-order constraints are imposed on X:

$$\mathrm{QE}(\exists X\,(X<0 \wedge ((\langle y:=2y+1\rangle y>0)\rightarrow X>y))) \equiv (\langle y:=2y+1\rangle y>0)\rightarrow y<0.$$

Modal formulas not containing elimination variable X can be handled by propositional abstraction in QE and remain unchanged. Syntactically, the reason for this is that d$\mathcal{L}$ rule applications on modal formulas that do not contain X will never produce formulas which do. The semantical reason for the same fact is a generalisation of the coincidence lemma to d$\mathcal{L}$, which says that values of variables that do not occur will neither affect the transition structure of a hybrid program nor the truth-value of formulas.

Lemma 2.6 (Coincidence lemma). *If the interpretations (and assignments and states, respectively) I,η,ν and J,ε,ω agree on all symbols that occur freely in the formula ϕ, then $val_{I,\eta}(\nu,\phi) = val_{J,\varepsilon}(\omega,\phi)$.*

Proof. The proof is by a simple structural induction using the definitions of valuation $val_{I,\eta}(\nu,\cdot)$ and $\rho_{I,\eta}(\cdot)$ in Definitions 2.5–2.7. □

2.5.3.4 Global Invertible Quantifier Rules

Rules i$\forall$ and i$\exists$ display an asymmetry. While i$\forall$ works locally on a branch, i$\exists$ needs to inspect all branches that contain X. The reason for this is that branches are implicitly combined conjunctively in sequent calculus, as all branches have to close simultaneously for a proof to succeed (Definition 2.11). Universal quantifiers can be handled separately for conjunctions by $\forall x\,(\phi\wedge\psi)\equiv\forall x\,\phi\wedge\forall x\,\psi$. Existential quantifiers, however, can only be dealt with separately for disjunctions but not for conjunctions: $\exists x\,(\phi\vee\psi)\equiv\exists x\,\phi\vee\exists x\,\psi$. In calculi with a disjunctive proof structure, the roles of i$\forall$ and i$\exists$ would be interchanged but the phenomenon remains.

Rule i$\exists$ can be applied to the full proof (i.e., all open goals) as a global closing substitution in the free-variable tableau calculus [122]; cf. App. A. By Lemma 2.6, however, rule i$\exists$ only needs to consider the set of all open goals $\Phi_i \vdash \Psi_i$ that actually contain X. Rule i$\exists$ resembles global closing substitutions in uninterpreted free-variable tableaux [134]. Both avoid the backtracking over closing substitutions that local closing substitutions require. Unlike closing substitutions, however, rule i$\exists$ uses the fixed semantics of function and predicate symbols of real arithmetic such that variables can be eliminated equivalently by QE before the proof completes. Applying i$\forall$ or i$\exists$ early does not necessarily close the proof. Instead, equivalent constraints on the remaining variables will be revealed, which can simplify the proof or help in deriving parametric constraints or invariants.

2.5.4 Verification Example

As a simple example to prove, recall the bouncing ball system (*ball*) from Example 2.5 on p. 45 and its d$\mathcal{L}$ specification from Example 2.7 on p. 48. Consider the intuitive property that the bouncing ball never bounces higher than initial height H when the precondition of property (2.3) holds initially:

$$(v^2 \leq 2g(H-h) \land h \geq 0 \land g > 0 \land H \geq 0 \land 1 > c \geq 0) \rightarrow [ball](0 \leq h \leq H). \quad (2.3^*)$$

The bouncing ball is very simple, but shows some interesting aspects of proofs. In order to simplify the proof notation, let us discard clock variable τ. Clock τ is not necessary for the property, and only used to ensure natural switching during the bounce to prevent the ball from bouncing multiple times while still on the ground (which would be *superdense* switching with multiple discrete switches at the exact same point in time).

For the proof, we define some abbreviations. Let ψ denote the general assumptions in the precondition about parameters that do not change during bouncing ball runs, and let ϕ denote the state-dependent part of the precondition, that is:

$$\psi \equiv g > 0 \land H \geq 0 \land 1 > c \geq 0,$$
$$\phi \equiv v^2 \leq 2g(H-h) \land h \geq 0.$$

The d$\mathcal{L}$ proof for the bouncing ball property (2.3) is shown in Fig. 2.18.

The proof starts with the property (2.3) at the bottom (goal). After normalising to sequent form with rules $\rightarrow$r,$\land$l, the proof follows an induction (using the rule *ind*$'$ from p. 86) with invariant ϕ. Rule *ind*$'$ produces two other proof subgoals that are not shown in Fig. 2.18: the proof goal that the precondition $\psi \land \phi$ implies the invariant ϕ (i.e., $\psi, \phi \vdash \phi$) and the proof goal that the invariant ϕ implies the postcondition, which gives $\psi, \phi \vdash \forall h \forall v\,(\phi \rightarrow 0{\leq}h{\leq}H)$. Both goals are trivial to prove by *ax* and $\forall$r,i$\forall$, respectively. The quantifiers $\forall h \forall v$ in the latter goal result from the universal closure $\forall^\alpha$ in rule *ind*$'$. Universal closures are not strictly necessary in this particular proof, because the premise only contains invariant ϕ and formula ψ about symbols that do not change during the hybrid program runs. Thus, the universal closure immediately disappears after applying $\forall$r. In general, however, universal closures in *ind* and the other global proof rules are critical for soundness; see Fig. 2.19a versus Fig. 2.19b.

After splitting the sequential composition by d$\mathcal{L}$ rule [;], the proof uses rule [$'$] with the solution $\langle h{:=}h+vt-\frac{g}{2}t^2, v{:=}v-gt\rangle$ of the differential equation system $h'=v, v'=-g$. We abbreviate this solution by $\langle\mathcal{S}_t\rangle$. Again, we use the simplified [$'$] rule from (2.12). QE cannot be applied to the result quantifier $\forall t{\geq}0\,\langle\mathcal{S}_t\rangle\ldots$, because the quantified variable t occurs in modalities to which QE is not applicable. Thus the proof uses $\forall$r to introduce a Skolem function s for the previously quantified variable t. But unlike for the proof in Fig. 2.14, we do not directly apply the resulting solution $\langle\mathcal{S}_s\rangle$ by rule $\langle{:=}\rangle$. The reason is the different system structure of the bouncing ball. In the bouncing ball program, the differential equation comes

$$
\begin{array}{ll}
 & * \\
\hline
\text{i}\forall & \psi,\phi,s{\geq}0,h+vs-\frac{g}{2}s^2=0 \vdash (-c(v-gs))^2 \leq 2g(H-(h+vs-\frac{g}{2}s^2)) \land h+vs-\frac{g}{2}s^2 \geq 0 \\
\hline
\langle:=\rangle & \psi,\phi,s{\geq}0,\langle\mathscr{S}_s\rangle h=0 \vdash \langle\mathscr{S}_s\rangle\langle v:=-cv\rangle\phi \\
\hline
[:=] & \psi,\phi,s{\geq}0,\langle\mathscr{S}_s\rangle h=0 \vdash \langle\mathscr{S}_s\rangle[v:=-cv]\phi \\
\hline
\rightarrow\text{r} & \psi,\phi,s{\geq}0 \vdash \langle\mathscr{S}_s\rangle(h=0\rightarrow[v:=-cv]\phi) \\
\hline
[?] & \psi,\phi,s{\geq}0 \vdash \langle\mathscr{S}_s\rangle[?h=0][v:=-cv]\phi \\
\hline
[;] & \psi,\phi,s{\geq}0 \vdash \langle\mathscr{S}_s\rangle[?h=0;v:=-cv]\phi
\end{array}
$$

$$
\begin{array}{ll}
 & * \\
\hline
\text{i}\forall & \psi,\phi,s{\geq}0,h+vs-\frac{g}{2}s^2>0 \vdash (v-gs)^2 \leq 2g(H-(h+vs-\frac{g}{2}s^2)) \land h+vs-\frac{g}{2}s^2 \geq 0 \\
\hline
\langle:=\rangle & \psi,\phi,s{\geq}0,\langle\mathscr{S}_s\rangle h>0 \vdash \langle\mathscr{S}_s\rangle\phi \\
\hline
\rightarrow\text{r} & \psi,\phi,s{\geq}0 \vdash \langle\mathscr{S}_s\rangle(h>0\rightarrow\phi) \\
\hline
[?] & \psi,\phi,s{\geq}0 \vdash \langle\mathscr{S}_s\rangle[?h>0]\phi
\end{array}
$$

$$
\begin{array}{ll}
 & \dots \qquad\qquad\qquad\qquad \dots \\
\hline
 & \psi,\phi,s{\geq}0 \vdash \langle\mathscr{S}_s\rangle[?h>0]\phi \qquad \psi,\phi,s{\geq}0 \vdash \langle\mathscr{S}_s\rangle[?h=0;v:=-cv]\phi \\
\hline
\land\text{r} & \psi,\phi,s{\geq}0 \vdash \langle\mathscr{S}_s\rangle([?h>0]\phi \land [?h=0;v:=-cv]\phi) \\
\hline
[\cup] & \psi,\phi,s{\geq}0 \vdash \langle\mathscr{S}_s\rangle[?h>0 \cup (?h=0;v:=-cv)]\phi \\
\hline
\rightarrow\text{r} & \psi,\phi \vdash s{\geq}0 \rightarrow \langle\mathscr{S}_s\rangle[?h>0 \cup (?h=0;v:=-cv)]\phi \\
\hline
\forall\text{r} & \psi,\phi \vdash \forall t{\geq}0\, \langle\mathscr{S}_t\rangle[?h>0 \cup (?h=0;v:=-cv)]\phi \\
\hline
['] & \psi,\phi \vdash [h''=-g][?h>0 \cup (?h=0;v:=-cv)]\phi \\
\hline
[;] & \psi,\phi \vdash [h''=-g;(?h>0 \cup (?h=0;v:=-cv))]\phi \\
\hline
ind' & \psi,\phi \vdash [(h''=-g;(?h>0 \cup (?h=0;v:=-cv)))^*](0{\leq}h{\leq}H) \\
\hline
\rightarrow\text{r},\land\text{l} & \vdash \psi\land\phi \rightarrow [(h''=-g;(?h>0 \cup (?h=0;v:=-cv)))^*](0{\leq}h{\leq}H)
\end{array}
$$

Fig. 2.18 Bouncing ball proof (no evolution domain)

$$
\begin{array}{c}
\text{unsound} \\
\hline
x\leq 0, x\leq 1 \vdash x+1\leq 1 \\
\hline
x\leq 0, x\leq 1 \vdash [x:=x+1]x\leq 1 \\
\hline
x\leq 0 \vdash x\leq 1 \rightarrow [x:=x+1]x\leq 1 \\
\hline
x\leq 0 \vdash [(x:=x+1)^*]x\leq 1
\end{array}
$$

Fig. 2.19a Unsound attempt of induction without universal closure $\forall^\alpha$

$$
\begin{array}{ll}
 & \text{not provable} \\
\hline
 & x\leq 0, y\leq 1 \vdash y+1\leq 1 \\
\hline
[:=],\langle:=\rangle & x\leq 0, y\leq 1 \vdash [y:=y+1]y\leq 1 \\
\hline
\rightarrow\text{r} & x\leq 0 \vdash y\leq 1 \rightarrow [y:=y+1]y\leq 1 \\
\hline
\forall\text{r} & x\leq 0 \vdash \forall x\,(x\leq 1 \rightarrow [x:=x+1]x\leq 1) \\
\hline
ind & x\leq 0 \vdash [(x:=x+1)^*]x\leq 1
\end{array}
$$

Fig. 2.19b Correct use of induction with universal closure $\forall^\alpha$, i.e., $\forall x$

first, and the discrete control equations are executed after that. Thus we keep the discrete jump set for the solution $\langle \mathcal{S}_s \rangle$ as an unmodified discrete jump prefix ($\langle \mathcal{J} \rangle$ in Definition 2.10) for the following rule applications and only apply the assignment with rule $\langle := \rangle$ to first-order formulas at the end of the proof.

On a side note: we could, in fact, just as well have used rule $\langle := \rangle$ here right away and substituted v, h inside the hybrid programs immediately, because there are no remaining loops or differential equations. That would clutter the notation, though, and we want to illustrate how discrete jump sets can be used as unmodified proof rule prefixes in Fig. 2.18.

Leaving prefix $\langle \mathcal{S}_s \rangle$ unchanged, the proof in Fig. 2.18 continues by splitting the choice between $h > 0$ and $h = 0$ with rule $[\cup]$ into two conjuncts, which split into two branches by rule $\wedge$r. On both branches, which are continued as indicated by the arrows, the test statements are turned into implications by rule $[?]$ and, ultimately, the accumulated discrete jump sets are applied (with rules $[:=], \langle := \rangle$) when the remaining formulas are simple. The last step of the proof is to reintroduce quantifiers for Skolem term s by rule i$\forall$ and apply quantifier elimination to the resulting first-order formulas on the left and right branches respectively:

$$\text{QE}\Big(\forall s\,(\psi \wedge \phi \wedge s \geq 0 \wedge h + vs - \frac{g}{2}s^2 > 0$$
$$\to (v - gs)^2 \leq 2g(H - (h + vs - \frac{g}{2}s^2)) \wedge h + vs - \frac{g}{2}s^2 \geq 0)\Big) \equiv \mathit{true};$$

$$\text{QE}\Big(\forall s\,(\psi \wedge \phi \wedge s \geq 0 \wedge h + vs - \frac{g}{2}s^2 = 0$$
$$\to (-c(v - gs))^2 \leq 2g(H - (h + vs - \frac{g}{2}s^2)) \wedge h + vs - \frac{g}{2}s^2 \geq 0)\Big) \equiv \mathit{true}.$$

In the proof of Fig. 2.18, we have used a bouncing ball without an evolution domain restriction. The bouncing ball property can also be proven with its evolution domain restricted to $h \geq 0$ on the differential equation system as in Fig. 2.2; see Fig. 2.20 for a proof. The proof is slightly more involved compared to Fig. 2.18, because of the extra constraints from the non-simplified rule $[']$. This time, for a change, we simply choose the full precondition as invariant, although the part marked in grey is still unaffected by the dynamics:

$$\phi \equiv v^2 \leq 2g(H - h) \wedge h \geq 0 \wedge g > 0 \wedge H \geq 0 \wedge 1 > c \geq 0.$$

Note, in particular, that there are many invariants that can be used to prove the same property. We just need to find one invariant that works.

$$
\dfrac{\dfrac{\dfrac{\dfrac{\dfrac{\dfrac{*}{\phi,s\geq 0,,h+vs-\frac{g}{2}s^2=0 \vdash (-c(v-gs))^2 \leq 2g(H-(h+vs-\frac{g}{2}s^2)) \land h+vs-\frac{g}{2}s^2\geq 0}\text{i}\forall}{\phi,s\geq 0,,\langle\mathscr{S}_s\rangle h=0 \vdash \langle\mathscr{S}_s\rangle\langle v:=-cv\rangle\phi}\langle:=\rangle}{\phi,s\geq 0,,\langle\mathscr{S}_s\rangle h=0 \vdash \langle\mathscr{S}_s\rangle[v:=-cv]\phi}[:=]}{\phi,s\geq 0,\langle\mathscr{S}_s\rangle h\geq 0 \vdash \langle\mathscr{S}_s\rangle(h=0\to[v:=-cv]\phi)}{\to}\text{r}}{\phi,s\geq 0,\langle\mathscr{S}_s\rangle h\geq 0 \vdash \langle\mathscr{S}_s\rangle[?h=0][v:=-cv]\phi}[?]}{\phi,s\geq 0,\langle\mathscr{S}_s\rangle h\geq 0 \vdash \langle\mathscr{S}_s\rangle[?h=0;v:=-cv]\phi}[;]
$$

$$
\dfrac{\dfrac{\dfrac{\dfrac{*}{\phi,s\geq 0,,h+vs-\frac{g}{2}s^2>0 \vdash (v-gs)^2 \leq 2g(H-(h+vs-\frac{g}{2}s^2)) \land h+vs-\frac{g}{2}s^2\geq 0}\text{i}\forall}{\phi,s\geq 0,,\langle\mathscr{S}_s\rangle h>0 \vdash \langle\mathscr{S}_s\rangle\phi}\langle:=\rangle}{\phi,s\geq 0,\langle\mathscr{S}_s\rangle h\geq 0 \vdash \langle\mathscr{S}_s\rangle(h>0\to\phi)}{\to}\text{r}}{\phi,s\geq 0,\langle\mathscr{S}_s\rangle h\geq 0 \vdash \langle\mathscr{S}_s\rangle[?h>0]\phi}[?]
$$

$$
\dfrac{\dfrac{\dfrac{\dfrac{\dfrac{\dfrac{\dfrac{\dfrac{\dfrac{\dfrac{\dots}{\dots \vdash \langle\mathscr{S}_s\rangle[?h>0]\phi} \quad \dfrac{\dots}{\phi,s\geq 0,\langle\mathscr{S}_s\rangle h\geq 0 \vdash \langle\mathscr{S}_s\rangle[?h=0;v:=-cv]\phi}}{\phi,s\geq 0,\langle\mathscr{S}_s\rangle h\geq 0 \vdash \langle\mathscr{S}_s\rangle([?h>0]\phi\land[?h=0;v:=-cv]\phi)}\land\text{r}}{\phi,s\geq 0,\langle\mathscr{S}_s\rangle h\geq 0 \vdash \langle\mathscr{S}_s\rangle[?h>0\cup(?h=0;v:=-cv)]\phi}[\cup]}{\phi,s\geq 0 \vdash \langle\mathscr{S}_s\rangle h\geq 0\to\langle\mathscr{S}_s\rangle[?h>0\cup(?h=0;v:=-cv)]\phi}{\to}\text{r}}{\phi \vdash s\geq 0\to(\langle\mathscr{S}_s\rangle h\geq 0\to\langle\mathscr{S}_s\rangle[?h>0\cup(?h=0;v:=-cv)]\phi)}{\to}\text{r}}{\phi \vdash \forall t\geq 0(\langle\mathscr{S}_t\rangle h\geq 0\to\langle\mathscr{S}_t\rangle[?h>0\cup(?h=0;v:=-cv)]\phi)}\forall\text{r}}{\phi \vdash [h''=-g\,\&\,h\geq 0][?h>0\cup(?h=0;v:=-cv)]\phi}[']}{\phi \vdash [h''=-g\,\&\,h\geq 0;(?h>0\cup(?h=0;v:=-cv))]\phi}[;]}{\phi \vdash [(h''=-g\,\&\,h\geq 0;(?h>0\cup(?h=0;v:=-cv)))^*](0\leq h\leq H)}ind'}{\vdash \phi\to[(h''=-g\,\&\,h\geq 0;(?h>0\cup(?h=0;v:=-cv)))^*](0\leq h\leq H)}{\to}\text{r}
$$

Fig. 2.20 Bouncing ball proof (with evolution domain)

2.6 Soundness

In this section, we prove that the d$\mathscr{L}$ calculus is a sound axiomatisation of the transition behaviour of hybrid systems. Whatever we can prove in the d$\mathscr{L}$ calculus is actually true.

The proof calculus for d$\mathscr{L}$ in Fig. 2.11 needs to fit to the semantics of differential dynamic logic from Sect. 2.3; otherwise, the proof rules would not be meaningful. Fortunately, every differential dynamic logic formula that can be derived in the d$\mathscr{L}$ calculus from Fig. 2.11 really is a valid formula! This property of the calculus is called *soundness* and is crucial, because it would be disastrous if a formula would be called "proven" when it is actually not valid, since we could not trust our proofs then. A calculus is sound iff every formula that can be derived in the calculus is also valid according to the semantics.

We prove that a successful deduction in the d$\mathcal{L}$ calculus always produces correct verification results about hybrid systems: The d$\mathcal{L}$ calculus is *sound*, i.e., all provable (closed) formulas are valid in all states of all interpretations. We can restrict our attention to closed formulas, i.e., formulas without free variables to begin with, because we can start with the universal closure of the formula for validity just as well. To reflect the interaction of free variables and Skolem terms, we adapt the notion of soundness for the liberalised δ^+-rule in free-variable tableau calculi [147] to sequent calculus.

A formula ϕ *has a model* [147] if there is an interpretation I and a state ν such that *for all* variable assignments η we have $I,\eta,\nu \models \phi$. Closed tableaux prove the unsatisfiability of the negated goal [147]. Sequent calculi work dually and show validity of the original proof goal. Consequently, we use the dual notion and say that formula ψ is a *consequence* of ϕ iff, for every I,ν *there is* an assignment η such that $I,\eta,\nu \models \psi$ provided that, for every I,ν, *there is* an assignment η such that $I,\eta,\nu \models \phi$. A proof rule that concludes Ψ from the premises Φ is *sound* if Ψ is, indeed, a consequence of Φ in the sense just defined. As usual, multiple branches in Ψ or Φ are combined conjunctively.

In this context, we think of free logical variables as being introduced by γ-rules, i.e., $\exists$r and $\forall$l (hence the implicit existential quantification of free logical variables by η). For closed formulas (without free logical variables), validity corresponds to being a consequence from an empty set of open goals. Hence, closed formulas that are provable with a sound deduction are *valid* (true in all states of all interpretations).

Theorem 2.1 (Soundness of d$\mathcal{L}$). *The* d$\mathcal{L}$ *calculus is sound.*

Proof. The calculus is sound if each rule instance is sound. All rules of the d$\mathcal{L}$ calculus except $\forall$r,$\exists$l and i$\exists$ are also *locally sound*, i.e., their conclusion is true at I,η,ν if all its premises are true in I,η,ν, which implies soundness. It is also easy to show that locally sound rules remain sound when adding contexts $\Gamma,\Delta,\langle\mathcal{J}\rangle$ as in Definition 2.10, since a discrete jump set $\langle\mathcal{J}\rangle$ characterises a unique state transition. Local soundness proofs of $\langle;\rangle$,$[;]$,$\langle\cup\rangle$,$[\cup]$,$\langle^{*n}\rangle$,$[^{*n}]$,$\langle?\rangle$,$[?]$ and propositional rules are as usual. Note that, for symmetric rules, local soundness implies that the premise and conclusion are *equivalent*, i.e., true in the same states. For an illustration of the dynamics behind the dynamic proof rules, we recall Fig. 2.12 from p. 83.

$\langle:=\rangle$ The rule $\langle:=\rangle$ is locally sound. Assume that the premise holds in I,η,ν, i.e., $I,\eta,\nu \models \phi_{x_1}^{\theta_1}\cdots_{x_n}^{\theta_n}$. We have to show that $I,\eta,\nu \models \langle x_1:=\theta_1,..,x_n:=\theta_n\rangle\phi$, i.e., $I,\eta,\omega \models \phi$ for a state ω with $(\nu,\omega)\in\rho_{I,\eta}(x_1:=\theta_1,..,x_n:=\theta_n)$. This follows directly from the Substitution Lemma 2.2 for admissible substitutions (Definition 2.8). The proof for rule $[:=]$ uses the fact that discrete jumps are deterministic.

$\langle;\rangle$ Rule $\langle;\rangle$ is locally sound. Assume that the premise holds in I,η,ν, i.e., $I,\eta,\nu \models \langle\alpha\rangle\langle\beta\rangle\phi$. We have to show that the conclusion holds in I,η,ν, i.e., $I,\eta,\nu \models \langle\alpha;\beta\rangle\phi$. By premise, $I,\eta,\nu \models \langle\alpha\rangle\langle\beta\rangle\phi$, we know that there is a state μ such that $(\nu,\mu)\in\rho_{I,\eta}(\alpha)$ and $I,\eta,\mu \models \langle\beta\rangle\phi$. Hence, there is a state ω such that $(\mu,\omega)\in\rho_{I,\eta}(\beta)$ and $I,\eta,\omega \models \phi$. Now, by the semantics of $\alpha;\beta$ (Definition 2.7), there is a transition from ν to ω (via intermediate state μ) along $\alpha;\beta$. Thus, $(\nu,\omega) \in \rho_{I,\eta}(\alpha;\beta)$ and $I,\eta,\omega \models \phi$, which implies $I,\eta,\nu \models \langle\alpha;\beta\rangle\phi$. The converse direction can be proven similarly to show equivalence and the local soundness of the dual rule $[;]$.

$\langle\cup\rangle$ Rule $\langle\cup\rangle$ is locally sound. Assume that the premise holds in I,η,ν, i.e., $I,\eta,\nu \models \langle\alpha\rangle\phi \vee \langle\beta\rangle\phi$. We have to show that the conclusion holds in I,η,ν, i.e., $I,\eta,\nu \models \langle\alpha\cup\beta\rangle\phi$. If the disjunction in the premise is true, then one of its disjuncts must hold in I,η,ν. Consider the case where $I,\eta,\nu \models \langle\alpha\rangle\phi$. Then there is a state ω such that $(\nu,\omega) \in \rho_{I,\eta}(\alpha)$ and $I,\eta,\omega \models \phi$. By the semantics of $\alpha\cup\beta$ in Definition 2.7, every transition of α is a transition of $\alpha\cup\beta$. Hence $(\nu,\omega) \in \rho_{I,\eta}(\alpha\cup\beta)$ and $I,\eta,\omega \models \phi$, which imply $I,\eta,\nu \models \langle\alpha\cup\beta\rangle\phi$. If, instead, the second disjunct $I,\eta,\nu \models \langle\beta\rangle\phi$ holds, then the proof is similar. Either way, we have $I,\eta,\nu \models \langle\alpha\cup\beta\rangle\phi$. The converse direction can be proven accordingly to show equivalence and the local soundness of the dual rule $[\cup]$.

$\langle^{*n}\rangle$ Rule $\langle^{*n}\rangle$ is locally sound. Assume that the premise holds in I,η,ν, i.e., assume $I,\eta,\nu \models \phi \vee \langle\alpha\rangle\langle\alpha^*\rangle\phi$. We have to show that the conclusion holds in I,η,ν, i.e., $I,\eta,\nu \models \langle\alpha^*\rangle\phi$. The disjunction in the premise holds; hence, one of the disjuncts holds. Consider the case where $I,\eta,\nu \models \phi$; then $I,\eta,\nu \models \langle\alpha^*\rangle\phi$ already holds with zero repetitions α^* for ϕ is true in the beginning. Consider the case where $I,\eta,\nu \models \langle\alpha\rangle\langle\alpha^*\rangle\phi$. Thus, there is an α-transition to a state μ such that $(\nu,\mu) \in \rho_{I,\eta}(\alpha)$ with $I,\eta,\mu \models \langle\alpha^*\rangle\phi$. Consequently, there is an α^*-transition to a state ω with $(\mu,\omega) \in \rho_{I,\eta}(\alpha^*)$ and $I,\eta,\omega \models \phi$. Obviously, every α-transition also is an α^*-transition, because repetitions may choose to repeat only once. In particular, by chaining the α-transition $(\nu,\mu) \in \rho_{I,\eta}(\alpha) \subset \rho_{I,\eta}(\alpha^*)$ with the α^*-transition $(\mu,\omega) \in \rho_{I,\eta}(\alpha^*)$, we obtain a longer α^*-transition $(\nu,\omega) \in \rho_{I,\eta}(\alpha^*)$ by the transition semantics in Definition 2.7. Hence, in either case, we conclude $I,\eta,\nu \models \langle\alpha^*\rangle\phi$. The converse direction can be proven accordingly to show equivalence and the local soundness of the dual rule $[^{*n}]$.

$\langle?\rangle$ Rule $\langle?\rangle$ is locally sound. Assume that the premise holds in I,η,ν, i.e., $I,\eta,\nu \models \chi\wedge\phi$. We have to show that the conclusion holds in I,η,ν, i.e., $I,\eta,\nu \models \langle?\chi\rangle\phi$. We have to show that there is a transition along $?\chi$ to a state where ϕ holds. By the semantics in Definition 2.7, there is only a transition along hybrid program $?\chi$ if $I,\eta,\nu \models \chi$ and the state is not changed by $?\chi$ transitions. Now the premise implies $I,\eta,\nu \models \chi$ and $I,\eta,\nu \models \phi$, which, together, imply $I,\eta,\nu \models \langle?\chi\rangle\phi$. Since this is the only case where $?\chi$ can make a transition to a state satisfying ϕ, it shows equivalence. Local soundness of the dual rule $[?]$ follows from this.

$\langle'\rangle$ The rule $\langle'\rangle$ is locally sound. Let $y_1,\dots,y_n$ be a solution for the differential equation system $x_1' = \theta_1,\dots,x_n' = \theta_n$ with symbolic initial values $x_1,\dots,x_n$. Let further $\langle\mathscr{S}_t\rangle$ be the jump set $\langle x_1 := y_1(t),\dots,x_n := y_n(t)\rangle$. Assume I,η,ν are such that the premise is true: $I,\eta,\nu \models \exists t{\geq}0\,(\bar{\chi}\wedge\langle\mathscr{S}_t\rangle\phi)$ with $\forall 0{\leq}\tilde{t}{\leq}t\,\langle\mathscr{S}_{\tilde{t}}\rangle\chi$ abbreviated as $\bar{\chi}$. For any $\zeta\in\mathbb{R}$, we denote by η^{ζ} the assignment that agrees with η except that it assigns ζ to t. Then, by assumption, there is a real value $r \geq 0$ such that $I, \eta^r, v \models \bar{\chi} \wedge \langle \mathscr{S}_t \rangle \phi$. Abbreviate $x_1' = \theta_1, .., x_n' = \theta_n \,\&\, \chi$ by $\mathscr{D}$. We have to show that $I, \eta, v \models \langle \mathscr{D} \rangle \phi$. Equivalently, by Lemma 2.6, we show $I, \eta^r, v \models \langle \mathscr{D} \rangle \phi$, because t is a fresh variable that does not occur in $\mathscr{D}$ or ϕ. Let function $f : [0, r] \to \mathrm{Sta}(\Sigma)$ be defined such that $(v, f(\zeta)) \in \rho_{I,\eta^\zeta}(\mathscr{S}_t)$ for all $\zeta \in [0, r]$. By premise, $f(0)$ is identical to v and ϕ holds at $f(r)$. Thus it only remains to be shown that f respects the constraints of Definition 2.7 for $\mathscr{D}$. In fact, f obeys the continuity and differentiability properties of Definition 2.7 by the corresponding properties of the y_i. Moreover, $val_{I,\eta^r}(f(\zeta), x_i) = val_{I,\eta^r}(v, y_i(t))$ has a derivative of value $val_{I,\eta^r}(f(\zeta), \theta_i)$, because y_i is a solution of the differential equation $x_i' = \theta_i$ with corresponding initial value $v(x_i)$. Further, it can be shown that the evolution domain χ is respected along f as follows: By premise, $I, \eta^r, v \models \bar{\chi}$ holds for the initial state v; thus $val_{I,\eta^r}(f(\zeta), \chi) = true$ for all $\zeta \in [0, r]$. Combining these results, we can conclude that f is a witness for $I, \eta, v \models \langle \mathscr{D} \rangle \phi$. The converse direction can be shown accordingly to prove the dual rule [′] using Lemma 2.1.

∀r The proof is a sequent calculus adaptation of that in [147]. By contraposition, assume that there are I, v such that for all η it is the case that $I, \eta, v \not\models \forall x\, \phi(x)$; hence $I, \eta, v \models \exists x \neg \phi(x)$. We construct an interpretation I' that agrees with I except for the new function symbol s. Let $b_1, \ldots, b_n \in \mathbb{R}$ be arbitrary elements and let η^b assign b_i to the respective X_i for $1 \leq i \leq n$. As $I, \eta, v \models \exists x \neg \phi(x)$ holds for all η, we pick a witness d for $I, \eta^b, v \models \exists x \neg \phi(x)$ and choose $I'(s)(b_1, \ldots, b_n) = d$. For this interpretation I' and state v we have $I', \eta, v \not\models \phi(s(X_1, \ldots, X_n))$ for all assignments η by Lemma 2.6, as $X_1, .., X_n$ are all free variables determining the truth-value of $\phi(s(X_1, \ldots, X_n))$. To see that the contexts Γ, Δ of Definition 2.10 can be added to instantiate this rule, consider the following. Since s is new and does not occur in the context Γ, Δ, the latter do not change their truth-value by passing from I to I'. Likewise, s is rigid so that it does not change its value by adding jump prefix $\langle \mathscr{J} \rangle$ which concludes the proof. The proof of ∃l is dual.

i∀ i∀ is locally sound. Assume that $I, \eta, v \models \mathrm{QE}(\forall X\, (\Phi(X) \vdash \Psi(X)))$. Since QE yields an equivalence, we can conclude $I, \eta, v \models \forall X\, (\Phi(X) \vdash \Psi(X))$. Then if the antecedent of the conclusion is true, $I, \eta, v \models \Phi(s(X_1, \ldots, X_n))$, we conclude $I, \eta, v \models \Psi(s(X_1, \ldots, X_n))$ by choosing $val_{I,\eta}(v, s(X_1, \ldots, X_n))$ for X in the premise. By admissibility of substitutions, variables $X_1, \ldots, X_n$ are free at all occurrences of $s(X_1, \ldots, X_n)$, and hence their value is the same in all occurrences.

∃r ∃r is locally sound by a simplified version of the proof in [147]. For any I, η, v with $I, \eta, v \models \phi(X)$ we can conclude $I, \eta, v \models \exists x\, \phi(x)$ according to the witness $\eta(X)$. The proof of ∀l is dual.

i∃ For any I, v let η be such that $I, \eta, v \models \mathrm{QE}(\exists X \bigwedge_i (\Phi_i \vdash \Psi_i))$. Again, this implies $I, \eta, v \models \exists X \bigwedge_i (\Phi_i \vdash \Psi_i)$, because quantifier elimination yields an equivalence. We pick a witness $d \in \mathbb{R}$ for this existential quantifier. As X does not occur anywhere else in the proof, it disappears from all open premises of the proof by applying i$\exists$. Hence, by the Coincidence Lemma 2.6, the value of X does not change the truth-value of the premise of i$\exists$. Consequently, η can be extended to η' by changing the interpretation of X to the witness d such that $I,\eta',\nu \models \bigwedge_i(\Phi_i \vdash \Psi_i)$. Thus, η' extends I,η,ν to a simultaneous model of all conclusions.

$\langle\rangle$gen Rules $[]$*gen–con* are locally sound by a variation of the usual proofs [149] using universal closures for local soundness. $[]$*gen*,$\langle\rangle$*gen* are simple refinements of Lemma 2.6 using the fact that the universal closure $\forall^\alpha$ comprises all variables that change in α. Let $I,\eta,\nu \models \langle\alpha\rangle\phi$, i.e., let $(\nu,\nu') \in \rho_{I,\eta}(\alpha)$ with $I,\eta,\nu' \models \phi$. As α can only change its bound variables, which are quantified universally in the universal closure $\forall^\alpha$, the premise implies $I,\eta,\nu' \models \phi \to \psi$; thus $I,\eta,\nu' \models \psi$ and $I,\eta,\nu \models \langle\alpha\rangle\psi$. The proof of $[]$*gen* is similar.

ind For any I,η,ν with $I,\eta,\nu \models \forall^\alpha(\phi \to [\alpha]\phi)$, we know $I,\eta,\nu' \models \phi \to [\alpha]\phi$ for all ν' with $(\nu,\nu') \in \rho_{I,\eta}(\alpha)$. As these share the same η, we can further conclude $I,\eta,\nu \models \phi \to [\alpha^*]\phi$ by induction along the series of states ν' reached from ν by repeating α. The universal closure is necessary as, otherwise, the premise may yield different η in different states ν'.

con Assume that the antecedent and premise hold in I,η,ν. By premise, we have $I,\eta[v\mapsto d],\nu' \models v>0 \land \varphi(v) \to \langle\alpha\rangle\varphi(v-1)$ for all $d \in \mathbb{R}$ and all states ν' that are reachable by α^* from ν, because $\forall^\alpha$ comprises all variables that are bound by α, which are the same as those bound by α^*. By antecedent, there is a $d \in \mathbb{R}$ such that $I,\eta[v\mapsto d],\nu \models \varphi(v)$. Now, the proof is a well-founded induction on d. If $d \le 0$, we directly have $I,\eta,\nu \models \langle\alpha^*\rangle\exists v{\le}0\,\varphi(v)$ for zero repetitions. Otherwise, if $d > 0$, we have, by premise, that

$$I,\eta[v\mapsto d],\nu \models v>0 \land \varphi(v) \to \langle\alpha\rangle\varphi(v-1).$$

As $v>0 \land \varphi(v)$ holds true at $I,\eta[v\mapsto d],\nu$, we have for some ν' with $(\nu,\nu') \in \rho_{I,\eta[v\mapsto d]}(\alpha)$ that $I,\eta[v\mapsto d],\nu' \models \varphi(v-1)$. In particular, we can conclude that $I,\eta[v\mapsto d-1],\nu' \models \varphi(v)$ satisfies the induction hypothesis for a smaller d and a reachable ν', because $(\nu,\nu') \in \rho_{I,\eta}(\alpha)$ as v does not occur in α. The induction is well-founded, because d decreases by 1 up to the base case $d \le 0$. □

With this soundness theorem, we now know that everything we prove in the d$\mathcal{L}$ calculus accurately reflects reality, because the syntactic proofs built with Fig. 2.11 fit to the semantics defined in Sect. 2.3.

2.7 Completeness

In this section, we prove that the d$\mathcal{L}$ calculus is a sound and complete axiomatisation of the transition behaviour of hybrid systems relative to differential equations.

With Soundness Theorem 2.1, we have shown that all provable formulas are valid. So we know that will never prove something that does not even hold (is not valid). The converse question is whether all valid formulas are also provable, i.e., whether we will always be able to prove all formulas that are "true" (valid). Have we just been lucky with the successful proofs that we managed to show so far? Or is there a deeper reason for which we can know that, in principle, we could also find proofs for all other valid formulas?

2.7.1 Incompleteness

Theorem 2.1 shows that all provable closed d$\mathcal{L}$ formulas are valid. The converse question is whether the d$\mathcal{L}$ calculus is *complete*, i.e., all valid d$\mathcal{L}$ formulas are provable. Combining completeness for first-order logic [147] and decidability of real arithmetic [81], it is easy to see that our calculus is complete for closed formulas of first-order real arithmetic by chaining the quantifier rules ∀r,∃l,∃r,∀l with the respective inverse rules i∀,i∃, using propositional rules as needed to unfold the propositional structure. In the presence of modalities, however, d$\mathcal{L}$ is not axiomatisable and, unlike its basis of first-order *real* arithmetic, d$\mathcal{L}$ is undecidable. Both unbounded repetition in the discrete fragment and unbounded evolution in the continuous fragment cause incompleteness. Beyond hybrid dynamics, where reachability is known to be undecidable [156], we show that even the purely discrete and purely continuous parts of d$\mathcal{L}$ are not effectively axiomatisable. Hence, valid d$\mathcal{L}$ formulas are not always provable.

Theorem 2.2 (Incompleteness of d$\mathcal{L}$). *Both the discrete fragment and the continuous fragment of* d$\mathcal{L}$ *are* not effectively axiomatisable, *i.e., they have no sound and complete effective calculus, because natural numbers are definable in both fragments.*

Proof. We prove that natural numbers are definable among the real numbers of d$\mathcal{L}$ interpretations in both fragments. Then these fragments extend first-order *integer* arithmetic such that the incompleteness theorem of Gödel [137] applies. Gödel's incompleteness theorem shows that no logic extending first-order integer arithmetic can have a sound and complete effective calculus. Natural numbers are definable in the discrete fragment without continuous evolutions using repetitive additions:

$$nat(n) \leftrightarrow \langle x := 0; (x := x+1)^* \rangle\, x = n.$$

In the continuous fragment, an isomorphic copy of the natural numbers is definable using linear differential equations:

$$nat(n) \leftrightarrow \exists s \exists c \exists \tau\, (s = 0 \land c = 1 \land \tau = 0 \land \langle s' = c, c' = -s, \tau' = 1 \rangle (s = 0 \land \tau = n)).$$

These differential equations characterise sin and cos as unique solutions for s and c,

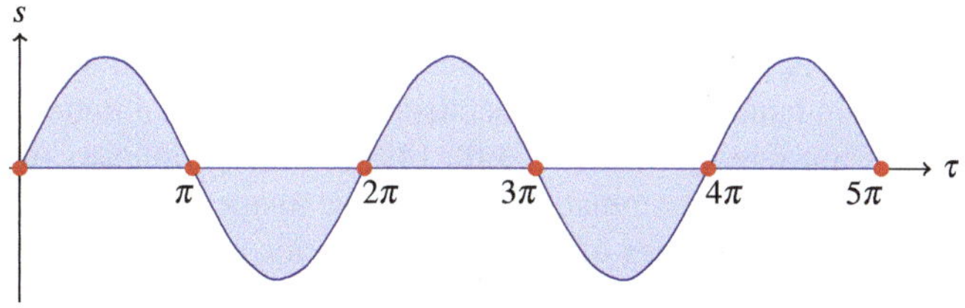

Fig. 2.21 Characterisation of $\mathbb{N}$ as zeros of solutions of differential equations

respectively. Their zeros, as detected by τ, correspond to an isomorphic copy of natural numbers, scaled by π, i.e., $nat(n)$ holds iff n is of the form $k\pi$ for a $k \in \mathbb{N}$; see Fig. 2.21. The initial values for s and c prevent the trivial solution identical to 0. □

In this context, note that hybrid programs contain a computationally complete sublanguage and that reachability of hybrid systems is undecidable [156].

2.7.2 *Relative Completeness*

The standard approach for showing adequacy of a calculus when its logic is not effectively axiomatisable is to analyse the deductive power of the calculus relative to a base logic or to an ineffective oracle rule for the base logic [87, 148, 149]. In calculi for discrete programs, completeness is proven relative to the handling of data [87, 148, 149]. For hybrid systems, this is inadequate: By Theorem 2.2, no sound calculus for $\mathsf{d}\mathcal{L}$ can be complete relative to its data (the reals), because its basis, first-order real arithmetic, is a perfectly decidable and axiomatisable theory [288]. If the $\mathsf{d}\mathcal{L}$ calculus itself would be complete relative to the data of first-order real arithmetic, then, since this is a decidable logic, the $\mathsf{d}\mathcal{L}$ calculus would be complete altogether, which would contradict Theorem 2.2. Thus, we need a different basis for a relative completeness argument. Unlike in classical discrete programs, the data is not where the complexity comes from. In hybrid dynamical systems, the complexity truly originates from the actual dynamics.

According to Theorem 2.2, both continuous evolutions and repetitive discrete transitions, as well as their interaction, cause non-axiomatisability of $\mathsf{d}\mathcal{L}$. Discrete transitions and repetition do not supersede the complexity of continuous transitions. Even relative to an oracle for handling properties of discrete jumps and repetition, the $\mathsf{d}\mathcal{L}$ calculus is not complete, simply because not all differential equations have solutions that are definable in first-order arithmetic so that rule $[']$ can be used. For instance, the solutions of $s' = c, c' = -s$ are trigonometric functions (like sin and cos), which are not first-order definable. The question is whether the converse is true, i.e., whether hybrid programs can be verified given that all required differential equations can be handled.

To calibrate the deductive power of the d$\mathcal{L}$ calculus in light of its inherent incompleteness, we analyse the quotient of reasoning about hybrid systems modulo differential equation handling. Using generalisations of the usual notions of relative completeness for discrete systems [87, 148, 149] to the hybrid case, we show that the d$\mathcal{L}$ calculus completely axiomatises d$\mathcal{L}$ relative to one single additional axiom about valid first-order properties of differential equations. Essentially, we drop the effectiveness requirement for one oracle axiom and show that the resulting d$\mathcal{L}$ calculus is sound and complete. We thus show that the d$\mathcal{L}$ calculus would be complete if only we had a complete replacement for $[']$,$\langle'\rangle$. Although repetitions and interactions of hybrid programs are more involved than purely continuous systems, this results emphasises the importance of studying approximations of this continuous oracle for the analysis of hybrid systems, as we do in Chap. 3.

As a basis, we define FOD as the *first-order logic of differential equations*, i.e., first-order real arithmetic augmented with formulas expressing properties of differential equations, that is, d$\mathcal{L}$ formulas of the form $[x_1' = \theta_1,\dots,x_n' = \theta_n]F$ with a first-order formula F. Dually, the diamond formula $\langle x_1' = \theta_1,\dots,x_n' = \theta_n\rangle F$ is expressible as $\neg[x_1' = \theta_1,\dots,x_n' = \theta_n]\neg F$.

Theorem 2.3 (Relative completeness of d$\mathcal{L}$). *The* d$\mathcal{L}$ *calculus is* complete relative to *FOD, i.e., every valid* d$\mathcal{L}$ *formula can be derived from FOD tautologies.*

Proof (Outline). The (constructive) proof, which, in full, is contained in the remainder of this section, adapts the techniques of Cook [87] and Harel [148, 149] to the hybrid case. The decisive step is to show that every valid property of a repetition α^* can be proven by rules *ind* or *con*, respectively, with a sufficiently strong invariant or variant that is expressible in d$\mathcal{L}$. For this, we show that d$\mathcal{L}$ formulas can be expressed equivalently in FOD, and that valid d$\mathcal{L}$ formulas can be derived from corresponding FOD axioms in the d$\mathcal{L}$ calculus. In turn, the crucial step is to construct a finite FOD formula that characterises the effect of unboundedly many repetitive hybrid transitions and just uses finitely many real variables. □

This main result completely aligns hybrid and continuous verification proof-theoretically. It gives a formal justification that reasoning about hybrid systems is possible to *exactly* the same extent to which it is possible to show properties of solutions of differential equations. Theorem 2.3 shows that superpositions or combinations of discrete jumps, continuous evolutions, and repetitions of hybrid processes can be verified whenever corresponding (intermediate) properties of differential equations are provable. Moreover, in a proof-theoretical sense, our calculus completely lifts all verification techniques for dynamical systems to hybrid systems perfectly. Summarising Theorems 2.1 and 2.3:

The d$\mathcal{L}$ calculus axiomatises the transition behaviour of hybrid systems completely relative to the handling of differential equations!

In the following subsections, we present a fully constructive proof of Theorem 2.3, which generalises the techniques of Harel [148, 149] and Cook [87] to the hybrid case. It shows that for every valid d$\mathcal{L}$ formula, there is a finite set of valid FOD formulas from which it can be derived in the d$\mathcal{L}$ calculus. Recall the proof outline of Theorem 2.3 for a road map of the proof.

Natural numbers are definable in FOD by Theorem 2.2. In this section, we abbreviate quantifiers over natural numbers, e.g., $\forall x\,(nat(x) \rightarrow \phi)$ by $\forall x\!:\!\mathbb{N}\ \phi$ and $\exists x\,(nat(x) \wedge \phi)$ by $\exists x\!:\!\mathbb{N}\ \phi$. Likewise, we abbreviate quantifiers over integers, e.g., $\forall x\,((nat(x) \vee nat(-x)) \rightarrow \phi)$ by $\forall x\!:\!\mathbb{Z}\ \phi$.

2.7.3 Characterising Real Gödel Encodings

As the central device for constructing a FOD formula that captures the effect of unboundedly many repetitive hybrid transitions and just uses finitely many real variables, we prove that a real version of Gödel encoding is definable in FOD. That is, we give a FOD formula that reversibly packs finite sequences of real values into a single real number.

Observe that a single differential equation system is *not* sufficient for defining these pairing functions as their solutions are differentiable, and yet, as a consequence of Morayne's theorem [213], there is no differentiable surjection $\mathbb{R} \rightarrow \mathbb{R}^2$, nor to any part of $\mathbb{R}^2$ of positive measure. We show that real sequences can be encoded nevertheless by chaining the effects of solutions of multiple differential equations and quantifiers.

Lemma 2.7 ($\mathbb{R}$-Gödel encoding). *The formula $at(Z,n,j,z)$, which holds iff Z is a real number that represents a Gödel encoding of a sequence of n real numbers with real value z at position j (for $1 \leq j \leq m$), is definable in FOD. For a formula $\phi(z)$ we abbreviate $\exists z\,(\mathrm{at}(Z,n,j,z) \wedge \phi(z))$ by $\phi(Z_j^{(n)})$.*

$$\sum_{i=0}^{\infty} \frac{a_i}{2^i} = a_0.a_1a_2\ldots \qquad \sum_{i=0}^{\infty} \frac{b_i}{2^i} = b_0.b_1b_2\ldots$$

$$\searrow\!\!\swarrow \quad \sum_{i=0}^{\infty} \left(\frac{a_i}{2^{2i-1}} + \frac{b_i}{2^{2i}}\right) = a_0b_0.a_1b_1a_2b_2\ldots$$

Fig. 2.22 Fractional encoding principle of $\mathbb{R}$-Gödel encoding by bit interleaving

Proof. The basic idea of the $\mathbb{R}$-Gödel encoding is to interleave the bits of real numbers as depicted in Fig. 2.22 (for a pairing of $n = 2$ numbers a and b). For defining $\mathrm{at}(Z,n,j,z)$, we use several auxiliary functions to improve readability; see Fig. 2.23. Note that these definitions need no recursion. Hence, as in the notation $\phi(Z_j^{(n)})$, we can consider occurrences of the function symbols as syntactic abbreviations for quantified variables satisfying the respective definitions.

$$
\begin{aligned}
\mathrm{at}(Z,n,j,z) &\leftrightarrow \forall i\!:\!\mathbb{Z}\ \mathrm{digit}(z,i) = \mathrm{digit}(Z,n(i-1)+j) \wedge nat(n) \wedge nat(j) \wedge n>0\\
\mathrm{digit}(a,i) &= \mathrm{intpart}(2\,\mathrm{frac}(2^{i-1}a))\\
\mathrm{intpart}(a) &= a - \mathrm{frac}(a)\\
\mathrm{frac}(a) = z &\leftrightarrow \exists i\!:\!\mathbb{Z}\ z = a - i \wedge -1 < z \wedge z < 1 \wedge az \geq 0\\
2^i = z &\leftrightarrow i \geq 0 \wedge \exists x \exists t\, (x=1 \wedge t=0 \wedge \langle x' = x\ln 2, t'=1\rangle (t=i \wedge x=z))\\
&\quad\ \vee\ i<0 \wedge \exists x \exists t\, (x=1 \wedge t=0 \wedge \langle x' = -x\ln 2, t'=-1\rangle (t=i \wedge x=z))\\
\ln 2 = z &\leftrightarrow \exists x \exists t\, (x=1 \wedge t=0 \wedge \langle x'=x, t'=1\rangle (x=2 \wedge t=z))
\end{aligned}
$$

Fig. 2.23 FOD definition characterising Gödel encoding of $\mathbb{R}$-sequences in one real number

The function symbol $\mathrm{digit}(a,i)$ gives the ith bit of $a \in \mathbb{R}$ when represented with basis 2. For $i > 0$, $\mathrm{digit}(a,i)$ yields fractional bits, and, for $i \leq 0$, it yields bits of the integer part. For instance, $\mathrm{digit}(a,1)$ yields the first fractional bit, $\mathrm{digit}(a,0)$ is the least-significant bit of the integer part of a. The function $\mathrm{intpart}(a)$ represents the integer part of $a \in \mathbb{R}$. The function $\mathrm{frac}(a)$ represents the fractional part of $a \in \mathbb{R}$, which drops all integer bits. The last constraint in its definition implies that $\mathrm{frac}(a)$ keeps the sign of a (or 0). Consequently, $\mathrm{intpart}(a)$ and $\mathrm{digit}(a,i)$ also keep the sign of a (or 0). Exponentiation 2^i is definable using differential equations, using an auxiliary characterisation of the natural logarithm $\ln 2$. The definition of 2^i splits into the case of exponential growth when $i \geq 0$ and a symmetric case of exponential decay when $i < 0$. □

2.7.4 Expressibility and Rendition of Hybrid Program Semantics

In order to show that d$\mathcal{L}$ is sufficiently expressive to state the invariants and variants that are needed for proving valid statements about loops with rules *ind* and *con*, we prove an expressibility result. We give a constructive proof that the state transition relation of hybrid programs is definable in FOD, i.e., there is a FOD formula $\mathcal{S}_\alpha(\vec{x},\vec{v})$ characterising the state transitions of hybrid program α from the state characterised by the vector $\vec{x}$ of variables to the state characterised by vector $\vec{v}$.

For this, we need to characterise hybrid processes equivalently by differential equations in FOD. Observe that the existence of such characterisations does *not* follow from results embedding Turing machines into differential equations [57, 140], because, unlike Turing machines, hybrid processes are not restricted to discrete values on a grid (such as $\mathbb{N}^k$) but work with continuous real values. Furthermore, Turing machines only have repetitions of discrete transitions on discrete data (e.g., $\mathbb{N}$). For hybrid programs, in contrast, we have to characterise repetitive interactions of discrete and continuous transitions in continuous space (some $\mathbb{R}^k$).

Lemma 2.8 (Hybrid program rendition). *For every hybrid program α with variables among $\vec{x} = x_1,\dots,x_k$, there is a FOD formula $\mathcal{S}_\alpha(\vec{x},\vec{v})$ with variables among the $2k$ distinct variables $\vec{x} = x_1,\dots,x_k$ and $\vec{v} = v_1,\dots,v_k$ such that*

$$\vDash \mathcal{S}_\alpha(\vec{x},\vec{v}) \leftrightarrow \langle\alpha\rangle \vec{x} = \vec{v}$$

or, equivalently, for every I,η,ν,

$$I,\eta,\nu \models \mathscr{S}_\alpha(\vec{x},\vec{v}) \text{ iff } (\nu,\nu[\vec{x}\mapsto val_{I,\eta}(\nu,\vec{v})]) \in \rho_{I,\eta}(\alpha).$$

$$\begin{aligned}
\mathscr{S}_{x_1:=\theta_1,\dots,x_k:=\theta_k}(\vec{x},\vec{v}) &\equiv \bigwedge_{i=1}^{k}(v_i=\theta_i)\\
\mathscr{S}_{x_1'=\theta_1,\dots,x_k'=\theta_k}(\vec{x},\vec{v}) &\equiv \langle x_1'=\theta_1,\dots,x_k'=\theta_k\rangle \vec{v}=\vec{x}\\
\mathscr{S}_{x_1'=\theta_1,\dots,x_k'=\theta_k\,\&\,\chi}(\vec{x},\vec{v}) &\equiv \exists t\,\big(t=0\wedge\langle x_1'=\theta_1,\dots,x_k'=\theta_k,t'=1\rangle\big(\vec{v}=\vec{x}\\
&\qquad\wedge[x_1'=-\theta_1,\dots,x_k'=-\theta_k,t'=-1](t\geq 0\rightarrow\chi)\big)\big)\\
\mathscr{S}_{?\chi}(\vec{x},\vec{v}) &\equiv \vec{v}=\vec{x}\wedge\chi\\
\mathscr{S}_{\beta\cup\gamma}(\vec{x},\vec{v}) &\equiv \mathscr{S}_\beta(\vec{x},\vec{v})\vee\mathscr{S}_\gamma(\vec{x},\vec{v})\\
\mathscr{S}_{\beta;\gamma}(\vec{x},\vec{v}) &\equiv \exists\vec{z}\,(\mathscr{S}_\beta(\vec{x},\vec{z})\wedge\mathscr{S}_\gamma(\vec{z},\vec{v}))\\
\mathscr{S}_{\beta^*}(\vec{x},\vec{v}) &\equiv \exists Z\,\exists n\!:\!\mathbb{N}\,\big(Z_1^{(n)}=\vec{x}\wedge Z_n^{(n)}=\vec{v}\\
&\qquad\wedge\forall i\!:\!\mathbb{N}\,(1\leq i<n\rightarrow\mathscr{S}_\beta(Z_i^{(n)},Z_{i+1}^{(n)}))\big)
\end{aligned}$$

Fig. 2.24 Explicit rendition of hybrid program transition semantics in FOD

Proof. By Lemma 2.6, interpretations of the vectors $\vec{x}$ and $\vec{v}$ characterise the input and output states, respectively, as far as α is concerned. These vectors are finite because α is finite. Vectorial equalities like $\vec{x}=\vec{v}$ or quantifiers $\exists\vec{v}$ are to be understood componentwise. The program rendition is defined inductively in Fig. 2.24. To simplify the notation, we assume that all variables $x_1,\dots,x_k$ are affected in discrete jumps and differential equations by adding vacuous $x_i:=x_i$, or $x_i'=0$ if x_i does not change in the respective statement.

Differential equations give FOD formulas; no further reduction is needed. Evolution along differential equations with evolution domain restrictions is definable by following the unique flow (Lemma 2.1) backwards. Continuous evolution is reversible, i.e., the transitions of $x_i'=-\theta$ are inverse to those of $x_i'=\theta$. Consequently, with an auxiliary variable t, all evolutions of $[x_1'=-\theta_1,\dots,x_k'=-\theta_k,t'=-1]$ follow the same flow as $\langle x_1'=\theta_1,\dots,x_k'=\theta_k,t'=1\rangle$, but backwards. By also reversing clock t, we ensure that, along the reverse flow, χ has been true at all times (because of the box modality) until starting time $t=0$; see Fig. 2.25.

To show reversibility, let $(\nu,\omega)\in\rho_{I,\eta}(x_1'=\theta_1,\dots,x_k'=\theta_k)$, that is, let $f:[0,r]\rightarrow \mathrm{Sta}(\Sigma)$ be a solution of $x_1'=\theta_1,\dots,x_k'=\theta_k$ starting in state ν and ending in ω. Then $g:[0,r]\rightarrow\mathrm{Sta}(\Sigma)$, defined as $g(\zeta)=f(r-\zeta)$, starts in ω and ends in ν. Thus, it only remains to show that g is a solution of $x_1'=-\theta_1,\dots,x_k'=-\theta_k$, which can be seen for $1\leq i\leq k$ as follows:

Fig. 2.25 Evolution domain checks along backwards flow over time t

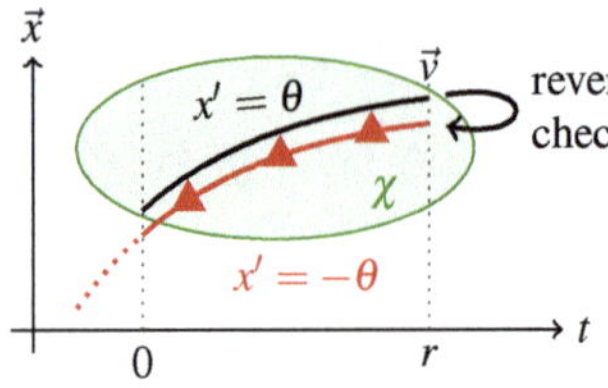

$$\frac{dg(t)(x_i)}{dt}(\zeta) = \frac{df(r-t)(x_i)}{dt}(\zeta) = \frac{df(u)(x_i)}{du}\frac{d(r-t)}{dt}(\zeta) = -\frac{df(u)(x_i)}{du}(\zeta)$$
$$= -val_{I,\eta}(f(\zeta),\theta_i) = val_{I,\eta}(f(\zeta),-\theta_i).$$

Unlike all other cases, case $\mathcal{S}_{x_1'=\theta_1,..,x_k'=\theta_k \,\&\, \chi}(\vec{x},\vec{v})$ in Fig. 2.24 uses nested FOD modalities. Nested modalities can be avoided in $\mathcal{S}_\alpha(\vec{x},\vec{v})$ using an equivalent FOD formula without them; see Fig. 2.25:

$$\exists t\, \exists r\, \big(t = 0 \wedge \langle x_1' = \theta_1,..,x_k' = \theta_k, t' = 1\rangle(\vec{v} = \vec{x} \wedge r = t) \wedge$$
$$\forall \vec{x}\, \forall t\, (\vec{x} = \vec{v} \wedge t = r \to [x_1' = -\theta_1,..,x_k' = -\theta_k, t' = -1](t \geq 0 \to \chi))\big).$$

With a finite formula, the characterisation of repetition $\mathcal{S}_{\beta^*}(\vec{x},\vec{v})$ in FOD needs to capture arbitrarily long sequences of intermediate real-valued states and the correct transition between successive states of such a sequence. To achieve this with first-order quantifiers, we use the real Gödel encoding from Lemma 2.7 in Fig. 2.24 to map unbounded sequences of real-valued states reversibly to a single real number Z, which can be quantified over in first-order logic. □

Using the program rendition from Lemma 2.8 to characterise modalities, we prove that every d$\mathcal{L}$ formula can be expressed equivalently in FOD by structural induction.

Lemma 2.9 (d$\mathcal{L}$ Expressibility). *Logic d$\mathcal{L}$ is* expressible *in FOD: for all d$\mathcal{L}$ formulas $\phi \in \mathrm{Fml}(\Sigma,V)$ there is a FOD formula $\phi^\# \in \mathrm{Fml}_{FOD}(\Sigma,V)$ that is equivalent, i.e., $\models \phi \leftrightarrow \phi^\#$. The converse holds trivially.*

Proof. The proof follows an induction on the structure of formula ϕ for which it is imperative to find an equivalent $\phi^\#$ in FOD. Observe that the construction of $\phi^\#$ from ϕ is effective.

0. If ϕ is a first-order formula, then $\phi^\# := \phi$ already is a FOD formula such that nothing has to be shown.
1. If ϕ is of the form $\varphi \vee \psi$, then by the induction hypothesis there are FOD formulas $\varphi^\#, \psi^\#$ such that $\models \varphi \leftrightarrow \varphi^\#$ and $\models \psi \leftrightarrow \psi^\#$, from which we can conclude by congruence that $\models (\varphi \vee \psi) \leftrightarrow (\varphi^\# \vee \psi^\#)$, giving $\models \phi \leftrightarrow \phi^\#$ by choosing $\varphi^\# \vee \psi^\#$ for $\phi^\#$. Similar reasoning addresses the other propositional connectives or quantifiers.
2. The case where ϕ is of the form $\langle\alpha\rangle\psi$ is a consequence of the characterisation of the semantics of hybrid programs in FOD. The expressibility conjecture holds

by the induction hypothesis using the equivalence of explicit hybrid program renditions from Lemma 2.8:

$$\vDash \ \langle\alpha\rangle\psi \leftrightarrow \exists\vec{v}\,(\mathscr{S}_\alpha(\vec{x},\vec{v}) \wedge \psi^{\#}{}^{\vec{v}}_{\vec{x}}).$$

3. The case where ϕ is $[\alpha]\psi$ is again a consequence of Lemma 2.8:

$$\vDash \ [\alpha]\psi \leftrightarrow \forall\vec{v}\,(\mathscr{S}_\alpha(\vec{x},\vec{v}) \rightarrow \psi^{\#}{}^{\vec{v}}_{\vec{x}})$$

□

The above proofs directly carry over to rich test $\mathsf{d}\mathscr{L}$, i.e., the logic where $\mathsf{d}\mathscr{L}$ formulas are allowed in tests $?\chi$ of hybrid programs and evolution domain restrictions χ of differential equations, when using $\chi^{\#}$ in place of χ in Fig. 2.24. Accordingly, nested modalities can be avoided in FOD by using the following formula for $\mathscr{S}_{x_1'=\theta_1,..,x_k'=\theta_k\,\&\,\chi}(\vec{x},\vec{v})$:

$$\begin{aligned}&\exists t\,\exists r\,\big(t=0 \wedge \langle x_1'=\theta_1,..,x_k'=\theta_k,t'=1\rangle(\vec{v}=\vec{x}\wedge r=t)\wedge\\&\qquad\forall\vec{z}\,\big(\exists\vec{x}\,\exists t\,(\vec{x}=\vec{v}\wedge t=r\wedge\langle x_1'=-\theta_1,..,x_k'=-\theta_k,t'=-1\rangle(t\geq 0\wedge\vec{z}=\vec{x}))\\&\qquad\quad\rightarrow \chi^{\#}{}^{\vec{z}}_{\vec{x}})\big).\end{aligned}$$

2.7.5 Relative Completeness of First-Order Assertions

As special cases of Theorem 2.3, we first prove relative completeness for first-order assertions about hybrid programs. These first-order cases constitute the basis for the general completeness proof for arbitrary formulas of differential dynamic logic.

In the following relative completeness proofs, we use the notation $\vdash_{\mathscr{D}} \phi$ to indicate that a $\mathsf{d}\mathscr{L}$ formula ϕ is derivable (Definition 2.11) from a set of FOD tautologies, which is equivalent to saying that ϕ is derivable in the $\mathsf{d}\mathscr{L}$ calculus augmented with a single *oracle axiom* $\mathscr{D}$ that gives all valid FOD instances. Likewise, we use the notation $\Gamma \vdash_{\mathscr{D}} \Delta$ to indicate that the sequent $\Gamma \vdash \Delta$ is derivable from $\mathscr{D}$.

For the completeness proof, we use several simplifications. For uniform proofs, we assume formulas to use a simplified vocabulary. A formula ϕ is valid iff it is true in *all* I,η,ν. In particular, we can assume valid ϕ to use Skolem constants (or state variables) instead of free logical variables. Existential quantifiers can be represented as modalities: $\exists x\,\phi \equiv \langle x'=1\rangle\phi \vee \langle x'=-1\rangle\phi$. For simplicity, we use cut (*cut*) and weakening to glue together subproofs propositionally. Weakening (i.e., from $\phi \vdash \psi$ infer $\phi_1,\phi \vdash \psi,\psi_1$) can be emulated using contexts Γ,Δ from Definition 2.10, and we use it implicitly together with rule *cut* in the following. Derivability of sequents and derivability of corresponding formulas are equivalent by the following lemma.

Lemma 2.10 (Derivability of sequents). $\vdash_{\mathscr{D}} \phi \rightarrow \psi$ *iff* $\phi \vdash_{\mathscr{D}} \psi$.

Proof. When we consider sequents as abbreviations for formulas, both sides are identical. Otherwise, let $\vdash_{\mathcal{D}} \phi \to \psi$ be derivable from $\mathcal{D}$. Using *cut* (and weakening) with $\phi \to \psi$, this derivation can be extended to one of $\phi \vdash_{\mathcal{D}} \psi$:

$$\text{cut}\,\dfrac{\dfrac{*}{\phi \vdash \phi \to \psi, \psi} \qquad \to\mathrm{l}\,\dfrac{\text{ax}\,\dfrac{*}{\phi \vdash \phi, \psi} \qquad \text{ax}\,\dfrac{*}{\psi, \phi \vdash \psi}}{\phi, \phi \to \psi \vdash \psi}}{\phi \vdash \psi}$$

The converse direction is by an application of $\to$r. □

Lemma 2.11 (Generalisation). *If $\vdash_{\mathcal{D}} \phi$ is provable without free logical variables, then so are $\vdash_{\mathcal{D}} \forall x\, \phi$ and $\vdash_{\mathcal{D}} \langle x_1 := \theta_1, \ldots x_n := \theta_n \rangle \phi$.*

Proof. For the second conjecture, let $\langle \mathscr{A} \rangle$ abbreviate $\langle x_1 := \theta_1, \ldots x_n := \theta_n \rangle$. We prefix each formula in the proof of ϕ with $\langle \mathscr{A} \rangle$ and show that this gives a proof of $\langle \mathscr{A} \rangle \phi$. i$\exists$ is not needed in the proof due to the absence of free logical variables. As an intermediate step, we first show that prefixing with $\langle \mathscr{A} \rangle$ gives an (extended) proof with rule applications generalised to allow for nested jump prefixes $\langle \mathscr{A} \rangle \langle \mathscr{J} \rangle$: By the argument in Theorem 4.1, it is easy to see for discrete jump sets $\langle \mathscr{A} \rangle$ and $\langle \mathscr{J} \rangle$ that the d$\mathcal{L}$ rules remain sound with nested jump prefix $\langle \mathscr{A} \rangle \langle \mathscr{J} \rangle$ in place of only a single prefix $\langle \mathscr{J} \rangle$ from Definition 2.10. Applicability conditions of rules do not depend on jump prefixes, as Definition 2.10 allows adding *any* jump prefix. Thus, we obtain a sound (extended) proof of $\langle \mathscr{A} \rangle \phi$ when replacing—with arbitrary unchanged context $\Gamma, \Delta, \langle \mathscr{J} \rangle$—every rule application of the form

$$\dfrac{\Gamma, \langle \mathscr{J} \rangle \Phi_1 \vdash \langle \mathscr{J} \rangle \Psi_1, \Delta \quad \ldots \quad \Gamma, \langle \mathscr{J} \rangle \Phi_n \vdash \langle \mathscr{J} \rangle \Psi_n, \Delta}{\Gamma, \langle \mathscr{J} \rangle \Phi_0 \vdash \langle \mathscr{J} \rangle \Psi_0, \Delta}$$

in the proof of ϕ by a rule application with the additional unchanged prefix $\langle \mathscr{A} \rangle$ for corresponding $\Gamma, \Delta, \langle \mathscr{J} \rangle$:

$$\dfrac{\Gamma, \langle \mathscr{A} \rangle \langle \mathscr{J} \rangle \Phi_1 \vdash \langle \mathscr{A} \rangle \langle \mathscr{J} \rangle \Psi_1, \Delta \quad \ldots \quad \Gamma, \langle \mathscr{A} \rangle \langle \mathscr{J} \rangle \Phi_n \vdash \langle \mathscr{A} \rangle \langle \mathscr{J} \rangle \Psi_n, \Delta}{\Gamma, \langle \mathscr{A} \rangle \langle \mathscr{J} \rangle \Phi_0 \vdash \langle \mathscr{A} \rangle \langle \mathscr{J} \rangle \Psi_0, \Delta} \tag{2.14}$$

Next, we show that these nested jump prefixes can be reduced to a single jump prefix as Definition 2.10 allows: Let $\langle \mathscr{A}\, \mathscr{J} \rangle$ denote the discrete jump set obtained by merging $\langle \mathscr{A} \rangle$ and $\langle \mathscr{J} \rangle$ using $\langle := \rangle$ as in Sect. 2.5.2. We replace each rule application (with nested prefixes) of the form (2.14) by the following derivation with only a single prefix (assuming $n = 1$ for notational convenience):

$$\langle := \rangle, \langle := \rangle\,\dfrac{\dfrac{\text{cut}\,\dfrac{\dfrac{\ldots}{\Gamma, \langle \mathscr{A} \rangle \langle \mathscr{J} \rangle \Phi_1 \vdash \langle \mathscr{A}\, \mathscr{J} \rangle \Psi_1, \Delta} \qquad \langle := \rangle\,\dfrac{\text{ax}\,\dfrac{*}{\Gamma, \langle \mathscr{A}\, \mathscr{J} \rangle \Phi_1 \vdash \langle \mathscr{A}\, \mathscr{J} \rangle \Phi_1, \Delta}}{\Gamma, \langle \mathscr{A}\, \mathscr{J} \rangle \Phi_1 \vdash \langle \mathscr{A} \rangle \langle \mathscr{J} \rangle \Phi_1, \Delta}}{\Gamma, \langle \mathscr{A}\, \mathscr{J} \rangle \Phi_1 \vdash \langle \mathscr{A}\, \mathscr{J} \rangle \Psi_1, \Delta}}{\Gamma, \langle \mathscr{A}\, \mathscr{J} \rangle \Phi_0 \vdash \langle \mathscr{A}\, \mathscr{J} \rangle \Psi_0, \Delta}}{\Gamma, \langle \mathscr{A} \rangle \langle \mathscr{J} \rangle \Phi_0 \vdash \langle \mathscr{A} \rangle \langle \mathscr{J} \rangle \Psi_0, \Delta}$$

The bottom most $\langle:=\rangle$ applications merge $\langle\mathscr{A}\rangle$ into $\langle\mathscr{J}\rangle$ in the antecedent and succedent, respectively. The unmarked rule applies the same rule that has been used in (2.14), which is applicable on $\Phi_0 \vdash \Psi_0$ for *any* context by Definition 2.10, including $\Gamma, \Delta, \langle\mathscr{A}\,\mathscr{J}\rangle$. The subsequent cut with $\langle\mathscr{A}\rangle\langle\mathscr{J}\rangle\Phi_1$ restores the form of the premise in (2.14). The left branch continues using a dual argument to turn succedent $\langle\mathscr{A}\,\mathscr{J}\rangle\Psi_1$ into $\langle\mathscr{A}\rangle\langle\mathscr{J}\rangle\Psi_1$, thereby yielding a set of non-extended rule applications with the same conclusions and premises as the extended rule application (2.14):

$$\dfrac{\Gamma, \langle\mathscr{A}\rangle\langle\mathscr{J}\rangle\Phi_1 \vdash \langle\mathscr{A}\rangle\langle\mathscr{J}\rangle\Psi_1, \Delta \qquad \dfrac{\dfrac{*}{\Gamma, \langle\mathscr{A}\,\mathscr{J}\rangle\Psi_1 \vdash \langle\mathscr{A}\,\mathscr{J}\rangle\Psi_1, \Delta}\,ax}{\Gamma, \langle\mathscr{A}\rangle\langle\mathscr{J}\rangle\Psi_1 \vdash \langle\mathscr{A}\,\mathscr{J}\rangle\Psi_1, \Delta}\,\langle:=\rangle}{\Gamma, \langle\mathscr{A}\rangle\langle\mathscr{J}\rangle\Phi_1 \vdash \langle\mathscr{A}\,\mathscr{J}\rangle\Psi_1, \Delta}\,cut$$

For reducing the first conjecture of this lemma to the second, let s be a Skolem constant for state variable x. By the above proof, we derive $\vdash_{\mathscr{D}} \langle x := s\rangle\phi$. Using $\forall$r, we continue this derivation to a proof of $\forall X\, \langle x := X\rangle\phi$, which we abbreviate as $\forall x\, \phi$ (see the text below Definition 2.8). Rule $\forall$r is applicable for Skolem constant s as no free logical variables occur in the proof. □

Now we prove two special cases of Theorem 2.3 for formulas of a special form.

Proposition 2.1 (Relative completeness of first-order safety). *For every hybrid program* $\alpha \in \mathrm{HP}(\Sigma, V)$ *and all FOD formulas* $F, G \in \mathrm{Fml}_{FOD}(\Sigma, V)$

$$\vDash F \to [\alpha]G \textit{ implies } \vdash_{\mathscr{D}} F \to [\alpha]G \textit{ (and } F \vdash_{\mathscr{D}} [\alpha]G \textit{ by Lemma 2.10).}$$

Proof. We generalise the relative completeness proof by Cook [87] to $\mathsf{d}\mathscr{L}$ and follow an induction on the structure of program α. In the following, *IH* is short for the induction hypothesis.

1. The cases where α is of the form $x_1 := \theta_1, \ldots, x_n := \theta_n$, $?\chi$, $\beta \cup \gamma$, or $\beta;\gamma$ are consequences of the soundness of the symmetric rules $[;]$,$[\cup]$,$[?]$,$\langle:=\rangle$,$[:=]$. Since these rules are symmetric, they perform equivalent transformations. Consequently, whenever their conclusion is valid, their premise is valid and of smaller complexity (the programs get simpler), and hence derivable by IH. Thus, we can derive $F \to [\alpha]G$ by applying the respective rule. We explicitly show the proof for $\beta;\gamma$ as it contains an extra twist.
2. $\vDash F \to [\beta;\gamma]G$, which implies $\vDash F \to [\beta][\gamma]G$. By Lemma 2.9, there is a FOD formula $G^\#$ such that $\vDash G^\# \leftrightarrow [\gamma]G$. From the validity of $\vDash F \to [\beta]G^\#$, we can conclude by IH that $F \vdash_{\mathscr{D}} [\beta]G^\#$ is derivable. Similarly, due to $\vDash G^\# \to [\gamma]G$, we conclude $\vdash_{\mathscr{D}} G^\# \to [\gamma]G$ by IH. Using Lemma 2.11, we conclude that also $\vdash_{\mathscr{D}} \forall^\beta (G^\# \to [\gamma]G)$. With an application of $[]gen$, the latter derivation can be extended to a derivation of $[\beta]G^\# \vdash_{\mathscr{D}} [\beta][\gamma]G$. Combining the above derivations propositionally by a cut with $[\beta]G^\#$, we can derive $F \vdash_{\mathscr{D}} [\beta][\gamma]G$, from which $[;]$ yields $F \vdash_{\mathscr{D}} [\beta;\gamma]G$ as desired (and Lemma 2.10 or $\to$r yield $\vdash_{\mathscr{D}} F \to [\beta;\gamma]G$).

3. $\vDash F \to [x_1' = \theta_1, \dots, x_n' = \theta_n]G$ is a FOD formula and hence derivable as a $\mathcal{D}$ axiom. Continuous evolution $x_1' = \theta_1, \dots, x_n' = \theta_n \,\&\, \chi$ with evolution domain restrictions is definable in FOD by Lemma 2.8, which we consider as an abbreviation in this proof.
4. $\vDash F \to [\beta^*]G$ can be derived by induction. For this, we define the invariant as a FOD encoding of the statement that all potential post-states of β^* satisfy G according to Lemma 2.9:

$$\phi \equiv ([\beta^*]G)^\# \equiv \forall \vec{v} \, (\mathcal{S}_{\beta^*}(\vec{x}, \vec{v}) \to G_{\vec{x}}^{\vec{v}}).$$

Since $F \to \phi$ and $\phi \to G$ are valid FOD formulas, they are derivable by $\mathcal{D}$; so is $F \vdash_{\mathcal{D}} \phi$ derivable by Lemma 2.10. By Lemma 2.11 and $[]gen$, $[\beta^*]\phi \vdash_{\mathcal{D}} [\beta^*]G$ is derivable. Likewise, $\phi \to [\beta]\phi$ is valid according to the semantics of repetition, and thus derivable by IH, since β is less complex. Using Lemma 2.11, we can derive $\vdash_{\mathcal{D}} \forall^\beta (\phi \to [\beta]\phi)$, from which *ind* yields $\phi \vdash_{\mathcal{D}} [\beta^*]\phi$. Combining the above derivations propositionally by a cut with $[\beta^*]\phi$ and ϕ yields $F \vdash_{\mathcal{D}} [\beta^*]G$.

□

Proposition 2.2 (Relative completeness of first-order liveness). *For each hybrid program $\alpha \in \mathrm{HP}(\Sigma, V)$ and all FOD formulas $F, G \in \mathrm{Fml}_{FOD}(\Sigma, V)$*

$$\vDash F \to \langle\alpha\rangle G \text{ implies } \vdash_{\mathcal{D}} F \to \langle\alpha\rangle G \text{ (and } F \vdash_{\mathcal{D}} \langle\alpha\rangle G \text{ by Lemma 2.10)}.$$

Proof. We generalise the arithmetic completeness proof by Harel [148] to the hybrid case. Most cases of the proof are simple adaptations of the corresponding cases in Proposition 2.1. What remains to be shown is the case of repetitions. Assume that $\vDash F \to \langle\beta^*\rangle G$. To derive this formula by *con*, we use a FOD formula $\varphi(n)$ as a variant expressing that, after n iterations, β can lead to a state satisfying G. This formula is obtained from Lemmas 2.8 and 2.9 as $(\langle\beta^*\rangle G)^\# \equiv \exists \vec{v} \, (\mathcal{S}_{\beta^*}(\vec{x}, \vec{v}) \wedge G_{\vec{x}}^{\vec{v}})$, *except* that the quantifier on the repetition count n is removed such that n becomes a free variable (plus index shifting to count repetitions):

$$\varphi(n-1) \equiv \exists \vec{v} \exists Z \, (Z_1^{(n)} = \vec{x} \wedge Z_n^{(n)} = \vec{v} \wedge \forall i{:}\mathbb{N} \, (1 \le i < n \to \mathcal{S}_\beta(Z_i^{(n)}, Z_{i+1}^{(n)})) \wedge G_{\vec{x}}^{\vec{v}}).$$

By Lemma 2.7, $\varphi(n)$ can only hold true if n is a natural number.

According to the loop semantics, $\vDash n > 0 \wedge \varphi(n) \to \langle\beta\rangle\varphi(n-1)$ is valid by construction: If $n > 0$ is a natural number then so is $n-1$, and if β reaches G after n repetitions, then, after executing β once, $n-1$ repetitions of β reach G. By IH, this formula is derivable, since β contains less loops. By Lemma 2.11, we extend this derivation to $\vdash_{\mathcal{D}} \forall^\beta \forall n{>}0 \, (\varphi(n) \to \langle\beta\rangle\varphi(n-1))$. Thus $\exists v \, \varphi(v) \vdash_{\mathcal{D}} \langle\beta^*\rangle \exists v{\le}0 \, \varphi(v)$ by *con*. It only remains to show that the antecedent is derivable from F and $\langle\beta^*\rangle G$ is derivable from the succedent. From our assumption, we conclude that the following are valid FOD formulas, hence $\mathcal{D}$ axioms:

- $\vDash F \to \exists v \, \varphi(v)$, because $\vDash F \to \langle\beta^*\rangle G$, and

- $\vDash (\exists v{\le}0\, \varphi(v)) \rightarrow G$, because $v \le 0$, and the fact, that by Lemma 2.7, $\varphi(v)$ only holds true for natural numbers, imply $\varphi(0)$. Further, $\varphi(0)$ entails G, because zero repetitions of β have no effect.

From the latter we derive $\vdash_{\mathscr{D}} \forall^{\beta}(\exists v{\le}0\, \varphi(v) \rightarrow G)$ by Lemma 2.11 and extend the derivation to $\langle\beta^*\rangle\exists v{\le}0\, \varphi(v) \vdash_{\mathscr{D}} \langle\beta^*\rangle G$ by $\langle\rangle gen$. From $\vdash_{\mathscr{D}} F \rightarrow \exists v\, \varphi(v)$ we conclude $F \vdash_{\mathscr{D}} \exists v\, \varphi(v)$ by Lemma 2.10. Now, the above derivations can be combined propositionally by a cut with $\langle\beta^*\rangle\exists v{\le}0\, \varphi(v)$ and with $\exists v\, \varphi(v)$ to yield $F \vdash_{\mathscr{D}} \langle\beta^*\rangle G$. □

2.7.6 Relative Completeness of the Differential Logic Calculus

Having succeeded with the proofs of the above statements we can finish the proof of Theorem 2.3, which is the central theoretical result of this chapter.

Proof (of Theorem 2.3). The proof follows a basic structure analogous to that of Harel's proof for the discrete case [148, Theorem 3.1]. We have to show that every valid d$\mathscr{L}$ formula ϕ can be proven from FOD axioms within the d$\mathscr{L}$ calculus: from $\vDash \phi$ we have to prove $\vdash_{\mathscr{D}} \phi$. The proof proceeds as follows: By propositional recombination, we inductively identify fragments of ϕ that correspond to $\phi_1 \rightarrow [\alpha]\phi_2$ or $\phi_1 \rightarrow \langle\alpha\rangle\phi_2$ logically. Next, we express subformulas ϕ_i equivalently in FOD by Lemma 2.9, and use Propositions 2.1 and 2.2 to resolve these first-order safety or liveness assertions. Finally, we prove that the original d$\mathscr{L}$ formula can be re-derived from the subproofs.

We can assume ϕ to be given in conjunctive normal form by appropriate propositional reasoning. In particular, we assume that negations are pushed inside over modalities using the dualities $\neg[\alpha]\phi \equiv \langle\alpha\rangle\neg\phi$ and $\neg\langle\alpha\rangle\phi \equiv [\alpha]\neg\phi$. The remainder of the proof follows an induction on a measure $|\phi|$ defined as the number of modalities in ϕ. For a simple and uniform proof, we assume quantifiers to be abbreviations for modal formulas: $\exists x\, \phi \equiv \langle x' = 1\rangle\phi \vee \langle x' = -1\rangle\phi$ and $\forall x\, \phi \equiv [x' = 1]\phi \wedge [x' = -1]\phi$.

0. $|\phi| = 0$; then ϕ is a first-order formula; hence derivable by $\mathscr{D}$.
1. ϕ is of the form $\neg\phi_1$; then ϕ_1 is first-order, as we assumed negations to be pushed inside. Hence, $|\phi| = 0$ and Case 0 applies.
2. ϕ is of the form $\phi_1 \wedge \phi_2$, then individually deduce the simpler proofs for $\vdash_{\mathscr{D}} \phi_1$ and $\vdash_{\mathscr{D}} \phi_2$ by IH, which can be combined by rule $\wedge$r.
3. ϕ is a disjunction and—without loss of generality—has one of the following forms (otherwise use associativity and commutativity to select a different order for the disjunction):

$$\phi_1 \vee [\alpha]\phi_2$$
$$\phi_1 \vee \langle\alpha\rangle\phi_2$$

As a unified notation for those cases we use $\phi_1 \vee \langle\![\alpha]\!\rangle \phi_2$. Then, $|\phi_2| < |\phi|$, since ϕ_2 has less modalities. Likewise, $|\phi_1| < |\phi|$ because $\langle\![\alpha]\!\rangle \phi_2$ contributes one modality to $|\phi|$ that is not part of ϕ_1.
According to Lemma 2.9 there are FOD formulas $\phi_1^\#, \phi_2^\#$ with $\vDash \phi_i \leftrightarrow \phi_i^\#$ for $i = 1,2$. By congruence, the validity $\vDash \phi$ yields $\vDash \phi_1^\# \vee \langle\![\alpha]\!\rangle \phi_2^\#$, which directly implies $\vDash \neg\phi_1^\# \rightarrow \langle\![\alpha]\!\rangle \phi_2^\#$. Then by Propositions 2.1 or 2.2, respectively, we can derive

$$\neg\phi_1^\# \vdash_{\mathcal{D}} \langle\![\alpha]\!\rangle \phi_2^\#. \tag{2.15}$$

Further $\vDash \phi_1 \leftrightarrow \phi_1^\#$ implies $\vDash \neg\phi_1 \rightarrow \neg\phi_1^\#$, which is derivable by IH, because $|\phi_1| < |\phi|$. By Lemma 2.10, we obtain $\neg\phi_1 \vdash_{\mathcal{D}} \neg\phi_1^\#$, which we combine with (2.15) by a cut with $\neg\phi_1^\#$ to

$$\neg\phi_1 \vdash_{\mathcal{D}} \langle\![\alpha]\!\rangle \phi_2^\#. \tag{2.16}$$

Likewise $\vDash \phi_2 \leftrightarrow \phi_2^\#$ implies $\vDash \phi_2^\# \rightarrow \phi_2$, which is derivable by IH, as $|\phi_2| < |\phi|$. We can extend the derivation of $\vdash_{\mathcal{D}} \phi_2^\# \rightarrow \phi_2$ to one of $\vdash_{\mathcal{D}} \forall^\alpha(\phi_2^\# \rightarrow \phi_2)$ by Lemma 2.11 and conclude $\langle\![\alpha]\!\rangle \phi_2^\# \vdash_{\mathcal{D}} \langle\![\alpha]\!\rangle \phi_2$ by $[]gen$–$\langle\rangle gen$. Finally we combine the latter derivation propositionally with (2.16) by a cut with $\langle\![\alpha]\!\rangle \phi_2^\#$ to derive $\neg\phi_1 \vdash_{\mathcal{D}} \langle\![\alpha]\!\rangle \phi_2$, from which $\vdash_{\mathcal{D}} \phi_1 \vee \langle\![\alpha]\!\rangle \phi_2$ can be obtained, again using *cut*, to complete the proof. □

This concludes the main theoretical proof of relative completeness of the d$\mathcal{L}$ calculus, i.e., of Theorem 2.3.

2.8 Relatively Semidecidable Fragments

To strengthen the completeness result from Theorem 2.3, we consider fragments of d$\mathcal{L}$ where the required FOD tautologies are sufficiently simple as differential equations have first-order definable flows and the required loop invariants (or variants) are expressible in first-order logic over the reals. In these fragments, the only difficulty is to find the required invariants and variants for the proof. Relative to an (ineffective) oracle that provides first-order invariants and variants for repetitions, the d$\mathcal{L}$ calculus can be used as a semidecision procedure. That is, when we assume the oracle to provide suitable (in)variants, validity of formulas can be proven in the d$\mathcal{L}$ calculus. If an imperfect oracle chooses inadequate (in)variants, applying the d$\mathcal{L}$ calculus rules results in goals that are not valid, which is again decidable by quantifier elimination in the d$\mathcal{L}$ calculus.

Theorem 2.4 (Relatively semidecidable fragment). *Relative to an oracle generating first-order invariants and variants, the* d$\mathcal{L}$ *calculus gives a backtracking-free semidecision procedure for (closed)* d$\mathcal{L}$ *formulas with differential equations having first-order definable flows.*

Proof (Outline). The (constructive) proof, which, in full, can be found in the remainder of this section, shows that there are always applicable d$\mathcal{L}$ rules that transform the formulas equivalently and that formulas in this d$\mathcal{L}$ proof descend along a well-founded order. For loops, we assume that suitable (in)variants are obtained from the oracle and we can guarantee termination when these (in)variants are first-order (or contain fewer loops). □

As a consequence, enumerating first-order invariants or variants gives a semidecision procedure for the fragment of Theorem 2.4. As a corollary to Theorems 2.2 and 2.4, there are valid d$\mathcal{L}$ formulas that need proper d$\mathcal{L}$ (or FOD) invariants to be provable and cannot be proven just using (in)variants of first-order real arithmetic. Similarly, the fragment with first-order definable flows and bounded loops is decidable: When loops α^* are annotated with natural numbers indicating the maximum number of repetitions of α, an effective oracle for Theorem 2.4 can be obtained by unrolling, e.g., by rule $\langle{}^{*n}\rangle$.

As an auxiliary result for proving Theorem 2.4, we show that, in d$\mathcal{L}$ proofs, Skolem symbols occur in a uniform way, i.e., a Skolem symbol s always occurs with the same list of arguments.

Lemma 2.12 (Uniform Skolem symbols). *Let ϕ be a d$\mathcal{L}$ formula without Skolem symbols. In any derivation of ϕ, Skolem symbols only occur with a unique list of free logical variables as arguments, provided that the formulas in cuts (rule cut) obey this restriction.*

Proof. The proof is by induction on the structure of proofs in the d$\mathcal{L}$ calculus. For derivations of length zero, the conjecture holds, because ϕ does not contain Skolem symbols. We show that the conjectured Skolem occurrence property is preserved in all subgoals when applying a rule to a goal that satisfies the conjecture.

- $\forall$r The symbols $s(X_1,\dots,X_n)$ introduced by rules $\forall$r,$\exists$l are of the required form as the X_i are precisely the free logical variables. In addition, the symbol $s(X_1,\dots,X_n)$ does not occur nested in other Skolem terms, because, by the induction hypothesis, the bound variable x does not occur in Skolem terms of the goal.
- i$\forall$ Rules i$\forall$ and i$\exists$ are only applicable to instances of first-order real arithmetic (Lemma 2.5), for which the equivalence transformations of quantifier elimination preserve the Skolem occurrence property, because they never introduce quantifiers to bind free variables.
- $\langle'\rangle$ Rule $\langle'\rangle$ preserves the property, as it only substitutes state variables $x_i \in \Sigma$, not logical variables $X_i \in V$.
- *cut* Cuts preserve the Skolem occurrence property, as we assumed the formulas that *cut* introduces to adhere to the Skolem occurrence property.
- – The other rules of the d$\mathcal{L}$ calculus preserve the property as they never replace arguments of Skolem function symbols (which are free variables by induction hypothesis). □

Proof (of Theorem 2.4). The proof is by well-founded induction. We prove that there is a well-founded strict partial order $\prec$ such that:

IH: For all non-atomic formulas occurring in the sequents during a proof, there is an applicable series of dℒ rules such that all resulting subgoals are simpler with respect to $\prec$ and have no additional free variables or function symbols, and their conjunction is equivalent to the conclusion (for suitable oracle choices).

By applying these dℒ rules exhaustively, we obtain a decision procedure relative to the oracle, because the subgoals descend along the well-founded order $\prec$, which has no infinite descending chain. Finally, validity of the remaining sequents with atomic formulas is decidable by evaluating ground instances (Definition 2.9), because, by IH, the resulting formulas have no free variables when the initial formula is closed (open formulas, in contrast, yield equivalent parameter constraints as results). We use the derived rules *ind′* and *con′* from p. 86 in place of *ind* and *con*; see Sect. 2.5.2. To obtain a backtracking-free procedure, we remove rules $\langle^{*n}\rangle$,$[^{*n}]$, $[]$*gen*,$\langle\rangle$*gen*,*ind*,*con* and *cut* from the calculus: If a calculus with less rules gives a decision procedure, then so does the full calculus.

We define the order $\prec$ as the lexicographical order of, respectively, the number of: loops, differential equations, sequential compositions, choices, modalities, quantifiers, number of different variables and Skolem function symbols, and the number of logical connectives. As a lexicographical order of natural numbers, $\prec$ is well-founded [99]. It lifts to sequents in rule applications (Definition 2.10) when all subgoals of all rule schemata are simpler than their goals with respect to $\prec$, which can be shown to retain well-foundedness as a multiset ordering [99].

Now the proof of IH is by induction along $\prec$. Let ϕ be a non-atomic formula of a sequent in an open branch of the proof. We assume ϕ to occur in the succedent; the respective proofs for the antecedent are dual. Hence, we consider the sequent to be of the form $\Gamma \vdash \phi, \Delta$.

1. If ϕ is of the form $\psi_1 \wedge \psi_2$, then rule $\wedge$r is applicable, yielding smaller sequents (with less logical connectives) that are equivalent. Other logical connectives are handled likewise using rules $\neg$r,$\vee$r,$\wedge$r,$\rightarrow$r, respectively.
2. If ϕ is of the form $[\alpha]\psi$ or $\langle\alpha\rangle\psi$ and α is of the form $?\chi$, $\beta;\gamma$ or $\beta \cup \gamma$, the corresponding rule $\langle;\rangle$,$[;]$,$\langle\cup\rangle$,$[\cup]$ or $\langle?\rangle$,$[?]$ is applicable, yielding a simpler yet equivalent formula.
3. If ϕ is of the form $[x_1' = \theta_1, \ldots, x_n' = \theta_n \,\&\, \chi]\psi$, then rule $[']$ is applicable, as we assumed differential equations to have first-order definable flows. The resulting formula is equivalent and simpler, because it contains fewer differential equations. It involves additional bound variables but not free variables. Case $\langle x_1' = \theta_1, \ldots, x_n' = \theta_n \,\&\, \chi\rangle\psi$ is similar, by rule $\langle'\rangle$.
4. If ϕ is of the form $[\alpha^*]\psi$, then rule *ind′* is applicable with a first-order invariant F obtained from the oracle. The resulting subgoals are simpler according to $\prec$, because they contain less loops (F does not contain loops). The resulting subgoals do not have additional free variables as all bound variables of α^* remain bound by the universal closure $\forall^\alpha$ in the respective premises. Finally, we assume the oracle to give an invariant such that the conjunction of the resulting subgoals is equivalent to the goal (otherwise we have nothing to show for

inadequate choices by the oracle). The case $\langle\alpha^*\rangle\psi$ is similar, using rule *con'* instead.

5. If ϕ is of the form $\langle x_1 := \theta_1, \ldots, x_n := \theta_n\rangle\psi$, there are two cases. If rule $\langle:=\rangle$ is applicable, it yields equivalent simpler sequents. Otherwise, we have

$$\psi \prec \langle x_1 := \theta_1, \ldots, x_n := \theta_n\rangle\psi.$$

Thus, by IH, there is a finite sequence of rule applications on ψ yielding equivalent sequents with atomic formulas. Prefixing the resulting proof with $\langle x_1 := \theta_1, \ldots, x_n := \theta_n\rangle$ yields a corresponding proof for deriving $\Gamma \vdash \phi, \Delta$ by Lemma 2.11. The formulas of the open branches of this proof resulting from ϕ are of the form $\langle x_1 := \theta_1, \ldots, x_n := \theta_n\rangle G$ for atomic formulas G, where, at the latest, rule $\langle:=\rangle$ is applicable, as substitutions are admissible on atomic formulas. Case $[x_1 := \theta_1, \ldots, x_n := \theta_n]\psi$ is similar, using rule $[:=]$ first.

6. If ϕ is of the form $\forall x\, \psi(x)$, we can apply rule $\forall$r giving $\psi(s(X_1, \ldots, X_n))$. Now, we have $\psi(s(X_1, \ldots, X_n)) \prec \forall x\, \psi(x)$; hence, by IH, $\psi(s(X_1, \ldots, X_n))$ can be transformed equivalently to a set of sequents of the form

$$\Phi_i(s(X_1, \ldots, X_n)) \vdash \Psi_i(s(X_1, \ldots, X_n))$$

with atomic formulas (without loss of generality, we can assume $s(X_1, \ldots, X_n)$ to occur in all branches). Hence, QE is defined for these atomic formulas and rule i$\forall$ can be applied on each branch, yielding $\mathrm{QE}(\forall s\,(\Phi_i(s) \vdash \Psi_i(s)))$. Consequently, the original sequent $\Gamma \vdash \forall x\, \psi(x), \Delta$ is equivalent to

$$\bigwedge_i \mathrm{QE}(\forall s\,(\Phi_i(s) \vdash \Psi_i(s)))$$

for the following reason: $\Gamma \vdash \psi(s(X_1, \ldots, X_n)), \Delta$ is equivalent to

$$\bigwedge_i (\Phi_i(s(X_1, \ldots, X_n)) \vdash \Psi_i(s(X_1, \ldots, X_n)))$$

by IH, using the equivalence $\mathrm{QE}(\forall s\,(F \wedge G)) \equiv \mathrm{QE}(\forall s\, F) \wedge \mathrm{QE}(\forall s\, G)$ and the fact that s does not occur in Γ, Δ. After applying rule i$\forall$, the result has no additional free symbols, although intermediate formulas do.

7. If ϕ is of the form $\exists x\, \psi(x)$, then rule $\exists$r is applicable giving $\psi(X)$ for a fresh logical variable X. Then $\psi(X) \prec \exists x\, \psi(x)$; hence, by IH, $\psi(X)$ can be transformed equivalently to a set of sequents $\Phi_i \vdash \Psi_i$ with atomic formulas. If no Skolem dependency on X occurs in $\Phi_i \vdash \Psi_i$, then QE is defined and rule i$\exists$ applicable, giving $\mathrm{QE}(\exists X \bigwedge_i(\Phi_i \vdash \Psi_i))$, which is equivalent to $\exists X \bigwedge_i(\Phi_i \vdash \Psi_i)$. By IH, this is equivalent to $\Gamma \vdash \exists X\, \psi(X), \Delta$, because X does not occur in Γ, Δ. Otherwise, if a Skolem term $s(X_1, \ldots, X, \ldots, X_n)$ occurs in a $\Phi_i \vdash \Psi_i$, then, by IH, the Skolem function s already occurred in $\psi(X)$. By Lemma 2.12, the Skolem term $s(X_1, \ldots, X, \ldots, X_n)$ itself must already have occurred in $\psi(X)$, which contradicts the fact that X is fresh and that bound variable x does not occur in Skolem

terms of $\exists x\, \psi(x)$, again by Lemma 2.12. After applying rule i$\exists$ the additional free variable X disappears. □

This completes the proof of Theorem 2.4, showing that the d$\mathcal{L}$ calculus can be used as a semidecision procedure for a particular set of (in)variants provided by an oracle. Consequently, these results show that, in a certain sense, finding (in)variants is the only challenge in hybrid systems' verification, because the d$\mathcal{L}$ calculus takes care of everything else. In Chap. 3 we revisit and strengthen this result, because we show that properties of differential equations can be proven by appropriate generalisations of (in)variants that we call differential invariants. Furthermore, we turn to the challenge of finding these (differential) invariants in Chap. 6.

2.9 Train Control Verification

In this section, we verify collision avoidance of the train control system presented in Sect. 2.4. Especially, we identify the constraints required for the free parameters of the system and discover the preconditions for safe driving.

2.9.1 *Finding Inductive Candidates*

Recall the d$\mathcal{L}$ formula from Sect. 2.4 that expresses that the simplified ETCS train control system ensures that trains always stay inside their movement authority m to ensure collision-freedom:

$$\psi \rightarrow [(ctrl; drive)^*]\, z \leq m \tag{2.7*}$$

We want to prove safety statement (2.7) of the simplified version of the European Train Control System. Note that this is a significantly simplified version showing only the true essentials of ETCS. We consider the ETCS cooperation protocol in more detail in Chap. 7.

Using parametric extraction techniques, we identify both the requirement ψ for safe driving and the induction hypothesis ϕ that is required for the proof. Similar to the proof in Fig. 2.16, which is dual to the proof in Fig. 2.14, an unwinding of the loop in (2.7) by rule $[^{*n}]$ can be used to extract a candidate for a parametric inductive hypothesis. It expresses that there is sufficient braking distance at current speed v, which basically corresponds to the controllability constraint for ETCS (as illustrated in Fig. 2.15 on p. 90):

$$\phi \equiv v^2 \leq 2b(m-z) \land b > 0 \land A \geq 0\ . \tag{2.17}$$

2.9.2 Inductive Verification

Using proof rule *ind* to prove $d\mathcal{L}$ formula (2.7) by induction, we show that (a) invariant ϕ holds initially, i.e., $\psi \vdash \phi$ (implying the antecedent of the conclusion of *ind*), (b) the invariant is sustained after each execution of $ctrl;drive$, and (c) invariant ϕ implies postcondition $z \leq m$. Case (c) holds by QE, as $0 \leq v^2 \leq 2b(m-z)$ and $b > 0$. The induction start (a) will be examined after the full proof, since we want to identify the prerequisite ψ for safe driving by proof analysis. In the proof of the induction step $\phi \to [ctrl;drive]\phi$, we omit condition $m - z \leq s$ from *ctrl*, because it is not used in the proof (braking remains safe with respect to $z \leq m$). The induction is provable in $d\mathcal{L}$ as follows (for notational convenience, we assume rule $\forall$r calls the Skolem constant for m again m, and so on, as there are no free logical variables):

$$
\dfrac{\dfrac{\dfrac{\dfrac{\dfrac{\dfrac{\dots}{\phi \vdash \langle a:=-b\rangle[drive]\phi} \qquad {\scriptstyle [?],\to\mathrm{r}}\;\dfrac{\dfrac{\dots}{\phi, m-z \geq s \vdash \langle a:=A\rangle[drive]\phi}}{\phi \vdash [?m-z\geq s; a:=A][drive]\phi}}{\phi \vdash [ctrl][drive]\phi}\;{\scriptstyle [\cup],\wedge\mathrm{r}}}{\phi \vdash [ctrl;drive]\phi}\;{\scriptstyle [;]}}{\vdash \phi \to [ctrl;drive]\phi}\;{\scriptstyle \to\mathrm{r}}}{\vdash \forall^{\alpha}(\phi \to [ctrl;drive]\phi)}\;{\scriptstyle \forall\mathrm{r}}}{\phi \vdash [(ctrl;drive)^*]\phi}\;{\scriptstyle ind}
$$

The differential equation system in *drive* is linear with a constant coefficient matrix M. Its solution can be obtained by symbolically computing the exponential series $e^{Mt}\eta$ with symbolic initial value $\eta = (z, v)$ and similar symbolic integration of the inhomogeneous part [297, §18.VI]; also see App. B.4. We abbreviate the solution $\langle z := -\frac{b}{2}t^2 + vt + z, v := -bt + v\rangle$ thus obtained by $\langle \mathcal{S}_t\rangle$. See Example B.3 in App. B for an explanation of why this is a solution of the differential equations. In this example, the evolution domain restrictions are convex; hence the constraint $\forall 0{\leq}\tilde{t}{\leq}t\,\langle\mathcal{S}_{\tilde{t}}\rangle\chi$ of rule $[']$ can be simplified to $\langle\mathcal{S}_t\rangle\chi$ as in (2.12) to save space. Further, we leave out conditions which are unnecessary for closing the above proof. In the left branch, the constrained evolution of clock τ is irrelevant and will be left out to save space (braking is the safest operation and can be continued indefinitely without extra risk). The left branch closes (marked $*$):

$$
\dfrac{\dfrac{\dfrac{\dfrac{\dfrac{\dfrac{\dfrac{*}{\phi, t\geq 0, -bt+v\geq 0 \vdash \langle\mathcal{S}_t\rangle\phi}\;{\scriptstyle \langle:=\rangle,\mathrm{i}\forall}}{\phi, t\geq 0, \langle v:=-bt+v\rangle v\geq 0 \vdash \langle\mathcal{S}_t\rangle\phi}\;{\scriptstyle \langle:=\rangle}}{\phi \vdash t\geq 0 \to (\langle v:=-bt+v\rangle v\geq 0 \to \langle\mathcal{S}_t\rangle\phi)}\;{\scriptstyle \to\mathrm{r},\to\mathrm{r}}}{\phi \vdash \forall t{\geq}0(\langle v:=-bt+v\rangle v\geq 0 \to \langle\mathcal{S}_t\rangle\phi)}\;{\scriptstyle \forall\mathrm{r}}}{\phi \vdash [z'=v, v'=-b\,\&\,v\geq 0]\phi}\;{\scriptstyle [']}}{\phi \vdash \langle a:=-b\rangle[drive]\phi}\;{\scriptstyle \langle:=\rangle}}{\phi \vdash [a:=-b][drive]\phi}\;{\scriptstyle [:=]}
$$

The right branch does not need the evolution domain constraint $v \geq 0$, because v does not decrease when accelerating. We again use $\langle \mathcal{S}_t \rangle$ as an abbreviation for the solution $\langle z := \frac{A}{2}t^2 + vt + z, v := At + v \rangle$.

$$
\begin{array}{rc}
 & \dots \\ \hline
 & \phi, m-z \geq s \vdash s \geq \frac{v^2}{2b} + \left(\frac{A}{b}+1\right)\left(\frac{A}{2}\varepsilon^2 + \varepsilon v\right) \\ \hline
\langle := \rangle, \mathrm{i}\forall & \phi, m-z \geq s, 0 \leq t \leq \varepsilon \vdash \langle \mathcal{S}_t \rangle \phi \\ \hline
\rightarrow\mathrm{r}, \langle := \rangle & \phi, m-z \geq s \vdash t \geq 0 \rightarrow (\langle \tau := t \rangle \tau \leq \varepsilon \rightarrow \langle \mathcal{S}_t \rangle \phi) \\ \hline
\forall\mathrm{r} & \phi, m-z \geq s \vdash \forall t {\geq} 0 (\langle \tau := t \rangle \tau \leq \varepsilon \rightarrow \langle \mathcal{S}_t \rangle \phi) \\ \hline
\langle := \rangle & \phi, m-z \geq s \vdash \langle \tau := 0 \rangle \forall t {\geq} 0 (\langle \tau := t + \tau \rangle \tau \leq \varepsilon \rightarrow \langle \mathcal{S}_t \rangle \phi) \\ \hline
\langle ' \rangle & \phi, m-z \geq s \vdash \langle \tau := 0 \rangle [z' = v, v' = A, \tau' = 1 \,\&\, \tau \leq \varepsilon] \phi \\ \hline
[:=] & \phi, m-z \geq s \vdash [\tau := 0][z' = v, v' = A, \tau' = 1 \,\&\, \tau \leq \varepsilon] \phi \\ \hline
\langle := \rangle & \phi, m-z \geq s \vdash \langle a := A \rangle [\tau := 0][z' = v, v' = a, \tau' = 1 \,\&\, \tau \leq \varepsilon] \phi \\ \hline
[;] & \phi, m-z \geq s \vdash \langle a := A \rangle [drive] \phi \\ \hline
[:=] & \phi, m-z \geq s \vdash [a := A][drive] \phi
\end{array}
$$

2.9.3 Parameter Constraint Discovery

The right branch only closes when the succedent of its open goal is guaranteed. That formula expresses that there will still be sufficient braking distance even after accelerating by $\leq A$ for up to ε seconds:

$$s \geq \frac{v^2}{2b} + \left(\frac{A}{b} + 1\right)\left(\frac{A}{2}\varepsilon^2 + \varepsilon v\right). \tag{2.18}$$

This constraint can be discovered automatically in the above proof by the indicated application of rule i$\forall$ using quantifier elimination with some simplifications. Constraint (2.18) is required to make sure invariant (2.17) still holds after accelerating. In fact, augmenting the case study with (2.18) makes the argument inductive, and the whole proof of the safety statement (2.7) closes when ψ is chosen identical to ϕ. Here, the conditions of ψ cannot be removed without leaving the proof open due to a counterexample, as the invariant (2.17) is a controllability constraint; see Sect. 2.5.3.1.

Quite unlike in the acceleration-free case [231], constraint (2.18) needs to be enforced dynamically as the affected variables change over time. That is, at the beginning of each *ctrl* cycle, s needs to be updated in accordance with (2.18), which admits complex behaviour as in Fig. 2.9b on p. 63. Further, this constraint can be used to find out how densely a track can be packed with trains in order to maximise ETCS throughput without compromising safety. The resulting provably safe train control system can be summarised as follows:

$$v^2 \leq 2b(m-z) \land b > 0 \land A \geq 0 \rightarrow [(ctrl; drive)^*] z \leq m \tag{2.19}$$

$$\begin{aligned} \text{where } ctrl &\equiv s := \frac{v^2}{2b} + \left(\frac{A}{b}+1\right)\left(\frac{A}{2}\varepsilon^2 + \varepsilon v\right); \\ &\quad (?m - z \le s; a := -b) \cup (?m - z \ge s; a := A) \\ drive &\equiv \tau := 0; (z' = v, v' = a, \tau' = 1 \,\&\, v \ge 0 \land \tau \le \varepsilon). \end{aligned}$$

Using the d$\mathcal{L}$ calculus, similar constraints can be derived (Sect. 4.8) to find out how early a train needs to start negotiation in order to minimise the risk of having to reduce speed when the MA is not extensible in time, which is the ST parameter of Fig. 2.8.

For the resulting ETCS system, liveness can be proven in the d$\mathcal{L}$ calculus by showing that the train can pass every point p by an appropriate choice of m by the RBC:

$$z = z_0 \land v = v_0 > 0 \land \varepsilon > 0 \land b > 0 \land A \ge 0 \to \forall p \exists m \langle (ctrl; drive)^* \rangle z \ge p. \quad (2.20)$$

For $A = 0$, the proof of property (2.20) uses the variant $\varphi(n) \equiv z + n\varepsilon v_0 \ge p \land v = v_0$ for rule *con*, which expresses that the speed does not decrease (until $n < 0$) and that the remaining distance from z to target p can be covered after at most n iteration cycles. This directly proves the property even when $A = 0$ for appropriate acceleration choices. For general $A \ge 0$, the following variant proves property (2.20) by *con*:

$$\varphi(n) \equiv ((z + n\varepsilon v_0 \ge p \land z_0 \le z \land v^2 \le v_0^2 + 2A(z - z_0) \land v \ge v_0 \land z \le p) \lor z \ge p) \land v \ge 0. \quad (2.21)$$

It expresses that, when $z \le p$, the remaining distance can be covered after at most n iterations while the train position and velocity increase, yet the velocity is bounded depending on the initial velocity v_0, acceleration A, and distance $z - z_0$. The appropriate choice of m to prove property (2.20) with this variant is

$$m \ge p + \frac{v_0^2 + 2A(p - z_0)}{2b} + \left(\frac{A}{b}+1\right)\left(\frac{A}{2}\varepsilon^2\right) + \varepsilon\sqrt{v_0^2 + 2A(p - z_0)},$$

which can be obtained by overapproximating braking condition (2.18) with the speed limit $v^2 \le v_0^2 + 2A(z - z_o)$ from the variant. We will analyse ETCS in more detail in Chap. 7.

In this example, we can see the effect of the d$\mathcal{L}$ calculus. It takes a specification of a hybrid system and successively identifies constraints on the parameters which are needed for correctness. These constraints can then be handled in a purely modular way by rules i$\forall$ and i$\exists$. As a typical characteristic of hybrid systems, further observe that intermediate formulas are significantly more complex than the original proof obligation, which can be expressed succinctly in the expressive language of d$\mathcal{L}$. This reflects the fact that the actual complexity of hybrid systems originates from hybrid interaction, not from a single transition. Still, using appropriate proof strategies (Chap. 5) for the d$\mathcal{L}$ calculus, the safety statement (2.7) with invariant (2.17) can be verified automatically in a theorem prover that invokes Mathematica for rules $\langle ' \rangle$,$[']$, i$\forall$, and i$\exists$. In fact, using the invariant computation techniques that we introduce in Chap. 6, the overall safety property (2.7) can be verified fully automatically even without providing an invariant manually.

2.10 Summary

We have introduced a first-order dynamic logic for hybrid programs, which are uniform operational models for hybrid systems with interacting discrete jumps and continuous evolutions along differential equations. For this differential dynamic logic, $d\mathcal{L}$, we have presented a concise generalised free-variable proof calculus over the reals.

Our sequent calculus for $d\mathcal{L}$ is a generalisation of classical calculi for discrete dynamic logic [35, 37, 149, 148] to the hybrid case. It is a compositional verification calculus for verifying properties of hybrid programs by decomposing them into properties of their parts. In order to handle interacting hybrid dynamics, we lift real quantifier elimination to the deductive calculus in a new modular way that is suitable for automation, using real-valued free variables, Skolem terms, and invertible quantifier rules over the reals.

As a fundamental result aligning hybrid and continuous reasoning proof-theoretically, we have proven our calculus to axiomatise the transition behaviour of hybrid systems completely relative to the handling of differential equations. To the best of our knowledge, this is the first relative completeness result for hybrid systems' verification. Moreover, we have demonstrated that our calculus is well suited for practical automatic verification in a realistic case study of a fully parametric version of the European Train Control System.

Dynamic logic can be augmented [37] to support reasoning about dynamically reconfiguring system structures, which we want to extend to hybrid systems in future work. While the $d\mathcal{L}$ calculus is complete relative to the continuous fragment, it is a subtle open problem whether a converse calculus can exist that is complete relative to various discrete fragments.

Chapter 3
Differential-Algebraic Dynamic Logic DAL

Contents

3.1	Introduction		**124**
	3.1.1	Related Work	128
	3.1.2	Structure of This Chapter	130
3.2	Syntax		**130**
	3.2.1	Terms	132
	3.2.2	Differential-Algebraic Programs	132
	3.2.3	Formulas	139
3.3	Semantics		**141**
	3.3.1	Transition Semantics of Differential-Algebraic Programs	141
	3.3.2	Valuation of Formulas	145
	3.3.3	Time Anomalies	145
	3.3.4	Conservative Extension	147
3.4	Collision Avoidance in Air Traffic Control		**148**
	3.4.1	Flight Dynamics	148
	3.4.2	Differential Axiomatisation	149
	3.4.3	Aircraft Collision Avoidance Manoeuvres	150
	3.4.4	Tangential Roundabout Manoeuvre	151
3.5	Proof Calculus		**152**
	3.5.1	Motivation	153
	3.5.2	Derivations and Differentiation	154
	3.5.3	Differential Reduction and Differential Elimination	160
	3.5.4	Proof Rules	162
	3.5.5	Deduction Modulo by Side Deduction	168
	3.5.6	Differential Induction with Differential Invariants	170
	3.5.7	Differential Induction with Differential Variants	181
3.6	Soundness		**185**
3.7	Restricting Differential Invariants		**188**
3.8	Differential Monotonicity Relaxations		**189**
3.9	Relative Completeness		**193**
3.10	Deductive Strength of Differential Induction		**194**
3.11	Air Traffic Control Verification		**197**
	3.11.1	Characterisation of Safe Roundabout Dynamics	197
	3.11.2	Tangential Entry Procedures	200
	3.11.3	Discussion	201
3.12	Summary		**201**

A. Platzer, Logical Analysis of Hybrid Systems,
DOI 10.1007/978-3-642-14509-4_3,

Synopsis We generalise dynamic logic to a logic for differential-algebraic programs, i.e., discrete programs augmented with first-order differential-algebraic formulas as continuous evolution constraints in addition to first-order discrete jump formulas. These programs characterise interacting discrete and continuous dynamics of hybrid systems elegantly and uniformly, including systems with disturbance and differential-algebraic equations. For our logic, we introduce a calculus over real arithmetic with discrete induction and a new *differential induction* with which differential-algebraic programs can be verified by exploiting their differential constraints algebraically without having to solve them. We develop the theory of differential induction and differential refinement and analyse their deductive power. As an example, we present parametric tangential roundabout manoeuvres in air traffic control and prove collision avoidance in our calculus.

3.1 Introduction

In Chap. 2, we have shown a verification logic for hybrid programs and proof rules for differential equations that are solvable in polynomial arithmetic. While the verification approach is, of course, more general, the primary application that we have seen so far is train control. The same proof techniques also work for many other systems, including car control where the continuous dynamics can be solved. Now we consider air traffic control scenarios as a motivating example for hybrid systems where differential equations can no longer be solved in decidable arithmetic, so that more advanced verification techniques for differential equations are needed.

Verification of Hybrid Systems

Flight manoeuvres in air traffic control [293, 196, 203, 104, 92, 238, 129, 171] give rise to hybrid systems with challenging dynamics. There the continuous dynamics results from continuous movement of aircraft in space, and the discrete dynamics is caused by the instantaneous switching of manoeuvring modes or by discrete aircraft controllers that decide when and how to initiate flight manoeuvres. Proper functioning of these systems is highly safety-critical for the spatial separation of aircraft during all flight manoeuvres, especially collision avoidance manoeuvres. Their analysis, however, is challenging due to the superposition of involved continuous flight dynamics with nontrivial discrete control, causing hybrid systems like these to be amenable neither to mere continuous reasoning nor to verification techniques for purely discrete systems. Since, especially in the presence of parameters, hybrid systems cannot be verified numerically [238, 85], we present a purely symbolic approach using combined deductive and algebraic verification techniques.

In practise, correctness of hybrid systems further depends on the choice of parameters that arise naturally from the degrees of freedom of how a part of the system can be instantiated or how a controller can respond to input [293, 97, 91, 238, 171]. For instance, correct angular velocities, proper timing, and compatible manoeuvre points are equally required for safe air traffic control [293, 238]. Additionally, relevant correctness properties for hybrid systems include safety, liveness, and mixed properties like reactivity (see Chap. 7), all of which can possibly involve (alternating) quantifiers or free variables for parameters. As a uniform approach for specifying and verifying these heterogeneous properties of hybrid systems with symbolic parameters, we introduce an extension of our first-order logic and dynamic logic for handling correctness statements about hybrid systems.

Logic for Hybrid Systems

The aim of this chapter is to present logical analysis techniques with which general hybrid systems with interacting discrete and continuous dynamics can be specified and verified in a coherent logical framework. To this end, we introduce the *differential-algebraic dynamic logic* (DA-logic or DAL for short) as the logic of general hybrid change. As an elegant and uniform operational model for hybrid systems in DAL, we introduce *differential-algebraic programs* (DA-programs). These programs combine *first-order discrete jump constraints* (DJ-constraints) to characterise discrete transitions with support for *first-order differential-algebraic constraints* (DA-constraints) to characterise continuous transitions. DA-constraints provide a convenient way for expressing continuous system evolution constraints and give a uniform semantics to differential evolutions, systems of differential equations [297], switched systems [55], invariant constraints [156, 97], triggers [55], differential-algebraic equations [132, 187], and differential inequalities [153, 297]. In DJ-constraints and DA-constraints, first-order quantifiers further give a natural and semantically well-founded way of expressing unbounded discrete or continuous nondeterminism in the dynamics, including nondeterminism resulting from internal choices or external disturbances. In interaction with appropriate control structure, DJ-constraints and DA-constraints can be combined to form DA-programs as uniform operational models for hybrid systems. With this, DA-programs are a generalised program notation for hybrid systems.

As a specification and verification logic for hybrid systems that are given as DA-programs, we design the first-order dynamic logic DAL as an extension of $\mathsf{d}\mathcal{L}$. In particular, we generalise discrete dynamic logic [149] to hybrid control and support DA-programs as actions of a first-order multi-modal logic [123], such that its modalities can be used to specify and verify correctness properties of more advanced hybrid systems. As in Chap. 2, the DAL formula $[\alpha]\phi$ expresses that all runs of DA-program α lead to states satisfying the DAL formula ϕ. Likewise, $\langle\alpha\rangle\phi$ says that there is at least one state reachable by DA-program α which satisfies ϕ. Similarly, $\exists p\,[\alpha]\langle\beta\rangle\phi$ says that there is a choice of parameter p such that for all possible behaviour of DA-program α there is a reaction of DA-program β that ensures ϕ.

Deductive Verification and Differential Induction

As a means for verifying hybrid systems by proving corresponding DAL formulas, we introduce a sequent calculus. It uses side deductions as a simple and concise, yet constructive, modular technique to integrate real quantifier elimination with calculus rules for modalities. For handling discrete transitions, we present a first-order generalisation of standard proof rules [149, 37]. Interacting continuous transitions are more involved. Formulas with very simple differential equations can be verified by using their solutions as in Chap. 2: Linear differential equations with nilpotent constant coefficients (i.e., $x' = Ax$ for a matrix A with $A^n = 0$ for some n) have polynomial solutions so that arithmetic formulas about these solutions can be verified by quantifier elimination [81]; see Apps. B and D.2 for details. This approach, however, does not scale to hybrid systems with more sophisticated differential constraints because their solutions do not support quantifier elimination (e.g., when they involve transcendental functions), cannot be given in closed form [297], are not computable [250], or do not even exist [297, 179]. As part of the descriptive power of differential equations, solutions of differential equations are much more complicated than the original equations and can become transcendental even for simple linear differential equations like $x' = -y, y' = x$, where the solutions will be trigonometric functions.

Instead, as a logical analysis technique for verifying DA-programs with more general differential-algebraic constraints, we introduce *first-order differential induction* as a fully algebraic form of proving logical statements about DA-constraints using their differential-algebraic constraints in a differential induction step instead of using their solutions in a reachability computation. Unlike in discrete induction, the invariant is a *differential invariant*, i.e., a property that is closed under total differentiation with respect to the differential constraints. There, the basic idea for showing invariance of a property F is to show that F holds initially and its total derivative F' holds always along the dynamics (with generalisations of total differentials to logical formulas and corresponding generalisations for quantified DA-constraints). This analysis considers all non-Zeno executions, i.e., where the system cannot switch its mode infinitely often in finite time. In addition, we introduce *differential strengthening* or differential cuts as a technique for refining the system dynamics by differential invariants until the property becomes provable for the refined dynamics, which we show to be crucial in practical applications.

Comparison

In Chap. 2 we have introduced a logic and calculus for verifying *hybrid programs*, which is the quantifier-free subclass of DA-programs without propositional connectives (see Table 3.1 for examples). Further, we have proven this calculus to be complete *relative to* the handling of differential equations (Theorem 2.3). Complementarily, in this chapter, we address the question how sophisticated differential constraints themselves can be specified and verified in a way that lifts to hybrid sys-

Table 3.1 Comparison of DAL with DA-programs versus $d\mathcal{L}$ with hybrid programs

	$d\mathcal{L}$/hybrid programs	DAL/DA-programs
expressive power	single assignments $x := 1$ differential equations $x_1' = d_1, x_2' = d_2$	propositional/quantified DJ-constraints $x > 0 \rightarrow \exists a\,(a < 5 \wedge x := a^2 + 1)$ propositional/quantified DA-constraints $\exists \omega \leq 1\,(d_1' = -\omega d_2 \wedge d_2' = \omega d_1) \vee d_1' \leq d_2' \leq 2d_1$
verification technology	substitutions *polynomial* solutions	quantifier elimination and substitutions first-order differential induction
quantifier integration	real-valued free variables, Skolemisation	side deductions
scope of applications	nilpotent dynamics, e.g., trains in $\mathbb{R}^1$	algebraic dynamics and polynomial differential constraints, e.g., curved aircraft flight

tems, and how these techniques can be integrated seamlessly into a logic. In this chapter, we thus show how properties of systems with more complicated dynamics can be expressed and proven.

To this end, we design differential-algebraic programs as the *first-order completion* of hybrid programs, and we augment both the logic and the calculus with means for handling DA-constraints. In particular, we extend our logic $d\mathcal{L}$ to the logic DAL with general first-order differential constraints plus first-order jump formulas and introduce differential induction for verifying differential-algebraic programs. Specifically, the continuous evolutions which can be handled by differential induction are strictly more expressive than those that previous calculi [306, 270, 97] or the $d\mathcal{L}$ calculus are able to handle. DAL even supports differential-algebraic equations [132, 187]. Consequently, the DAL calculus can verify much more general scenarios, including the dynamics of aircraft manoeuvres or the dynamics of systems with disturbances, which were out of scope for approaches that require polynomial solutions [125, 228]. Table 3.1 summarises the differences in syntactic expressiveness, discrete and continuous verification technology, arithmetic quantifier integration approach, and overall scope of applicability. The DAL extensions presented in this chapter are both complementary to and compatible with our $d\mathcal{L}$ calculus extensions for integrating arithmetic as presented in Chap. 2.

Contributions

The first contribution of this chapter is the generalised differential-algebraic dynamic logic DAL for differential-algebraic programs as the first-order completion of hybrid programs. DAL provides a uniform semantics and a concise language for specifying and verifying correctness properties of general hybrid systems with sophisticated (possibly quantified) first-order dynamics. The main contribution is a verification calculus for DAL including uniform proof rules for differential induction along first-order differential-algebraic constraints with differential invariants, differential variants, and differential strengthening. Our main theoretical contribution is our analysis of the deductive power of differential induction for classes of differential invariants. As an applied contribution, we introduce a generalised tangential roundabout manoeuvre in air traffic control and demonstrate the capabilities of our approach by verifying collision avoidance in the DAL calculus. To the best of our knowledge, this is the first formal proof for safety of the hybrid dynamics of an aircraft manoeuvre with curved flight dynamics and the first sound verification result for collision avoidance with curved aircraft dynamics.

3.1.1 Related Work

Most verification approaches for hybrid systems follow the paradigm of model checking for hybrid automata and use approximations or abstraction refinement, e.g., [156, 77, 21], because reachability is undecidable for hybrid automata [156]. We have shown in previous work [238] that even reachability problems for fairly restricted classes of single continuous transitions are not decidable using numerical computations. Thus, we follow a purely symbolic approach in this book.

Invariants of Hybrid Systems

Several authors [274, 269, 251, 252] argue that invariant techniques scale to more general dynamics than explicit reach-set computations or techniques that require solutions of the differential equations [125, 228, 231, 233]. Among them, there are model checking approaches [274, 269] that use equational polynomial invariants based on Gröbner basis computations. Still, the approach of Rodríguez-Carbonell and Tiwari [269] requires closed-form solutions and is restricted to linear dynamics. The major limitation of these approaches [274, 269], however, is that they only work for equational invariants of fully equation-definable hybrid systems, including equational initial sets and switching surfaces. Yet, this assumes highly regular systems without tolerances and only works for null sets. In practise, the set of initial states usually does not have measure zero, though. A thorough analysis of collision avoidance manoeuvres, for instance, should consider all initial flight paths in free flight instead of just a single restricted position corridor. In train control, the relevant constraints and initial regions are non-equational.

Prajna et al. [251, 252] have generalised Lyapunov functions to barrier certificates, i.e., a function B decreasing along the dynamics whose zero set separates initial from unsafe states. Further, they focus on stochastic extensions. DAL provides barrier certificates as a special case using $B \leq 0$ as a differential invariant. In a similar vein, criticality functions [91] generalise Lyapunov-functions from stability to safety, which DAL provides as a special case of differential invariants.

We generalise purely equational invariants [274, 269] and single polynomial expressions [274, 251, 252, 91] to general differential induction with real arithmetic formulas. In practise, such more general differential invariants are needed for verifying sophisticated hybrid systems, including aircraft manoeuvres. Further, unlike other approaches [274, 269, 251, 252], DAL leverages the full deductive power of logic, combining differential induction with discrete induction to lift these proof techniques uniformly to hybrid systems. In addition, dynamic logic can be used to prove sophisticated statements involving quantifier and modality alternations for parametric verification [231]. Finally, the DAL calculus supports combinations with differential variants for liveness properties and combinations with differential strengthening, which we show to be crucial in verifying realistic aircraft manoeuvres.

Air Traffic Control Verification

In air traffic control, Tomlin et al. [293] analyse competitive aircraft manoeuvres game-theoretically using Hamilton-Jacobi-Isaacs partial differential equations. They derive saddle solutions for purely angular or purely linear control actions. They propose roundabout manoeuvres and give bounded-time verification results for trapezoidal straight-line approximations. Our symbolic techniques avoid exponential state space discretisations that are required for complicated PDEs and are thus more scalable for automation. Further, we handle fully parametric cases, even for more complicated curved flight dynamics.

Hwang et al. [171] have presented a straight-line aircraft conflict avoidance manoeuvre that involves optimisation over complicated trigonometric computations, and validate it on random numerical simulation. They show examples where the decisions of the manoeuvre change only slightly for small perturbations. Hwang et al. do not, however, prove that their proposed manoeuvre is safe with respect to actual hybrid flight dynamics.

Dowek et al. [104] and Galdino et al. [129] consider straight-line manoeuvres and formalise geometrical proofs in PVS. As in the work of Hwang et al. [171], they do not, however, consider curved flight paths nor verify actual hybrid dynamics but work with geometrical meta-level reasoning instead.

In all these approaches [104, 129, 171], it remains to be proven separately that the geometrical meta-level considerations actually fit the hybrid dynamics and flight equations. In contrast, our approach directly works for the hybrid flight dynamics and we verify roundabout manoeuvres with curves instead of straight-line manoeuvres with non-flyable instant turns only. A few approaches [92, 203] have been undertaken to model check discretisations of roundabout manoeuvres, which indicate avoidance of orthogonal collisions. However, the counterexamples found by our model checker in previous work [238] show for these manoeuvres that collision avoidance does not extend to other initial flight paths.

3.1.2 Structure of This Chapter

In Sects. 3.2 and 3.3, we introduce syntax and semantics of the differential-algebraic logic DAL. In Sect. 3.4, we introduce tangential roundabout manoeuvres in air traffic control as a case study and running example. Further, we introduce a sequent calculus with differential induction for DAL in Sect. 3.5 and prove soundness in Sect. 3.6. We show extensions of differential induction techniques in Sect. 3.7. We exploit differential induction techniques for differential monotonicity relaxations in Sect. 3.8. We prove relative completeness of the DAL calculus in Sect. 3.9 and compare the deductive strength of differential invariants in Sect. 3.10. Using the DAL calculus, we prove, in Sect. 3.11, safety of the tangential roundabout manoeuvre in air traffic control. Finally, we draw conclusions and discuss future work in Sect. 3.12.

3.2 Syntax of Differential-Algebraic Logic

In this section, we introduce the *differential-algebraic logic* (DAL) as a specification and verification logic for *differential-algebraic programs* (DA-programs). DA-programs constitute an elegant and uniform model for hybrid systems with expressive dynamics, including differential-algebraic equations, differential inequalities, and disturbance in the dynamics. We start with an informal introduction that motivates the definitions to come. DA-programs have three basic characteristics, as follows.

Discrete jump constraints. Discrete transitions, which can possibly lead to discontinuous change, are represented as discrete jump constraints (*DJ-constraints*), i.e., first-order formulas with instantaneous assignments of values to state variables as additional atomic formulas. DJ-constraints specify what new values the respective state variables assume by an instant change. For instance, $d_1 := -d_2$ specifies that the value of variable d_1 is changed to the value of $-d_2$. Multiple discrete changes can be combined conjunctively ($\wedge$) with simultaneous effect, for instance, $d_1 := -d_2 \wedge d_2 := d_1$, which assigns the previous value of $-d_2$ to d_1 and, simultaneously, the previous value of d_1 to d_2. This operation instantly rotates the vector $d = (d_1, d_2)$ by $\pi/2$ to the left. Using $d := d^\perp$ as a short vectorial notation for this jump, the DJ-constraint $(d_1 > 0 \to d := d^\perp) \wedge (d_1 \leq 0 \to d := -d^\perp)$ specifies that the direction of the rotation depends on the initial value of d_1. Finally, the DJ-constraint $\exists a\,(\omega := a^2 \wedge a < 5)$ assigns the square of *some* number a less than 5 to ω.

Differential-algebraic constraints. Continuous dynamics is represented with differential-algebraic constraints (*DA-constraints*) as evolution constraints, i.e., first-order formulas with differential symbols x', e.g., in differential equations or inequalities. DA-constraints specify how state variables change continuously over time. For instance, $x_1' = d_1 \wedge x_2' = d_2$ says that the system evolves continuously by moving the vector $x = (x_1, x_2)$ in direction $d = (d_1, d_2)$ along the differential equation system $(x_1' = d_1, x_2' = d_2)$. Likewise, $d_1' = -\omega d_2 \wedge d_2' = \omega d_1 \wedge d_1 \geq 0$ specifies that the vector d is rotating continuously with angular velocity ω, so that (in conjunction with $x_1' = d_1 \wedge x_2' = d_2$), the direction in which point x is heading changes over time. By adding $d_1 \geq 0$ conjunctively to the DA-constraint, we express that the curving will only be able to continue while $d_1 \geq 0$. This evolution will have to stop before $d_1 < 0$. The evolution is impossible altogether if $d_1 \geq 0$ already fails to hold initially. DA-constraints are first-order and can have quantifiers.The quantified DA-constraint $\exists \omega\, (d_1' = -\omega d_2 \wedge d_2' = \omega d_1 \wedge -1 \leq \omega \leq 1)$ characterises rotation with *some* angular velocity $-1 \leq \omega \leq 1$, which may even change over time, in contrast to $d_1' = -\omega d_2 \wedge d_2' = \omega d_1 \wedge -1 \leq \omega \leq 1 \wedge \omega' = 0$ or constraint $d_1' = -\omega d_2 \wedge d_2' = \omega$, where ω is not allowed to change.

Differential-algebraic programs. As an operational model for hybrid systems, DJ-constraints and DA-constraints, which represent general discrete and continuous transitions, respectively, can be combined to form a DA-program using regular expression operators ($\cup, ^*, ;$) of regular discrete dynamic logic [149] as control structure. For example, $\omega := 1 \cup \omega := -1$ describes a controller that can choose to set angular velocity ω either to a left or to a right curve by a nondeterministic choice ($\cup$). Similarly, sequential composition $\omega := \omega + 1; d_1' = -\omega d_2 \wedge d_2' = \omega d_1$ says that the system first increases its angular velocity ω by a discrete transition and then switches to a mode in which it follows a continuous rotation with this angular velocity.

Discussion

Not all constraints involving $x := \theta$ or x' qualify as reasonable ways of characterising elementary system transitions. Unlike positive occurrences, negative occurrences of assignments such as in $\neg(x := 5)$ are somewhat pointless, because they impose no meaningful transition constraints on what new value x actually assumes (but only on what value is *not* assigned to x). Likewise, negative occurrences of differential constraints as in $\neg(x' = 5)$ would be pointless as they do not constrain the overall evolution but allow arbitrary transitions.

Further, we disallow duplicate constraints that constrain the same variable in incompatible ways at the same time as, e.g., in $x := 2 \wedge x := 3$ or $x' = 2 \wedge x' = 3$. At any state during a system evolution, variable x can only assume one value at a time, not both 2 and 3 at once. Similarly, variables cannot evolve with contradictory slopes at the same time for any positive duration.

Finally, $\forall a\, (x := a)$ would be equivalent to false, because it is impossible to assign all possible choices for a (hence all reals) simultaneously to x, which can only assume one value at a time. Likewise, no interesting evolutions are possible along $\forall a\, (x' = a)$, because x' can only equal one real value at a time. Dually, $\exists a\, (a := \theta)$ is equivalent to true, because the DJ-constraint imposes no constraints, nor has any visible effects (the scope of the quantified a ends with the DJ-constraint). The situation with $\exists a\, (a' = \theta)$ is similar.

Even though semantics and proof rules for all these cases could still be defined, the respective transitions are degenerate and their technical handling is not very illuminating. Hence, we define DJ-constraints and DA-constraints to avoid these insignificant cases altogether. Note that the syntactical restrictions are non-essential but simplify the presentation by allowing us to focus on the interesting cases instead.

3.2.1 Terms

To simplify the presentation, we use side deduction rules for quantifiers in this chapter. The free-variable calculus rules from Chap. 2 are compatible with the findings in this chapter, but we refrain from using them formally in this chapter to keep things simple. Consequently, we do not need to distinguish between free logical variables from V and free state variables from Σ. Thus, we do not distinguish Σ and V here at all.

The formulas of DAL are built over a signature Σ of real-valued function and predicate symbols. The signature Σ contains the usual function and predicate symbols for real arithmetic, $+,-,\cdot,/,=,\leq,<,\geq,>$, and number symbols such as $0,1$. State variables are represented as real-valued function symbols of arity zero (constants) in Σ. These state variables are *flexible* [37], i.e., their interpretation can change from state to state while following the transitions of a DA-program. Observe that there is no need to distinguish between discrete and continuous variables in DAL. The set $\mathrm{Term}(\Sigma)$ of *terms* is defined as in classical first-order logic, yielding rational expressions over the reals. The set of formulas of first-order logic is defined as usual (Definition 2.2 and App. A), giving first-order real arithmetic.

Although we are primarily interested in polynomial cases, our techniques generalise to the presence of division. Yet to avoid partiality in the semantics, we only allow use of p/q when $q \neq 0$ is present or ensured. Essentially, we assume constraints ϕ containing a term of the form p/q to mean an appropriate constraint like $\phi \land \neg(q=0)$. Note that, in a certain sense, divisions cause less difficulties for the calculus than for the semantics. Particularly, our calculus uses indirect means of differential induction to conclude properties of solutions of DA-constraints, thereby avoiding the need to handle singularities in these solutions explicitly, as caused by divisions by zero.

3.2.2 Differential-Algebraic Programs

We build DA-programs with first-order discrete jumps and first-order differential-algebraic constraints as primitive operations, which interact using regular control structure by regular-expression-style operators ($;\ \cup\ ,^*$). Reflecting the discussion before Sect. 3.2.1, we characterise reasonable occurrences for changes like $x := \theta$ or x'

in these constraints as follows. We call a formula G an *affirmative subformula* of a first-order formula F iff:

1. G is a positive subformula of F, i.e., it occurs with an even number of negations, and
2. no variable y that occurs in G is in the scope of a universal quantifier $\forall y$ of a positive subformula of F (or in the scope of an existential quantifier $\exists y$ of a negative subformula of F).

Discrete Jump Constraints

Definition 3.1 (Discrete jump constraint). A *discrete jump constraint (or DJ-constraint)* is a formula $\mathscr{J}$ of first-order real arithmetic over Σ with additional atomic formulas of the form $x := \theta$ where $x \in \Sigma$, $\theta \in \mathrm{Term}(\Sigma)$. The latter are called assignments and are only allowed in affirmative subformulas of DJ-constraints that are not in the scope of a quantifier for x of $\mathscr{J}$. A DJ-constraint without assignments is called *jump-free*. A variable x is (possibly) *changed* in $\mathscr{J}$ iff an assignment of the form $x := \theta$ occurs in $\mathscr{J}$.

The effect of $(x_1 := \theta_1 \wedge .. \wedge x_n := \theta_n \wedge x_1 > 0) \vee (x_1 := \vartheta_1 \wedge .. \wedge x_n := \vartheta_n \wedge x_1 < 0)$ is to simultaneously change the interpretations of the variables x_i to the respective θ_i if $x_1 > 0$, and to change the x_i to ϑ_i if $x_1 < 0$. If neither case applies ($x_1 = 0$), the DJ-constraint evaluates to false as no disjunct applies, so no jump is possible at all, which will prevent the system from continuing any further. In particular, a jump-free DJ-constraint such as $x \geq y$ corresponds to a test. It completes without changing the state if, in fact, $x \geq y$ holds true in the current state, and it aborts system evolution otherwise (deadlock). Especially, unlike the assignment $x := \theta$, which changes the value of x to that of θ, the test $x = \theta$ fails by aborting the system evolution if x does not already happen to have the value θ. If cases overlap, as in $(x := x-1 \wedge x \geq 0) \vee x := 0$, either disjunct can be chosen to take effect by a nondeterministic choice.

Quantifiers within DJ-constraints express *unbounded discrete nondeterministic choices*. For instance, the following quantified DJ-constraint assigns *some* vector $u \in \mathbb{R}^2$ to e such that the rays spanned by $d = (d_1, d_2)$ and $u = (u_1, u_2)$ intersect:

$$\exists u_1 \exists u_2 \, (e_1 := u_1 \wedge e_2 := u_2 \wedge \exists \lambda{>}0 \, \exists \mu{>}0 \, (\lambda d_1 = \mu u_1 \wedge \lambda d_2 = \mu u_2)).$$

We informally use *vectorial notation* when no confusion arises. Using vectorial quantifiers, equations, arithmetic, and assignments, the latter DJ-constraint simplifies to:

$$\exists u \, (e := u \wedge \exists \lambda{>}0 \, \exists \mu{>}0 \, \lambda d = \mu u).$$

Example 3.1 (DJ-constraints). Examples of DJ-constraints include:

- $d_1 := -d_2$

- $d_1 := -d_2 \land d_2 := d_1$ — simultaneous effect is an instant left rotation by $\frac{\pi}{2}$
- $(d_1 > 0 \to d_1 := -d_2 \land d_2 := d_1) \land (d_1 \leq 0 \to d_1 := d_2 \land d_2 := -d_1)$ — effect depends on the sign of d_1
- $\exists a\,(\omega := a^2 \land a < 5)$ — quantified nondeterministic effect
- $d_1 > 0 \to \exists a\,(a < 5 \land d_1 := a^2 + 1)$ — but this DJ-constraint is inhomogeneous, it does not specify what happens if $d_1 \leq 0$.

The following cases are not allowed as DJ-constraints:

- $d_1 := -d_2 \land d_1 := 0$ — incompatible jump, because d_1 cannot assume both values at once
- $\neg(d_1 := 5)$ — does not specify what really *is* assigned to d_1, only what is not
- $\forall a\,(\omega := a^2)$ — ω cannot assume all those values a^2 at once for all a
- $\exists a\,(a := d_1)$ — is just equivalent to *true*, because it has no visible effects or constraints (the scope of the quantified a ends with the DJ-constraint)

It is not impossible to give a reasonable semantics to these disallowed cases and handle them in proof rules. Yet it would complicate the technical treatment unnecessarily and would distract from the important aspects. □

Differential-Algebraic Constraints

Definition 3.2 (Differential-algebraic constraints). A *differential-algebraic constraint (DA-constraint)* is a formula $\mathscr{D}$ of first-order real arithmetic over $\Sigma \cup \Sigma'$ in which symbols of Σ' only occur in affirmative subformulas that are not in the scope of a quantifier of $\mathscr{D}$ for that symbol. Here, Σ' is the set of all *differential symbols* $x^{(n)}$ with $n \in \mathbb{N}$ for state variables $x \in \Sigma$. A DA-constraint without differential symbols $x^{(n)}$ for $n \geq 1$ is called *non-differential.* A variable x is (possibly) *changed* in $\mathscr{D}$ iff $x^{(n)}$ occurs in $\mathscr{D}$ for an $n \geq 1$.

Syntactically, $x^{(n)}$ is like an ordinary function symbol of arity 0, but only allowed to occur within DA-constraints, and not in any other formula. The intended semantics of a differential symbol $x^{(n)}$ is to denote the nth time derivative of x, which is used to form differential equations (or differential inequalities). We write x' for $x^{(1)}$ and x'' for $x^{(2)}$ and, sometimes, $x^{(0)}$ for the non-differential symbol x. The (partial) *order* $\mathrm{ord}_x \mathscr{D}$ of a DA-constraint $\mathscr{D}$ *in* x is the highest order $n \in \mathbb{N}$ of a differential symbol $x^{(n)}$ occurring in $\mathscr{D}$, or is not defined if no such $x^{(n)}$ occurs. The notion of order is accordingly for terms.

The effect of a DA-constraint $\mathscr{D}$ is an ongoing continuous evolution respecting the differential and non-differential constraints of $\mathscr{D}$ during the whole evolution. For instance, the effect of $(x' = \theta \land x > 0) \lor (x' = -x^2 \land x < 0)$ is that the system evolves along $x' = \theta$ while $x > 0$, yet evolves along $x' = -x^2$ when $x < 0$. This evolution can stop at any time but is never allowed to enter the region where neither case applies anymore ($x = 0$). There also is no transition at all if $x = 0$ holds in the beginning, because no disjunct applies in the initial state then.

More generally, the differential constraints of $\mathscr{D}$ describe how the valuations of the respective state variables change continuously over time while following $\mathscr{D}$. The non-differential constraints of $\mathscr{D}$ can be understood to express evolution domain restrictions or invariant regions of these evolutions for which the differential equations apply or within which the evolution resides. For instance, in the DA-constraint $d_1' = -\omega d_2 \wedge d_2' = \omega d_1 \wedge d_1 \geq 0$, differential equations $d_1' = -\omega d_2$ and $d_2' = \omega d_1$ describe the change and $d_1 \geq 0$ describes the invariance region or maximal domain of evolution. Overlapping cases are resolved as in DJ-constraints, i.e., by nondeterministic choice. Likewise, a DA-constraint where no case applies aborts the system evolution as it does not satisfy the DA-constraint. Hence, non-differential DA-constraints and jump-free DJ-constraints are both equivalent to pure tests ($?\chi$ from Chap. 2). Except for such tests, we need to distinguish DA-constraints from DJ-constraints: Only DA-constraints can have evolutions of nonzero duration and only DJ-constraints can lead to discontinuous changes.

Quantifiers within DA-constraints express *continuous nondeterministic choices*. For example, constraint $\exists u\,(d_1' = -(\omega+u)d_2 \wedge d_2' = (\omega+u)d_1 \wedge -0.1 \leq u \leq 0.1)$ expresses that the system follows a continuous evolution in which, at each time, the differential equations are respected for *some* choice of u in $-0.1 \leq u \leq 0.1$. In particular, the choice of u can be different at each time so that u amounts to a bounded nondeterministic disturbance during the rotation in the above DA-constraint.

Example 3.2 (DA-constraints). Examples of DA-constraints include:

- $x_1' = d_1 \wedge x_2' = d_2$
- $x_1' = d_1 \wedge x_2' = d_2 \wedge d_1' = -\omega d_2 \wedge d_2' = \omega d_1$
- $d_1' = -\omega d_2 \wedge d_2' = \omega d_1 \wedge d_1 \geq 0$
- $(d_1 > 0 \to d_1' = -d_2 \wedge d_2' = d_1) \wedge (d_1 \leq 0 \to d_1' = d_2 \wedge d_2' = -d_1)$
- $x_1' \leq d_1 \wedge x_2' \leq d_2$
- $x_1' = d_1 \wedge x_2' \leq d_2 \wedge d_1' < 2 \wedge d_2' \geq 0$
- $\exists \omega\,(d_1' = -\omega d_2 \wedge d_2' = \omega d_1 \wedge -1 \leq \omega \leq 1)$
- $\exists \omega\,(d_1' = -\omega d_2 \wedge d_2' = \omega d_1 \wedge -1 \leq \omega \leq 1) \vee (d_1' \leq d_2' \leq 2d_1)$
- $d_1 > 0 \to x_1' = d_1 \wedge x_2' = d_2 \wedge d_1' = -\omega d_2$ — but this DA-constraint is inhomogeneous, it does not specify what happens if $d_1 \leq 0$

The following cases are not allowed as DA-constraints:

- $d_1' = -d_2 \wedge d_1' = 1$ — incompatible slope, because d_1 cannot evolve with two different slopes at once
- $\neg(d_1' = 5)$ — does not specify what the slope of d_1 really *is*, only what it is not
- $\forall \omega\,(d_1' = -\omega d_2 \wedge d_2' = \omega d_1)$ — d_1 cannot have all those slopes at once
- $\exists a\,(a' = d_1)$ — is just equivalent to *true*, because it has no visible effects or constraints (the scope of the quantified a ends with the DA-constraint)

It is not impossible to give a reasonable semantics to these disallowed cases and handle them in proof rules. But it would complicate the technical treatment unnecessarily. □

When using constraint formulas to characterise system transitions, we face the usual frame problem: Typically, one does not expect variables to change their values unless the respective constraint explicitly specifies how. In this chapter, we indicate constant variables explicitly so that no confusion arises. In practical applications, however, it can be quite cumbersome to have to specify $z := z$ or $z' = 0$ explicitly for all variables z that are not supposed to change in a DJ-constraint or DA-constraint, respectively. To account for that, we will define the DAL semantics so that variables that are not changed by a DJ-constraint or DA-constraint keep their value. Since free nondeterministic change of variable y is expressible using $\exists a\,(y := a)$ or $\exists a\,(y' = a)$, respectively, we expect the changes of all changed variables to be specified explicitly in all cases of the constraints to improve readability:

Definition 3.3 (Homogeneous constraints). A DA-constraint or DJ-constraint $\mathscr{C}$ is called *homogeneous* iff, in each of the disjuncts of a disjunctive normal form of $\mathscr{C}$, every changed variable of $\mathscr{C}$ is changed exactly once.

Note that Lemma 3.3 from Sect. 3.5.2 will show that DA-constraints are equivalent to their disjunctive normal forms and, hence, the notion of homogeneity is well-defined. Throughout the chapter, we assume that all DA-constraints and DJ-constraints are homogeneous, thereby ensuring that all changed variables receive a new value in all cases of the respective constraint (or stay constant because they are changed nowhere in the constraint) and that no change conflicts occur.

Hence, variable y does not change during the DA-constraint $x' = -x \wedge x \geq y$ but works as a constant lower bound for the evolution of x, because no differential symbol $y^{(n)}$ with $n \geq 1$ occurs so that $y' = 0$ is assumed implicitly. If y is intended to vary, but its variation is not specified by a differential equation, because y varies according to some algebraic relation with x, then quantified DA-constraints can be used to represent such *differential-algebraic equations* [132]. For instance, the differential-algebraic equation $x' = -x, y^2 = x$, in which $y^2 = x$ is an algebraic variational constraint specifying how y changes over time, is expressible as the DA-constraint $x' = -x \wedge \exists u\,(y' = u \wedge y^2 = x)$. There, the quantified differential constraint on y essentially says that y can change arbitrarily (with arbitrary disturbance u as slope), but only so that it always respects the relation $y^2 = x$.

Differential-Algebraic Programs

Now we can define DA-programs as regular combinations of DJ-constraints and DA-constraints.

Definition 3.4 (Differential-algebraic programs). The set DA-program(Σ) of *differential-algebraic programs*, with typical elements α, β, is inductively defined as the smallest set such that:

- If $\mathscr{J}$ is a DJ-constraint over Σ, then $\mathscr{J} \in$ DA-program(Σ).
- If $\mathscr{D}$ is a DA-constraint over $\Sigma \cup \Sigma'$, then $\mathscr{D} \in$ DA-program(Σ).
- If $\alpha, \beta \in$ DA-program(Σ) then $(\alpha \cup \beta) \in$ DA-program(Σ).

Table 3.2 Statements and (informal) effects of differential-algebraic programs

DA-program	Operation	Effect
$\mathscr{J}$	discrete jump	jump constraint with assignments holds for discrete jump
$\mathscr{D}$	diff.-alg. flow	differential-algebraic constraint holds during continuous flow
$\alpha;\beta$	seq. composition	DA-program β starts after DA-program α finishes
$\alpha\cup\beta$	nondet. choice	choice between alternative DA-programs α or β
α^*	nondet. repetition	repeats DA-program α n-times for any $n\in\mathbb{N}$

- If $\alpha,\beta \in$ DA-program(Σ) then $(\alpha;\beta) \in$ DA-program(Σ).
- If $\alpha \in$ DA-program(Σ) then $(\alpha^*) \in$ DA-program(Σ).

Choices $\alpha\cup\beta$ are used to express behavioural alternatives between α and β, i.e., the system follows either α or β. In particular, the difference between the DA-constraint $\mathscr{D}\vee\mathscr{E}$ and the DA-program $\mathscr{D}\cup\mathscr{E}$ is that the system has to commit to one choice of either $\mathscr{D}$ or $\mathscr{E}$ in $\mathscr{D}\cup\mathscr{E}$, but it can switch back and forth multiple times between disjunct $\mathscr{D}$ and $\mathscr{E}$ in $\mathscr{D}\vee\mathscr{E}$. The sequential composition $\alpha;\beta$ says that DA-program β starts executing after α has finished (β never starts if α does not terminate, e.g., due to a failed test in α). Observe that, as with repetitions, continuous evolutions within α can take more or less time. This nondeterminism is inherent in hybrid systems and as such reflected in DA-programs. Additional restrictions on the permitted duration of evolutions can simply be specified using auxiliary clocks, i.e., variables of derivative $\tau'=1$. For instance, $\tau:=0; x'=-x^2\wedge\tau'=1\wedge\tau\le 5; ?\tau\ge 2$ specifies that the system only follows those evolutions along $x'=-x^2$ that take at most five (conjunct $\tau\le 5$) but at least two time units (subsequent test $?\tau\ge 2$). Repetition α^* is used to express that the hybrid process α repeats any number of times, including zero.

Table 3.2 summarises the statements and (informal) effects of DA-programs. DA-programs still form a Kleene algebra [182] like hybrid programs from Chap. 2, but the primitive programs are more powerful. Further tests $?\chi$ are not necessary, because they can be defined by a jump-free DJ-constraint χ or by a non-differential DA-constraint χ.

Example 3.3 (Train control with disturbance). Recall the hybrid program for a train control system from d$\mathcal{L}$ formula (2.7) in Sect. 2.4:

$$\begin{aligned} &\psi \to [(ctrl;drive)^*]z\le m \qquad (2.7^*)\\ \text{where } ctrl &\equiv (?m-z\le s; a:=-b)\cup(?m-z\ge s; a:=A),\\ drive &\equiv \tau:=0; (z'=v, v'=a, \tau'=1\,\&\,v\ge 0\wedge\tau\le\varepsilon). \end{aligned}$$

With a minor notational variation, this hybrid program is a DA-program:

$$\begin{aligned} &\psi \to [(ctrl;drive)^*]z\le m\\ \text{where } ctrl &\equiv (m-z\le s\wedge a:=-b)\cup(m-z\ge s\wedge a:=A),\\ drive &\equiv \tau:=0; (z'=v\wedge v'=a\wedge\tau'=1\wedge v\ge 0\wedge\tau\le\varepsilon). \end{aligned}$$

Note also that we have decided to merge the test $?m - z \leq s$ with the subsequent assignment $a := -b$ here to form the DJ-constraint $m - z \leq s \wedge a := -b$. This simplification is non-essential and just used to show the expressiveness of DA-programs. In fact, we could also replace *ctrl* by a single DJ-constraint with the same semantics:

$$(m - z \leq s \wedge a := -b) \vee (m - z \geq s \wedge a := A).$$

The simple program assumes ideal-world dynamics with perfect control of the acceleration a, which takes effect exactly by the differential equation $z' = v, v' = a$. Yet this is not quite realistic. In reality, there are more aspects to train control that have an important impact on the dynamics, including wind, track conditions, slope of the track, mass, etc. In order to take these effects into account without having to build a full physical model of all parts, we replace the differential equation by a differential inequality, which allows for minor discrepancies. We thus replace the differential equation DA-constraint $z' = v \wedge v' = a \wedge v \geq 0$ with the differential inequality DA-constraint

$$z' = v \wedge a - l \leq v' \leq a + u \wedge v \geq 0 \tag{3.1}$$

for bounds $l > 0$ and $u \geq 0$ on how much the actual acceleration may deviate from the set acceleration a. See Fig. 3.1 for an illustration of how the dynamics can deviate from the ideal-world dynamics in the presence of a disturbance that can change over time. With this generalisation the program can only be represented as a DA-

Fig. 3.1 Controllability violated in the presence of disturbance

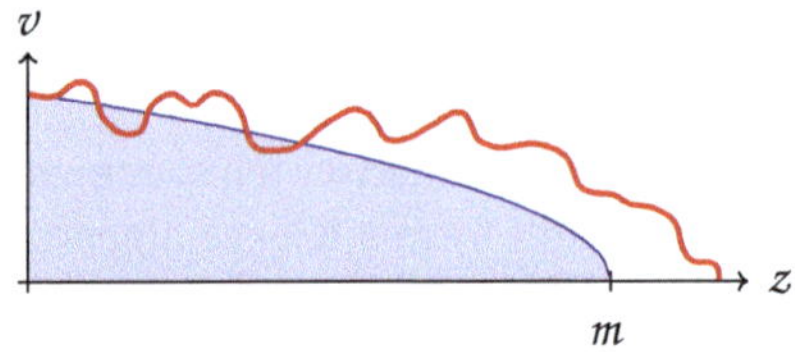

program, and no longer as a hybrid program. Now, the formula from (2.7) will no longer be a d$\mathcal{L}$ formula, because DA-programs are not allowed in d$\mathcal{L}$. Instead, we will see that it is a well-formed DAL formula in Sect. 3.2.3.

Let us continue this line and replace the original differential equation by a quantified DA-constraint with a quantified disturbance as input to the dynamics:

$$z' = v \wedge \exists d\, (v' = d \wedge a - l \leq d \wedge d \leq a + u) \wedge v \geq 0. \tag{3.2}$$

This DA-constraint expresses that, at every point in time, there is a disturbance d affecting the actual dynamics. Yet this disturbance is bounded at any point by $a - l$ and $a + u$. Again, we obtain a DA-program instead of a hybrid program and again the resulting formula is no longer a d$\mathcal{L}$ formula. One question is what the relation is of those two different ways of generalising the ideal-world dynamics. It will turn out in Sect. 3.5.3 that the DA-constraints (3.1) and (3.2) are, in fact, equivalent, and so are the resulting DA-programs. □

Classification of Differential-Algebraic Programs

DA-programs give rise to an elegant syntactic hierarchy of discrete, continuous, and hybrid systems; see Table 3.3. Purely conjunctive DA-constraints correspond to continuous dynamical systems [279]. DA-constraints with disjunctions correspond to switched continuous dynamical systems [55]. DA-programs without DA-constraints correspond to discrete dynamical systems or, when restricted to domain $\mathbb{N}$ (which is definable in DAL), to discrete while programs [149]. Regular combinations of DJ-constraints form a complete basis of discrete programs [149]. Finally, general DA-programs correspond to (first-order generalisations of) hybrid dynamical systems [55, 156, 97].

Table 3.3 Classification of differential-algebraic programs and correspondence to dynamical systems

DA-program class	Dynamical systems class
conjunctive DA-constraints	continuous dynamical systems
DA-constraints	switched continuous dynamical systems
no DA-constraints	discrete dynamical systems
no DA-constraints, over $\mathbb{N}$	discrete while programs
general DA-programs	hybrid dynamical systems
	+ (generalised to first-order dynamics)

3.2.3 Formulas of Differential-Algebraic Logic

The set of *formulas* of DAL is defined as common in first-order dynamic logic [149]. They are built using propositional connectives and, in addition, if α is a DA-program and ϕ is a DAL formula, then $[\alpha]\phi, \langle\alpha\rangle\phi$ are DAL formulas. The intuitive reading of $[\alpha]\phi$ is that every run of DA-program α leads to states satisfying ϕ. Dually, $\langle\alpha\rangle\phi$ expresses that there is at least one run of DA-program α leading to such a state.

Definition 3.5 (DAL formulas). The set $\text{Fml}(\Sigma)$ of DAL *formulas*, with typical elements ϕ, ψ, is inductively defined as the smallest set with:

- If $\theta_1, \theta_2 \in \text{Term}(\Sigma)$ are terms, then $(\theta_1 \geq \theta_2) \in \text{Fml}(\Sigma)$. The definition is accordingly for $=, \leq, <, >$.
- If $\phi, \psi \in \text{Fml}(\Sigma)$, then $\neg\phi, (\phi \wedge \psi), (\phi \vee \psi), (\phi \to \psi) \in \text{Fml}(\Sigma)$.
- If $\phi \in \text{Fml}(\Sigma)$ and $\alpha \in \text{DA-program}(\Sigma)$, then $[\alpha]\phi, \langle\alpha\rangle\phi \in \text{Fml}(\Sigma)$.

Quantifiers in DAL formulas are definable in terms of DA-constraints or quantified DJ-constraints. We consider quantifiers as abbreviations:

$$\forall x\,\phi \equiv [\exists a\, x := a]\phi \equiv [x' = 1 \vee x' = -1]\phi$$
$$\exists x\,\phi, \equiv \langle\exists a\, x := a\rangle\phi \equiv \langle x' = 1 \vee x' = -1\rangle\phi.$$

Table 3.4 Operators and meaning in differential-algebraic dynamic logic (DAL)

DAL	Operator	Meaning
$\theta_1 \geq \theta_2$	comparison	true iff denotation of θ_1 is greater or equal that of θ_2
$\neg\phi$	negation / not	true if ϕ is false
$\phi \wedge \psi$	conjunction / and	true if both ϕ and ψ are true
$\phi \vee \psi$	disjunction / or	true if ϕ is true or if ψ is true
$\phi \rightarrow \psi$	implication / implies	true if ϕ is false or ψ is true
$\phi \leftrightarrow \psi$	bi-implication / equivalent	true if ϕ and ψ are both true or both false
$\forall x\,\phi$	universal quantifier / for all	true if ϕ is true for all values of variable x
$\exists x\,\phi$	existential quantifier / exists	true if ϕ is true for some values of variable x
$[\alpha]\phi$	$[\cdot]$ modality / box	true if ϕ is true after all runs of DA-program α
$\langle\alpha\rangle\phi$	$\langle\cdot\rangle$ modality / diamond	true if ϕ is true after at least one run of DA-program α

The DAL formula $[\exists a\,x\,{:=}\,a]\phi$ considers *all* possibilities of assigning *some* value a to x, which amounts to universal quantification. Likewise, $\langle\exists a\,x\,{:=}\,a\rangle\phi$ considers some such choice, which is existential quantification. Similarly, the indeterminate continuous evolution $x' = 1 \vee x' = -1$ reaches all values, which amounts to the respective quantifier when combined with the appropriate modality. Note here that we can define quantifiers in terms of both DJ-constraints and DA-constraints although the former are discrete and the latter are continuous. The reason for this is that the time that passed during the continuous evolution is not observable in the system as no clock variable is included in the DA-constraint that could measure the progress of time.

With these abbreviations, we summarise the operators of differential-algebraic dynamic logic in Table 3.4. The structure of the logic DAL is mostly identical to the structure of $d\mathcal{L}$ except for the significantly more expressive DA-programs inside modalities.

Example 3.4 (Train control with disturbance). In Example 3.3 we have replaced the differential equation of the train example by more general DA-constraints: the differential inequality (3.1) and the quantified DA-constraint (3.2), respectively. The generalised counterpart of the $d\mathcal{L}$ formula (2.7) for its hybrid program is the following DAL formula for its DA-program:

$$\begin{aligned} &\psi \rightarrow [(ctrl_d; drive_d)^*]\, z \leq m &(3.3)\\ \text{with } ctrl_d &\equiv (?m-z \leq s; a:=-b) \cup (?m-z \geq s; a:=A)\\ drive_d &\equiv \tau:=0;\\ &\quad (z' = v \wedge \exists d\,(v' = d \wedge a-l \leq d \wedge d \leq a+u) \wedge \tau' = 1 \wedge v \geq 0 \wedge \tau \leq \varepsilon). \end{aligned}$$

DAL formula (3.3) is more general than $d\mathcal{L}$ formula (2.7), because it makes a safety claim about a more general dynamics, with nondeterministic quantified input d as disturbance. If we prove DAL formula (3.3) then this implies validity of $d\mathcal{L}$ formula (2.7), because all behaviour of the hybrid program is a special case of the behaviour of the DA-program (with constant disturbance $d = 0$). Yet, how do we prove the more general DAL formula to be valid? Certainly, we can no longer solve its differential equations, because the DA-constraints now depend on quantified input or differential inequalities. □

One common pattern for representing safety statements about hybrid control loops is to use DAL formulas of the form $\phi \to [(controller; plant)^*]\psi$ for specifying that the system satisfies property ψ whenever the initial state satisfies ϕ. There, the system repeats a controller plant feedback loop, with a DA-constraint *plant* describing the continuous plant dynamics and a discrete DA-program *controller* describing the control decisions. The controller plant interaction repeats as indicated by the repetition star. Still, more general forms of systems and properties can be formulated and verified in DAL as well.

3.3 Semantics of Differential-Algebraic Logic

The semantics of DAL is a Kripke semantics with possible states of a hybrid system as possible worlds, where the accessibility relation between worlds is generated by the discrete or continuous transitions of DA-programs. A potential behaviour of a hybrid system corresponds to a succession of states that contain the observable values of system variables during its hybrid evolution.

3.3.1 Transition Semantics of Differential-Algebraic Programs

Since in this chapter we do not distinguish between free logical variables and constants, the semantics does not need to distinguish states and variable assignments.

A *state* is a map $\nu : \Sigma \to \mathbb{R}$; the set of all states is denoted by $\mathrm{State}(\Sigma)$. The function and predicate symbols of real arithmetic are interpreted as usual.

Definition 3.6 (Valuation of terms). The *valuation* $val(\nu,\cdot)$ *of terms* with respect to state ν is defined by

1. $val(\nu,x) = \nu(x)$ if $x \in \Sigma$ is a variable.
2. $val(\nu,\theta_1+\theta_2) = val(\nu,\theta_1)+val(\nu,\theta_2)$.
3. $val(\nu,\theta_1-\theta_2) = val(\nu,\theta_1)-val(\nu,\theta_2)$.
4. $val(\nu,\theta_1\cdot\theta_2) = val(\nu,\theta_1)\cdot val(\nu,\theta_2)$.
5. $val(\nu,\theta_1/\theta_2) = val(\nu,\theta_1)/val(\nu,\theta_2)$ if $val(\nu,\theta_2) \neq 0$.

Note that we do not need the semantics of θ_1/θ_2 for $val(\nu,\theta_2) = 0$, because we have assumed the presence of constraints ensuring $\neg(\theta_2 = 0)$ for divisions.

Discrete Jump Constraints

The interpretation of discrete jump constraints is defined as in first-order real arithmetic, with the addition of an interpretation for assignment formulas.

Definition 3.7 (Interpretation of discrete jump constraints). The *interpretation* $(\nu,\omega) \models \mathscr{J}$ of DJ-constraint $\mathscr{J}$ for the pair of states (ν,ω) is defined as follows, where $\omega(z) = val(\omega,z) = val(\nu,z) = \nu(z)$ for all variables z that are not changed in $\mathscr{J}$:

1. $(\nu,\omega) \models x := \theta$ iff $val(\omega,x) = val(\nu,\theta)$.
2. $(\nu,\omega) \models \theta_1 \geq \theta_2$ iff $val(\nu,\theta_1) \geq val(\nu,\theta_2)$, and accordingly for $=,\leq,<,>$.
3. $(\nu,\omega) \models \phi \wedge \psi$ iff $(\nu,\omega) \models \phi$ and $(\nu,\omega) \models \psi$. Accordingly for $\neg,\vee,\rightarrow$.
4. $(\nu,\omega) \models \phi \vee \psi$ iff $(\nu,\omega) \models \phi$ or $(\nu,\omega) \models \psi$.
5. $(\nu,\omega) \models \neg\phi$ iff it is not the case that $(\nu,\omega) \models \phi$.
6. $(\nu,\omega) \models \phi \rightarrow \psi$ iff it is not the case that $(\nu,\omega) \models \phi$ or it is the case that $(\nu,\omega) \models \psi$.
7. $(\nu,\omega) \models \forall x\,\phi$ iff $(\nu_x,\omega) \models \phi$ for all states ν_x that agree with ν except for the value of x.
8. $(\nu,\omega) \models \exists x\,\phi$ iff $(\nu_x,\omega) \models \phi$ for some state ν_x that agrees with ν except for the value of x.

Differential-Algebraic Constraints

To give a semantics to DA-constraints, differential symbols $x' \in \Sigma'$ must get a meaning. However, a DA-constraint like $d_1' = -\omega d_2 \wedge d_2' = \omega d_1$ cannot be interpreted in a single state ν, because derivatives are not defined in isolated points. Instead, DA-constraints are constraints that have to hold for an evolution of states over time. Along such a *flow* function $\varphi : [0,r] \to \mathrm{State}(\Sigma)$, DA-constraints can again be interpreted locally by assigning to the formal differential symbol d_1' the analytic time derivative of the value of d_1 along φ at the respective points in time. As we assumed DA-constraints to avoid zero divisions, analytic derivatives are well-defined for $r > 0$ as $\mathrm{State}(\Sigma)$ is isomorphic to a finite-dimensional real space with respect to the finitely many differential symbols occurring in the DA-constraint. We give a uniform definition for all durations $r \geq 0$ and defer the discussion of the understanding for $r = 0$ until the DA-constraint semantics has been presented in full. The philosophy behind hybrid systems is to isolate discontinuities in discrete transitions. Thus we assume that state variables (and their differential symbols, if present) always vary continuously along continuous evolutions over time.

Definition 3.8 (Differential state flow). A function $\varphi : [0,r] \to \mathrm{State}(\Sigma)$ is called *state flow* of duration $r \geq 0$ if φ is componentwise continuous on $[0,r]$, i.e., for all $x \in \Sigma$, $\varphi(\zeta)(x)$ is continuous in ζ. Then, the *differentially augmented state* $\bar{\varphi}(\zeta)$ of φ at $\zeta \in [0,r]$ agrees with $\varphi(\zeta)$ except that it further assigns values to some of the differential symbols $x^{(n)} \in \Sigma'$: If $\varphi(t)(x)$ is n times continuously differentiable in t at ζ, then $\bar{\varphi}(\zeta)$ assigns the nth time derivative $\frac{\mathrm{d}^n \varphi(t)(x)}{\mathrm{d}t^n}(\zeta)$ of x at ζ to differential symbol $x^{(n)} \in \Sigma'$; otherwise the value of $x^{(n)} \in \Sigma'$ is not defined in $\bar{\varphi}(\zeta)$.

For a DA-constraint $\mathscr{D}$, a state flow φ of duration r is called *state flow of the order of* $\mathscr{D}$ iff the value of each differential symbol occurring in $\mathscr{D}$ is defined on $[0,r]$, i.e., $\varphi(\zeta)(x)$ is n times continuously differentiable in ζ on $[0,r]$ for $n = \mathrm{ord}_x \mathscr{D}$.

Example 3.5 (Differential state flow). We want to give a semantics to the DA-constraint $x' = x^3$, for which we have to specify where the dynamics of this DA-constraint evolves to when evolution starts in a state ν. Clearly, we can say what the semantics of the right-hand side of the differential equation is, and how the term x^3 evaluates in any state ν. But what should x' evaluate to? And how do we define when the terms x' and x^3 are equal. Remember: we also have to define when one term is less than the other for giving a semantics to differential inequalities like $x' < x^3$ or $x' \leq x^3$. At a single state ν, which is just an isolated point, derivatives are not defined, so that we cannot really give a meaning to the DA-constraints $x' = x^3$ or $x' \leq x^3$ when looking only at a single point.

Yet along a continuous state flow φ, we can make sense of x'. At any time ζ during the continuous state flow φ, we have values for all the variables, and can thus evaluate $val(\bar{\varphi}(\zeta), x^3)$ based on the value of variable $\bar{\varphi}(\zeta)(x)$ and of any other variables that occur. At time ζ along the flow, however, we can also give a value to x' as the value of the derivative of the value $val(\bar{\varphi}(t), x)$ of x by time t at the point in time ζ; see illustration in Fig. 3.2. We assign this value of the derivative at ζ to

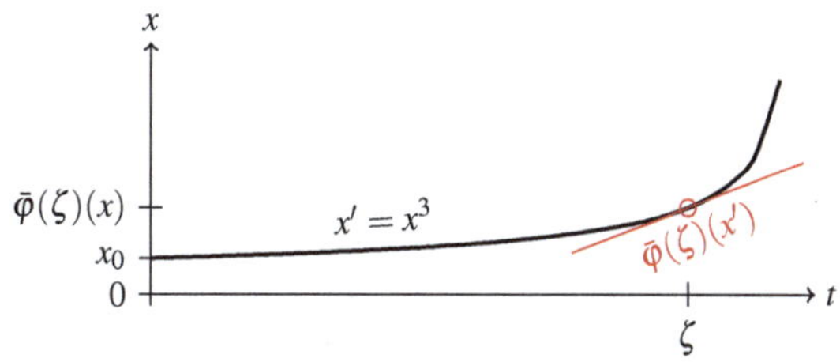

Fig. 3.2 Differential state flow

$\bar{\varphi}(\zeta)(x')$ in the differentially augmented state $\bar{\varphi}$ at time ζ. Intuitively, $\bar{\varphi}(\zeta)(x')$ is determined by considering how the value $val(\bar{\varphi}(\zeta), x)$ of x changes along the flow φ when we change time ζ "only a little bit". With this definition of the differentially augmented states $\bar{\varphi}$, we can easily evaluate DA-constraints like $x' = x^3$ or $x' \leq x^3$ at every point in time ζ by checking if $\bar{\varphi}(\zeta) \models x' = x^3$ or $\bar{\varphi}(\zeta) \models x' \leq x^3$, respectively, hold in the standard valuation of first-order logic, which is the same as checking if $val(\bar{\varphi}(\zeta), x') = val(\bar{\varphi}(\zeta), x^3)$ or $val(\bar{\varphi}(\zeta), x') \leq val(\bar{\varphi}(\zeta), x^3)$, respectively. The same principle can be used to give a semantics to more complicated DA-constraints, just by evaluating the first-order logic (including quantifiers) locally at every point in time along the differentially augmented state flow $\bar{\varphi}(\zeta)$. □

Definition 3.9 (Interpretation of differential-algebraic constraints). The *interpretation of DA-constraint $\mathscr{D}$* with respect to a state flow φ of the order of $\mathscr{D}$ and duration $r \geq 0$ is defined by: $\varphi \models \mathscr{D}$ iff, for all $\zeta \in [0, r]$,

1. $\bar{\varphi}(\zeta) \models_{\mathbb{R}} \mathscr{D}$ using the standard semantics $\models_{\mathbb{R}}$ of first-order real arithmetic (App. A), and
2. $val(\bar{\varphi}(\zeta), z) = val(\bar{\varphi}(0), z)$ for all variables z that are not changed by $\mathscr{D}$.

Observe that, along the state flows for a DA-constraint $\mathscr{D}$, only those variables whose differential symbols occur in $\mathscr{D}$ have to be continuously differentiable to the appropriate order. Quantified variables can change more arbitrarily (even discontinuously) during the evolution. The reason is that the semantics does not directly relate the value of a quantified variable like u in $\exists u\, x' = u^2$ at time ζ with the values that u assumes at later times. In particular, the value chosen for the quantified variable u can be different at every point in time, thereby giving a semantics of disturbance. Quantified variables may be constrained indirectly by their relations, though: In $\exists u\, x' = u^2$, the value of u^2 (but not that of u) also varies continuously over time, because x' varies continuously.

As a consequence of Picard-Lindelöf's theorem, a.k.a. the Cauchy-Lipschitz theorem (Theorem B.2), and using the fact that DAL terms are continuously differentiable on the open domain where divisors are nonzero, the flows of explicit quantifier-free, conjunctive DA-constraints of the form $x_1' = \theta_1 \wedge \cdots \wedge x_n' = \theta_n \wedge \chi$ with non-differential constraint χ are unique (as long as they exist): For each duration and initial value, there is at most one state flow φ respecting the DA-constraint (see Lemma 2.1). Yet, this is *not* the case for disjunctive DA-constraints, differential inequalities, quantified DA-constraints, or DA-constraints in implicit form such as $x'^2 - 1 = 0$, which has solutions $x(t) = x(0) + t$ and $x(t) = x(0) - t$. Finally, a non-differential constraint χ imposes no change but only tests whether χ holds. Hence, without differential constraints, a non-differential DA-constraint χ itself only has constant flows (if any), i.e., $\varphi(\zeta) = \varphi(0)$ for all ζ.

Restrictions of differential state flows to a prefix are again state flows. In particular, for all differential equations, the restriction to the point interval $[0,0]$ yields a trivial flow of no effect. For such point duration $r = 0$, however, derivatives and differentiability are not defined. To admit trivial flows nevertheless, the understanding of a DA-constraint is that its differential terms take no effect for flows of zero duration. That is, for trivial flows, atomic formulas with differential symbols are defined to evaluate to *true* as they occur only positively in DA-constraints. Thus, only the non-differential constraints of $\mathscr{D}$ impose constraints for trivial flows. A state flow of duration zero satisfying $\mathscr{D}$ and starting in some state ν exists iff ν satisfies the non-differential part of $\mathscr{D}$, which acts as a test condition.

Differential-Algebraic Programs

Based on the semantics of DJ-constraints and DA-constraints that we have defined above, we can now define the transition semantics of DA-programs. We define the transition semantics, $\rho(\alpha)$, of a DA-program α, compositionally and denotationally in terms of the semantics of its parts. The semantics of a DA-program is captured by the discrete or continuous transitions that are possible by following this DA-program. For DJ-constraints this transition relation holds for pairs of states that satisfy the jump constraints. For DA-constraints, the transition relation holds for pairs of states that can be interconnected by a differential state flow respecting the DA-constraint.

Definition 3.10 (Transition semantics of differential-algebraic programs). The *valuation, $\rho(\alpha)$ of a DA-program α* is a *transition relation* on states. It specifies which state ω is reachable from a state ν by operations of the hybrid system α and is defined as:

1. $(\nu,\omega) \in \rho(\mathscr{J})$ iff $(\nu,\omega) \models \mathscr{J}$ according to Definition 3.7 when $\mathscr{J}$ is a DJ-constraint.
2. $\rho(\mathscr{D}) = \{(\varphi(0),\varphi(r)) \ : \ \varphi$ is a differential state flow of the order of $\mathscr{D}$ and some duration $r \geq 0$ such that $\varphi \models \mathscr{D}\}$ when $\mathscr{D}$ is a DA-constraint; see Definition 3.9.
3. $\rho(\alpha \cup \beta) = \rho(\alpha) \cup \rho(\beta)$.
4. $\rho(\alpha;\beta) = \{(\nu,\omega) : (\nu,\mu) \in \rho(\alpha), (\mu,\omega) \in \rho(\beta)$ for some state $\mu\}$.
5. $(\nu,\omega) \in \rho(\alpha^*)$ iff there is $n \in \mathbb{N}$ and there are $\nu = \nu_0,\dots,\nu_n = \omega$ such that $(\nu_i,\nu_{i+1}) \in \rho(\alpha)$ for all $0 \leq i < n$.

3.3.2 Valuation of Formulas

Now, the interpretation of DAL formulas is defined as usual for first-order modal logic [123, 149], with the transition semantics, $\rho(\alpha)$, of DA-programs for modalities. The semantics of formulas is compositional and denotational, that is, the semantics of a complex formula is defined as a simple function of the semantics of its subformulas. The definition is, in fact, an equivalent reformulation of the definition of the semantics of $\mathsf{d}\mathscr{L}$ in Sect. 2.3, yet using the semantics of DA-programs instead of that of hybrid programs.

Definition 3.11 (Interpretation of DAL formulas). The *interpretation* $\models$ of DAL formulas with respect to state ν is defined as

1. $\nu \models \theta_1 \geq \theta_2$ iff $val(\nu,\theta_1) \geq val(\nu,\theta_2)$, and accordingly for $=,\leq,<,>$.
2. $\nu \models \phi \land \psi$ iff $\nu \models \phi$ and $\nu \models \psi$.
3. $\nu \models \phi \lor \psi$ iff $\nu \models \phi$ or $\nu \models \psi$.
4. $\nu \models \neg\phi$ iff it is not the case that $\nu \models \phi$.
5. $\nu \models \phi \to \psi$ iff it is not the case that $\nu \models \phi$ or it is the case that $\nu \models \psi$.
6. $\nu \models [\alpha]\phi$ iff $\omega \models \phi$ for all states ω with $(\nu,\omega) \in \rho(\alpha)$.
7. $\nu \models \langle\alpha\rangle\phi$ iff $\omega \models \phi$ for some state ω with $(\nu,\omega) \in \rho(\alpha)$.

The semantics of quantifiers is defined, because we considered them as abbreviations. In particular:

5. $\nu \models \forall x\,\phi$ iff $\omega \models \phi$ for all states ω that agree with ν except for the value of x.
6. $\nu \models \exists x\,\phi$ iff $\omega \models \phi$ for some ω that agrees with ν except for the value of x.

3.3.3 Time Anomalies

Hybrid systems evolve along piecewise continuous trajectories, which consist of a sequence of continuous flows interrupted by (possibly discontinuous) discrete jumps. A common phenomenon in hybrid system models is that their semantics and analysis are more controversial when discrete and continuous behaviour are allowed to interact without certain regularity assumptions [176, 270, 97, 156]. Zeno-anomalies occur when the hybrid system is allowed to take infinitely many discrete transitions in finite time.

Consider the DA-program $(a' = -1 \land d \leq a; d := d/2)^*$ starting in a state where $a > d > 0$ and a and d progress towards goal 0. The (inverse) clock variable a decreases continuously, yet d bounds the maximum duration of each continuous evolution phase. At the latest when $a = d$, variable d decreases by a discrete transition. This Zeno system generates infinitely many transitions in finite time and it is impossible for clock a to finally reach 0, because $a \geq d > 0$ will always remain true; see the dynamics in Fig. 3.3. Yet this behaviour is, in a certain sense,

Fig. 3.3 Zeno system run

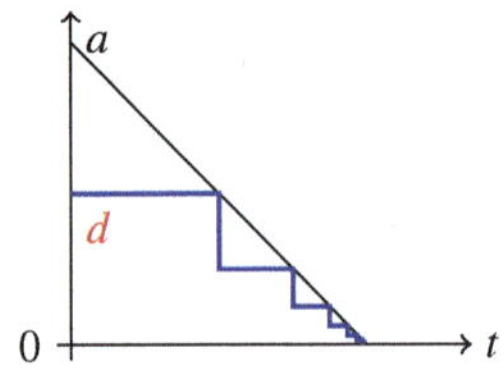

counterfactual, because it fails to obey divergence of time: Real time diverges, whereas clock a converges to 0. Further, systems with Zeno-anomalies cannot be realised [176, 270, 97, 156] so that corresponding regularity assumptions can be justified for practical purposes.

Another example for a Zeno system is the bouncing ball from Fig. 2.2 on p. 45. The bouncing ball will bounce infinitely often in finite time unless the damping coefficient c decreases to 0 at some point.

To avoid pitfalls of time anomalies, we define the DAL semantics so that it only refers to well-defined system behaviour with finitely many transitions in finite time: We restrict the semantics of DA-constraints and disallow infinite numbers of switches between differential equations in bounded time. With DA-constraint $\mathscr{D}$ defined as, say, $(x \geq 0 \to x'' = -1) \land (x < 0 \to x'' = 1) \land y' = 1$, the DAL formula

$$\exists e \, \langle \mathscr{D} \rangle [\mathscr{D}] (y > e \to x \leq d)$$

expresses that, after some time, the system can stabilise such that it always remains within the region $x \leq d$ when $y > e$ for some choice of e. For such a stability property, we do not analyse what happens *after* there have been infinitely many switches from $x'' = 1$ to $x'' = -1$ within the first second. Instead, our semantics is such that our calculus reveals what happens for *any finite* number of switches. Accordingly, we restrict the semantics of DA-constraints to only accept non-Zeno evolutions:

Definition 3.12. A differential state flow φ for a DA-constraint $\mathscr{D}$ is called *non-Zeno* if there only is a finite number of points in time where some variable needs to obey another differential constraint of $\mathscr{D}$ than those before the respective point in time: Let $\mathscr{D}_1 \lor \cdots \lor \mathscr{D}_n$ be a disjunctive normal form of $\mathscr{D}$; then flow

$\varphi : [0,r] \to \text{State}(\Sigma)$ is non-Zeno iff there are $m \in \mathbb{N}$ and $0 = \zeta_0 < \zeta_1 < \cdots < \zeta_m = r$ and indices $i_1, \ldots, i_m \in \{1, \ldots, n\}$ such that φ respects $\mathscr{D}_{i_k}$ on the interval $[\zeta_{k-1}, \zeta_k]$, i.e., $\varphi|_{[\zeta_{k-1},\zeta_k]} \models \mathscr{D}_{i_k}$ for all $k \in \{1, \ldots, m\}$.

The semantics of DA-programs entails that runs with non-Zeno state flows are non-Zeno, because α^* does not accept infinitely many switches.

3.3.4 Conservative Extension

The following result shows that $\mathsf{d}\mathscr{L}$ formulas with hybrid programs can be embedded syntactically into the extension of DAL by DA-programs without changing the meaning of the $\mathsf{d}\mathscr{L}$ formulas. That is, the semantics of DAL formulas given in Definition 3.11 and 3.10 is equivalent to the semantics given in Definitions 2.6 and 2.7 for the sublogic $\mathsf{d}\mathscr{L}$ of dTL using the syntactic embedding of hybrid programs into DA-programs from Table 3.5.

Table 3.5 Embedding hybrid programs as DA-programs

Hybrid program	DA-program
assignment / discrete jump set	DA-constraint
$x_1 := \theta_1, \ldots, x_n := \theta_n$	$x_1 := \theta_1 \wedge \cdots \wedge x_n := \theta_n$
differential equation system	DA-constraint
$x_1' = \theta_1, \ldots, x_n' = \theta_n \,\&\, \chi$	$x_1' = \theta_1 \wedge \cdots \wedge x_n' = \theta_n \wedge \chi$
test	DJ-constraint / DA-constraint
$?\chi$	χ

Proposition 3.1 (Conservative extension). *The logic* DAL *is a* conservative extension *of* $\mathsf{d}\mathscr{L}$*, i.e., the set of valid* $\mathsf{d}\mathscr{L}$ *formulas is the same with respect to the transition semantics of hybrid programs (Definition 2.7) as with respect to the transition semantics of DA-programs (Definition 3.10).*

Proof. The valuation of formulas of $\mathsf{d}\mathscr{L}$ and DAL is directly compatible (Definition 3.11 and 2.6, respectively). By comparing Definition 2.7 with Definition 3.10, it is easy to see that we only need to show that differential equations generate the same transitions. Using vectorial notation, let $x' = \theta \,\&\, \chi$ be a differential equation with evolution domain restriction χ. Let $(v, \omega) \in \rho(x' = \theta \,\&\, \chi)$ according to a flow φ of duration r as a witness due to Definition 2.7. Then φ is a differential state flow of the order of $x' = \theta$ and $\varphi(0) = v, \varphi(r) = \omega$ and $\varphi \models \chi$ and the value of variables z other than x remains constant. Assume $r > 0$ as there is nothing else to show otherwise. By Definition 2.7, we know that $\varphi \models x' = \theta$ holds on the interval $(0,r)$ and have to show that there is a continuation of φ so that $\varphi \models x' = \theta$ holds on $[0,r]$.

The right-hand side θ of the differential equation assumes values that are defined along φ, because $\varphi \models \chi$ and χ guards against zeros of denominators. Hence, the image of φ remains in the domain of definition of θ. Further, φ is continuous on $[0,r]$;

hence, as a compact image of a continuous map, its image is compact. Thus, by the continuation theorem for solutions of differential equations (Proposition B.1), φ can be continued to a solution of $x' = \theta$ on $[0,r]$.

Conversely, it is easy to see that $(v,\omega) \in \rho(x' = \theta \wedge \chi)$ according to Definition 3.10 directly implies $(v,\omega) \in \rho(x' = \theta \,\&\, \chi)$ according to Definition 2.7. □

Clearly, the DAL calculus will *not* be a conservative extension of the $d\mathscr{L}$ calculus, because it contains more powerful proof rules for verifying properties of differential equations. We will see that there are $d\mathscr{L}$ formulas that can be proven in the DAL calculus but not in the $d\mathscr{L}$ calculus. With our differential induction techniques in Sects. 3.5.6 and 3.5.7, the DAL calculus has better handling of differential equations than the solution-based proof rules presented in Chap. 2.

3.4 Collision Avoidance in Air Traffic Control

As a case study, which will serve as a running example, we show how succinctly collision avoidance manoeuvres in air traffic control can be described in DAL. In Sect. 3.11, we will verify such manoeuvres in the DAL calculus.

3.4.1 Flight Dynamics

Assuming, for simplicity, aircraft remain at the same altitude, an aircraft is described by its planar position $x = (x_1, x_2) \in \mathbb{R}^2$ and angular orientation ϑ. The dynamics of an aircraft is determined by its linear velocity $v \in \mathbb{R}$ and angular velocity ω; see Fig. 3.4 (depicted with $\vartheta = 0$ in the illustration). When neglecting wind, gravitation,

Fig. 3.4 Aircraft dynamics

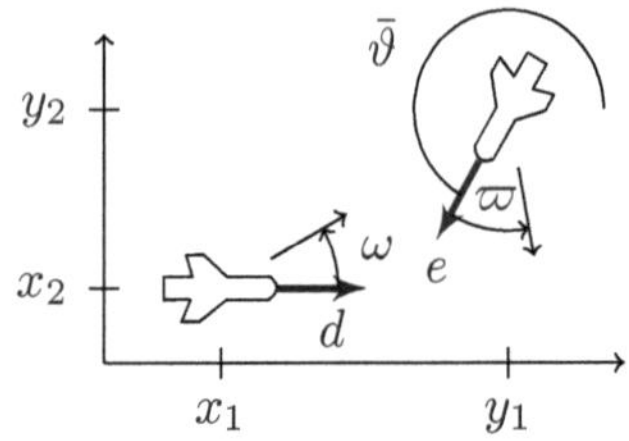

and so on, which is appropriate for analysing cooperation in air traffic control [293, 196, 203, 92, 238], the in-flight dynamics of an aircraft at x can be described by the following differential equation system; see [293] for details:

$$x_1' = v\cos\vartheta \qquad x_2' = v\sin\vartheta \qquad \vartheta' = \omega. \tag{3.4}$$

That is, the linear velocity v of the aircraft changes both positions x_1 and x_2 in the (planar) direction corresponding to the orientation ϑ the aircraft is currently heading toward. Further, the angular velocity ω of the aircraft changes the orientation ϑ of the aircraft.

3.4.2 Differential Axiomatisation

Unlike for straight-line flight ($\omega = 0$), the nonlinear dynamics in (3.4) is difficult to analyse [293, 196, 203, 92, 238] for curved flight ($\omega \neq 0$), especially due to the trigonometric expressions which are generally undecidable. Solving (3.4) requires the Floquet theory of differential equations with periodic coefficients [297, Theorem 18.X] and yields mixed polynomial expressions with multiple trigonometric functions. A true challenge, however, is the need to verify properties of the states that the aircraft reach by following these solutions, which requires proving that complicated formulas with mixed polynomial arithmetic and trigonometric functions hold true for all values of state variables and all possible evolution durations. However, quantified arithmetic with trigonometric functions is undecidable: By Gödel's incompleteness theorem [137], the resulting first-order real arithmetic with trigonometric functions is not semidecidable, because the roots of sin characterise an isomorphic copy of natural numbers (Theorem 2.2).

To obtain polynomial dynamics, we axiomatise the trigonometric functions in the dynamics differentially and reparametrise the state correspondingly. Instead of angular orientation ϑ and linear velocity v, we use the linear speed vector

$$d = (d_1, d_2) := (v\cos\vartheta, v\sin\vartheta) \in \mathbb{R}^2$$

which describes both the linear speed $\|d\| := \sqrt{d_1^2 + d_2^2} = v$ and the orientation of the aircraft in space; see Figs. 3.4 and 3.5. Substituting this coordinate change into

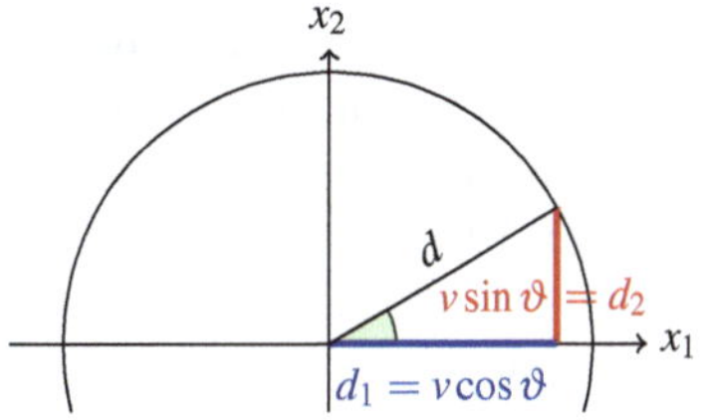

Fig. 3.5 Reparametrise for differential axiomatisation

differential equations (3.4), we immediately have $x_1' = d_1$ and $x_2' = d_2$. With the coordinate change, we further obtain differential equations for d_1, d_2 from differential equation system (3.4) by simple symbolic differentiation:

$$d_1' = (v\cos\vartheta)' = v'\cos\vartheta + v(-\sin\vartheta)\vartheta' = -(v\sin\vartheta)\omega = -\omega d_2,$$

$$d_2' = (v\sin\vartheta)' = v'\sin\vartheta + v(\cos\vartheta)\vartheta' \quad = \quad (v\cos\vartheta)\omega = \quad \omega d_1.$$

The middle equality holds for constant linear velocity ($v' = 0$), which we assume, because only limited variations in linear speed are possible and cost-effective during the flight [293, 196] so that angular velocity ω is the primary control parameter in air traffic control. Hence, equations (3.4) can be restated as the following DA-constraint $\mathscr{F}(\omega)$:

$$x_1' = d_1 \wedge x_2' = d_2 \wedge d_1' = -\omega d_2 \wedge d_2' = \omega d_1 \qquad (\mathscr{F}(\omega))$$
$$y_1' = e_1 \wedge y_2' = e_2 \wedge e_1' = -\varpi e_2 \wedge e_2' = \varpi e_1 \qquad (\mathscr{G}(\varpi))$$

DA-constraint $\mathscr{F}(\omega)$ expresses that position $x = (x_1, x_2)$ changes according to the linear speed vector $d = (d_1, d_2)$, which in turn rotates according to ω. Simultaneous movement together with a second aircraft at $y \in \mathbb{R}^2$ having linear speed $e \in \mathbb{R}^2$ (also indicated with angle $\bar{\vartheta}$ in Fig. 3.4) and angular velocity ϖ corresponds to the DA-constraint $\mathscr{F}(\omega) \wedge \mathscr{G}(\varpi)$. DA-constraints capture simultaneous dynamics of multiple traffic agents succinctly using conjunction.

By this *differential axiomatisation*, we thus obtain polynomial differential equations. Note, however, that their solutions still involve the same complicated non-linear trigonometric expressions so that solutions still give undecidable arithmetic (see Example B.4 in App. B). Our proof calculus in this chapter works with the differential equations themselves and not with their solutions, so that differential axiomatisation helps. Since the solutions involve trigonometric functions, previous approaches [306, 156, 125, 270, 97, 228] were not able to handle such dynamics.

3.4.3 Aircraft Collision Avoidance Manoeuvres

Due to possible turbulence or collisions, a flight configuration is unsafe if another aircraft is within a protected zone of radius p, i.e., $\|x - y\|^2 < p^2$. Guiding aircraft by collision avoidance manoeuvres to automatically resolve conflicting flight paths that would lead to possible loss of separation, is a major challenge for both air traffic control and verification [293, 196, 203, 104, 92, 238, 129, 171]. Several different classes of collision avoidance manoeuvres for air traffic control have been suggested [293, 196, 203, 104, 129, 171]. The classical traffic alert and collision avoidance system (TCAS) [196] directs one aircraft on climbing routes and the other on descending routes to resolve conflicts at different altitudes but keeps otherwise unmodified straight-line flight paths. While the simplistic TCAS manoeuvre has several benefits, it does not scale up easily to multiple aircraft or dense traffic situations near airports. As a more scalable alternative, Tomlin et al. [293] suggested roundabout manoeuvres on circular paths (see Fig. 3.6a), where, even at the same altitude, several aircraft can participate in collision avoidance manoeuvres. Because the continuous dynamics of curved flights with $\omega \neq 0$ is quite intricate, Tomlin et al. [293] and Massink and De Francesco [203] have analysed trapezoidal straight-line ($\omega = 0$)

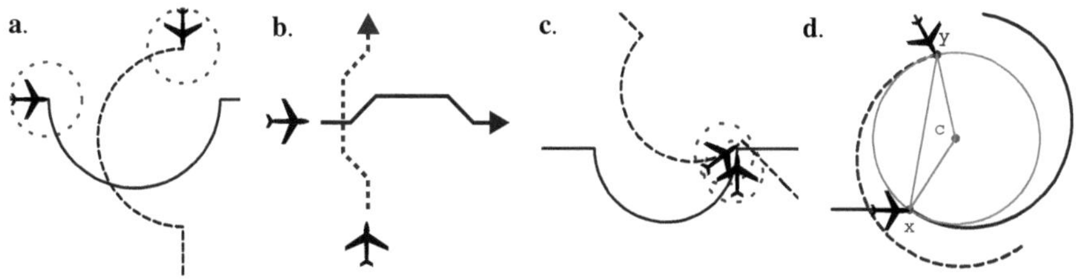

Fig. 3.6 Flight manoeuvres for collision avoidance in air traffic control

approximations of roundabouts instead, which consist only of a series of two to five straight line segments connected by several instant turns (Fig. 3.6b). Unfortunately, the discontinuities in instant turns are not flyable by aircraft.

As a more realistic model, we investigate curved roundabout manoeuvres proposed by Tomlin et al. [293]. Roundabouts have proper flight curves with nonzero angular velocities ω (Fig. 3.6a). We have shown previously [238] that classical roundabout manoeuvres with fixed turns [293, 196, 203, 92] are unsafe for non-orthogonal initial flight paths (see Fig. 3.6c for a counterexample that our model checker found), and we have proposed a tangential roundabout manoeuvre [238] with position-dependent evasive actions to overcome these deficiencies. However, because of general limits of numerical approximation techniques [238, 85], we could not actually verify the tangential roundabout manoeuvre numerically.

In this chapter, we introduce a generalised class of tangential roundabout manoeuvres with curved flight paths (Fig. 3.6d) and formally verify separation properties of this manoeuvre in the purely symbolic DAL calculus. Our main motivation for studying roundabouts are their curved flight paths, which constitute a substantial challenge for verification of hybrid systems with nontrivial dynamics and an important part of realistic flight manoeuvres.

3.4.4 Tangential Roundabout Manoeuvre

In the tangential roundabout manoeuvre, sketched in Fig. 3.6d, the idea is that the aircraft agree on some common angular velocity ω and common centre c around which both can circle safely without coming closer to each other (their linear velocities can differ, though, to compensate for different cruise speeds). Note that neither c nor ω needs to be discovered by complicated online trajectory predictions. Instead, we present a simple characterisation of safe choices for the parameters of the tangential roundabout manoeuvre in Sect. 3.11 and determine safety of the resulting flight paths using formal proofs in the DAL calculus.

In Fig. 3.7, we introduce the DAL model for the tangential roundabout manoeuvre, which is a simplified and more uniform generalisation of our previous work [238]. Observe how concisely complicated aircraft manoeuvres can be specified in DAL. There, safety property ψ for aircraft manoeuvres expresses that pro-

$$\begin{aligned}
\psi &\equiv \phi \rightarrow [trm^*]\phi \\
\phi &\equiv \|x-y\|^2 \geq p^2 \equiv (x_1-y_1)^2+(x_2-y_2)^2 \geq p^2 \\
trm &\equiv free;\ tang;\ \mathscr{F}(\omega) \wedge \mathscr{G}(\omega) \\
free &\equiv \exists \omega\, \mathscr{F}(\omega) \wedge \exists \varpi\, \mathscr{G}(\varpi) \wedge \phi \\
tang &\equiv \text{will be derived in Sect. 3.11}
\end{aligned}$$

Fig. 3.7 Flight control with tangential roundabout collision avoidance manoeuvres

tected zones are respected during the flight (specified by the separation property ϕ). The flight controller (*trm**) performs collision avoidance manoeuvres by tangential roundabouts and repeats these manoeuvres any number of times, as needed, as indicated by the * repetition operator. During each *trm* phase, the aircraft first perform arbitrary free flight (*free*) by (repeatedly) independently adjusting their angular velocities ω and ϖ (by $\exists\omega$ and $\exists\varpi$) while the aircraft are safely separated. This is expressed by conjunct ϕ of the DA-constraint. Observe that, unlike in $\exists u\,(\omega := u);\, \mathscr{F}(\omega)$, angular velocities can be (re)adjusted continuously during free flight in $\exists\omega\,\mathscr{F}(\omega)$, rather than just once. In particular, *free* includes piecewise constant choices as in $\big(\exists u\,(\omega := u) \wedge \exists u\,(\varpi := u);\, \mathscr{F}(\omega) \wedge \mathscr{G}(\varpi)\big)^*$. Due to evolution domain ϕ of *free*, the tangential roundabout manoeuvre must be initiated (by a tangential initiation controller *tang*) before the flight paths become unsafe. Then, the tangential roundabout manoeuvre itself is carried out by the DA-constraint $\mathscr{F}(\omega) \wedge \mathscr{G}(\omega)$ according to some common angular velocity ω determined by *tang*. Finally, the collision avoidance roundabouts can be left again by repeating the loop *trm** and entering arbitrary free flight at any time. When further conflicts occur during free flight, the controller in Fig. 3.7 again enters roundabout conflict resolution manoeuvres.

In summary, property ψ of Fig. 3.7 expresses that the aircraft remain safe during the flight, especially during evasive roundabout manoeuvres. In Sect. 3.11, we will determine a constraint on the parameter adjustment by *tang* such that the roundabout manoeuvre is safe, and we give a simple choice for *tang* respecting this parameter constraint.

3.5 Proof Calculus for Differential-Algebraic Logic

In this section, we introduce a sequent calculus for proving DAL formulas. The basic idea is to symbolically compute the effects of DA-programs and successively transform them into simpler logical formulas describing their effects by symbolic decomposition. The calculus consists of standard propositional rules, dedicated rules for handling DA-program modalities, including differential induction rules for sophisticated differential constraints, and side deduction rules for integrating real quantifier elimination.

For our calculus, recall the definition of substitutions: The result of applying to ϕ the substitution that replaces x by θ is defined as usual (Sect. 2.5.1); it is denoted by ϕ_x^θ. Likewise, in a simultaneous substitution $\phi_{x_1}^{\theta_1} \dots {}_{x_n}^{\theta_n}$ the x_i are replaced simultaneously by the respective θ_i.

3.5.1 Motivation

DA-constraints are a very expressive formalism and can represent rich forms of continuous dynamics, including differential equations, differential-algebraic equations, differential inequalities, and differential equations with quantified nondeterminism and disturbance input. Even for quite simple differential equations, formal verification of properties of their behaviour is challenging and, in general, undecidable (Theorem 2.2). Even the linear differential equations $x' = a_1x + a_2y \land y' = a_3x + a_4y$ have trigonometric solutions that fall outside decidable classes of arithmetic for appropriate coefficients $a_i \in \mathbb{R}$.

Working with solutions of differential equations as in $d\mathcal{L}$ rules $\langle'\rangle$,$[']$ of Chap. 2 is, thus, not a very promising verification approach. Instead, we are looking for a verification principle that works implicitly based on the local dynamics and the differential equation itself, not on its solution. Solutions of differential equations are usually much more complicated than the differential equations themselves, which makes differential equations representationally powerful—and even more so for the more expressive DA-constraints.

Intuitively, the global solution of a differential equation, while helpful, is not really required for deciding a property. Consider Fig. 3.8, which shows the vector fields of a differential equation system (intuitively, at each point it shows the vector of the right-hand side, i.e., the direction into which the differential equations dynamics points locally). Geometrically, a solution is obtained by following the

Fig. 3.8 Vector field and a solution of a differential equation

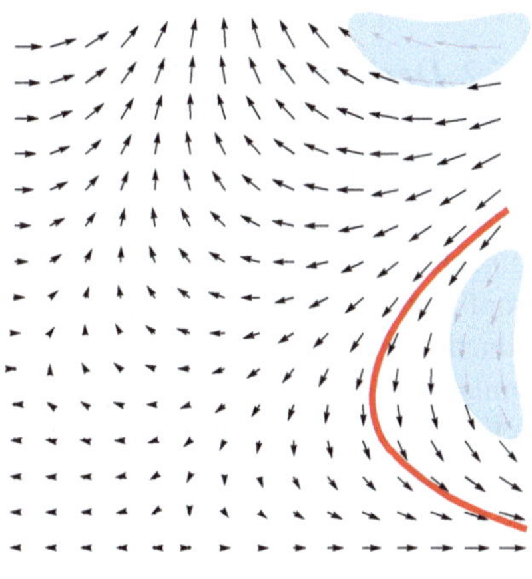

direction of the dynamics "at every point" along the vector field. The solid curve shows one global solution starting at a particular point as initial value. Suppose the two shaded transparent regions represent the set of unsafe states for a system. In principle, verification could proceed by considering each global solution starting at any initial point (typically uncountably many) and show that no such solution enters the unsafe region at any time (uncountably many points in time). Of course, literally following this enumeration of solutions and enumerating all points in time will not work because there are uncountably many initial states and points in time. This principle is what proof rules $\langle ' \rangle$,$[']$ from the d$\mathscr{L}$ calculus in Fig. 2.11 on p. 79 make formally rigorous and sound using symbolic polynomial solutions and quantifier elimination in real-closed fields. It is generally not very scalable and it can be quite difficult to ensure soundness for more general dynamics.

The point is, however, that proving does not need to use solutions of differential equations! In Fig. 3.8 the question about whether the system can ever enter an unsafe state (in the two transparent shaded regions) can be answered by inspection of the local dynamics of the vector field alone. If the vector field never points from a safe state into the unsafe region, then, intuitively, the system can never end unsafe if it starts safe and always follows the local dynamics of the vector field. Similarly, for a reachability property, we do not necessarily need a solution for the differential equations to decide if the system is able to reach a target region. Intuitively, it can also be answered in the positive if the dynamics is always making good progress towards the target region locally.

The challenge is how to turn this geometrical intuition into a sound proof principle. In fact, it turns out throughout the course of this chapter that this is a surprisingly subtle matter and easy to get wrong. In this section, we develop a sound reasoning framework and a proof calculus for DAL that follows these principles of proving by local dynamics without leading astray into the unsound. We first develop appropriate notions from differential algebra that will be invaluable for our formal development of sound proof rules.

3.5.2 Derivations and Differentiation

As a purely algebraic device for proving properties about continuous evolutions in our calculus, we define syntactic derivations of terms and show that their valuation coincides with analytic differentiation (the *total differential*). With this, we can build proof rules for verifying DA-programs fully algebraically by a differential form of induction without the need to carry out analytic reasoning about analytic limits or similar concepts that would require higher-order logic. The advantage of our syntactic notions that are inspired from differential algebra is that syntactic algebraic reasoning can be carried out with a calculus that is still suitable for automatic theorem proving.

Derivations

We define a syntactic total derivation and prove that its valuation along differential flows coincides with analytic differentiation.

Definition 3.13 (Derivation). The operator $D : \mathrm{Term}(\Sigma \cup \Sigma') \to \mathrm{Term}(\Sigma \cup \Sigma')$ that is defined as follows is called *syntactic (total) derivation*:

$$D(r) = 0 \qquad \text{if } r \in \mathbb{Q} \text{ is a rational number} \tag{3.5a}$$

$$D(x^{(n)}) = x^{(n+1)} \qquad \text{if } x \in \Sigma \text{ is a state variable}, n \geq 0 \tag{3.5b}$$

$$D(a+b) = D(a) + D(b) \tag{3.5c}$$

$$D(a-b) = D(a) - D(b) \tag{3.5d}$$

$$D(a \cdot b) = D(a) \cdot b + a \cdot D(b) \tag{3.5e}$$

$$D(a/b) = (D(a) \cdot b - a \cdot D(b))/b^2 \tag{3.5f}$$

For a first-order formula F, we define the following abbreviations:

$$D(F) \equiv \bigwedge_{i=1}^{m} D(F_i) \qquad \text{where } \{F_1, \ldots, F_m\} \text{ is the set of all literals of } F;$$

$$D(a \geq b) \equiv D(a) \geq D(b) \quad \text{and accordingly for } <, >, \leq, = \text{ or negative literals.}$$

Recall that a *literal* is a logical formula with no logical operators other than $\neg$. A literal is negative if it has an odd number of $\neg$ operators, and is positive if it has an even number. More than one $\neg$ is not needed because $\neg\neg F_i \equiv F_i$.

To illustrate the naturalness of this definition, we briefly align it in terms of the algebraic structures from differential algebra [179]. Case (3.5a) defines number symbols as *differential constants*, which do not change during continuous evolution. Their total derivative is zero. Equation (3.5c) and the *Leibniz* or *product rule* (3.5e) are defining conditions for *derivation operators on rings*. The derivative of a sum is the sum of the derivatives (additivity or a homomorphic property with respect to addition) according to equation (3.5c). Furthermore, the derivative of a product is the derivative of one factor times the other factor plus the one factor times the derivative of the other factor as in (3.5e). Equation (3.5d) is a derived rule for subtraction according to $a - b = a + (-1) \cdot b$ and again expresses a homomorphic property, now with respect to subtraction. In addition, equation (3.5b) uniquely defines operator D on the *differential polynomial algebra* spanned by the *differential indeterminates* $x \in \Sigma$. It says that we understand the higher-order differential symbol $x^{(n+1)}$ as the derivative of the symbol $x^{(n)}$ for all state variables $x \in \Sigma$ and orders $n \geq 0$. Equation (3.5f) canonically extends D to the *differential field of quotients* by the usual *quotient rule*. As the base field $\mathbb{R}$ has no zero divisors, the right-hand side of (3.5f) is defined whenever the original division a/b can be carried out, which, as we assumed, is guarded by $b \neq 0$. The resulting structure $\mathrm{Term}(\Sigma \cup \Sigma')$, together with the derivation D, corresponds to the *differential field of rational fractions* with state variables as *differential indeterminates* over $\mathbb{R}$ and with rational numbers as *differential constants*.

The conjunctive definition of the formula $D(F)$ in Definition 3.13 corresponds to the joint total derivative of all atomic subformulas of F and will be an important tool for differential induction rules of our calculus.

Example 3.6 (Total differential of aircraft separation). Consider the separation property ϕ for aircraft manoeuvres from Fig. 3.7 on p. 152, which expresses that two aircraft at positions $x = (x_1, x_2)$ and $y = (y_1, y_2)$, respectively, have at least distance p:

$$F \equiv (x_1 - y_1)^2 + (x_2 - y_2)^2 \geq p^2.$$

The total differential of formula F is formed by applying the syntactic total derivation $D(\cdot)$ on the terms of each literal, which gives:

$$\begin{aligned} D(F) &\equiv D((x_1 - y_1)^2) + D((x_2 - y_2)^2) \geq D(p^2) \\ &\equiv 2(x_1 - y_1)D(x_1 - y_1) + 2(x_2 - y_2)D(x_2 - y_2) \geq 2pD(p) \\ &\equiv 2(x_1 - y_1)(D(x_1) - D(y_1)) + 2(x_2 - y_2)(D(x_2) - D(y_2)) \geq 2pD(p) \\ &\equiv 2(x_1 - y_1)(x_1' - y_1') + 2(x_2 - y_2)(x_2' - y_2') \geq 2pp'. \end{aligned}$$

An interesting question that we will answer in the following is, what can we conclude about the truth-value of F along the evolution of the system by evaluating its total derivative $D(F)$? □

The following central lemma, which is the differential counterpart of the substitution lemma, establishes the connection between syntactic derivation of terms and semantic differentiation as an analytic operation to obtain analytic derivatives of valuations along differential state flows. It will allow us to draw analytic conclusions about the behaviour of a system along differential equations from the truth of purely algebraic formulas obtained by syntactic derivation. In a nutshell, the following lemma shows that, along a flow, analytic derivatives of valuations coincide with valuations of syntactic derivations. For comparison, the classical substitution lemma (Lemma 2.2) shows that the valuation $val(\sigma^* \nu, \phi)$ in the semantically modified state $\sigma^* \nu$ equals valuation of the syntactic substitution $val(\nu, \phi_x^\theta)$, where $\sigma^* \nu$ is like ν except that x is interpreted as $val(\nu, \theta)$. The derivation lemma has the same form, using derivation instead of substitution.

Lemma 3.1 (Derivation lemma). *The valuation of* DAL *terms is a* differential homomorphism*: Let $\theta \in \mathrm{Term}(\Sigma)$ and let $\varphi : [0, r] \to \mathrm{State}(\Sigma)$ be any state flow of the order of $D(\theta)$ and of duration $r > 0$ along which the value of θ is defined (as no divisions by zero occur). Then we have for all $\zeta \in [0, r]$ that*

$$\frac{\mathrm{d}\,val(\varphi(t), \theta)}{\mathrm{d}t}(\zeta) = val(\bar{\varphi}(\zeta), D(\theta)).$$

In particular, $val(\varphi(t), \theta)$ is continuously differentiable (where θ is defined) and its derivative exists on $[0, r]$.

Proof. The proof is an inductive consequence of the correspondence of the semantics of differential symbols and analytic derivatives in state flows (Definition 3.8). It uses the assumption that the flow φ remains within the domain of definition of θ and is continuously differentiable in all variables of θ. In particular, all denominators are nonzero during φ.

- If θ is a variable x, the conjecture holds immediately by Definition 3.8:

$$\frac{\mathrm{d}\,val(\varphi(t),x)}{\mathrm{d}t}(\zeta) = \frac{\mathrm{d}\,\varphi(t)(x)}{\mathrm{d}t}(\zeta) = \bar{\varphi}(\zeta)(x') = val(\bar{\varphi}(\zeta),D(x)).$$

  There, the derivative exists because the state flow is of order 1 in x and, thus, (continuously) differentiable for x.
- If θ is of the form $a+b$, the desired result can be obtained by using the properties of derivatives, derivations (Definition 3.13), and valuations (Definition 3.6):

$$\begin{aligned}
&\frac{\mathrm{d}}{\mathrm{d}t}(val(\varphi(t),a+b))(\zeta) && \\
&= \frac{\mathrm{d}}{\mathrm{d}t}(val(\varphi(t),a)+val(\varphi(t),b))(\zeta) && val(\nu,\cdot) \text{ homomorphic for } + \\
&= \frac{\mathrm{d}}{\mathrm{d}t}(val(\varphi(t),a))(\zeta)+\frac{\mathrm{d}}{\mathrm{d}t}(val(\varphi(t),b))(\zeta) && \frac{\mathrm{d}}{\mathrm{d}t} \text{ is a (linear) derivation} \\
&= val(\bar{\varphi}(\zeta),D(a))+val(\bar{\varphi}(\zeta),D(b)) && \text{by induction hypothesis} \\
&= val(\bar{\varphi}(\zeta),D(a)+D(b)) && val(\nu,\cdot) \text{ homomorphic for } + \\
&= val(\bar{\varphi}(\zeta),D(a+b)) && D(\cdot) \text{ is a syntactic derivation}
\end{aligned}$$

- The case where θ is of the form $a\cdot b$ or $a-b$ is similar, using Leibniz product rule (3.5e) or subtractivity (3.5d) of Definition 3.13, respectively.
- The case where θ is of the form a/b uses (3.5f) of Definition 3.13 and further depends on the assumption that $b \neq 0$ along φ. This holds as the value of θ is assumed to be defined all along state flow φ.
- The values of numbers $r \in \mathbb{Q}$ do not change during a state flow (in fact, they are not affected by the state at all); hence their derivative is $D(r)=0$. □

Differential Transformations

The substitution property Lemma 2.3 can be lifted to differential equations, i.e., differential equations can be used for equivalent substitutions along differential state flows respecting the corresponding differential constraints. In a nutshell, the following lemma can be used to substitute right-hand sides of differential equations for the left-hand side derivatives for flows along which these differential equations hold. For comparison, the classical substitution property says that equals can be substituted for equals, i.e., left-hand sides of equations can be substituted by right-hand sides of equations within formulas in which the equations hold.

Lemma 3.2 (Differential substitution property). *If φ is a state flow satisfying $\varphi \models x_1' = \theta_1 \wedge \cdots \wedge x_n' = \theta_n \wedge \chi$, then $\varphi \models \mathscr{D} \leftrightarrow (\chi \rightarrow \mathscr{D}_{x_1'}^{\theta_1} \cdots {}_{x_n'}^{\theta_n})$ holds for all DA-constraints $\mathscr{D}$.*

Proof. The proof is by using the Substitution Lemma 2.2 for first-order logic on the basis of $val(\bar{\varphi}(\zeta), x_i') = val(\bar{\varphi}(\zeta), \theta_i)$ and $\bar{\varphi}(\zeta) \models \chi$ at each time ζ in the domain of φ. □

Example 3.7 (Differential substitution for aircraft dynamics). Continuing the aircraft scenario from Example 3.6, recall the differential equations ($\mathscr{F}(\omega)$) for flight:

$$x_1' = d_1 \wedge x_2' = d_2 \wedge d_1' = -\omega d_2 \wedge d_2' = \omega d_1 \qquad (\mathscr{F}(\omega)^*)$$

While the aircraft follow this differential equation, all corresponding differential state flows φ satisfy

$$\varphi \models x_1' = d_1 \wedge x_2' = d_2 \wedge d_1' = -\omega d_2 \wedge d_2' = \omega d_1$$

Consequently, by Lemma 3.2, along φ we can substitute in the right-hand sides of these differential equations for the left-hand sides. When we perform this substitution for $D(F)$ from Example 3.6, we get

$$D(F)_{x_1'}^{d_1}{}_{x_2'}^{d_2}{}_{d_1'}^{-\omega d_2}{}_{d_2'}^{\omega d_1} \equiv 2(x_1 - y_1)(d_1 - y_1') + 2(x_2 - y_2)(d_2 - y_2') \geq 2pp'.$$

If we also substitute in the differential equations ($\mathscr{G}(\varpi)$) for the second aircraft and assume $p' = 0$, we obtain

$$D(F)_{x_1'}^{d_1}{}_{x_2'}^{d_2}{}_{d_1'}^{-\omega d_2}{}_{d_2'}^{\omega d_1}{}_{y_1'}^{e_1}{}_{y_2'}^{e_2}{}_{y_1'}^{-\varpi e_2}{}_{e_2'}^{\varpi e_1}{}_{p'}^{0} \equiv 2(x_1 - y_1)(d_1 - e_1) + 2(x_2 - y_2)(d_2 - e_2) \geq 0.$$

Expressions like these will play a very important role in our proof calculus, because we will use them to show (in)variance of properties without the need for knowing solutions of differential equations. From the truth of such a differential substitution of a total derivative, we will be able to conclude invariance of F by the derivation lemma. □

The following lemma captures that the semantics of DA-constraints is not sensitive to how the DA-constraint is presented. It also plays its part in the soundness proof of our calculus, because it immediately makes all implicational and equivalence transformations of real arithmetic available for DA-constraints.

Lemma 3.3 (Differential transformation principle). *Let $\mathscr{D}$ and $\mathscr{E}$ be two DA-constraints (with the same changed variables). If $\mathscr{D} \rightarrow \mathscr{E}$ is a tautology of (non-differential) first-order real arithmetic (that is, when considering $x^{(n)}$ as a new variable independent of x), then $\rho(\mathscr{D}) \subseteq \rho(\mathscr{E})$.*

Proof. Let the first-order formulas ϕ and ψ be obtained from $\mathscr{D}$ and $\mathscr{E}$, respectively, by replacing all x' with new variable symbols X (accordingly for higher-order differential symbols $x^{(n)}$). Using vectorial notation, we write $\phi_X^{x'}$ for the formula obtained from ϕ by substituting all variables X by x'. Thus, $\phi_X^{x'}$ is $\mathscr{D}$ and $\psi_X^{x'}$ is $\mathscr{E}$. Let $\phi \to \psi$ be valid in (non-differential) real arithmetic. Let $(\nu, \omega) \in \rho(\mathscr{D})$ according to a state flow φ. Then φ also is a state flow for $\mathscr{E}$ that justifies $(\nu, \omega) \in \rho(\mathscr{E})$: For any $\zeta \in [0,r]$, we have $\bar{\varphi}(\zeta) \models \mathscr{D}$. Hence $\bar{\varphi}(\zeta) \models \mathscr{E}$, because $\bar{\varphi}(\zeta) \models \phi_X^{x'}$ immediately implies $\bar{\varphi}(\zeta) \models \psi_X^{x'}$ by validity of $\phi \to \psi$. The assumption of $\mathscr{D}$ and $\mathscr{E}$ having the same set of changed variables is only required for compatibility with condition 2 of Definition 3.9, which enforces that unchanged variables z remain constant. It can be established easily by adding constraints of the form $z' = 0$ as required. □

DA-constraints $\mathscr{D}$ and $\mathscr{E}$ are *equivalent* iff $\rho(\mathscr{D}) = \rho(\mathscr{E})$. In particular, the semantics of DA-programs is preserved when replacing a DA-constraint by another DA-constraint that is equivalent in non-differential first-order real arithmetic (similarly for DJ-constraints).

Example 3.8. Lemma 3.3 is a very powerful tool for analysing DA-constraints, because it lifts equivalence or implication-preserving transformations from first-order real arithmetic to DA-constraints. In general, it may be difficult to determine the relationship of the reachability relations of differential equations or DA-constraints. With Lemma 3.3, however, we can use the first-order arithmetic structure to find out. For example, the dynamics of DA-constraint $x' = 5x \wedge y' = 1 - x \wedge x > 0$ can be overapproximated safely by the dynamics of $x' = 5x \wedge y' \leq 1 \wedge x > 0$, because the following implication is valid in first-order logic, where we consider $x^{(1)}$ and $y^{(1)}$ as new variables independent of x and y:

$$x^{(1)} = 5x \wedge y^{(1)} = 1 - x \wedge x > 0 \to x^{(1)} = 5x \wedge y^{(1)} \leq 1 \wedge x > 0.$$

□

Example 3.9 (Disturbance in the train dynamics). Similarly, we have a formally precise way to prove why the transition relation $\rho(\cdot)$ of train dynamics $z' = v \wedge v' = a$ from Chap. 2 is contained in the transition relation of the differential inequality $z' = v \wedge v' \geq a - l \wedge v' \leq a + u$ for bounds $l > 0$ and $u \geq 0$, which, in turn, is contained in the transition relation of $z' = v \wedge v' \geq a - l$. The first differential equation characterises ideal-world dynamics, where the chosen acceleration a takes effect exactly. The second differential inequality characterises dynamics with a bounded disturbance on the actual acceleration a. By the nature of the differential inequality, the actual disturbance may vary quite arbitrarily over time, but only inside the bounds $-l$ and u. The last differential inequality is a variant that is only bounded in one direction.

The proof of inclusion of the above three DA-constraints only amounts to checking the validity of the following formulas in first-order real arithmetic, again considering $z^{(1)}$ and $v^{(1)}$ as new variables:

$$z^{(1)} = v \wedge v^{(1)} = a \to z^{(1)} = v \wedge v^{(1)} \geq a - l \wedge v^{(1)} \leq a + u,$$

$$z^{(1)} = v \land v^{(1)} \geq a - l \land v^{(1)} \leq a + u \rightarrow z^{(1)} = v \land v^{(1)} \geq a - l.$$

Similarly, we prove that the differential inequality $z' = v \land v' \geq a - l \land v' \leq a + u$ and the following quantified DA-constraint have equivalent reachability relations:

$$z' = v \land \exists d\, (v' = d \land a - l \leq d \land d \leq a + u) \land v \geq 0. \qquad (3.2^*)$$

By Lemma 3.3, we only need to check equivalence of the corresponding formulas in first-order real arithmetic, considering $z^{(1)}$ and $v^{(1)}$ as new variables:

$$(z^{(1)} = v \land v^{(1)} \geq a - l \land v^{(1)} \leq a + u)$$
$$\leftrightarrow (z^{(1)} = v \land \exists d\, (v^{(1)} = d \land a - l \leq d \land d \leq a + u) \land v \geq 0).$$

□

Counterexample 3.10 (Same changed variables). For correctness of Lemma 3.3, we have to assume that $\mathscr{D}$ is not allowed to have more differential variables than $\mathscr{E}$, because of our implicit assumption that variables without differential constraints are constant. At first sight, it may look like the transition relation $z' = v \land v' = a \land t' = 1$ may be overapproximated by $z' = v \land t' = 1$, because the following arithmetic formula is valid:

$$z^{(1)} = v \land v^{(1)} = a \land t^{(1)} = 1 \rightarrow z^{(1)} = v \land t^{(1)} = 1.$$

But this does not work, because of our convention that assumes variables do not change if they are not mentioned in a differential equation. Thus in $z' = v \land t' = 1$, we silently assume $v' = 0$, because no differential constraint for v is mentioned. Then the transition relations are very different, however, because v may change in the former but will stay constant in the latter. If, instead, we make the implicit assumption $v' = 0$ explicit in the DA-constraint, so that both DA-constraints have the same changed variables, then this subtlety does not happen, because the following formula is *not* valid (unless $a = 0$, in which case both DA-constraints are, indeed, equivalent):

$$z^{(1)} = v \land v^{(1)} = a \land t^{(1)} = 1 \rightarrow z^{(1)} = v \land t^{(1)} = 1 \land v^{(1)} = 0.$$

□

3.5.3 Differential Reduction and Differential Elimination

Using the expressive power of DA-constraints, several reductions can be performed to simplify the syntactic form of DA-constraints. With quantified DA-constraints, we can reduce differential inequalities to quantified differential equations equivalently:

Lemma 3.4 (Differential inequality elimination). *DA-constraints admit* differential inequality elimination, *i.e., with each DA-constraint $\mathscr{D}$, an equivalent DA-constraint without differential inequalities can be effectively associated that has no other free variables.*

Proof. Let $\mathscr{E}$ be obtained from $\mathscr{D}$ by replacing all differential inequalities $\theta_1 \leq \theta_2$ by a quantified differential equation $\exists u\,(\theta_1 = \theta_2 - u \wedge u \geq 0)$ with a new variable u for the quantified disturbance (accordingly for $\geq, >, <$). By Lemma 3.3, the equivalence of $\mathscr{D}$ and $\mathscr{E}$ is a simple consequence of the corresponding equivalences in first-order real arithmetic. □

Example 3.11. Lemma 3.4 can turn differential inequalities to DA-constraints with quantifiers and equations in place of the differential inequalities. Let $l > 0$. We can transform the differential inequality $z' = v \wedge v' \geq a - l \wedge v \geq 0$ to an equivalent quantified DA-constraint

$$\exists d\,(z' = v \wedge v' = a - d \wedge d \leq l \wedge v \geq 0).$$

Both DA-constraints are equivalent. The latter DA-constraint has a structural advantage: we only have differential *equations* and quantifiers, while the inequalities are isolated in the non-differential evolution domain constraints $d \leq l \wedge v \geq 0$. Thus we can handle differential inequalities if we can handle quantified disturbance. In fact, we could even go on and replace *all* inequalities (differential or not) by equivalent equations and quantifiers with Lemma 3.4:

$$\exists d\,(z' = v \wedge v' = a - d \wedge \exists u_1\, d = l - u_1^2 \wedge \exists u_2\, v = u_2^2).$$

But, unlike differential inequalities, those inequalities that only occur in the non-differential part $d \leq l \wedge v \geq 0$ are not complicated for proofs, because they can be handled easily by quantifier elimination in real-closed fields [288, 81]; also see App. D. □

In this book, we assume this transformation has been applied such that we can focus on DA-constraints with differential equations, i.e., where differential symbols only occur in differential equations, and where inequalities do not contain differential symbols. Yet, the DA-constraint resulting from Lemma 3.4 could become inhomogeneous when multiple differential equations are produced for the same variable from multiple differential inequalities. For instance, $\theta_1 \leq x' \leq \theta_2$ produces $\exists u\, \exists v\,(x' = \theta_1 + u \wedge x' = \theta_2 - v \wedge u \geq 0 \wedge v \geq 0)$, which has two differential equations for x. To homogenise this DA-constraint again, we use the following:

Lemma 3.5 (Differential equation normalisation). *DA-constraints admit* differential equation normalisation, *i.e., with each DA-constraint $\mathscr{D}$, an equivalent DA-constraint with at most one differential equation for each differential symbol can be effectively associated that has no other free variables. Furthermore, this differential equation is* explicit, *i.e., of the form $x^{(n)} = \theta$ where* $\mathrm{ord}_x\,\theta < n$.

Proof. For each differential symbol $x^{(n)} \in \Sigma'$ occurring in $\mathscr{D}$, we introduce a new non-differential variable $X_n \in \Sigma$. Let $\mathscr{D}^{X_n}_{x^{(n)}}$ denote the result of substituting X_n for $x^{(n)}$ in $\mathscr{D}$. By Lemma 3.3, the equivalence of $\mathscr{D}$ and $\exists X_n\,(x^{(n)} = X_n \wedge \mathscr{D}^{X_n}_{x^{(n)}})$ is a simple consequence of the corresponding equivalence in first-order logic. Proceeding inductively for all such $x^{(n)} \in \Sigma'$ in $\mathscr{D}$ gives the desired result. □

Example 3.12. Directly using the construction in Lemma 3.4 on $\theta_1 \leq x' \leq \theta_2$ we obtain an equivalent DA-constraint $\exists u\,\exists v\,(x' = \theta_1 + u \wedge x' = \theta_2 - v \wedge u \geq 0 \wedge v \geq 0)$ without differential inequalities but with inhomogeneous parts (two separate differential equations for x). After normalisation with Lemma 3.5, we obtain the equivalent DA-constraint $\exists X_1\,\exists u\,\exists v\,(x' = X_1 \wedge X_1 = \theta_1 + u \wedge X_1 = \theta_2 - v \wedge u \geq 0 \wedge v \geq 0)$. □

Similarly, higher-order differential constraints reduce to first-order constraints by introducing new non-differential auxiliary variables X_n for each of the higher-order differential symbols $x^{(n)}$. For $1 \leq \mathrm{ord}_x\,\theta < n$, we can replace a higher-order differential equation $x^{(n)} = \theta$ with:

$$x' = X_1 \wedge X_1' = X_2 \wedge \ldots \wedge X_{n-2}' = X_{n-1} \wedge X_{n-1}' = \theta^{X_1}_{x'} \cdots {}^{X_{n-1}}_{x^{(n-1)}} {}^{X_{n-1}'}_{x^{(n)}}\,.$$

3.5.4 Rules of the Calculus for Differential-Algebraic Logic

Sequents and substitutions are defined as in Sect. 2.5.2. We again assume bound variable renaming as needed. In the DAL calculus, only admissible substitutions are applicable, which is crucial for soundness.

Definition 3.14 (Admissible substitution). An application of a substitution σ is *admissible* if no replaced variable x occurs in the scope of a quantifier or modality binding x or a variable of the replacement $\sigma(x)$. A modality *binds* x if its DA-program (possibly) changes x, i.e., if it contains a DJ-constraint containing $x := \theta$ or a DA-constraint containing $x^{(n)} \in \Sigma'$ for an $n \geq 1$.

As usual in sequent calculus—although the direction of entailment is from *premises* (above rule bar) to *conclusion* (below)—the order of reasoning and reading is *goal-directed* in practise: Rules are applied backwards, that is, starting from the desired conclusion at the bottom (goal) to the resulting premises (subgoals). To highlight the logical essence of the DAL calculus, Fig. 3.9 provides rule *schemata* with which the following definition associates the calculus rules that are applicable during a DAL proof. The calculus inherits the propositional rules from Fig. 2.11 on p. 79 and further consists of first-order quantifier rules (r∀,l∀,r∃,l∃), rules for dynamic modalities (⟨;⟩–*DS*), and global rules ([]*gen*,⟨⟩*gen*,*ind*,*con*,*DI*,*DV*). The DAL rules that have the same name as corresponding d$\mathcal{L}$ rules are actually identical, except for minor syntactic variations and generalisations. Rule ⟨;⟩, for instance, is identical in the d$\mathcal{L}$ calculus in Fig. 2.11 and the DAL calculus in Fig. 3.9. We just repeat it here for convenience to have a comprehensive representation of the DAL calculus.

The definition of rules is a simplified version of that in Definition 2.10, with side deductions in place of the free-variable quantifier rules from Chap. 2. Further, we can simplify the presentation by avoiding update prefixes (which would also be sound here when allowing conjunctive DA-constraints as prefixes). Note that this choice generally requires more complicated invariants and variants than in the d$\mathscr{L}$ calculus of Chap. 2, where discrete jump set prefixes are allowed for rule applications so that more information about the pre-state is retained automatically.

Definition 3.15 (Rules). The *rule schemata* in Fig. 3.9—in which *all* substitutions need to be admissible—induce *calculus rules* by:

1. If

$$\frac{\Phi_1 \vdash \Psi_1 \quad \dots \quad \Phi_n \vdash \Psi_n}{\Phi_0 \vdash \Psi_0}$$

is an instance of one of the rule schemata in Fig. 3.9, then

$$\frac{\Gamma, \Phi_1 \vdash \Psi_1, \Delta \quad \dots \quad \Gamma, \Phi_n \vdash \Psi_n, \Delta}{\Gamma, \Phi_0 \vdash \Psi_0, \Delta}$$

can be applied as a proof rule of the DAL calculus, where Γ, Δ are arbitrary finite sets of context formulas (including empty sets).

2. Symmetric schemata can be applied on either side of the sequent. If

$$\frac{\phi_1}{\phi_0}$$

is an instance of one of the symmetric rule schemata (dynamic rules $\langle;\rangle$–$[D]$) in Fig. 3.9, then

$$\frac{\Gamma \vdash \phi_1, \Delta}{\Gamma \vdash \phi_0, \Delta} \qquad \text{and} \qquad \frac{\Gamma, \phi_1 \vdash \Delta}{\Gamma, \phi_0 \vdash \Delta}$$

can both be applied as proof rules of the DAL calculus, where Γ, Δ are arbitrary finite sets of context formulas (including empty sets).

Propositional Rules

For propositional logic, we reuse the standard propositional rules $\neg$r–*cut* from the d$\mathscr{L}$ calculus in Fig. 2.11.

First-Order Quantifier Rules

Unlike in uninterpreted first-order logic [122, 123], quantifier rules have to respect the specific semantics of real arithmetic. Thus, our rules handle real quantifiers

$$(\mathrm{r}\forall)\ \frac{\mathrm{QE}(\forall x \bigwedge_i(\Gamma_i \vdash \Delta_i))}{\Gamma \vdash \Delta, \forall x\, \phi}{}^{1} \qquad (\mathrm{r}\exists)\ \frac{\mathrm{QE}(\exists x \bigwedge_i(\Gamma_i \vdash \Delta_i))}{\Gamma \vdash \Delta, \exists x\, \phi}{}^{1}$$

$$(\mathrm{l}\forall)\ \frac{\mathrm{QE}(\exists x \bigwedge_i(\Gamma_i \vdash \Delta_i))}{\Gamma, \forall x\, \phi \vdash \Delta}{}^{1} \qquad (\mathrm{l}\exists)\ \frac{\mathrm{QE}(\forall x \bigwedge_i(\Gamma_i \vdash \Delta_i))}{\Gamma, \exists x\, \phi \vdash \Delta}{}^{1}$$

$$(\langle;\rangle)\ \frac{\langle\alpha\rangle\langle\beta\rangle\phi}{\langle\alpha;\beta\rangle\phi} \qquad (\langle\exists\rangle)\ \frac{\exists x\,\langle\mathscr{J}\rangle\phi}{\langle\exists x\,\mathscr{J}\rangle\phi} \qquad (\langle:=\rangle)\ \frac{\chi \wedge \phi_{x_1}^{\theta_1} \cdots {}_{x_n}^{\theta_n}}{\langle x_1 := \theta_1 \wedge .. \wedge x_n := \theta_n \wedge \chi\rangle\phi}{}^{3}$$

$$([;])\ \frac{[\alpha][\beta]\phi}{[\alpha;\beta]\phi} \qquad ([\exists])\ \frac{\forall x\,[\mathscr{J}]\phi}{[\exists x\,\mathscr{J}]\phi} \qquad ([:=])\ \frac{\chi \rightarrow \phi_{x_1}^{\theta_1} \cdots {}_{x_n}^{\theta_n}}{[x_1 := \theta_1 \wedge .. \wedge x_n := \theta_n \wedge \chi]\phi}{}^{3}$$

$$(\langle\cup\rangle)\ \frac{\langle\alpha\rangle\phi \vee \langle\beta\rangle\phi}{\langle\alpha\cup\beta\rangle\phi} \qquad (\langle J\rangle)\ \frac{\langle\mathscr{J}_1 \cup \ldots \cup \mathscr{J}_n\rangle\phi}{\langle\mathscr{J}\rangle\phi}{}^{2} \qquad (\langle D\rangle)\ \frac{\langle(\mathscr{D}_1 \cup \ldots \cup \mathscr{D}_n)^*\rangle\phi}{\langle\mathscr{D}\rangle\phi}{}^{4}$$

$$([\cup])\ \frac{[\alpha]\phi \wedge [\beta]\phi}{[\alpha\cup\beta]\phi} \qquad ([J])\ \frac{[\mathscr{J}_1 \cup \ldots \cup \mathscr{J}_n]\phi}{[\mathscr{J}]\phi}{}^{2} \qquad ([D])\ \frac{[(\mathscr{D}_1 \cup \ldots \cup \mathscr{D}_n)^*]\phi}{[\mathscr{D}]\phi}{}^{4}$$

$$([DR])\ \frac{\vdash [\mathscr{E}]\phi}{\vdash [\mathscr{D}]\phi}{}^{5} \qquad (\langle DR\rangle)\ \frac{\vdash \langle\mathscr{D}\rangle\phi}{\vdash \langle\mathscr{E}\rangle\phi}{}^{5} \qquad (DS)\ \frac{\vdash [\mathscr{D}]\chi \quad \vdash [\mathscr{D} \wedge \chi]\phi}{\vdash [\mathscr{D}]\phi}$$

$$([]gen)\ \frac{\vdash \forall^\alpha(\phi \rightarrow \psi)}{[\alpha]\phi \vdash [\alpha]\psi} \qquad (\langle\rangle gen)\ \frac{\vdash \forall^\alpha(\phi \rightarrow \psi)}{\langle\alpha\rangle\phi \vdash \langle\alpha\rangle\psi}$$

$$(ind)\ \frac{\vdash \forall^\alpha(\phi \rightarrow [\alpha]\phi)}{\phi \vdash [\alpha^*]\phi} \qquad (con)\ \frac{\vdash \forall^\alpha \forall v{>}0\,(\varphi(v) \rightarrow \langle\alpha\rangle\varphi(v-1))}{\exists v\,\varphi(v) \vdash \langle\alpha^*\rangle \exists v{\leq}0\,\varphi(v)}{}^{6}$$

$$(DI)\ \frac{\vdash \forall^\alpha \forall y_1 .. \forall y_k\,(\chi \rightarrow F'{}_{x_1'}^{\theta_1} \cdots {}_{x_n'}^{\theta_n})}{[\exists y_1 .. \exists y_k\,\chi]F \vdash [\exists y_1 .. \exists y_k\,(x_1' = \theta_1 \wedge .. \wedge x_n' = \theta_n \wedge \chi)]F}{}^{7}$$

$$(DV)\ \frac{\vdash \exists \varepsilon{>}0\,\forall^\alpha \forall y_1 .. y_k\,(\neg F \wedge \chi \rightarrow (F' \geq \varepsilon)_{x_1'}^{\theta_1} \cdots {}_{x_n'}^{\theta_n})}{[\exists y_1 .. y_k\,(x_1' = \theta_1 \wedge .. \wedge x_n' = \theta_n \wedge {\sim}F)]\chi \vdash \langle\exists y_1 .. y_k\,(x_1' = \theta_1 \wedge .. \wedge x_n' = \theta_n \wedge \chi)\rangle F}{}^{8}$$

[1] $\Gamma_i \vdash \Delta_i$ are obtained from the subgoals of side deduction ($\star$) in Fig. 3.10, in which x is assumed to occur in first-order formulas only, as QE is then applicable. The side deduction starts from goal $\Gamma \vdash \Delta, \phi$ at the bottom (or $\Gamma, \phi \vdash \Delta$ for l$\forall$ and l$\exists$), where x does not occur in Γ, Δ using renaming.

[2] $\mathscr{J}_1 \vee \cdots \vee \mathscr{J}_n$ is a disjunctive normal form of the DJ-constraint $\mathscr{J}$.

[3] Rule applicable for any reordering of the conjuncts of the DJ-constraint where χ is jump-free.

[4] $\mathscr{D}_1 \vee \cdots \vee \mathscr{D}_n$ is a disjunctive normal form of the DA-constraint $\mathscr{D}$.

[5] $\mathscr{D}$ implies $\mathscr{E}$, i.e., satisfies the assumptions of Lemma 3.3.

[6] Logical variable v does not occur in α.

[7] Applicable for any reordering of the conjuncts where χ is non-differential. F is first-order without negative equalities, and F' abbreviates $D(F)$, with z' replaced with 0 for unchanged variables.

[8] Like DI, but F contains no equalities and the differential equations are Lipschitz continuous.

Fig. 3.9 Rule schemata of the proof calculus for differential-algebraic dynamic logic

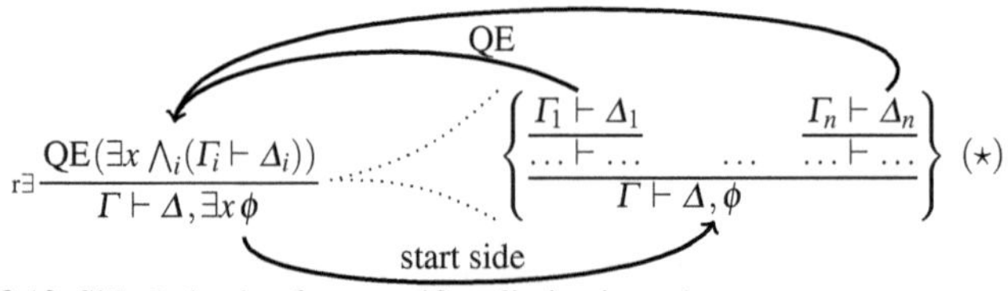

Fig. 3.10 Side deduction for quantifier elimination rules

using quantifier elimination (QE) over the reals [81]. Unfortunately, QE is only defined in first-order real arithmetic and cannot handle DAL modalities, where variables evolve along hybrid trajectories over time. We establish compatibility with dynamic modalities using (cut-like) side deductions for the quantifier rules, as illustrated in Sect. 3.5.5. Alternatively, the quantifier rules in Fig. 3.9 can be replaced by the quantifier rules of Chap. 2, which generalise free variables, Skolemisation, and Deskolemisation to real arithmetic for integrating quantifier elimination with modal rules. Instead, here, we use side deductions that we have introduced in previous work [231] as a very intuitive and simple approach for handling real quantifiers.

Dynamic Rules

The dynamic rules transform a DA-program into simpler logical formulas. Rules $\langle;\rangle$,$[;]$,$\langle\cup\rangle$,$[\cup]$ are as in discrete dynamic logic [149, 37] and in the $d\mathcal{L}$ calculus in Fig. 2.11. The new rules $\langle J\rangle$ and $[J]$ normalise DJ-constraints to their disjunctive normal form such that the jump alternatives can be read off easily. They also turn the disjuncts of the disjunctive normal form into nondeterministic choices so that $\langle\cup\rangle$,$[\cup]$ will reason separately for each case. Rules $\langle\exists\rangle$ and $[\exists]$ lift quantified choices in DJ-constraints to DAL quantifiers, which are, in turn, handled by quantifier rules. They use the fact that *all choices of* assigning *some value* to x correspond to the universal quantifier (rule $[\exists]$), whereas *some choice of* assigning *some value* to x corresponds to the existential quantifier (rule $\langle\exists\rangle$).

Rules $\langle:=\rangle$ and $[:=]$ use generalised simultaneous substitutions for handling discrete change by substituting the respective new values θ_i for all the affected variables x_i at once (for all $i = 1,\dots,n$) and checking the jump-free constraint χ. In fact, this check or assumption of the jump-free constraint χ is the only essential difference in rules $\langle:=\rangle$,$[:=]$ in Fig. 3.9 compared to those in Fig. 2.11. Notice however, that as in rules $\langle?\rangle$,$[?]$ from Fig. 2.11, the presence of χ already makes $[x:=\theta\wedge\chi]\phi$ and $\langle x:=\theta\wedge\chi\rangle\phi$ non-equivalent. Thus, for $\langle x:=\theta\wedge\chi\rangle\phi$ rule $\langle:=\rangle$ is used to prove that χ holds true (otherwise there is no transition and thus the reachability property is false) and that ϕ holds after replacing x with its new value θ. Rule $[:=]$ for $[x:=\theta\wedge\chi]\phi$ instead assumes that χ holds true (otherwise there is no transition and thus nothing to show) and that ϕ holds after replacing x with its new value θ.

Similarly to rules $\langle J\rangle$,$[J]$, rules $\langle D\rangle$,$[D]$ normalise DA-constraints to a form where their differential evolution alternatives are readily identifiable. They turn a DA-constraints into its disjunctive normal form and split disjuncts into nondeterministic choices. Unlike for $\langle J\rangle$,$[J]$, however, continuous evolutions take time so that the system can switch back and forth (repeatedly) between the various cases of the DA-constraint during one continuous evolution, and hence the repetition. Observe that finitely many repetitions are sufficient for non-Zeno flows (Definition 3.12), which can only switch finitely often in finite time.

Rules $[DR]$,$\langle DR\rangle$, and DS are differential weakening, differential refinement, and differential strengthening rules for DA-constraints, respectively. In rules $[DR]$ and $\langle DR\rangle$, DA-constraint $\mathscr{D}$ implies $\mathscr{E}$ in real arithmetic according to Lemma 3.3, which is easy to decide by QE in practise. Note that $[DR]$ and $\langle DR\rangle$ are sound for *any* such combination of $\mathscr{D}$ and $\mathscr{E}$. Their primary practical purpose is to use $[DR]$ for overapproximating individual variable evolutions by a weaker version (*differential weakening*) and $\langle DR\rangle$ for refining nondeterministic variable evolutions to specific differential equations (*differential refinement*). For instance, rule $[DR]$ can weaken $[x' = 5x \wedge y' = 1 - x \wedge x > 0]\phi$ to $[x' = 5x \wedge y' \leq 1 \wedge x > 0]\phi$, because the former DA-constraint implies the latter DA-constraint according to Lemma 3.3. Such overapproximations have the advantage of decoupling differential equations. In particular, we use $[DR]$ onto project conjunctive differential constraints $\mathscr{D}$ to their non-differential constraints. As we illustrate in Sect. 3.11, this gives a powerful verification technique in combination with differential strengthening (*DS*), which can refine the system dynamics by auxiliary constraints. Rule $\langle DR\rangle$ can refine nondeterministic variable evolutions into specific cases, e.g., $\langle z' = v \wedge v' \geq -b\rangle\phi$ to the borderline case $\langle z' = v \wedge v' = -b\rangle\phi$. The latter refines the former, because it is a special case (the latter DA-constraint implies the former DA-constraint in the sense of Lemma 3.3).

While rule $[DR]$ can be used to remove information from a DA-constraint, the differential strengthening rule *DS* does the opposite and can be used to add information. But, of course, *DS* can only add constraints χ (to the right premise) that have been proven to be invariant (in the left premise); otherwise it would alter the reachable set incorrectly. With the left premise proven, rule *DS* changes the dynamics of the DA-program and restricts the DA-constraint $\mathscr{D}$ to remain within evolution domain χ. But this restriction is a pseudo restriction, because the left premise shows that χ actually is an invariant of the previous dynamics $\mathscr{D}$. In fact, assuming the left premise $[\mathscr{D}]\chi$ to be valid, the right premise $\vdash [\mathscr{D} \wedge \chi]\phi$ and conclusion $\vdash [\mathscr{D}]\phi$ are actually equivalent (as we will show in Proposition 6.2). The differential strengthening rule *DS* is essentially a *differential cut*. We address the problem of automatically determining the respective strengthenings χ that actually help the proof in Chap. 6, where we derive automatic verification algorithms from the results presented in this chapter. Furthermore, $[DR]$ and $\langle DR\rangle$ make all equivalence transformations on DA-constraints from Sect. 3.5.3 available as proof rules, including index reduction techniques for differential-algebraic equations [132].

Note that DAL does not need rules for handling negation in DA-constraints or DJ-constraints, as—possibly after applying $\langle J\rangle$,$[J]$ or $\langle D\rangle$,$[D]$, respectively—negations only occur in jump-free or non-differential parts, because assignments and differential symbols only occur positively by Definitions 3.1 and 3.2. Similarly, no rules for universal quantifiers within DA-constraints or DJ-constraints are needed. Like other propositional operators or quantifiers, negation and universal quantifiers are allowed without restriction in non-differential or jump-free χ and are then handled by $\langle :=\rangle$,$[:=]$ or *DI*,*DV*.

Global Rules

The global rules $[]gen$,$\langle\rangle gen$,*ind*,*con*,*DI*,*DV* depend on the truth of their premises in all states, which is ensured by the universal closure with respect to all bound variables of the respective DA-program α (see Definition 3.14). In particular, the rules $[]gen$,$\langle\rangle gen$,*ind*,*con* are as in the $\mathsf{d}\mathcal{L}$ calculus of Fig. 2.11. The differential induction rules *DI* and *DV* are new.

Rule *DI* is a rule for *differential induction*, which is a continuous form of induction along differential constraints. The induction rules *ind* and *DI* (or *con* and *DV* respectively) differ in the way the invariant is sustained (or in the way the variant makes progress). Rule *ind* uses the inductive nature of repetition and, ultimately, follows an induction on the number of repetitions. Rule *DI*, in contrast, uses continuity of evolution and the differential equation for a continuous induction step with the *differential invariant* F: If F holds initially (antecedent of conclusion) and its total differential F' satisfies the same relations when taking into account the differential constraints (premise), then F itself is sustained differentially (succedent of conclusion). Formula F' abbreviates $D(F)$ (see Definition 3.13) with z' replaced with 0 for all variables z that are unchanged by the DA-constraint, i.e., that are distinct from $\{x_1,\dots,x_n\}$, because these are assumed constant in the semantics. By bound variable renaming, the y_i do not occur in F. Thus, an important difference between the operating principles of rules *ind* and *DI* is that *ind* uses induction over natural numbers for repetitions of a loop, whereas *DI* uses induction in the continuous domain along the vector fields of differential equations, and is based on differential algebra (Sects. 3.5.2 and 3.5.3). A notable special instance of rule *DI* is the following case for quantifier-free DA-constraints:

$$\frac{\vdash \forall^\alpha(\chi \to F'{}^{\theta_1}_{x_1'}\cdots{}^{\theta_n}_{x_n'})}{[\chi]F \vdash [x_1' = \theta_1 \land .. \land x_n' = \theta_n \land \chi]F}$$

Rule *DV* is a *differential variant* rule where the variant F is finally reached differentially (with some minimal progress ε), rather than sustained as in *DI*. Differential induction, the requirement of the differential equations for *DV* to be Lipschitz continuous, and the notations $F' \geq \varepsilon$ and $\sim F$ will be illustrated in more detail in Sects. 3.5.6 and 3.5.7 after side deductions for quantifiers have been explained in Sect. 3.5.5. Finally, global rules can be combined with generalisation ($[]gen$,$\langle\rangle gen$) to strengthen postconditions as needed, similarly to rules *ind'* and *con'* from p. 86. A notable special instance of rule *DV* is the following case for quantifier-free DA-constraints:

$$\frac{\vdash \exists\varepsilon{>}0\,\forall^\alpha(\neg F \land \chi \to (F' \geq \varepsilon)^{\theta_1}_{x_1'}\cdots{}^{\theta_n}_{x_n'})}{[x_1' = \theta_1 \land .. \land x_n' = \theta_n \land {\sim}F]\chi \vdash \langle x_1' = \theta_1 \land .. \land x_n' = \theta_n \land \chi\rangle F}$$

Derivability and Proofs

Provability can be defined as a simplified version of Definition 2.11, because all DAL rules have only one conclusion; so a proof will be inductively defined as a tree, not as an acyclic graph (as was necessary for the quantifier rules from Chap. 2).

Definition 3.16 (Provability). A formula ψ is *provable* from a set Φ of formulas, denoted by $\Phi \vdash_{\mathrm{DAL}} \psi$, iff there is a finite set $\Phi_0 \subseteq \Phi$ for which the sequent $\Phi_0 \vdash \psi$ is derivable. Derivability is inductively defined so that a sequent $\Phi \vdash \Psi$ is *derivable* iff there is a proof rule of the DAL calculus (Definition 3.15) with conclusion $\Phi \vdash \Psi$ such that all premises of the rule are derivable.

3.5.5 Deduction Modulo by Side Deduction

The quantifier rules constitute a purely modular interface to mathematical reasoning. They can use any theory that admits quantifier elimination and has a decidable ground theory, e.g., the theory of first-order real arithmetic, which is equivalent to the theory of real-closed fields [81]. Unlike in deduction modulo approaches of Dowek et al. [103] and Tinelli [290], the information given to the background prover is not restricted to ground formulas [290] or atomic formulas [103], and the effect of modalities has to be taken into account.

Integrating quantifier elimination to deal with statements about real quantities is quite challenging in the presence of modalities that influence the values of variables and terms. Real quantifier elimination cannot be applied to formulas with mixed quantifiers and modalities such as $\exists x\,[x' = -x; x := 2x]x \leq 5$. To find out which first-order constraints are actually imposed on x by this DAL formula, we have to take into account how x evolves from $\exists x$ to $x \leq 5$ along the hybrid system dynamics. Hence, our calculus first unveils the first-order constraints on x before applying QE. To achieve this in a concise and simple way, we use side deductions that we have introduced in previous work [231].

The effect of a side deduction is as follows. First, the DAL calculus discovers all relevant first-order constraints from modal formulas using a side deduction. Secondly, these constraints are re-imported into the main proof and equivalently reduced using QE, and then the main proof continues. For instance, an application of r$\exists$ to a sequent $\Gamma \vdash \Delta, \exists x\,\phi$ starts a side deduction (marked $(\star)$ in Fig. 3.10) with the unquantified kernel $\Gamma \vdash \Delta, \phi$ as a goal at the bottom. This side deduction is carried out in the DAL calculus until x no longer occurs within modal formulas of the remaining open branches $\Gamma_i \vdash \Delta_i$ of $(\star)$. Once all occurrences of x are in first-order formulas, the resulting subgoals $\Gamma_i \vdash \Delta_i$ of $(\star)$ are copied back to the main proof and QE is applied to eliminate x altogether (which determines the resulting subgoal of rule r$\exists$ on the upper left side of Fig. 3.10). The remaining modal formulas not containing x can be considered as atoms for this purpose, as they do not impose constraints on x. Formally, this can be proven using the coincidence lemma 2.6. When

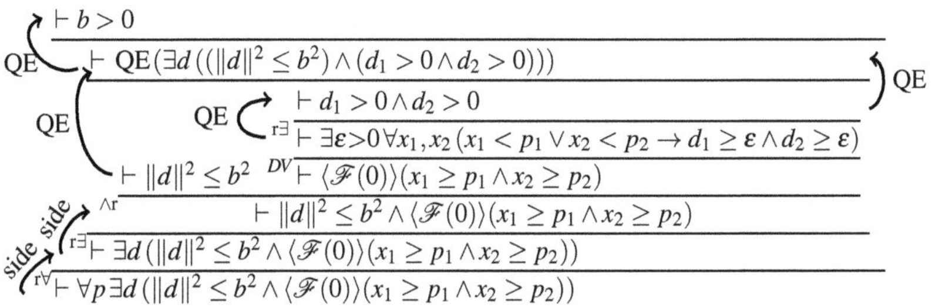

Fig. 3.11 Nested side deductions and differential variants for progress property

several quantifiers are nested, side deductions will be nested in a cascade, as they can again spawn further side deductions. According to the applicability conditions of quantifier rules, inner nested side deductions need to be completed by QE before outer deductions continue. For instance, further side deductions started within ($\star$) of Fig. 3.10 will be completed before ($\star$) continues and the quantifier elimination result of ($\star$) is returned to the main r∃ application.

Example 3.13 (Aircraft progress). To illustrate how our calculus combines arithmetic with dynamic reasoning using side deductions, we look at an aircraft example. Using the notation from Sect. 3.4 where $\mathscr{F}(\omega)$ denotes the flight equation with angular velocity ω, the following DAL formula expresses a simple progress property about aircraft: The aircraft at x can finally fly beyond any point $p \in \mathbb{R}^2$ by adjusting its speed vector d appropriately, using only speed vectors $d \in \mathbb{R}^2$ of bounded speed $\|d\| \le b$, i.e., $\|d\|^2 \le b^2 \equiv d_1^2 + d_2^2 \le b^2$:

$$\forall p\, \exists d\,(\|d\|^2 \le b^2 \wedge \langle \mathscr{F}(0) \rangle (x_1 \ge p_1 \wedge x_2 \ge p_2)). \tag{3.6}$$

There, point p is constant during the evolution, i.e., $p_1' = p_2' = 0$ and $b' = 0$. The DAL proof in Fig. 3.11 proves this property using nested side deductions for nested quantifiers and differential variant induction *DV*. Applying r∀ in the main branch yields a side deduction for quantifier $\forall p$, which, in turn, yields another side deduction by applying r∃ for the nested quantifier $\exists d$. □

These nested side deductions in Fig. 3.11 are inlined and indicated by indenting the side deductions, with arrows marked "side" pointing to the start of the respective inner side deduction and arrows marked "QE" pointing back to the continuation of the outer deduction (as in Fig. 3.10). The two branches for the side deduction for r∃ recombine conjunctively and, after quantifiers are re-added, quantifier elimination yields $b > 0$, which reveals the parameter constraint on the speed bound b. Consequently, property (3.6) holds true and the proof closes for all positive speed bounds $b > 0$. The right branch of the inner r∃ side deduction uses differential variant induction *DV*, as will be illustrated in Sect. 3.5.7. There, the quantifiers for x_1, x_2 result from the universal closure $\forall^\alpha$ in *DV*. The subsequent innermost r∃ side deduction can be abbreviated by directly applying QE, because the affected formula already is first-order. That is, the side deduction for $\exists\varepsilon$ is the identity proof with zero rule applications.

As with the other aircraft examples in this chapter, formula (3.6) is provable in our theorem prover KeYmaera within a few seconds, despite the complicated underlying aircraft dynamics.

3.5.6 Differential Induction with Differential Invariants

The purpose of *DI* and *DV* is to prove properties about continuous evolutions by differential induction using differential invariants or differential variants, respectively. They work with the differential constraints directly instead of with the complicated (possibly undecidable) arithmetic of their solutions. Unlike approaches using solutions [125, 228, 231, 233, 235], differential induction can even be used to verify systems with nondeterministic quantified input, which would otherwise cause quantified higher-order functions for the time-dependent input of the solutions. Solutions of differential equations lead to a quantifier over time (see $\mathsf{d}\mathscr{L}$ rules $[']$,$\langle'\rangle$ in Fig. 2.11). In addition to their dependency on time, solutions of quantified differential equations or differential equations with input disturbance depend on a function $u : [0,\infty) \to \mathbb{R}$ that specifies how this input $u(t)$ changes over time t. When generalising solution-based proof rules for these situations, quantification over disturbances would thus lead to quantifiers over functions in higher-order logic, as in $\forall u : \mathbb{R} \to \mathbb{R}\,\forall t{\geq}0\,\langle x{:=}y_u(t)\rangle\phi$, where $y_u(t)$ is the solution at time t depending on the input disturbance function u. With the theory of differential induction, we avoid these higher-order quantifiers in the proof calculus. Further, unlike in discrete induction, differential induction proof rules exploit continuity of evolution and knowledge of the differential constraints for a continuous induction step. We demonstrate the capabilities and the necessity of the requirements of differential induction rules in a series of examples and counterexamples.

Differential Invariants

Rule *DI* uses differential induction to prove that F is a *differential invariant*, i.e., F is closed under total differentiation (Definition 3.13) relative to the differential constraints. For this, the premise of *DI* shows that the total differential F'—i.e., $D(F)$ with z' replaced by 0 for unchanged variables z— holds within evolution domain χ when substituting the differential equations into F'. Now, if F holds initially (antecedent of conclusion), then F itself is sustained (succedent of conclusion). Intuitively, the premise expresses that, within χ, the total derivative F' along the differential constraints is pointing inwards or transversally to F but never outwards to $\neg F$; see Fig. 3.12. Intuitively, if we start in F and, as indicated by F', the local dynamics never points outside F, then the system always stays in F when following the dynamics. At this point, it is important to note that, even though meta-proofs about

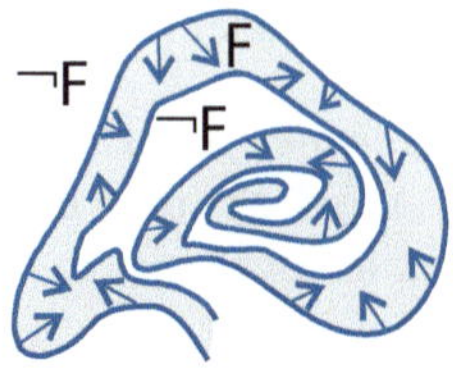

Fig. 3.12 Differential invariants

DAL involve analytic reasoning, proofs within the DAL calculus are fully algebraic, including the handling of differential constraints by *DI*. Further observe that the premise of *DI* is a well-formed DAL formula, because all differential symbols are replaced by non-differential terms when forming $F'^{\theta_1}_{x'_1} \dots {}^{\theta_n}_{x'_n}$.

Example 3.14 (Quartic dynamics). As a first simple example, consider the differential equation $x' = x^4$. It is not so easy to see the solution of this differential equation. Still, with implicit means of differential induction, we can establish easily that the solution always stays above $\frac{1}{4}$ whenever the dynamics initially starts above $\frac{1}{4}$:

$$\text{DI}\;\dfrac{\text{r}\forall\;\dfrac{*}{\vdash \forall x\,(x^4 \geq 0)}}{x \geq \frac{1}{4} \vdash [x' = x^4]x \geq \frac{1}{4}}$$

This deduction proves the invariance of $x \geq \frac{1}{4}$ along the differential equation $x' = x^4$ by differential induction and without having to solve the differential equation. To apply the differential induction rule *DI*, we form the total derivative of the differential invariant $F \equiv x \geq \frac{1}{4}$ and obtain the differential expression $F' \equiv D(x \geq \frac{1}{4}) \equiv x' \geq 0$. Now, the differential induction rule *DI* takes into account that the derivative of state variable x along the dynamics is known (the trick, of course, is to show why this intuitive reasoning is sound, which we will prove in Sect. 3.6). Substituting the differential equation $x' = x^4$ into the inequality above yields $F'^{x^4}_{x'} \equiv x^4 \geq 0$, which is a valid formula and closes by quantifier elimination with r$\forall$. Observe how elegantly differential induction establishes the desired result indirectly by working with the differential equation itself in an algebraic way instead of requiring its solution.

Even more so, for the differential equations $x' = x^2 + x^4$ or $x' = x^2 - 4x + 6$, solutions are hard to obtain both symbolically and numerically. With differential induction, however, we directly establish the following result about their dynamics:

$$\text{DI}\;\dfrac{\text{r}\forall\;\dfrac{*}{\vdash \forall x\,(3(x^2 + x^4) \geq 0)}}{3x \geq \frac{1}{4} \vdash [x' = x^2 + x^4]3x \geq \frac{1}{4}}$$

$$\text{DI}\;\dfrac{\text{r}\forall\;\dfrac{*}{\vdash \forall x\,(3(x^2 - 4x + 6) \geq 0)}}{3x \geq \frac{1}{4} \vdash [x' = x^2 - 4x + 6]3x \geq \frac{1}{4}}$$

For the latter proof, note that $3(x^2-4x+6) = 3((x-2)^2+2) \geq 0$. □

Example 3.15 (Cubic dynamics). Similarly, differential induction can easily prove that $\frac{1}{3} \leq 5x^2$ is an invariant of the cubic dynamics $x' = x^3$; see the proof in Fig. 3.13a for the dynamics in Fig. 3.13b. To apply the differential induction rule *DI*, we again

$$\dfrac{\dfrac{*}{\vdash \forall x\,(0 \leq 5\cdot 2x(x^3))}\,\text{r}\forall}{\frac{1}{3} \leq 5x^2 \vdash [x' = x^3]\,\frac{1}{3} \leq 5x^2}\,DI$$

Fig. 3.13a Cubic dynamics proof

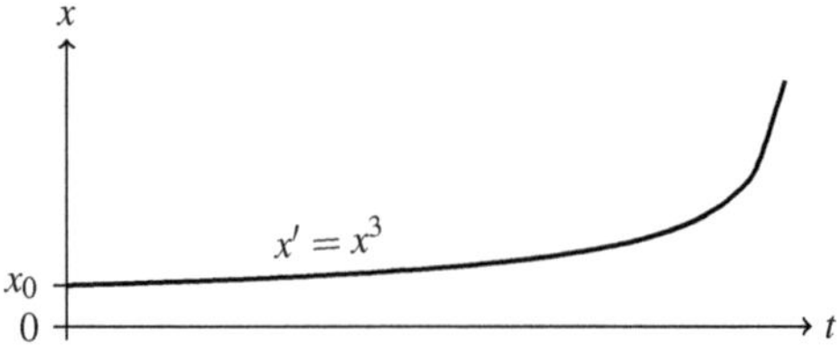

Fig. 3.13b Cubic dynamics

form the total derivative of the differential invariant $F \equiv \frac{1}{3} \leq 5x^2$, which gives the differential expression $F' \equiv D(\frac{1}{3} \leq 5x^2) \equiv 0 \leq 5\cdot 2xx'$. Now, the differential induction rule *DI* takes into account that the derivative of state variable x along the dynamics is known. Substituting the differential equation $x' = x^3$ into the inequality yields $F'^{x^3}_{x'} \equiv 0 \leq 5\cdot 2xx^3$, which is a valid formula and closes by quantifier elimination with r∀. □

Example 3.16 (Linear versus angular speed). Consider the following simple proof, which shows that the speed v of the aircraft with position x is maintained even when it changes its angular velocity ω nondeterministically during the flight (as in mode *free* of Fig. 3.7). Again, recall the flight equation with angular velocity ω:

$$x_1' = d_1 \wedge x_2' = d_2 \wedge d_1' = -\omega d_2 \wedge d_2' = \omega d_1 \qquad (\mathscr{F}(\omega)^*)$$

$$\begin{array}{ll}
 & * \\
\hline
\text{QE} & \vdash \text{QE}(\forall x_1,x_2\,\forall d_1,d_2\,\forall\omega\,(2d_1(-\omega d_2)+2d_2\omega d_1 = 0)) \\
\hline
\text{r}\forall & \vdash \forall x_1,x_2\,\forall d_1,d_2\,\forall\omega\,(2d_1(-\omega d_2)+2d_2\omega d_1 = 0) \quad \text{QE} \\
\hline
DI & d_1^2+d_2^2 = v^2 \vdash [\exists\omega\,\mathscr{F}(\omega)]\,d_1^2+d_2^2 = v^2 \\
\hline
\to\text{r} & \vdash d_1^2+d_2^2 = v^2 \to [\exists\omega\,\mathscr{F}(\omega)]\,d_1^2+d_2^2 = v^2 \quad (\text{side}) \\
\hline
\text{r}\forall & \vdash \forall v\,(d_1^2+d_2^2 = v^2 \to [\exists\omega\,\mathscr{F}(\omega)]\,d_1^2+d_2^2 = v^2)
\end{array}$$

The total derivative of the property $F \equiv d_1^2+d_2^2 = v^2$ for differential induction with *DI* is $F' \equiv D(d_1^2+d_2^2 = v^2) \equiv 2d_1d_1' + 2d_2d_2' = 2vv'$. Substituting the differential equations $\mathscr{F}(\omega)$ of flight yields $F'^{-\omega d_2\ \omega d_1\ 0}_{d_1'\ \ d_2'\ \ v'} \equiv 2d_1(-\omega d_2) + 2d_2\omega d_1 = 0$, which is valid and closes by quantifier elimination. This example shows the difference of differential continuous evolution (of d_1, d_2) and nondeterministic continuous evolution (of ω). The DA-constraint specifies how the d_i evolve along differential equations;

hence d_i' is substituted in F'. For ω, in contrast, the DA-constraint is nondeterministic ($\exists\omega$) and does not specify how ω changes precisely. In particular, there is no equation for ω' that could be used for substitution. Yet such an equation is not even needed for forming the premise of *DI*, because, after bound variable renaming, ω cannot occur in F, since the scope of $\exists\omega$ ends with the DA-constraint and does not extend to postcondition F. In the proof, the quantifiers for x_i and d_i result from the universal closure $\forall^\alpha$ in *DI*. The quantifier for ω is introduced by *DV* and ensures that *all* possible evolutions of ω are taken into account as there is no specific equation for ω'. After all, ω is a quantified input and we cannot know a specific value for its slope, but have to expect any ($\forall\omega$) choice. Finally, note that in such cases without existential variables, side deductions can be inlined; see Chap. 2 for formal details on proof rules with Skolem function terms. □

Requirements of Differential Invariants

Next, we illustrate why the requirements formulated for the proof rule *DI* are necessary in general.

Counterexample 3.17. For soundness of differential induction, it is crucial that Definition 3.13 defines the total derivative $D(F \vee G)$ of a disjunction conjunctively as $D(F) \wedge D(G)$ instead of as $D(F) \vee D(G)$. From an initial state ν which satisfies $\nu \models F$, and hence $\nu \models F \vee G$, the formula $F \vee G$ only is sustained differentially if F itself is a differential invariant, not if G is. For instance, $x_1 \geq 0 \vee d_1^2 + d_2^2 = v^2$ is no differential invariant of $\exists\omega\, \mathscr{F}(\omega)$, because $x_1 \geq 0$ can be invalidated by appropriate curved flights along $\mathscr{F}(\omega)$; see formula (3.6). In practise, splitting differential induction proofs over disjunctions can be useful.

Note, however, that $D(F \vee G) \equiv D(F) \wedge D(G)$ is not the only definition that works. We could just as well define $D(F \vee G) \equiv (F \wedge D(F)) \vee (G \wedge D(G))$ instead. It is a simple exercise to show soundness of this modification. □

Counterexample 3.18 (Restricting differential invariance). It may be tempting to suspect that in *DI* the differential invariant F only needs to be differentially inductive at the states where F actually holds true. After all, the rule is used to prove that F stays true all the time, and in discrete loop induction (*ind*), the invariant ϕ can also be assumed to hold when showing the induction step in the premise. So why did we not assume F when proving F' in the premise of rule *DI*? See Fig. 3.14 for a tempting attempt of what a proof rule could be. But the differential induction needs

Fig. 3.14 Unsound restriction of differential invariance

$$\frac{\vdash \forall^\alpha (F \wedge \chi \to F'^{\theta_1}_{x_1'} \cdots ^{\theta_n}_{x_n'})}{[\chi]F \vdash [x_1' = \theta_1 \wedge .. \wedge x_n' = \theta_n \wedge \chi]F}$$

to hold in a neighbourhood, such that adding F (or even just the border of F) to the assumptions in the premise of *DI* would be unsound! Consider the counterexample in Fig. 3.15a, where the differential invariance restricted to F would seem to indicate the region $x^2 \leq 0$ was never left when following the dynamics $x' = 1$. This is, of course, completely counterfactual as the dynamics in Fig. 3.15b shows, where region $x^2 \leq 0$ is actually left immediately when following the dynamics $x' = 1$.

$$\dfrac{\dfrac{* \text{ (unsound)}}{\vdash \forall x\,(x^2 \leq 0 \rightarrow 2x \leq 0)}}{x^2 \leq 0 \vdash [x' = 1]x^2 \leq 0}$$

Fig. 3.15a Restricting differential invariance

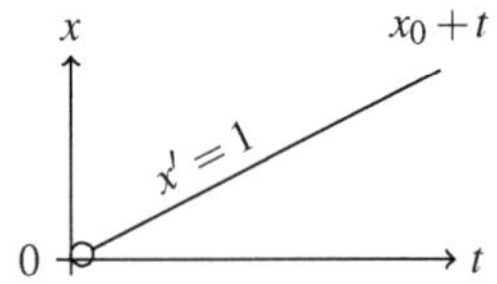

Fig. 3.15b Linear dynamics

Thus, although restricting the domain where proof rules require differential invariance appears to be a tempting idea, we have to be very careful to ensure the proof rules are sound in this subtle domain. In fact, the same counterexample in Fig. 3.15a also demonstrates unsoundness of other approaches that have been proposed to handle nontrivial differential equations [251, 145]. With a minor variation (replacing assumption $x^2 \leq 0$ in the premise by $x^2 = 0$), Fig. 3.15a also shows that it is not sufficient to restrict the differential to the border of F. The problem causing this unsoundness is circular reasoning and the fact that derivatives are only defined in domains with non-empty interior. Strictly speaking, in the beginning we only know invariant F to hold at a single point (antecedent of conclusion of *DI*). Now if we assume F to hold in some domain for the induction step (premise of *DI*), then, initially, we only know that F holds in a region with an empty interior. This is not sufficient to conclude anything based on following derivatives, because these are not defined on regions with an empty interior. Thus, the reasoning in Fig. 3.15a assumes more than it has proven already, which explains the unsoundness, and the unsoundness of the rule in Fig. 3.14. □

This counterexample illustrates that differential invariance is a powerful but also subtle matter and that we have to prove soundness of the proof rules very carefully. We prove soundness of the DAL calculus and its differential induction principles in Sect. 3.6. In particular, we cannot generally use differential invariants for proving their continuous induction step. If, however, F describes an open set (e.g., F only involves strict inequalities), then *DI* is sound even when adding F to the assumptions of the premise as in Fig. 3.14. Likewise, F can be added to the assumptions of the premise when strengthening F' to strict inequalities. We will prove both refinements in Sect. 3.7. Furthermore, differential strengthening (*DS*) can be an extraordinarily successful proof technique for successively enriching evolution domain restrictions by derived invariants until F itself becomes differentially inductive, as we illustrate in Sects. 3.10 and 3.11. Also if polynomial solutions exist, they can be used as differential invariants.

Counterexample 3.19 (Negative equations). It is crucial for soundness of differential induction that F be not allowed to contain negative equations. In the following counterexample, variable x can reach $x = 0$ without its derivative ever being 0; again, see Fig. 3.15b for the dynamics.

$$\dfrac{\dfrac{*\ \text{(unsound)}}{\vdash \forall x\,(1 \neq 0)}}{x \neq 0 \vdash [x' = 1]x \neq 0}$$

If, instead, both inequalities $x < 0$ and $x > 0$ are differential invariants of a system (e.g., of the differential equation $x' = x$), then $x \neq 0$ can be proven indirectly by representing it equivalently as $x < 0 \vee x > 0$. We pursue this further in Example 3.29 in Sect. 3.7. □

Differential Weakening

A useful special case of the differential refinement rule $[DR]$ is the following derived weakening rule, with which we can assume the evolution domain constraint χ to hold for all reachable states.

Lemma 3.6 (Differential weakening). *The following is a derived proof rule:*

$$([DR'])\quad \dfrac{\vdash \forall^{\alpha}\forall y_1 \ldots \forall y_k\,(\chi \to \phi)}{\vdash [\exists y_1 \ldots \exists y_k\,(x_1' = \theta_1 \wedge \cdots \wedge x_n' = \theta_n \wedge \chi)]\phi}$$

Proof. Rule $[DR']$ is sound, because χ is true along all state flows of the DA-constraint and ϕ is a consequence of χ in all reachable states (as overapproximated by $\forall^{\alpha}$) by premise. Rule $[DR']$ can be derived as follows:

$$\begin{array}{rl}
 & \vdash \mathrm{QE}(\forall y_1 \ldots \forall y_k \forall d_1 \ldots \forall d_n\,(\chi_{x_1}^{d_1}\cdots{}_{x_n}^{d_n} \to \phi_{x_1}^{d_1}\cdots{}_{x_n}^{d_n}))\\ \hline
\mathrm{QE} & \quad \vdash \chi_{x_1}^{d_1}\cdots{}_{x_n}^{d_n} \to \phi_{x_1}^{d_1}\cdots{}_{x_n}^{d_n}\\ \hline
\text{side}\ [:=] & \quad \vdash [x_1 := d_1 \wedge \cdots \wedge x_n := d_n \wedge \chi_{x_1}^{d_1}\cdots{}_{x_n}^{d_n}]\phi\\ \hline
\mathrm{r}\forall & \vdash \forall y_1 \ldots \forall y_k \forall d_1 \ldots \forall d_n\,([x_1 := d_1 \wedge \cdots \wedge x_n := d_n \wedge \chi_{x_1}^{d_1}\cdots{}_{x_n}^{d_n}]\phi)\\ \hline
[\exists] & \vdash [\exists y_1 \ldots \exists y_k \exists d_1 \ldots \exists d_n\,(x_1 := d_1 \wedge \cdots \wedge x_n := d_n \wedge \chi_{x_1}^{d_1}\cdots{}_{x_n}^{d_n})]\phi\\ \hline
[DR] & \vdash [\exists y_1 \ldots \exists y_k \exists d_1 \ldots \exists d_n\,(x_1' = d_1 \wedge \cdots \wedge x_n' = d_n \wedge \chi)]\phi\\ \hline
[DR] & \vdash [\exists y_1 \ldots \exists y_k\,(x_1' = \theta_1 \wedge \cdots \wedge x_n' = \theta_n \wedge \chi)]\phi
\end{array}$$

The second application of $[DR]$ uses the fact that fully nondeterministic continuous state change is equivalent to fully nondeterministic discrete state change, as they generate the same transitions. Finally, $\chi \to \phi$ can be obtained by bound variable renaming. □

Differential Invariants for Disturbance

So far, we have used differential induction rule *DI* for various differential equations. Yet proof rule *DI* and the technique of differential induction is more general than that. The rule also works for DA-constraints with quantifiers, differential inequalities, or systems with disturbance in the dynamics.

Example 3.20 (Disturbance in the train dynamics). Differential induction can be used to handle differential inequalities and disturbance in differential equations. Continuing Example 3.9 from p. 159, let us abbreviate the differential inequality

$$z' = v \land a - l \le v' \le a + u \land v \ge 0 \tag{3.1*}$$

succinctly by $z''{\approx}a$. This differential inequality characterises that the train at position z with velocity v follows approximately the chosen acceleration a, with a deviation from a that is bounded by lower bound $-l \le 0$ and upper bound $u \ge 0$. Now, for the choice of $a := -b$ for braking, we can prove that the train that we considered in Sect. 2.5.3 always stays inside its movement authority m, despite the differential inequality disturbance in the dynamics, when starting in an initial state that satisfies:

$$\phi \equiv v^2 \le 2(b-u)(m-z) \land b > u \ge 0 \land l \ge 0.$$

See Fig. 3.16 for a proof. The proof uses differential strengthening rule *DS* to aug-

$$
{\scriptstyle \to r}\dfrac{
{\scriptstyle DS}\dfrac{
{\scriptstyle [DR]}\dfrac{
{\scriptstyle DI}\dfrac{
{\scriptstyle r\forall}\dfrac{*}{\phi \vdash \forall^{\alpha}\forall d\,(-l \le d \le u \land v{\ge}0 \to 2v(d-b) \le -2v(b-u))}
}{\phi \vdash [\exists d\,(-l \le d \le u \land z'' = -b + d \land v \ge 0)]\phi}
}{\phi \vdash [z''{\approx}-b \land v \ge 0]\phi}
\qquad
{\scriptstyle [DR']}\dfrac{
{\scriptstyle r\forall}\dfrac{*}{\vdash \forall^{\alpha}(v{\ge}0 \land \phi \to z \le m)}
}{\vdash [z''{\approx}-b \land v{\ge}0 \land \phi]z{\le}m}
}{\phi \vdash [z''{\approx}-b \land v \ge 0]z \le m}
}{\vdash \phi \to [z''{\approx}-b \land v \ge 0]z \le m}
$$

Fig. 3.16 Proof of MA-safety in braking mode with disturbance

ment the differential inequality $z''{\approx}-b$ with ϕ as an auxiliary invariant. The resulting right branch where *DS* added ϕ to the DA-constraint can then be proven easily using differential weakening $[DR']$ to show that the newly augmented evolution domain region $v \ge 0 \land \phi$ implies the postcondition $z \le m$. In the left branch, we use differential induction rule *DI* to prove that the auxiliary invariant ϕ that we assumed in the right branch is actually an invariant of the dynamics and can thus be added soundly to the evolution domain restriction in the right branch, because ϕ is a pseudo restriction. Before the differential induction *DI* on the left branch, we use differential refinement $[DR]$ for an equivalence transformation that replaces the differential inequality $z''{\approx}-b$ from (3.1) with a quantified DA-constraint with a quantifier for the disturbance d that is restricted to the domain $-l \le d \le u$:

$$\exists d\,(-l \le d \le u \land z' = v \land v' = -b + d \land v \ge 0). \tag{3.7}$$

The equivalence of the quantified DA-constraint (3.7) with disturbance and the differential inequality (3.1) that is required for applying $[DR]$ can be established easily using Lemma 3.3. More detailed examples for properties of dynamics with disturbance, even equivalence properties, can be found in Sect. 7.4. □

Differential Invariants and Oscillation

Differential invariants are an interesting proof technique also for proving properties of dynamical systems with oscillation. In these systems, solutions of the differential equations cannot (really) be used for verification, because the solutions of oscillating processes involve trigonometric functions of undecidable classes of arithmetic.

Example 3.21 (Damped oscillator). Recall the damped oscillator from Example 1.3 on p. 8, which corresponds to the DA-constraint

$$x' = y \land y' = -\omega^2 x - 2d\omega y \tag{3.8}$$

with parameters $\omega \geq 0$ (for the undamped angular frequency) and $d \geq 0$ (for the damping ratio). See Fig. 3.17 for one example of an evolution along this continuous dynamics. Figure 3.17 shows a trajectory in the x, y space on the left, and an evolu-

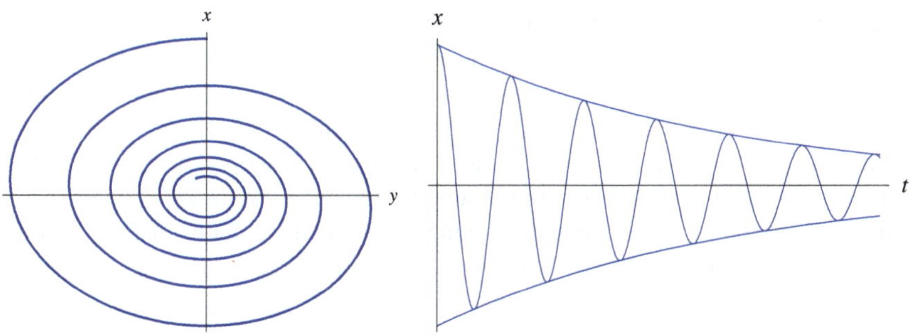

Fig. 3.17 Trajectory and evolution of a damped oscillator

tion of x over time t on the right. For the damped oscillator dynamics, we can easily prove the following invariance property:

$$\dfrac{\dfrac{*}{\omega \geq 0, d \geq 0 \vdash \forall x \forall y\, (2\omega^2 xy - 2\omega^2 xy - 4d\omega y^2 \leq 0)}\, r\forall}{\omega \geq 0, d \geq 0, \omega^2 x^2 + y^2 \leq c^2 \vdash [x' = y \land y' = -\omega^2 x - 2d\omega y]\, \omega^2 x^2 + y^2 \leq c^2}\, DI$$

The total derivative of the property $F \equiv \omega^2 x^2 + y^2 \leq c^2$ for differential induction with DI is $F' \equiv D(\omega^2 x^2 + y^2 \leq c^2) \equiv 2\omega^2 x x' + 2\omega\omega' x^2 + 2yy' \leq 0$. Substituting with the differential equations (3.8) gives $F'^{\,y}_{x'}{}^{-\omega^2 x - 2d\omega y}_{y'}{}^{0}_{\omega'}{}^{0}_{d'}{}^{0}_{c'}$, which is equivalent to:

$$2\omega^2xy + 2y(-\omega^2x - 2d\omega y) \leq 0 \equiv 2\omega^2xy - 2\omega^2xy - 4d\omega y^2 \leq 0 \equiv -4d\omega y^2 \leq 0.$$

This formula is valid and closes by quantifier elimination (r∀), because $d\omega \geq 0$. □

Example 3.22 (Switched damped oscillators). Recall the switched damped oscillator from Example 1.3, which switches between two different damped oscillators of the form (3.8) for different choices of ω and d. The switched damped oscillator switches from one damped oscillator to the other at the diagonals of the state space. See Fig. 3.18 for an illustration of one possible evolution of the system. The hybrid system switches between two different damped oscillators at the switching surfaces depicted by the diagonal lines in Fig. 1.6 on p. 8 from one mode (shown in solid curves, with smaller ω) to the other mode (shown in dashed curves, with greater ω) and back again. At the solid diagonal, the system switches to the dynamics shown by a solid curve (with a smaller ω value), and at the dashed diagonal, it switches to the dynamics shown by a dashed curve (with greater ω value). We model a generalised parametric version of a corresponding switched damped oscillator in the first line of Fig. 3.19, with subsequent abbreviations. More generally, we assume the system

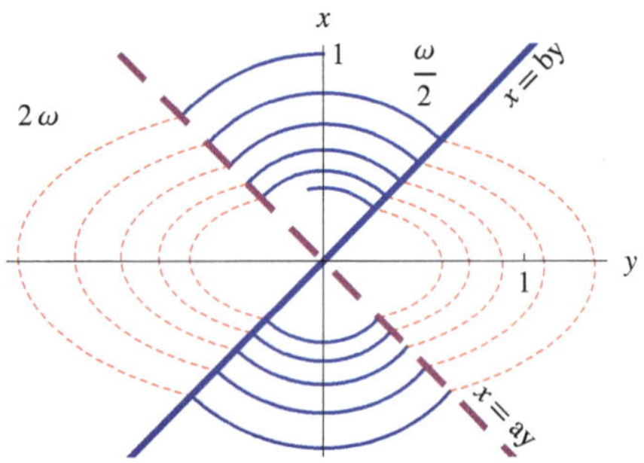

Fig. 3.18 Trajectory switching between two damped oscillators

$$\phi \to [(\mathscr{D}; (?T \cup (?x = ay; \omega := 2\omega ..) \cup (?x = by; \omega := \frac{\omega}{2} ..)))^*]\phi$$

$$\text{where} \quad \phi \equiv \omega \geq 0 \land d \geq 0 \land \omega^2x^2 + y^2 \leq c$$

$$\mathscr{D} \equiv x' = y \land y' = -\omega^2x - 2d\omega y$$

$$?T \equiv ?true$$

$$(?x = ay; \omega := 2\omega ..) \equiv ?x = ay; \omega := 2\omega;\ d := \frac{d}{2}; c := c\frac{(2\omega)^2 + 1^2}{\omega^2 + 1^2}$$

$$(?x = by; \omega := \frac{\omega}{2} ..) \equiv ?x = by; \omega := \frac{\omega}{2};\ d := 2d; c := c\frac{\omega^2 + 1^2}{(2\omega)^2 + 1^2}$$

Fig. 3.19 Parametric switched damped oscillator system

switches on a diagonal of the form $x = ay$ (dashed diagonal) to the faster angular frequency ($\omega := 2\omega$), and switches to the slower angular frequency($\omega := \frac{\omega}{2}$) at a diagonal of the form $x = by$ (solid diagonal). When the undamped angular frequency ω changes, the damping factor d is changed conversely and the range coefficient c adapts to the current ω.

Note that the system in Fig. 3.19 does not force the system to switch modes eagerly (there is no evolution domain restriction on the differential equations). Yet, while the system does not force switching, it only allows switching at the respective diagonals, when the tests $?x = ay$ or $?x = by$ are successful, which is all that we need. The intuition why we do not need to enforce switching to obtain a safe model is that the previous example, Example 3.21, showed that the continuous DA-constraint is safe and only switching could make it unsafe. But comparing the safe (and stable) evolution in Fig. 3.18 with the unsafe (and even divergent) evolution in Fig. 3.20, we find that the choice of a and b for the switching surfaces is crucial,

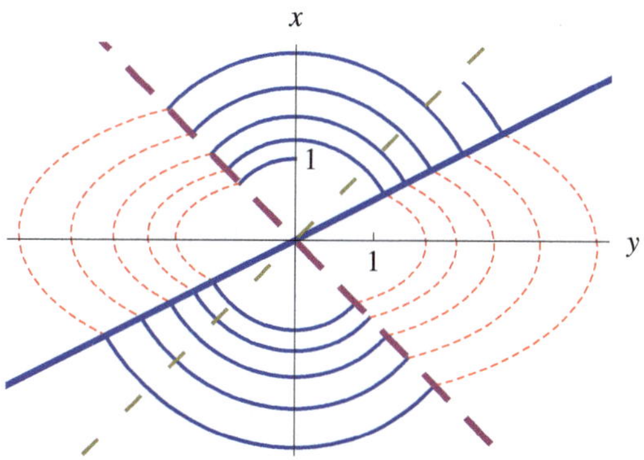

Fig. 3.20 Instable trajectory switching between two damped oscillators

which is the only difference between both evolutions. How do we find out which choices of parameters a and b are safe?

With differential induction (and differential strengthening and differential weakening, respectively), we can analyse the safety property in Fig. 3.19 of the switched damped oscillators, which conjectures that the system always stays within the region $\omega^2 x^2 + y^2 \leq c$. We analyse this property in the DAL proof in Fig. 3.21, using the abbreviations from Fig. 3.19.

The proof follows an induction by proof rule *ind* with invariant ϕ and then uses differential strengthening *DS* to show that ϕ is an auxiliary invariant of the DA-constraint $\mathscr{D}$, which can be proven by differential induction *DI* in the left branch immediately. Consequently, differential strengthening rule *DS* adds the auxiliary invariant ϕ. This rule adds auxiliary invariant ϕ into the evolution domain restrictions on the right branch after it has been proven $\in$ on the left branch to be invariant under the dynamics by *DI*. The addition of ϕ on the right branch then is a pseudo restriction of the dynamics, because ϕ is an invariant. The right branch now uses the (derived) differential weakening rule $[DR']$ to overapproximate the dynamics to the new evolution domain restriction ϕ. From ϕ, the remaining switching dynamics of the DA-program can be proven by splitting it into the three cases resulting from

$$
\dfrac{\dfrac{\dfrac{*}{\phi \vdash [\mathscr{D}]\phi}\,DI \qquad \dfrac{\dfrac{\dfrac{\dfrac{\dfrac{\dfrac{\dfrac{\dfrac{*}{\phi \vdash \phi}\,ax}{\phi \vdash [?T]\phi}\,[?] \quad \dfrac{\phi \vdash -2 \le a \le 2 \qquad \phi \vdash b^2 \ge \frac{1}{3}}{\phi \vdash [(?x = ay; \omega := 2\omega..)]\phi \wedge [(?x = by; \omega := \frac{\omega}{2}..)]\phi}}{\phi \vdash [?T]\phi \wedge [(?x = ay; \omega := 2\omega..)]\phi \wedge [(?x = by; \omega := \frac{\omega}{2}..)]\phi}\,\wedge r}{\phi \vdash [(?T \cup (?x = ay; \omega := 2\omega..) \cup (?x = by; \omega := \frac{\omega}{2}..))]\phi}\,[\cup]}{\phi \vdash \phi \to [(?T \cup (?x = ay; \omega := 2\omega..) \cup (?x = by; \omega := \frac{\omega}{2}..))]\phi}\,\to r}{\phi \vdash \forall^\alpha(\phi \to [(?T \cup (?x = ay; \omega := 2\omega..) \cup (?x = by; \omega := \frac{\omega}{2}..))]\phi)}\,\forall r}{\phi \vdash [\mathscr{D} \wedge \phi][(?T \cup (?x = ay; \omega := 2\omega..) \cup (?x = by; \omega := \frac{\omega}{2}..))]\phi}\,[DR']}{\phi \vdash [\mathscr{D}][(?T \cup (?x = ay; \omega := 2\omega..) \cup (?x = by; \omega := \frac{\omega}{2}..))]\phi}\,DS}{\phi \vdash [\mathscr{D}; (?T \cup (?x = ay; \omega := 2\omega..) \cup (?x = by; \omega := \frac{\omega}{2}..))]\phi}\,[;]}{\phi \vdash [(\mathscr{D}; (?T \cup (?x = ay; \omega := 2\omega..) \cup (?x = by; \omega := \frac{\omega}{2}..)))^*]\phi}\,ind}{\vdash \phi \to [(\mathscr{D}; (?T \cup (?x = ay; \omega := 2\omega..) \cup (?x = by; \omega := \frac{\omega}{2}..)))^*]\phi}\,\to r
$$

Fig. 3.21 Parametric switched damped oscillator proof

the three cases of the nondeterministic choice by rules $[\cup]$,$\wedge$r. The leftmost of those branches is provable by the axiom rule *ax* directly, the rightmost two cases need a few more rule applications, including quantifier elimination, and then lead to the two branches indicated on the top right:

$$\phi \vdash -2 \le a \le 2 \qquad \text{and} \qquad \phi \vdash b^2 \ge \frac{1}{3}$$

The left case corresponds to the choice $(?x = ay; \omega := 2\omega..)$, and the right case to the choice $(?x = by; \omega := \frac{\omega}{2}..)$. From this proof, we can thus identify the constraints on the switching regions a and b that we need for the safety property to hold — quite similarly to our approach in Sect. 2.9 and, later, in Sect. 3.11. The resulting constraints we identify by combining the missing requirements are:

$$-2 \le a \le 2 \wedge b^2 \ge \frac{1}{3}. \tag{3.9}$$

In particular, when we add these assumptions to ϕ, the proof in Fig. 3.21 closes successfully, showing that the switched damped oscillator does not diverge but stays inside its region under these constraints.

Observe that, even though the system switches between two safe and stable continuous modes (damped oscillators), it is not at all evident that the switched damped oscillator is safe. In fact, the counterexample in Fig. 3.20 shows that surprisingly small variations of a and/or b make the switched system unsafe and instable, even make it diverge to infinity. The discovered constraints (3.9) characterise the different situations. For the evolution in Fig. 3.18, we have chosen the switching conditions $a = -1, b = 1$, and for Fig. 3.20 we have chosen $a = -1, b = 0.5$. □

Structural Properties of Differential Invariants

Differential invariants enjoy structural closure properties. They are closed under conjunction (because of the conjunctive definition in Definition 3.13) and the next lemma shows that they are closed under differentiation, which we summarise as:

$$F, G \text{ differential invariants then } F \wedge G \text{ differential invariant (of same system)}$$
$$F \text{ differential invariant then } D(F) \text{ differential invariant (of same system)}$$

Lemma 3.7 (Closure properties of differential invariants). *Differential invariants are closed under differentiation: The total derivative of a differential invariant is an invariant of the same DA-constraint.*

Proof. Let F be a differential invariant, i.e., satisfy *DI* for some DA-constraint of the form $\exists y\,(x' = \theta \wedge \chi)$, using vectorial notation for x and y. Hence, the premise of *DI* is provable: $\forall x \forall y\,(\chi \to F'^{\theta}_{x'})$ where the quantifier for x results from the universal closure $\forall^{\alpha}$. We have to show that the derivative $F'^{\theta}_{x'}$ is invariant and extend the proof to a proof of $[\exists y\,(x' = \theta \wedge \chi)]F'^{\theta}_{x'}$ by weakening (Lemma 3.6):

$$\dfrac{\dfrac{*}{\vdash \mathrm{QE}(\forall x \forall y\,(\chi \to F'^{\theta}_{x'}))}\,\mathrm{r}\forall}{\vdash [\exists y\,(x' = \theta \wedge \chi)]F'^{\theta}_{x'}}\,[DR']$$

□

3.5.7 *Differential Induction with Differential Variants*

The differential induction techniques with differential invariants (rule *DI*) can prove invariance properties. There is a dual proof technique of differential induction with differential *variants* (rule *DV*) to prove reachability or attractor properties.

Differential Variants

Unlike the differential induction rule *DI* for differential invariants, rule *DV* uses differential induction to prove that F is a *differential variant*, which is reached differentially as an attractor region rather than sustained differentially as in *DI*. The essential difference between *DV* and *DI* thus is the progress condition $F' \geq \varepsilon$ in the premise, saying that the total differential of F along the DA-constraint is positive and at least some $\varepsilon > 0$. There, $F' \geq \varepsilon$ is a mnemonic notation for replacing all occurrences of inequalities $a \geq b$ in F' with $a \geq b + \varepsilon$ and $a > b$ by $a > b + \varepsilon$ (accordingly for $\leq, >, <$). Intuitively, the premise expresses that, wherever χ holds but F does not yet hold, the total derivative is pointing towards F; see Fig. 3.22.

Fig. 3.22 Differential variants

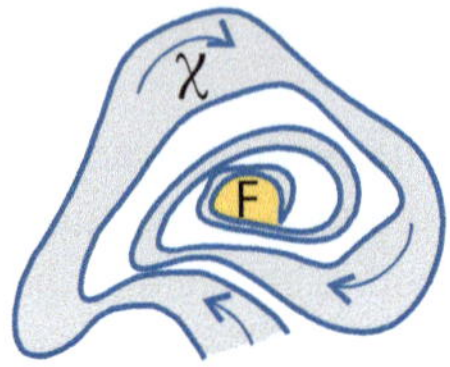

Especially, $F' \geq \varepsilon$ guarantees a minimum progress rate of ε towards F along the dynamics. To further ensure that the continuous evolution towards F remains within χ, the antecedent of the conclusion shows that χ holds *until* F is attained, which can again be proven using *DI*. Overall, the premise of rule *DV* shows that the dynamics makes progress (at least ε) toward F, and the antecedent shows that the dynamics does not leave evolution domain restriction χ on the way to F. In this context, $\sim F$ is a shorthand notation for *weak negation*, i.e., the operation that behaves like $\neg$, except that $\sim(a \geq b) \equiv b \geq a$ and $\sim(a > b) \equiv a \leq b$. Unlike negation, weak negation retains the border of F, which is required in *DV* as χ needs to continue to hold (including the border of F) until F is reached. Especially, for rule *DV*, invariant χ is not required to hold after F has been reached successfully. The operations $F' \geq \varepsilon$ and $\sim F$ are defined accordingly for other inequalities (in rule *DV*, we do not permit F to contain equalities; see Counterexample 3.25 below). Again, we demonstrate differential induction and the necessity of its prerequisites in a series of examples.

Example 3.23. As a very simple example for using differential induction with differential variants, consider the property where we want to prove that $\langle x' = a \rangle x \geq b$, i.e., we can finally reach region $x \geq b$, when we follow the dynamics $x' = a$ long enough. We analyse this DAL formula in the following DAL proof:

$$\cfrac{\cfrac{\vdash a > 0}{\vdash \exists \varepsilon{>}0\, \forall x\, (x \leq b \rightarrow a \geq \varepsilon)}\,r\exists}{\vdash \langle x' = a \rangle x \geq b}\,DV$$

As the proof reveals, the property is valid if only $a > 0$, which makes sense, because the system dynamics is then evolving towards $x \geq b$; otherwise it is evolving away from $x \geq b$ (if $a < 0$) or is constant ($a = 0$). For the above proof, we do not need to solve the differential equations. Solving the differential equation would be trivial here for a constant a, but is more involved when a is an arbitrary term. Thus, instead, we just form the total differential of $F \equiv x \geq b$, which gives $F' = x' \geq b'$. When we substitute in the differential equations $x' = a$ and assume that $a' = 0, b' = 0$, we obtain $F'^{a\,0\,0}_{x'\,a'\,b'} \equiv a \geq 0$. Consequently, $(F' \geq \varepsilon)^{a\,0\,0}_{x'\,a'\,b'}$ gives $a \geq \varepsilon$. If we can prove $a \geq \varepsilon$ holds for one common minimum progress $\varepsilon > 0$, then the system makes some minimum progress towards the goal and will reach it in finite time. This even holds if we restrict the progress condition to all x that have not yet reached $x \geq b$ or are on the border of $x \geq b$, which is the assumption $x \leq b$ in the premise. □

Example 3.24 (Aircraft progress). Recall the aircraft progress property that we proved in Fig. 3.11 on p. 169. In the rightmost side deduction, *DV* is used to prove that $F \equiv x_1 \geq p_1 \wedge x_2 \geq p_2$ is finally reached. There, the total derivative is $F' \equiv x_1' \geq 0 \wedge x_2' \geq 0$, which yields $d_1 \geq 0 \wedge d_2 \geq 0$ when substituting the flight equations $\mathscr{F}(\omega)$, because $x_1' = d_1, x_2' = d_2, p_1' = p_2' = 0$. Consequently, for $\omega = 0$, $(F' \geq \varepsilon)_{x_1'\,x_2'\,d_1'\,\,\,d_2'\,\,p_1'\,p_2'}^{d_1\,d_2\,-\omega d_2\,\omega d_1\,0\,\,0}$ is identical to $(F' \geq \varepsilon)_{x_1'\,x_2'\,p_1'\,p_2'}^{d_1\,d_2\,0\,\,0}$, giving $d_1 \geq \varepsilon \wedge d_2 \geq \varepsilon$. Similarly, the proof for formula (3.6) can be generalised to differential inequalities, again assuming $d_1' = d_2' = p_1' = p_2' = 0$ and $b' = 0$:

$$\forall p\, \exists d\, (\|d\|^2 \leq b^2 \wedge \langle x_1' \geq d_1 \wedge x_2' \geq d_2 \rangle (x_1 \geq p_1 \wedge x_2 \geq p_2)).$$

Using differential refinement rule $\langle DR \rangle$ and Lemma 3.4, the differential inequalities, which express lower bounds on the evolution of x_1 and x_2, can be reduced equivalently to differential equations with quantified disturbance $u \in \mathbb{R}^2$:

$$\forall p\, \exists d\, ..\langle \exists u\, (x_1' = d_1 + u_1 \wedge x_2' = d_2 + u_2 \wedge u_1 \geq 0 \wedge u_2 \geq 0)) \rangle (x_1 \geq p_1 \wedge x_2 \geq p_2).$$

The proof for this DAL formula is identical to that in Fig. 3.11, except that *DV* now yields $\forall x \forall u\, ((x_1 < p_1 \vee x_2 < p_2) \wedge u_1 \geq 0 \wedge u_2 \geq 0 \rightarrow d_1 + u_1 \geq \varepsilon \wedge d_2 + u_2 \geq \varepsilon)$. □

Requirements of Differential Variants

Like the proof rule *DI* for differential invariants, the proof rule *DV* for differential variants has requirements on the formulas. We illustrate why the requirements formulated for the proof rule *DV* are necessary in general.

Counterexample 3.25 (Equational differential variants). Rule *DV* is not applicable for equations like $x = y$. Even though $x = y$ can be encoded as $F \equiv x \leq y \wedge x \geq y$, the corresponding $F' \geq \varepsilon \equiv x' + \varepsilon \leq y' \wedge x' \geq y' + \varepsilon$ is equivalent to false for $\varepsilon > 0$. Indeed, assuming $a' = b' = 0$, the validity of a formula like $\langle x' = a \wedge y' = b \rangle x = y$ depends on a more involved relationship of the initial values of x and y and the constants a and b: It is true, iff $(x - y)(a - b) < 0 \vee x = y$ holds initially.

More generally, differential variants cannot (directly) verify conjunctive equations as in $\langle x' = a \wedge y' = b \rangle (x = 0 \wedge y = 0)$ because differential variants guarantee that a target region F will be reached, but not when precisely. In particular, differential variants cannot guarantee that $x = 0$ and $y = 0$ would be reached simultaneously. In fact, for $a, b \neq 0$, the above reachability property is only valid iff $bx = ay \wedge ax < 0$ initially. □

Counterexample 3.26 (Minimal progress requirement). Unlike in discrete domains, strictly monotonic sequences can converge in $\mathbb{R}$. Thus, the premise $F' \geq \varepsilon$ for an $\varepsilon > 0$ of *DV* cannot be weakened to $F' > 0$. To see why, consider the counterexample in Fig. 3.23a, in which x converges monotonically to 0 along the dynamics shown in Fig. 3.23b. This counterexample illustrates that differential variance is a

$$\dfrac{\dfrac{*\ (\text{unsound})}{\vdash \forall x\,(x>0\to -x<0)}}{\vdash \langle x'=-x\rangle\, x\le 0}$$

Fig. 3.23a Monotonically decreasing convergent counterexample

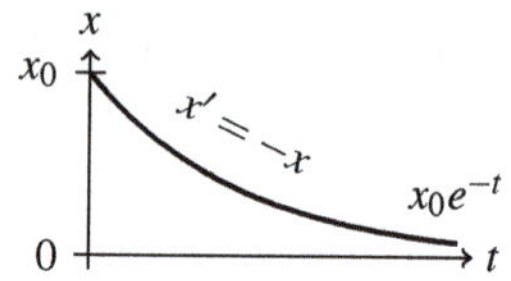

Fig. 3.23b Convergent descent dynamics

$$\dfrac{\dfrac{\mathit{false}}{\vdash \forall x\,(-x\ge 0)}\ \text{r}\forall}{x\ge 0\vdash [x'=-x]\,x\ge 0}\ DI$$

Fig. 3.23c Non-inductive property in convergent descent

matter that is no less subtle than differential invariance, and we have to prove soundness of the proof rules very carefully. Moreover, this example demonstrates that, in the presence of convergent dynamics, a property like $x\ge 0$ can be invariant even though it is not differentially invariant; see Fig. 3.23c. □

Counterexample 3.27 (Lipschitz continuity requirement). As the counterexample in Fig. 3.24a shows, Lipschitz continuity (or at least the existence of a solution of sufficient duration) is, in fact, a necessary prerequisite for *DV*. For $x=y=0$ initially, the solution of the differential equations in Fig. 3.24a is $x(t)=t$ and $y(t)=\tan t$. In explosive examples like the corresponding dynamics in Fig. 3.24b, where solution y grows unbounded in finite time, the duration of existence of solutions is limited so that the target region $x\ge 6$ is physically unreachable. More precisely, the dynamics

$$\dfrac{\dfrac{*\ (\text{unsound})}{\vdash \exists \varepsilon{>}0\,\forall x\,\forall y\,(x<6\to 1\ge \varepsilon)}}{\vdash \langle x'=1\wedge y'=1+y^2\rangle\, x\ge 6}$$

Fig. 3.24a Counterexample of unbounded dynamics without Lipschitz continuity

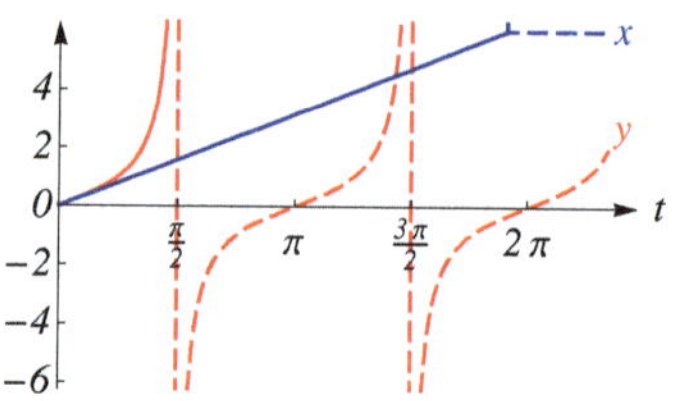

Fig. 3.24b Explosive dynamics with limited duration of solutions

is not well-posed beyond the explosive point of unbounded growth at the singularity $\frac{\pi}{2}$ and is non-physical beyond that singularity. The reason this can happen is that the differential equation is not (globally) Lipschitz continuous but only locally Lipschitz continuous and disobeys divergence of time (Sect. 3.3). The condition of (global) Lipschitz continuity is directly expressible as a formula for *DV*:

$$\exists L \forall y_1 \ldots \forall y_k \forall x_1 \ldots \forall x_n \forall \tilde{y}_1 \ldots \forall \tilde{y}_k \forall \tilde{x}_1 \ldots \forall \tilde{x}_n$$
$$(\theta_1 - \tilde{\theta}_1)^2 + \cdots + (\theta_n - \tilde{\theta}_n)^2 \leq L^2((x_1 - \tilde{x}_1)^2 + \cdots + (x_n - \tilde{x}_n)^2)$$

where $\tilde{\theta}_i$ denotes the result of substituting all x_j in θ_i with the corresponding $\tilde{x}_j$, and the y_j with $\tilde{y}_j$. Observe that, besides Lipschitz continuity, any other condition can be used for rule *DV* that ensures the existence of a solution of sufficient duration. □

3.6 Soundness

In this section we prove that verification with the DAL calculus always produces correct results about DA-programs, i.e., the DAL calculus is sound. In light of the various subtleties and sources of unsoundness we have pointed out in the previous sections, we will devote our attention to a very careful soundness proof.

Theorem 3.1 (Soundness of DAL). *The* DAL *calculus is* sound, *i.e., every* DAL *formula that can be derived in the* DAL *calculus is valid (true in all states).*

Proof. The calculus is sound if each rule instance is sound. The rules of the DAL calculus are even *locally sound*, i.e., their conclusion is true in state ν if all its premises are true in ν. Local soundness implies soundness. The local soundness proofs of $\langle;\rangle$,$[;]$,$\langle\cup\rangle$,$[\cup]$ and the propositional rules are as in Theorem 2.1. Similarly, rules *ind* and *con* are local versions of induction schemes, and the proof is as in Theorem 2.1; likewise for rules $[]gen$,$\langle\rangle gen$. The local soundness of $\langle:=\rangle$,$[:=]$ is a generalisation of the proofs for update rules [37] to first-order DJ-constraints. The proofs for $\langle\exists\rangle$,$[\exists]$,$\langle J\rangle$,$[J]$ are simple. Finally, our results from Theorem 2.1 can be lifted to show that locally sound rules are closed under addition of the Γ,Δ context in Definition 3.16. Soundness would even be closed under the addition of conjunctive DJ-constraints as rule prefixes as in Definition 2.11. For soundness, however, conjunctive DJ-constraints are crucial here [37, 235] as these are deterministic.

r$\exists$ Rule r$\exists$ is locally sound: Let ν be a state in which the premise is true, i.e.,

$$\nu \models \mathrm{QE}(\exists x \bigwedge_i (\Gamma_i \vdash \Delta_i)).$$

We have to show that the conclusion is true in this state. Using the fact that quantifier elimination (QE) yields an equivalence, we see that state ν also satisfies $\exists x \bigwedge_i (\Gamma_i \vdash \Delta_i)$ prior to the quantifier elimination. Hence, for some state ν_x that agrees with ν except for the value of x, we obtain:

$$\nu_x \models \bigwedge_i (\Gamma_i \vdash \Delta_i).$$

As the side deduction $(\star)$ in Fig. 3.10 is inductively shown to be locally sound, we can conclude that $\nu_x \models (\Gamma \vdash \Delta, \phi)$. Hence, $\nu \models \exists x (\Gamma \vdash \Delta, \phi)$.

Now the conjecture can be obtained using standard reasoning with quantifiers and the absence of x in Γ,Δ by rewriting the conclusion with local equivalences:

$$\exists x\,(\Gamma \vdash \Delta,\phi) \equiv \exists x\,(\neg\Gamma \vee \Delta \vee \phi) \equiv \neg\Gamma \vee \Delta \vee \exists x\,\phi \equiv \Gamma \vdash \Delta, \exists x\,\phi\,. \tag{3.10}$$

The soundness proof for r$\forall$ is similar since (3.10) holds for any quantifier, as x does not occur in Γ,Δ. The proofs of l$\exists$ and l$\forall$ can be derived using duality of quantifiers.

$[D]$ By Lemma 3.3, there is an equivalent disjunctive normal form $\mathscr{D}_1 \vee \cdots \vee \mathscr{D}_n$ of $\mathscr{D}$. Thus, it only remains to show that $\rho(\mathscr{D}) \subseteq \rho((\mathscr{D}_1 \cup \ldots \cup \mathscr{D}_n)^*)$ as the converse inclusion is obvious. Let φ be a differential state flow for a transition $(\nu,\omega) \in \rho(\mathscr{D})$. We assume that φ is non-Zeno according to Definition 3.12. Thus, there is a finite number, m, of switches between the disjuncts $\mathscr{D}_i$, say $\mathscr{D}_{i_1}, \mathscr{D}_{i_2}, \ldots, \mathscr{D}_{i_m}$. Then, the transition (ν,ω) belonging to φ can be simulated piecewise by m repetitions of $\mathscr{D}_1 \cup \ldots \cup \mathscr{D}_n$, where each piece selects the respective part $\mathscr{D}_{i_j}$. The proof for $\langle D\rangle$ is similar.

$[DR]$ Local soundness of the rules $[DR]$ and $\langle DR\rangle$ is an immediate consequence of Lemma 3.3 and the respective semantics of modalities.

DS Rule *DS* can be proven locally sound using the fact that the left premise implies that every flow φ that satisfies $\mathscr{D}$ also satisfies χ *all along* the flow. Thus, $\varphi \models \mathscr{D}$ implies $\varphi \models \mathscr{D} \wedge \chi$ so that the right premise entails the conclusion.

DI Let ν satisfy the premise and the antecedent of the conclusion as, otherwise, there is nothing to show. Because $D(F)$ is defined in terms of the literals of F, we can assume F to be in disjunctive normal form (also see Lemma 3.3). Consider any disjunct G of F that is true at ν. In order to show that F is sustained during the continuous evolution, it is sufficient to show that each conjunct of G is. We can assume these conjuncts to be of the form $c \geq 0$ (or $c > 0$ where the proof is similar). Finally, using vectorial notation, we write $x' = \theta$ for the differential equation system and $\exists y$ for the chain of quantifiers. Now let $\varphi : [0,r] \to \text{State}(\Sigma)$ be any state flow with $\varphi \models \exists y\,(x' = \theta \wedge \chi)$ beginning in $\varphi(0) = \nu$. In particular, $\varphi \models \exists y\,\chi$, which, by antecedent, implies $\nu \models F$, i.e., $c \geq 0$ holds at ν. We assume duration $r > 0$, because the other case is immediate ($\nu \models c \geq 0$ already holds). We show that $c \geq 0$ holds all along the flow φ, i.e., $\varphi \models c \geq 0$.
Suppose there was a $\zeta \in [0,r]$ where $\varphi(\zeta) \models c < 0$; this will lead to a contradiction. Then the function $h : [0,r] \to \mathbb{R}$ defined as $h(t) = val(\varphi(t),c)$ satisfies $h(0) \geq 0 > h(\zeta)$, because $\nu \models c \geq 0$ by antecedent. Clearly, φ is of the order of $D(c)$, because: φ is of order 1 for all variables in vector x, and trivially of order ∞ for variables that do not change during the DA-constraint. Further, by bound variable renaming, $D(c)$ cannot contain the quantified variables y; hence, φ is not required to be of any order in y. The value of c is defined all along φ, because we have assumed χ to guard against zeros of denominators. Thus, by Lemma 3.1, h is continuous on $[0,r]$ and

differentiable at every $\xi \in (0,r)$. The mean value theorem implies that there is a $\xi \in (0,\zeta)$ such that $\frac{dh(t)}{dt}(\xi)\cdot(\zeta-0) = h(\zeta)-h(0) < 0$. In particular, since $\zeta \geq 0$, we can conclude that $\frac{dh(t)}{dt}(\xi) < 0$. Now Lemma 3.1 implies that $\frac{dh(t)}{dt}(\xi) = val(\bar{\varphi}(\xi), D(c)) < 0$. The latter equals[1] $val(\bar{\varphi}(\xi)_y^u, D(c)_{x'}^{\theta})$ by Lemma 3.2, because $\varphi \models \exists y\,(x' = \theta \wedge \chi)$ so that $\bar{\varphi}(\xi)_y^u \models x' = \theta \wedge \chi$ for some $u \in \mathbb{R}$ and because y' does not occur and $y \notin c$. This, however, is a contradiction, because the premise implies that $\varphi \models \forall y\,(\chi \to D(c)_{x'}^{\theta} \geq 0)$ as $\forall^\alpha$ comprises all variables that change during the flow φ along $x' = \theta$, i.e., the vector x. In particular, as $\bar{\varphi}(\xi)_y^u \models \chi$ holds, we have $\bar{\varphi}(\xi)_y^u \models D(c)_{x'}^{\theta} \geq 0$.

DV First, we consider the quantifier free case, again using vectorial notation. Let ν be any state satisfying the premise and the antecedent of the conclusion. Since ν satisfies the premise and, after bound variable renaming, ε is a fresh variable, we can assume ν to satisfy $\nu \models \forall^\alpha(\neg F \wedge \chi \to (F' \geq \varepsilon)_{x'}^{\theta})$. For *DV*, we required $x' = \theta$ to be Lipschitz continuous so that the global Picard-Lindelöf theorem (Theorem B.2 or its corollary Corollary B.1) ensures the existence of a global solution of arbitrary duration $r \geq 0$, which is all we need here. Let φ be a state flow corresponding to a solution of the differential equation $x' = \theta$ starting in ν of some duration $r \geq 0$. If there is a point in time ζ at which $\varphi(\zeta) \models F$, then by antecedent, until (and including, because $\sim F$ contains the closure of $\neg F$) the first such point, χ holds true during φ. Hence, the restriction of φ to $[0,\zeta]$ is a state flow witnessing $\nu \models \langle x' = \theta \wedge \chi \rangle F$. If, otherwise, there is no such point, then we show that extending φ by choosing a larger r will inevitably make F true. We thus have $\varphi \models \neg F \wedge \chi$ and, by premise, $\varphi \models F'^{\theta}_{x'} \geq \varepsilon$, because $\forall^\alpha$ comprises the variables x that change during φ. By Definition 3.13, $F'^{\theta}_{x'} \geq \varepsilon$ is a conjunction. Consider one of its conjuncts, say $c'^{\theta}_{x'} \geq \varepsilon$ belonging to a literal $c \geq 0$ of F (the other cases are similar). Again, φ is of the order of $D(c)$ and the value of c is defined along φ, because $\varphi \models \chi$ and χ is assumed to guard against zeros. Hence, by the mean value theorem, Lemma 3.1, and Lemma 3.2, we conclude for each $\zeta \in [0,r]$ that

$$val(\varphi(\zeta),c) - val(\varphi(0),c) = val(\bar{\varphi}(\xi), c'^{\theta}_{x'})(\zeta - 0) \geq \zeta val(\varphi(0),\varepsilon)$$

for some $\xi \in (0,\zeta)$. Now as $val(\varphi(0),\varepsilon) > 0$ we have for all $\zeta > -\frac{val(\varphi(0),c)}{val(\varphi(0),\varepsilon)}$ that $\varphi(\zeta) \models c \geq 0$ and $\varphi(r) \models c \geq 0$, and even $\varphi(r) \models c > 0$. By enlarging r sufficiently, we have that all literals $c \geq 0$ of one conjunct of F are true, which concludes the proof, because, until F finally holds, $\varphi \models \chi$ is implied by the antecedent as shown earlier.

In the presence of quantifiers ($\exists y$ with vectorial notation), rule *DV* implies a slightly stronger statement, because y is quantified universally in the premise (and antecedent): F can be reached for *all* choices of y that respect χ

[1] For $u \in \mathbb{R}$ let $\bar{\varphi}(\xi)_y^u$ denote the (augmented) state that agrees with $\bar{\varphi}(\xi)$ except that the value of y is u.

(rather than just for one). By antecedent, there is a $u \in \mathbb{R}$ such that $\nu_y^u \models \chi$. Hence, ν_y^u satisfies the assumptions of the above quantifier-free case. Thus, $\nu_y^u \models \langle x' = \theta \wedge \chi\rangle F$, which entails that $\nu \models \langle \exists y\,(x' = \theta \wedge \chi)\rangle F$ using u constantly as the value for the quantified variable y during the evolution. □

Consequently, we know that all formulas provable in the DAL calculus are valid and hence reflect true properties.

3.7 Restricting Differential Invariants

Example 3.18 on p. 173 shows that differential invariant F cannot generally be assumed to hold in the premise of *DI* without losing soundness. Nevertheless, we present two corresponding refinements of *DI* that are indeed sound, even though they assume the differential invariant F to hold in the induction step.

Proposition 3.2 (Open differential induction). *Using the notation of the proof rulesDI,DV, the following variations of differential induction rule DI are sound (in DI′, F describes an open set):*

$$(DI')\ \frac{\vdash \forall^\alpha \forall y_1 \ldots \forall y_k\,(F \wedge \chi \to F'^{\theta_1}_{x'_1} \cdots ^{\theta_n}_{x'_n})}{[\exists y_1 \ldots \exists y_k\,\chi]F \vdash [\exists y_1 \ldots \exists y_k\,(x'_1 = \theta_1 \wedge \cdots \wedge x'_n = \theta_n \wedge \chi)]F}$$

$$(DI'')\ \frac{\vdash \forall^\alpha \forall y_1 \ldots \forall y_k\,(F \wedge \chi \to (F' > 0)^{\theta_1}_{x'_1} \cdots ^{\theta_n}_{x'_n})}{[\exists y_1 \ldots \exists y_k\,\chi]F \vdash [\exists y_1 \ldots \exists y_k\,(x'_1 = \theta_1 \wedge \cdots \wedge x'_n = \theta_n \wedge \chi)]F}$$

Proof. The proof that rule *DI′* is sound is similar to the soundness proof for *DI* in Theorem 3.1, except that assuming $\varphi(\zeta) \models \neg F$ only yields $h(0) \geq 0 \geq h(\zeta)$, which does not lead to a contradiction. However, by using the fact that F is open, the distance to the border of F is positive in the initial state $\varphi(0)$, which yields the inequality $h(0) > 0 \geq h(\zeta)$, and the contradiction arises accordingly.

The soundness of rule *DI″* needs more adaptation. Repeating the argument for *DI*, we can assume F to be of the form $c \geq 0$. Suppose there was a $\iota \in [0,r]$ where $\varphi(\iota) \models c < 0$, which will lead to a contradiction. Let $\zeta \in [0,r]$ be the infimum of these ι; hence, $\varphi(\zeta) \models c = 0$ by continuity. Then the function $h : [0,r] \to \mathbb{R}$ defined as $h(t) = val(\varphi(t), c)$ satisfies $h(0) \geq 0 \geq h(\zeta)$, because $\nu \models c \geq 0$ by antecedent. By repeating the argument with Lemma 3.1 as in the proof for *DI*, h is continuous on $[0,r]$ and differentiable at every $\xi \in (0,r)$ with a derivative of $\frac{dh(t)}{dt}(\xi) = val(\bar\varphi(\xi), D(c))$, which in turn equals $val(\bar\varphi(\xi), D(c)^\theta_{x'})$, as $\varphi \models x' = \theta$. Now, the mean value theorem implies that there is a $\xi \in (0,\zeta)$ such that

$$\frac{dh(t)}{dt}(\xi) \cdot (\zeta - 0) = h(\zeta) - h(0) \leq 0.$$

In particular, as $\zeta \geq 0$, we can conclude that $\frac{dh(t)}{dt}(\xi) = val(\bar\varphi(\xi), D(c)^\theta_{x'}) \leq 0$. This, however, contradicts the fact that the premise implies $\bar\varphi(\xi) \models D(c)^\theta_{x'} > 0$, as the flow satisfies $\varphi \models \chi$ and $\varphi(\xi) \models c \geq 0$, because $\zeta > \xi$ is the infimum of the counterexamples ι with $\varphi(\iota) \models c < 0$. □

Example 3.28. Consider the differential equation $x' = x^3$. With either rule *DI'* or rule *DI''*, we can establish easily that the system stays above $\frac{1}{4}$ whenever the dynamics starts above $\frac{1}{4}$ (refer to Fig. 3.13b on p. 172 for the dynamics):

$$
DI'\;\dfrac{r\forall\;\dfrac{*}{\vdash \forall x\,(x > \frac{1}{4} \rightarrow x^3 > 0)}}{x > \frac{1}{4} \vdash [x' = x^3]x > \frac{1}{4}}
$$

Observe that this property is not provable with rule *DI* directly, because $x^3 > 0$ does not hold for all x, but only for those where the invariant $x > \frac{1}{4}$ is true already. □

Example 3.29 (Negative equations splitting). As an example with a negative equality consider the DAL formula $x \neq 0 \rightarrow [x' = x]x \neq 0$ with the negative equality $x \neq 0$. In examples like these, differential induction does not work directly; see Counterexample 3.19. Yet when we replace $x \neq 0$ by the equivalent $x < 0 \vee x > 0$, we can prove invariance of this formula along the dynamics $x' = x$ separately by reasoning by cases; see Fig. 3.25. In the left branch, we strengthen the differential

$$
\vee l\;\dfrac{DS\;\dfrac{DI'\;\dfrac{r\forall\;\dfrac{*}{\vdash \forall x\,(x < 0 \rightarrow x < 0)}}{x < 0 \vdash [x' = x]x < 0} \qquad [DR']\;\dfrac{r\forall\;\dfrac{*}{\vdash \forall x\,(x < 0 \rightarrow x \neq 0)}}{x < 0 \vdash [x' = x \wedge x < 0]x \neq 0}}{x < 0 \vdash [x' = x]x \neq 0} \qquad DS\;\dfrac{*\ \text{similarly}}{x > 0 \vdash [x' = x]x \neq 0}}{x < 0 \vee x > 0 \vdash [x' = x]x \neq 0}
$$

Fig. 3.25 Differential induction splitting over disjunctions for negative equations

equations with $x < 0$ by *DS* and prove that $x < 0$ is a differential invariant (by rule *DI'* or *DI''*). In the middle branch, we show that the auxiliary evolution domain $x < 0$ implies the postcondition $x \neq 0$ by differential weakening $[DR']$. On the right branch, there is a corresponding proof for strengthening with $x > 0$ by *DS* and proving differential invariance by rule *DI'* and differential weakening by $[DR']$ to show that $x > 0$ also implies $x \neq 0$. Note in particular that two different ways to use differential strengthening are required on both branches here when splitting differential induction over disjunctions. □

3.8 Differential Monotonicity Relaxations

Evolution domain constraints of DA-constraints are helpful for differential induction rule *DI*, because they provide stronger assumptions for the premise. In fact, the whole purpose of the differential strengthening rule *DS* is to enrich evolution domain constraints of DA-constraints in order to weaken the subgoals of *DI*.

In contrast, evolution domain constraints are more demanding for differential variant induction rule *DV*, because the antecedent of its goal requires that evolution domain region χ be shown to remain true throughout the evolution (after all, evolution domains χ of DA-constraints $\mathscr{D}$ are restrictions that can be used as assumptions for $[\mathscr{D} \wedge \chi]\phi$ but have to be shown to hold true for $\langle \mathscr{D} \wedge \chi \rangle \phi$). Similarly, for evolution rules that are based on solutions of differential equations—rules $\langle ' \rangle$,$[']$ from the d$\mathscr{L}$ calculus in Fig. 2.11 on p. 79—evolution domain regions make the subgoal formulas much more complex (even though they lead to weaker subgoals for rule $[']$). In particular, evolution domain regions increase the number of quantifier alternations in $\langle ' \rangle$,$[']$, which have the predominant influence on the complexity of quantifier elimination [94].

For simplifying non-differential evolution domain region χ from $[\mathscr{D} \wedge \chi]\phi$ with a DA-constraint $\mathscr{D}$, we can simply use the differential weakening rule $[DR]$ to drop χ:

$$[DR]\ \frac{\vdash [\mathscr{D}]\phi}{\vdash [\mathscr{D} \wedge \chi]\phi}$$

Less conservatively, we can approximate the assumption $\forall 0 \le \tilde{t} \le t\, \langle \mathscr{S}_{\tilde{t}} \rangle \chi$ on the solution $\mathscr{S}_t$ in the subgoal of rule $[']$ from Fig. 2.11 by $\langle \mathscr{S}_t \rangle \chi$ (or by $\chi \wedge \langle \mathscr{S}_t \rangle \chi$) in a sound yet incomplete way. If every evolution of the solution $\mathscr{S}_t$ satisfying χ at the end satisfies ϕ, then every evolution along $\mathscr{D}$ satisfying χ all the time must satisfy ϕ even more so. Thus, the following variant of $[']$ is sound (but less complete) where $\mathscr{S}_t$ is the solution of the differential equation as for rule $[']$:

$$([']')\ \frac{\forall t \ge 0\, (\chi \to \langle \mathscr{S}_t \rangle (\chi \to \phi))}{[x_1' = \theta_1 \wedge .. \wedge x_n' = \theta_n \wedge \chi]\phi}$$

For simplifying non-differential evolution domain region χ from $\langle \mathscr{D} \wedge \chi \rangle \phi$, we can use the dual of the differential strengthening rule *DS* and show that χ remains true throughout the evolution along $\mathscr{D}$ (left subgoal, which can be handled using *DI*) so that only some evolution along $\mathscr{D}$ remains to be found that actually reaches ϕ (right subgoal):

$$DS\ \frac{\vdash [\mathscr{D}]\chi \quad \vdash \langle \mathscr{D} \rangle \phi}{\vdash \langle \mathscr{D} \wedge \chi \rangle \phi}$$

When using *DI* to prove the left subgoal $[\mathscr{D}]\chi$ by showing validity of a formula of the form $\forall^{\mathscr{D}} \chi'^{\theta}_{x'}$, we actually show invariance of χ along $\mathscr{D}$.

More generally, we can use differential-algebraic techniques similar to differential induction to prove the weaker property of monotonicity instead of invariance of χ.

Definition 3.17 (Monotonicity derivation). Let α be a DA-program. For a first-order formula F, the following formula is called *monotonicity derivation* of F, where the syntactic derivative $D(a)$ is defined according to Definition 3.13:

$$M^{\alpha}(F) \equiv \bigwedge_{i=1}^{m} M^{\alpha}(F_i) \quad \text{where } \{F_1,\dots,F_m\} \text{ is the set of all literals of } F;$$

$$M^{\alpha}(a \sim b) \equiv \forall^{\alpha}(D(a) \geq D(b)) \vee \forall^{\alpha}(D(a) \leq D(b)) \quad \text{where } \sim \in \{\leq,\geq,<,>,=\}.$$

Proposition 3.3 (Differential monotonicity). *Let $\langle \mathscr{S}_t \rangle$ be the DJ-constraint for the solution at time t of the symbolic initial value problem for the differential equation $\mathscr{D}$ defined as $x_1' = \theta_1 \wedge \dots \wedge x_n' = \theta_n$ as in rule $\langle ' \rangle$ of Fig. 2.11. Let the non-differential constraint χ be a conjunction of atomic formulas without negative equalities; then the following is a sound proof rule:*

$$(\langle ' \rangle') \; \frac{\vdash \exists t{\geq}0\,(\chi \wedge \langle \mathscr{S}_t \rangle(\chi \wedge \phi)) \qquad \vdash M^{\mathscr{D}}(\chi)_{x_1'}^{\theta_1} \cdots {}_{x_n'}^{\theta_n}}{\vdash \langle x_1' = \theta_1 \wedge \dots \wedge x_n' = \theta_n \wedge \chi \rangle \phi}$$

Proof. Local soundness is a simple consequence of the well-known fact that, for a differentiable function f, monotonic increasing of f on an interval $[c,d]$ is equivalent to $f'(z) \geq 0$ on (c,d). With this, the right subgoal implies that, for any conjunct $a \geq b$ of χ (likewise for $\leq,<,>,=$), the value of $a-b$ is either monotonically increasing or monotonically decreasing along the flow. Either way, if χ holds in the beginning *and* the end of a flow of some duration t (as implied by the left subgoal), monotonicity implies that χ holds all along the flow, so that the subgoals imply the conclusion as in case $\langle ' \rangle$ of the proof of Theorem 2.1.

Formally, let φ be a state flow of an appropriate duration r following solution $\mathscr{S}_t$ according to the left subgoal as in Theorem 2.1. By the left subgoal we have $\varphi(r) \models \phi$ (when r is the witness for $\exists t{\geq}0$) and we only need to show that $\varphi \models \chi$. Consider a conjunct $a \geq b$ of χ and consider the case where the right subgoal implies $\varphi(0) \models \forall^{\mathscr{D}}(D(a) \leq D(b))_{x'}^{\theta}$, using vectorial notation for x and θ. Then $\varphi \models (a' \leq b')_{x'}^{\theta}$, because the universal closure $\forall^{\mathscr{D}}$ comprises all variables that change during the flow φ along $\mathscr{D}$. Thus $\varphi \models a' \leq b'$ by Lemma 3.2. Let $h : [0,r] \to \mathbb{R}$ be the function defined as $h(t) = val(\varphi(t), a-b)$. Again, φ is of the order of $a'-b'$ (φ is of the order 1 in each x_i and of arbitrary order for other variables) and the value of $a-b$ is defined all along φ, because χ guards against zeros in χ. Thus, Lemma 3.1 is applicable and h is differentiable at every $\xi \in (0,r)$. For any $\zeta \in [0,r]$, we have to show that $\varphi(\zeta) \models \chi$. By the mean value theorem, there is a $\xi \in (\zeta, r)$ such that, when using Lemma 3.1, we have

$$h(r) - h(\zeta) = h'(\xi) \cdot (r-\zeta) = val(\bar{\varphi}(\xi), a'-b') \cdot (r-\zeta) \leq 0$$

because $\bar{\varphi}(\xi) \models a' \leq b'$. Thus, we have $h(\zeta) \geq h(r) \geq 0$, since $\varphi(r) \models \chi$, which implies that $\varphi(r) \models a-b \geq 0$ and $\varphi(r) \models a \geq b$.

The other case where the right subgoal implies $\varphi(0) \models \forall^{\alpha}(a' \geq b')$ is simpler using the fact that $\varphi(0) \models \chi$ and is, in fact, a direct consequence of the proof of *DI* in Theorem 3.1. The other conjuncts of the form $a \leq b, a < b, a > b$, and $a = b$ are almost identical, because the monotonicity argument for $a-b$ carries over easily, as the respective conjunct holds before and after the continuous evolution. □

Example 3.30 (Monotonic invariants in train control). Consider the differential constraints for train control (equation (2.7) on p. 62 in Sect. 2.4). For the DA-constraint $z' = v \land v' = a \land \tau' = 1 \land v \geq 0 \land \tau \leq \varepsilon$, evolution domain region $v \geq 0 \land \tau \leq \varepsilon$ can be shown to be monotonic or convex with respect to the dynamics. That is, if it holds in the beginning and at the end of an evolution, the invariant also holds in between. Monotonicity is easy to prove with the above proof rule using the following symbolic computations for the right subgoal (where $\forall^\alpha$ is $\forall z \forall v \forall \tau$):

$$\begin{aligned}
&((\forall^\alpha (v' \geq 0) \lor \forall^\alpha (v' \leq 0)) \land (\forall^\alpha (\tau' \geq \varepsilon') \lor \forall^\alpha (\tau' \leq \varepsilon')))_{z'\,v'\,\tau'\,\varepsilon'}^{v\;a\;1\;0} \\
&\equiv (\forall^\alpha (a \geq 0) \lor \forall^\alpha (a \leq 0)) \land (\forall^\alpha (1 \geq 0) \lor \forall^\alpha (1 \leq 0)) \\
&\equiv \mathit{true}.
\end{aligned}$$

Observe that the invariant domain will not be a differential invariant, here, because $v \geq 0$ is only an invariant of $z' = v \land v' = a$ for $a \geq 0$. For any a, however, $v \geq 0$ will be either a monotonically increasing (if $a \geq 0$ constantly) or a monotonically decreasing (if $a \leq 0$ constantly) property, one of which is true for every constant a. Thus, if $v \geq 0$ has been true *before and after* an evolution along $z' = v \land v' = a$, it must have been true throughout this evolution. Likewise, $\tau \leq \varepsilon$ never is a differential invariant of $\tau' = 1$, because the passing of time along $\tau' = 1$ will inevitably violate $\tau \leq \varepsilon$ sooner or later. Still, it is a monotonically decreasing property. Consequently, the monotonicity relaxation of Proposition 3.3 applies for the train control example, thereby simplifying proofs with evolution domain regions considerably, because the invariant only needs to be checked before and after rather than throughout the evolution. For instance, this simplifies the proof of property (2.20) on p. 121. □

Similarly, the right subgoal of rule $\langle ' \rangle'$ is a sufficient condition to ensure that rule $[']'$ is a complete replacement for rule $[']$.

Counterexample 3.31 (Disjunctive monotonicity). For soundness of the differential monotonicity relaxations, it is crucial that rule $\langle ' \rangle'$ only accept conjunctive evolution domain regions. As the counterexample in Fig. 3.26a with the dynamics in Fig. 3.26b shows, differential monotonicity relaxations do not hold for disjunctive evolution domain regions, because the same disjunct has to hold before and after the evolution for monotonicity arguments to be sound. Let χ abbreviate the disjunctive evolution domain region $x \leq 1 \lor x \geq 2$. Then the differential monotonicity criterion $\forall x (1 \leq 0) \lor \forall x (1 \geq 0)$ would be fulfilled, but a different disjunct holds at the initial state $x = 0$ than at the target $x \geq 3$ so that monotonicity implies neither that $x \leq 1$ nor that $x \geq 2$ holds in between. □

Counterexample 3.32 (Negative equalities). A similar counterexample shows why rule $\langle ' \rangle'$ does not allow negative equalities. Along the dynamics $x' = 1 \land x \neq 2$ we cannot conclude from the truth of $x \neq 2$ before and after the evolution that, on the basis of a condition on the derivative $x' \neq 0$, $x \neq 2$ held true throughout the evolution. A continuous evolution from $x = 0$ to $x = 3$ still leaves $x \neq 2$ in between. □

$$\dfrac{\dfrac{*\ (\text{unsound})}{\vdash \exists t\geq 0\,\big(\chi\wedge\langle x:=x+t\rangle(\chi\wedge x\geq 3)\big)}}{x=0\vdash\langle x'=1\wedge(x\leq 1\vee x\geq 2)\rangle x\geq 3}$$

Fig. 3.26a Counterexample for disjunctive monotonicity

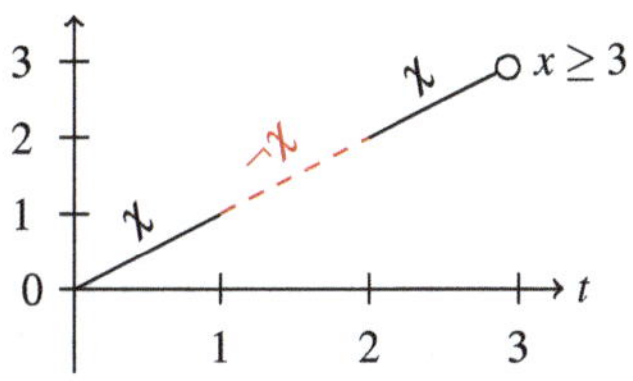

Fig. 3.26b Interrupted dynamics

3.9 Relative Completeness

As a consequence of the Incompleteness Theorem 2.2 for d$\mathscr{L}$ and the fact that DAL is a conservative extension of d$\mathscr{L}$ (Proposition 3.1), the DAL calculus is not effectively axiomatisable (yet even just reachability is undecidable for hybrid systems [156]).

It is easy to see that the relative completeness proof for d$\mathscr{L}$ (Theorem 2.3) generalises to DAL with only minor modifications when using the first-order logic of DA-constraints as a basis in place of FOD (again, nested modalities can be avoided when using quantifiers). The first-order logic of DA-constraints results from FOD by allowing DA-constraints in place of differential equations inside modalities.

Theorem 3.2 (Relative completeness of DAL). *The* DAL *calculus is* complete relative to *DA-constraints, i.e., every valid* DAL *formula can be derived from tautologies of the first-order logic of DA-constraints.*

Proof. The proof is a simple adaptation of the proof of Theorem 2.3 for d$\mathscr{L}$ in Sect. 2.7.2: In the proof of the program rendition Lemma 2.8, we replace all cases for continuous evolutions along differential equations or for discrete jumps by the following cases for DA-constraints $\mathscr{D}$ or DJ-constraints $\mathscr{J}$, respectively:

$$\mathscr{S}_{\mathscr{D}}(\vec{x},\vec{v}) \equiv \langle\mathscr{D}\rangle\vec{v}=\vec{x},$$
$$\mathscr{S}_{\mathscr{J}}(\vec{x},\vec{v}) \equiv \mathscr{J}^{v_i=\theta_i}_{x_i:=\theta_i}.$$

The first-order formula $\mathscr{J}^{v_i=\theta_i}_{x_i:=\theta_i}$ results from $\mathscr{J}$ by replacing all assignments of any form $x_i := \theta_i$ with equations $v_i = \theta_i$. The rest of the relative completeness proof generalises immediately using the fact that the respective rules ($\langle\exists\rangle$,$[:=]$) for DJ-constraints are symmetric, and hence equivalent, and their premises are of smaller complexity. □

Note that, for generalising the relative completeness proof in the most simple way, we formally need to allow update prefixes in DAL proofs as in Definition 2.10, which is easily seen to be sound for deterministic DJ-constraints.

3.10 Deductive Strength of Differential Induction

We analyse the deductive power of differential induction with respect to classes of formulas that are allowed as differential invariants. For purely equational differential invariants, the deductive power is not affected by allowing or disallowing propositional operators in differential invariants:

Proposition 3.4 (Equational deductive power). *The deductive power of differential induction with atomic equations is identical to the deductive power of differential induction with propositional combinations of polynomial equations: Formulas are provable with propositional combinations of equations as differential invariants iff they are provable with only atomic equations as differential invariants.*

Proof. We show that every differential invariant that is a propositional combination ϕ of polynomial equations is expressible as a single atomic polynomial equation (the converse inclusion is obvious). We assume ϕ to be in negation normal form and reduce ϕ inductively using the following transformations:

- If ϕ is of the form $p_1 = p_2 \vee q_1 = q_2$, then ϕ is equivalent to the single equation $(p_1 - p_2)(q_1 - q_2) = 0$. Further, $\phi' \equiv p_1' = p_2' \wedge q_1' = q_2'$ directly implies

$$((p_1 - p_2)(q_1 - q_2))' = 0 \equiv (p_1' - p_2')(q_1 - q_2) + (p_1 - p_2)(q_1' - q_2') = 0.$$

- If ϕ is of the form $p_1 = p_2 \wedge q_1 = q_2$, then ϕ is equivalent to the single equation $(p_1 - p_2)^2 + (q_1 - q_2)^2 = 0$. Further, $\phi' \equiv p_1' = p_2' \wedge q_1' = q_2'$ implies

$$\left((p_1 - p_2)^2 + (q_1 - q_2)^2\right)' = 0 \equiv 2(p_1 - p_2)(p_1' - p_2') + 2(q_1 - q_2)(q_1' - q_2') = 0.$$

- If ϕ is of the form $\neg(p_1 = p_2)$, then ϕ does not qualify as a differential invariant, because it contains a negative equality, which is disallowed for *DI* according to the conditions in Fig. 3.9. □

Observe, however, that the required polynomial degree of atomic equations is larger than for propositional combinations, which can have computational disadvantages for quantifier elimination.

For general differential invariants, where inequalities are allowed, the situation is different: We show that, in general, the deductive power of differential induction depends on which class of formulas is allowed as differential invariants! Some DAL formulas cannot by proven by a differential induction step with only atomic formulas but no propositional operators as differential invariants, while they are provable immediately using unrestricted differential invariants.

Theorem 3.3 (Deductive power). *The deductive power of differential induction with arbitrary formulas exceeds the deductive power of differential induction with atomic formulas: All* DAL *formulas that are provable using atomic differential invariants are provable using general differential invariants, but not vice versa!*

Proof. The inclusion is obvious. Conversely, we have to show that there are DAL formulas that are provable with general differential invariants but not with atomic differential invariants. Consider the following example, which is provable using rule DI', i.e., the variant of DI for open sets (Sect. 3.7), with the non-atomic formula $x>0 \land y>0$ as differential invariant:

$$\dfrac{\dfrac{*}{\vdash \forall x \forall y\,(x>0 \land y>0 \rightarrow xy>0 \land xy>0)}\,\text{r}\forall}{x>0 \land y>0 \vdash [x'=xy \land y'=xy](x>0 \land y>0)}\,DI'$$

First, we show that this formula is not provable by a differential induction step with only atomic formulas as differential invariants. Suppose there was a single polynomial $p(x,y)$ in variables x,y such that $p(x,y)>0$ is a differential invariant proving the above formula, which will lead to a contradiction. The conditions for differential invariants (DI or DI') imply that the following formulas have to be valid:

1. $x>0 \land y>0 \rightarrow p(x,y)>0$, as differential invariants have to hold in the prestate according to the antecedent of DI (or DI').
2. $p(x,y)>0 \rightarrow x>0 \land y>0$, as the differential invariant has to imply the postcondition (when using $[]gen$ to show that the differential invariant implies the postcondition).

In particular, $x>0 \land y>0 \leftrightarrow p(x,y)>0$ is valid, and p is not the zero polynomial. Thus, p enjoys the property:

$$p(x,y) \geq 0 \text{ for } x \geq 0, y \geq 0, \text{ and, otherwise, } p(x,y) \leq 0. \tag{3.11}$$

Assume p has minimal total degree with property (3.11). Now, $p(x,0)$ is a univariate polynomial in x with zeros at all $x>0$; thus $p(x,0)=0$ is the zero polynomial, hence y divides $p(x,y)$. Similarly, $p(0,y)=0$ for all y, and hence x divides $p(x,y)$. Thus, xy divides p. But by comparing the signs (cf. Fig. 3.27), we see that the polynomial $\frac{-p(-x,-y)}{xy}$ also satisfies property (3.11) with a smaller total degree than p, which is a contradiction. In detail: $p(x,y)$ satisfies the sign conditions (3.11) indicated in the outer part of Fig. 3.27. It is divisible by xy, but $\frac{p(x,y)}{xy}$ satisfies different sign conditions (indicated in the middle part of Fig. 3.27). Yet, by flipping the signs, $\frac{-p(-x,-y)}{xy}$ again satisfies the same sign conditions (3.11) as $p(x,y)$ but with a smaller total degree (indicated in the inner part of Fig. 3.27), which is a contradiction.

Similarly, there is no polynomial p such that $x>0 \land y>0 \leftrightarrow p(x,y)=0$, because only the zero polynomial is zero on the full quadrant $(0,\infty)^2$. Finally, property $x>0 \land y>0 \leftrightarrow p(x,y) \geq 0$ is impossible for continuity reasons, which imply that $p(0,0)=0$, which is a contradiction. More generally, the same argument holds for any other sign condition that is supposed to characterise one quadrant of $\mathbb{R}^2$ uniquely.

Observe that, so far, the argument does not depend on the actual dynamics and is, thus, still valid in the presence of arbitrary differential weakening ($[DR]$).

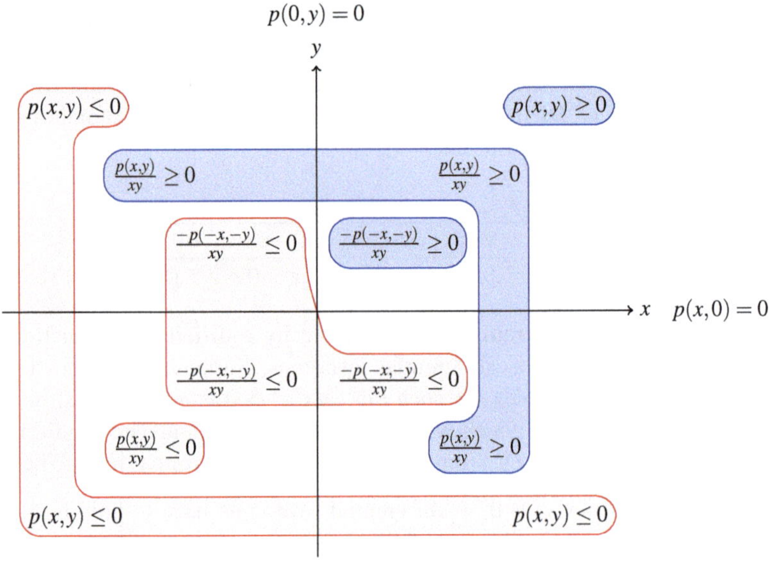

Fig. 3.27 Quadrant sign selection regions of differential invariant

Next, to see that the above example cannot be proven indirectly after differential strengthening (*DS*), we use the fact that, inductively, the strengthening χ itself needs to be a differential invariant: Ultimately, the left subgoal of *DS* can only be shown using differential induction. The above example, however, is built such that, as $x' = xy$ is the differential equation, $xy > 0$ is required for $x > 0$ to be a differential invariant (which thus also requires $y > 0$). Conversely, due to $y' = xy$, formula $xy > 0$ is a prerequisite for the differential invariance of $y > 0$ (which thus also needs $x > 0$). Yet, for differential invariance of $xy > 0$, we have to prove $xy > 0 \rightarrow (y+x)xy > 0$ for DI', because $(xy)'^{xy\,xy}_{x'\,y'}$ gives $(x'y+yx')^{xy\,xy}_{x'\,y'}$, i.e., $xyy+yxy$. But the property $xy > 0 \rightarrow (y+x)xy > 0$ is, again, equivalent to $x \geq 0 \vee y \geq 0$, and thus equivalent to $\neg(-x > 0 \wedge -y > 0)$, which cannot be proven by atomic differential induction (or differential weakening) according to the first part of this proof. Thus, the required atomic differential invariants have

Fig. 3.28 Circular dependencies for differential strengthening

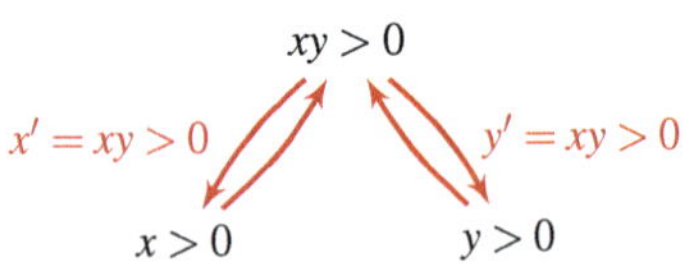

circular dependencies for differential strengthening by $x > 0$, $y > 0$, and $xy > 0$, respectively; see Fig. 3.28. These cannot be resolved in any proof tree without simultaneous differential induction using non-atomic differential invariants, because differential strengthenings have to be ordered totally along each proof branch. □

As a special case, this result implies that differential induction in DAL is deductively stronger than approaches using barrier certificates [251, 252], criticality functions [91], or polynomial invariant equations [274, 269]. On top of that, the DAL calculus adds differential strengthening and weakening techniques, which add further deductive power. The roundabout manoeuvre that we verify in the next section is a practical example where differential induction with mixed non-atomic formulas and successive differential strengthening turns out to be decisive.

3.11 Air Traffic Control Verification

In this section we verify that the tangential roundabout manoeuvre for collision avoidance in air traffic control that we presented in Sect. 3.4 is collision-free. That is, the manoeuvre directs aircraft on flight paths with global minimal distance $p > 0$. We determine a corresponding parameter constraint on the *tang* procedure from the roundabout manoeuvre in Fig. 3.7. Using differential induction and differential strengthening, we can verify the flight manoeuvre despite the complicated hybrid flight dynamics of aircraft whose solutions fall into undecidable classes of arithmetic. Recall the differential flight equations from Sect. 3.4.2:

$$x_1' = d_1 \land x_2' = d_2 \land d_1' = -\omega d_2 \land d_2' = \omega d_1 \qquad (\mathscr{F}(\omega)^*)$$
$$y_1' = e_1 \land y_2' = e_2 \land e_1' = -\varpi e_2 \land e_2' = \varpi e_1 \qquad (\mathscr{G}(\varpi)^*)$$

3.11.1 Characterisation of Safe Roundabout Dynamics

Property ϕ in Fig. 3.7 defines *safe states* as those with separation $\|x-y\| \geq p$. This does *not*, however, characterise the states with *safe dynamics*: Several states that satisfy ϕ will not remain safe when following curved roundabout flight manoeuvres; see Fig. 3.6c on p. 151 for a counterexample violating ϕ after some time. In particular, the angular velocity ω and initial speed vectors d and e must fit to the relative positioning of the aircraft x and y. Otherwise the aircraft dynamics will not remain safe from safely separated initial states. In order to discover the required parametric constraints for safety of the roundabout manoeuvre, we analyse the DAL formula $\phi \to [trm^*]\phi$ in the DAL calculus and identify a corresponding parameter constraint $\mathscr{T}$. For notational convenience, we inline side deductions and slightly simplify the universal closure notation $\forall^\alpha$ by taking free variables as universally quantified here, as with Skolem terms in Chap. 2, because the following DAL proof needs no existential variables.

$$
\dfrac{\dfrac{\dfrac{\dfrac{\dfrac{\dfrac{\dfrac{*}{\phi \vdash \forall x,y,d,e(\phi \to \phi)}\,{\scriptstyle r\forall}}{\phi \vdash [free]\phi}\,{\scriptstyle [DR']} \qquad \dfrac{\dfrac{\cdots}{\phi \vdash [tang](\phi \land \mathscr{T})} \quad \dfrac{\cdots}{\phi \land \mathscr{T} \vdash [\mathscr{F}(\omega) \land \mathscr{G}(\omega)]\phi}}{\phi \vdash [tang; \mathscr{F}(\omega) \land \mathscr{G}(\omega)]\phi}\,{\scriptstyle []gen}}{\phi \vdash [free][tang; \mathscr{F}(\omega) \land \mathscr{G}(\omega)]\phi}\,{\scriptstyle []gen}}{\phi \vdash [trm]\phi}\,{\scriptstyle [;]}}{\vdash \forall^{\alpha}(\phi \to [trm]\phi)}\,{\scriptstyle r\forall,\to r}}{\phi \vdash [trm^*]\phi}\,{\scriptstyle ind}}{\vdash \phi \to [trm^*]\phi}\,{\scriptstyle \to r}
$$

The left branch closes, because postcondition ϕ is the evolution domain restriction in free flight such that its DA-constraint can be weakened by Lemma 3.6. In the other branches, $\mathscr{T}$ is the parameter constraint that *tang* needs to establish in addition to ϕ (middle branch) for the roundabout dynamics to be safe (right branch). Hence condition $\mathscr{T}$ mediates between the middle and right branches. Using successive quantifier elimination on the right branch, we derive the following constraint $\mathscr{T}$ as a prerequisite for ϕ to be differentially inductive. It is the decisive constraint that characterises configurations with safely controllable dynamics in curved roundabout manoeuvres (using vectorial notation and orthogonal complements $d^{\perp}$ from Sect. 3.2):

$$
\begin{aligned}
\mathscr{T} &\equiv d - e = \omega(x-y)^{\perp} \quad \left(\text{or, equivalently } (d-e)^{\perp} = -\omega(x-y)\right) \qquad (3.12)\\
&\equiv d_1 - e_1 = -\omega(x_2 - y_2) \land d_2 - e_2 = \omega(x_1 - y_1).
\end{aligned}
$$

This formula expresses that the relative speed vector $d-e$ is orthogonal to the relative position $x-y$ and compatible with the angular velocity ω and tangential orientation of d and e. Figure 3.29a illustrates the symmetric case of $\mathscr{T}$ with identical linear speed $\|d\| = \|e\|$. Figures 3.29b and 3.29c show asymmetric cases with distinct linear speeds $\|d\| \neq \|e\|$, which is possible as well. Condition $\mathscr{T}$ gives the decisive

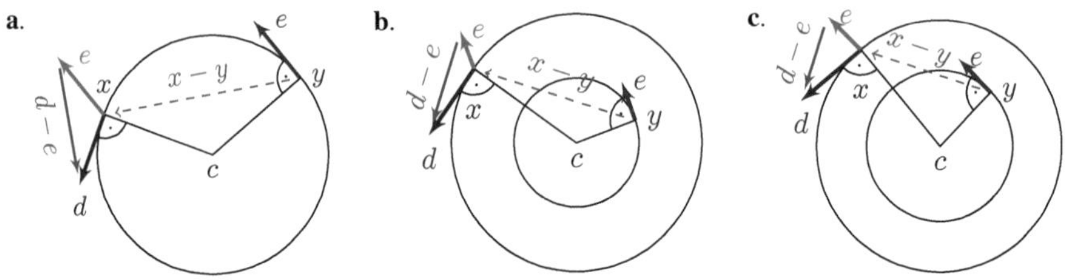

Fig. 3.29 Tangential construction for characteristics $\mathscr{T}$ of roundabout dynamics

handle for an inductive characterisation of safe tangential roundabout configurations: For the right branch of the above proof, we need to show that the tangential configuration $\mathscr{T}$ is sufficient for ϕ to be sustained during curved evasive actions. In the following, we prove that the relative speed vector configuration $\mathscr{T}$ is itself differentially inductive (rule *DI* in left branch). We use differential strengthening with *DS* as a differential cut to augment the dynamics with $\mathscr{T}$ as a derived invariant for proving that the actual safety property ϕ is sustained (right branch), again by differential induction rule *DI*:

$$\dfrac{\dfrac{\dfrac{*}{\vdash \forall^\alpha(\mathscr{T}'_{\mathscr{F}(\omega)\wedge\mathscr{G}(\omega)})}\,\mathrm{r}\forall}{\phi,\mathscr{T}\vdash[\mathscr{F}(\omega)\wedge\mathscr{G}(\omega)]\mathscr{T}}\,DI \qquad \dfrac{\dfrac{*}{\vdash \forall^\alpha(\mathscr{T}\to\phi'_{\mathscr{F}(\omega)\wedge\mathscr{G}(\omega)})}\,\mathrm{r}\forall}{\phi\vdash[\mathscr{F}(\omega)\wedge\mathscr{G}(\omega)\wedge\mathscr{T}]\phi}\,DI}{\dfrac{\phi,\mathscr{T}\vdash[\mathscr{F}(\omega)\wedge\mathscr{G}(\omega)]\phi}{\phi\wedge\mathscr{T}\vdash[\mathscr{F}(\omega)\wedge\mathscr{G}(\omega)]\phi}\,\wedge\mathrm{l}}\,DS$$

Observe that differential strengthening by rule *DS* is crucial for the proof, because neither ϕ nor $\mathscr{T}\wedge\phi$ is differentially inductive for $\mathscr{F}(\omega)\wedge\mathscr{G}(\omega)$! Instead, the tangential configuration $\mathscr{T}$ itself is differentially inductive relative to $\mathscr{F}(\omega)\wedge\mathscr{G}(\omega)$ (left branch) and strong enough to make ϕ differentially inductive relative to the augmented DA-constraint $\mathscr{F}(\omega)\wedge\mathscr{G}(\omega)\wedge\mathscr{T}$ (right branch). For readability, we use a slightly weaker rule for differential induction, with ϕ rather than $[\mathscr{T}]\phi$ in the antecedent of the conclusion. This variant can be derived easily using a cut and will again be called *DI*. The differential induction *DI* on the left and right branch close using quantifier elimination in r$\forall$. The arithmetic is also provable by the following algebraic equational reasoning ($\mathscr{T}'_{\mathscr{F}(\omega)\wedge\mathscr{G}(\omega)}$ is a short notation for substituting the differential equations from $\mathscr{F}(\omega)\wedge\mathscr{G}(\omega)$ into $D(\mathscr{T})$; see Lemma 3.2):

$$\begin{aligned}
\mathscr{T}'_{\mathscr{F}(\omega)\wedge\mathscr{G}(\omega)} &\equiv ((d_1-e_1)'=-\omega(x_2-y_2)'\wedge(d_2-e_2)'=\omega(x_1-y_1)')_{\mathscr{F}(\omega)\wedge\mathscr{G}(\omega)}\\
&\equiv (d_1'-e_1'=-\omega(x_2'-y_2')\wedge d_2'-e_2'=\omega(x_1'-y_1'))_{\mathscr{F}(\omega)\wedge\mathscr{G}(\omega)}\\
&\equiv -\omega d_2+\omega e_2=-\omega(d_2-e_2)\wedge\omega d_1-\omega e_1=\omega(d_1-e_1) \quad \equiv \mathit{true}\\
\phi'_{\mathscr{F}(\omega)\wedge\mathscr{G}(\omega)} &\equiv (2(x_1-y_1)(x_1-y_1)'+2(x_2-y_2)(x_2-y_2)'\geq 0)_{\mathscr{F}(\omega)\wedge\mathscr{G}(\omega)}\\
&\equiv (2(x_1-y_1)(x_1'-y_1')+2(x_2-y_2)(x_2'-y_2')\geq 0)_{\mathscr{F}(\omega)\wedge\mathscr{G}(\omega)}\\
&\equiv 2(x_1-y_1)(d_1-e_1)+2(x_2-y_2)(d_2-e_2)\geq 0\\
(\text{using } \mathscr{T}) &\equiv 2(x_1-y_1)(-\omega(x_2-y_2))+2(x_2-y_2)\omega(x_1-y_1)=0\geq 0 \quad \equiv \mathit{true}.
\end{aligned}$$

Altogether, we have shown that *every* tangential roundabout evasion manoeuvre respecting $\mathscr{T}$ is safe. Further, the middle branch of the above proof reveals the parameter constraint imposed on *tang* for safe roundabouts, which concludes the proof of the following result.

Theorem 3.4 (Safety of tangential roundabout manoeuvre). *For every choice of the tangential entry procedure that satisfies $\phi\to[tang](\phi\wedge\mathscr{T})$, the tangential roundabout flight manoeuvre in Fig. 3.7 safely avoids collisions, i.e., it directs aircraft on flight paths with minimal horizontal aircraft separation at least $p>0$.*

This result can be proven in our theorem prover [242] in two seconds including user interactions for rules *ind* and *DS*. Its proof does not need rule $[]gen$, which we only used here to shorten the proof presentation. Theorem 3.4 expresses unbounded-time safety for fully parametric tangential roundabouts with arbitrary choices for the free parameters. The proof of Theorem 3.4 generalises to roundabouts entered by more than two participants when ϕ and $\mathscr{T}$ are augmented similarly. For instance, using our automatic proof procedure from Chap. 6, our theorem prover can prove mutual collision avoidance for five aircraft fully automatically; see Chaps. 6 and 8. Likewise, rules *DI* and *DS* can be used to prove that external separation to all other sufficiently far points is maintained during the roundabout manoeuvre. In particular, the manoeuvre only needs bounded space:

Proposition 3.5 (External separation of roundabout manoeuvres). *Separation of aircraft x to all external points $u \in \mathbb{R}^2$ of distance beyond the roundabout diameter $2r$ is maintained, because the following* DAL *formula is provable:*

$$r \geq 0 \land (r\omega)^2 = \|d\|^2 \to \forall u\,(\|x-u\|^2 > (2r+p)^2 \to [\mathscr{F}(\omega)](\|x-u\|^2 > p^2)).$$

3.11.2 Tangential Entry Procedures

As a simple choice for the tangential initiation procedure *tang* satisfying property $\mathscr{T}$, consider the following operation which chooses an arbitrary angular velocity ω and an arbitrary centre $c \in \mathbb{R}^2$ for the roundabout manoeuvre, and adjusts d and e tangentially:

$$tang \equiv \exists u\,\omega := u;\ \exists c\,(d := \omega(x-c)^\perp \land e := \omega(y-c)^\perp). \tag{3.13}$$

This formula expresses that the speed vectors d and e of both aircraft at x and y, respectively, are tangential and of the same angular velocity ω relative to the intended centre c of the roundabout, with the same orientation (Fig. 3.29). For this choice, the assumption of Theorem 3.4 can be proven after rule [:=] substitutes the corresponding terms for d and e in $\mathscr{T}$, using rule r$\forall$ (or linearity of $d^\perp$):

$$
{\scriptstyle [;],[\exists],[\exists]}\dfrac{{\scriptstyle \text{r}\forall,\text{r}\forall}\dfrac{{\scriptstyle [:=]}\dfrac{{\scriptstyle \land\text{r}}\dfrac{{\scriptstyle \text{ax}}\dfrac{*}{\phi \vdash \phi} \qquad \dfrac{*}{\phi \vdash \omega(x-c)^\perp - \omega(y-c)^\perp = \omega(x-y)^\perp}}{\phi \vdash \phi \land \omega(x-c)^\perp - \omega(y-c)^\perp = \omega(x-y)^\perp}}{\phi \vdash [d := \omega(x-c)^\perp \land e := \omega(y-c)^\perp](\phi \land \mathscr{T})}}{\phi \vdash \forall\omega\forall c\,[d := \omega(x-c)^\perp \land e := \omega(y-c)^\perp](\phi \land \mathscr{T})}}{\phi \vdash [tang](\phi \land \mathscr{T})}
$$

It can also be shown that $\exists c\,(d = \omega(x-c)^\perp \land e = \omega(y-c)^\perp)$ is equivalent to $\mathscr{T}$ for nonzero ω. With choice (3.13), the tangential roundabout manoeuvre in Fig. 3.7 is safe and has been significantly simplified and generalised in comparison to our prior work [238].

3.11.3 Discussion

Our tangential roundabout manoeuvre leaves open the questions of how and when precisely the collision avoidance manoeuvre is initiated or left. For instance, (3.13) does not restrict c and ω but accepts any choice including choices optimising secondary objectives such as fuel consumption. Furthermore, as specified in Fig. 3.7 and proven in this section, the roundabout manoeuvre can be left safely with arbitrary free flight by repeating the loop at any time: The roundabout manoeuvre will simply be initiated again during free flight when necessary. As a special case, this open policy includes free flight, enabling the aircraft to leave the roundabout in their original direction. While the simple choice (3.13) is possibly discontinuous in d and e, it is comparably easy to see that there are fully curved entry and exit procedures that remain safe when the entry procedure is initiated with sufficient distance by using the separation limit of Proposition 3.5. We refine the roundabout collision avoidance manoeuvre and develop a corresponding entry procedure in Chap. 8. Our proof shows that the tangential roundabout manoeuvre is safe for every such entry procedure. In particular, the control parameters c and ω of (3.13) can also be chosen such that the resulting speed vectors d and e are in a bounded range meeting external speed requirements of the aircraft, which can be proven in the DAL calculus easily:

$$\forall v\,(\phi \rightarrow \langle tang\rangle(\phi \wedge \mathscr{T} \wedge \|d\|^2 = \|e\|^2 = v^2)). \tag{3.14}$$

3.12 Summary

We have introduced a first-order dynamic logic for differential-algebraic programs with interacting first-order discrete jump constraints and first-order differential-algebraic constraints. For this differential-algebraic logic, DAL, we have presented a calculus for verifying hybrid systems given as differential-algebraic programs.

In differential-algebraic programs, both internal choices and disturbances during continuous evolutions and nondeterminism in discrete operations can be described uniformly by quantifiers. Most importantly, we have introduced first-order differential induction with differential invariants and differential variants for proving correctness statements with first-order differential-algebraic constraints purely algebraically, using the differential constraints themselves instead of their solutions. In combination with successive differential strengthening or differential cuts for refining the system dynamics by auxiliary differential invariants, we obtain a powerful verification calculus for systems with challenging dynamics. We have compared the deductive strength for classes of differential invariants and have shown that the deductive power of general differential induction exceeds the deductive power of atomic differential invariants.

We have demonstrated that our calculus can be used successfully for verifying fully parametric roundabout manoeuvres in air traffic control. To the best of our knowledge, this is the first formal proof for unbounded safety of hybrid aircraft dynamics in curved collision avoidance manoeuvres for air traffic control. Moreover, we argue that our fully formal proof about aircraft gives more confidence in flight manoeuvres than informal approaches that do not consider the actual hybrid flight dynamics [171, 104, 129] or results that only prevent orthogonal collisions in discretisations of the system [92, 203]. Our logic DAL is also more convenient, because hybrid systems like the tangential roundabout manoeuvre can be specified and verified uniformly within a single logic. Despite challenging flight dynamics, the DAL formulas about aircraft and roundabout manoeuvres that we presented in this chapter can be proven in our theorem prover KeYmaera within a few seconds.

While this work answers the open issues (1), (3), and (4) raised in the work of Piazza et al. [228], we are interested in extending differential-algebraic methods to address further questions about hybrid systems. In Chap. 6, we investigate algorithms for constructing differential invariants automatically on the basis of our DAL calculus presented here. Interesting future work for the aircraft case study is to find a fully curved manoeuvre that achieves collision avoidance by joint horizontal and vertical evasive actions.

Chapter 4
Differential Temporal Dynamic Logic dTL

Contents

4.1 Introduction **204**
 4.1.1 Related Work 205
 4.1.2 Structure of This Chapter 206
4.2 Syntax **206**
 4.2.1 Hybrid Programs 207
 4.2.2 State and Trace Formulas 207
4.3 Semantics **210**
 4.3.1 Trace Semantics of Hybrid Programs 210
 4.3.2 Valuation of State and Trace Formulas 213
 4.3.3 Conservative Temporal Extension 215
4.4 Safety Invariants in Train Control **216**
4.5 Proof Calculus **217**
 4.5.1 Proof Rules 218
 4.5.2 Verification Example 221
4.6 Soundness **221**
4.7 Completeness **223**
 4.7.1 Incompleteness 223
 4.7.2 Relative Completeness 224
 4.7.3 Expressibility and Rendition of Hybrid Trace Semantics . 225
 4.7.4 Modular Relative Completeness Proof 226
4.8 Verification of Train Control Safety Invariants **227**
4.9 Liveness by Quantifier Alternation **228**
4.10 Summary **230**

A. Platzer, Logical Analysis of Hybrid Systems,
DOI 10.1007/978-3-642-14509-4_4,

Synopsis We combine first-order dynamic logic for reasoning about the possible behaviour of hybrid systems with temporal logic for reasoning about the temporal behaviour during their operation. Our logic supports verification of hybrid programs with first-order definable flows and provides a uniform treatment of discrete and continuous evolution. For our combined logic, we generalise the semantics of dynamic modalities to refer to hybrid traces instead of final states. Further, we prove that this gives a conservative extension of our dynamic logic for hybrid systems. On this basis, we provide a modular verification calculus that reduces correctness of temporal behaviour of hybrid systems to nontemporal reasoning, and prove that we obtain a complete axiomatisation relative to the nontemporal base logic. Using this calculus, we analyse safety invariants in a train control system and symbolically synthesise parametric safety constraints.

4.1 Introduction

Correctness of real-time and hybrid systems depends on a safe operation throughout *all* states of all possible trajectories, and the behaviour at intermediate states is highly relevant [90].

Temporal logics (TLs) use temporal operators to talk about intermediate states [247, 114, 115, 6, 284]. In addition to having successful uses in model checking [78, 6, 159, 156, 217], temporal logics have been used in deductive approaches to prove validity of formulas in calculi [97, 96]. Among other shortcomings and difficulties discussed in Chap. 1, the major drawback of TL calculi for our purpose is that TL formulas cannot generally characterise the operations of a specific hybrid system.

Like model checking, *dynamic logic* (DL) [149] can directly analyse the behaviour of actual system models. However, DL only considers the behaviour at the final states, which is insufficient for verifying safety invariants that have to hold all the time, throughout the execution of the system.

We close this gap of expressivity by combining first-order dynamic logic [149] with temporal logic [247, 114, 115]. We use the generalisation of operational system models and semantics to hybrid systems from Chap. 2. In this chapter, we introduce a temporal dynamic logic dTL, which provides modalities for quantifying over traces of hybrid systems based on differential dynamic logic. We equip dTL with temporal operators to state what is true all along a trace or at some point during a trace. In this chapter, we modify the semantics of the dynamic modality $[\alpha]$ to refer to all *traces* of α instead of all final states reachable with α (similarly for $\langle\alpha\rangle$). For instance, the formula $[\alpha]\Box\phi$ expresses that ϕ is true at each state during all traces of the hybrid system α. With this, dTL can also be used to verify temporal statements about the behaviour of α at intermediate states during system runs. As in our nontemporal dynamic logic $d\mathcal{L}$, we use hybrid programs as an operational model for hybrid systems, since they admit a uniform compositional treatment of interacting discrete and continuous evolution in logic.

As a semantical foundation for combined temporal dynamic formulas, we introduce a hybrid trace semantics for dTL. We prove that dTL is a conservative extension of $d\mathcal{L}$, that is, for nontemporal specifications, trace semantics is equivalent to the nontemporal transition semantics of $d\mathcal{L}$ from Chap. 2.

As a means for verification, we introduce a sequent calculus for dTL that successively reduces temporal statements about traces of hybrid programs to nontemporal $d\mathcal{L}$ formulas. In this way, we make the intuition formally precise that temporal safety invariants can be checked by augmenting proofs with appropriate assertions about intermediate states. As in Chap. 2, our calculus works compositionally: It decomposes correctness statements about hybrid programs structurally into corresponding statements about its parts by symbolic transformation. Observe that this is somewhat challenging for hybrid systems, because even a single elementary system operation of continuous evolution exhibits temporal behaviour as it assumes several different states as time passes.

Contributions

Our approach combines the advantages of dynamic logic in reasoning about the behaviour of (multiple and parametric) operational system models with those of temporal logic to verify temporal statements about traces. Our first contribution is the logic dTL, which provides a coherent foundation for reasoning about the temporal behaviour of operational models of hybrid systems with symbolic parameters. The main contribution in this chapter is our calculus for deductively verifying temporal statements about hybrid systems, which is a complete axiomatisation relative to nontemporal $d\mathcal{L}$.

4.1.1 Related Work

Based on [254], Beckert and Schlager [38] added separate trace modalities to dynamic logic and presented a relatively complete calculus for discrete while programs. Their approach only handles discrete state spaces. In contrast, dTL works for hybrid programs with continuous state spaces. There, a particular challenge is that invariants may change their truth-value multiple times during a single continuous evolution; hence relevant temporal behaviour even occurs during single transitions.

Davoren and Nerode [97] extended the propositional modal μ-calculus with a semantics in hybrid systems and examine topological aspects. In [96], Davoren et al. gave a semantics in general flow systems for a generalisation of CTL* [115]. In both cases, the authors of [97] and [96] provided Hilbert-style calculi to prove formulas that are valid for all systems simultaneously using abstract actions.

As discussed in Sect. 1.2, the strength of our logic primarily is that it is a first-order dynamic logic and handles actual hybrid programs like $x := x+1; x' = 2y$ rather than only abstract actions of unknown effect.

4.1.2 Structure of This Chapter

After introducing syntax and semantics of the differential temporal dynamic logic dTL in Sects. 4.2 and 4.3, we introduce a modular sequent calculus for dTL in Sect. 4.5 that extends our previous calculi with temporal proof rules in a completely modular way. We prove soundness and relative completeness in Sects. 4.6 and 4.7, respectively. In Sect. 4.8, we use our calculus to analyse safety invariants in the train control system from Sect. 4.4. We further present extensions for quantifier alternation and liveness in Sect. 4.9. We draw conclusions and discuss future work in Sect. 4.10.

4.2 Syntax of Temporal Dynamic Logic for Hybrid Systems

The *temporal differential dynamic logic* dTL extends dynamic logic [149] with three concepts for verifying temporal specifications of hybrid systems:

Hybrid programs. The behaviour of hybrid systems can be described by hybrid programs (Sect. 2.2.2), which generalise real-time programs [159] to hybrid change. The distinguishing feature of hybrid programs in this context is that they provide uniform discrete jumps and continuous evolutions along differential equations, which can be combined by regular control operations. While hybrid automata [156] can be embedded, program structures are more amenable to compositional symbolic processing by calculus rules.

Modal operators. Modalities of dynamic logic express statements about all possible behaviour ($[\alpha]\pi$) of a system α, or about the existence of a trace ($\langle\alpha\rangle\pi$), satisfying condition π. Unlike in standard dynamic logic, α is a modal of a hybrid system. We use hybrid programs to describe α as in Chap. 2. Yet, unlike in standard dynamic logic [149] or $d\mathcal{L}$, π is a *trace formula* in dTL, and π can refer to all states that occur *during* a trace using temporal operators.

Temporal operators. For dTL, the temporal trace formula $\Box\phi$ expresses that the formula ϕ holds all along a trace selected by $[\alpha]$ or $\langle\alpha\rangle$. For instance, the state formula $\langle\alpha\rangle\Box\phi$ says that the state formula ϕ holds at every state along at least one trace of α. Dually, the trace formula $\Diamond\phi$ expresses that ϕ holds at some point during such a trace. It can occur in a state formula $\langle\alpha\rangle\Diamond\phi$ to express that there is such a state in some trace of α, or as $[\alpha]\Diamond\phi$ to say that along each trace there is a state satisfying ϕ. In this chapter, the primary focus of attention is on homogeneous combinations of path and trace quantifiers like $[\alpha]\Box\phi$ or $\langle\alpha\rangle\Diamond\phi$.

4.2.1 Hybrid Programs

The formulas of dTL are built from a non-empty set Σ of real-valued variables and function and predicate symbols. Signature Σ is assumed to contain the usual function and predicate symbols for real arithmetic: $0, 1, +, \cdot, =, \leq, <, \geq, >$. For simplicity, we do not distinguish between logical variables in V and state variables from Σ. The set $\mathrm{Trm}(\Sigma)$ of *terms* is defined as in classical first-order logic.

The hybrid programs allowed in dynamic modalities of dTL are the same as those of $d\mathcal{L}$; see Definition 2.3 in Sect. 2.2.2. They are built from elementary discrete jumps and continuous evolutions using a regular control structure. The set $\mathrm{HP}(\Sigma)$ of hybrid programs with variables in Σ is defined in Definition 2.3. Similarly, differential-algebraic programs from Chap. 3 can be allowed when using DAL as a basis instead of $d\mathcal{L}$, giving *differential-algebraic temporal dynamic logic* DATL.

4.2.2 State and Trace Formulas

The formulas of dTL are defined similarly to first-order dynamic logic [149]. However, the modalities $[\alpha]$ and $\langle\alpha\rangle$ accept trace formulas that refer to the temporal behaviour of *all* states along a trace. Inspired by CTL and CTL* [114, 115], we distinguish between state formulas, which are true or false in states, and trace formulas, which are true or false for system traces. The sets $\mathrm{Fml}(\Sigma)$ of state formulas and $\mathrm{Fml}_T(\Sigma)$ of trace formulas with variables in Σ are simultaneously inductively defined in Definition 4.1.

Definition 4.1 (dTL formulas). The set $\mathrm{Fml}(\Sigma)$ of *(state) formulas* is simultaneously inductively defined as the smallest set such that:

1. If $p \in \Sigma$ is a predicate of arity $n \geq 0$ and if $\theta_1, \ldots, \theta_n \in \mathrm{Trm}(\Sigma)$ are terms, then $p(\theta_1, \ldots, \theta_n) \in \mathrm{Fml}(\Sigma)$.
2. If $\phi, \psi \in \mathrm{Fml}(\Sigma)$, then $\neg\phi, (\phi \wedge \psi), (\phi \vee \psi), (\phi \rightarrow \psi) \in \mathrm{Fml}(\Sigma)$.
3. If $\phi \in \mathrm{Fml}(\Sigma)$ and $x \in \Sigma$, then $\forall x\, \phi, \exists x\, \phi \in \mathrm{Fml}(\Sigma)$.
4. If $\pi \in \mathrm{Fml}_T(\Sigma)$ and $\alpha \in \mathrm{HP}(\Sigma)$, then $[\alpha]\pi, \langle\alpha\rangle\pi \in \mathrm{Fml}(\Sigma)$.

The set $\mathrm{Fml}_T(\Sigma)$ of *trace formulas* is the smallest set with:

1. If $\phi \in \mathrm{Fml}(\Sigma)$, then $\phi \in \mathrm{Fml}_T(\Sigma)$.
2. If $\phi \in \mathrm{Fml}(\Sigma)$, then $\Box\phi, \Diamond\phi \in \mathrm{Fml}_T(\Sigma)$.

Formulas without $\Box$ and $\Diamond$, i.e., without Case 2 of the trace formulas, are *nontemporal* $d\mathcal{L}$ *formulas* (Chap. 2). Unlike in CTL, state formulas are true on a trace (Case 1) if they hold for the *last* state of a trace, not for the first. Thus, $[\alpha]\phi$ expresses that ϕ is true at the end of each trace of α. In contrast, $[\alpha]\Box\phi$ expresses that ϕ is true all along all states of every trace of α. This combination gives a smooth embedding of nontemporal $d\mathcal{L}$ into dTL and makes it possible to define

Table 4.1 Operators and meaning in differential temporal dynamic logic (dTL)

dTL Notation	Operator	Meaning
$p(\theta_1,\ldots,\theta_n)$	atomic predicate	true iff predicate p holds for $(\theta_1,\ldots,\theta_n)$
$\neg\phi$	negation / not	true if ϕ is false
$\phi \wedge \psi$	conjunction / and	true if both ϕ and ψ are true
$\phi \vee \psi$	disjunction / or	true if ϕ is true or if ψ is true
$\phi \rightarrow \psi$	implication / implies	true if ϕ is false or ψ is true
$\phi \leftrightarrow \psi$	bi-implication / equivalent	true if ϕ and ψ are both true or both false
$\forall x\, \phi$	universal quantifier / for all	ϕ is true for all values of variable x
$\exists x\, \phi$	existential quantifier / exists	ϕ is true for some values of variable x
$[\alpha]\phi$	$[\cdot]$ modality / box	ϕ is true after all runs of HP α
$\langle\alpha\rangle\phi$	$\langle\cdot\rangle$ modality / diamond	ϕ is true after at least one run of HP α
$[\alpha]\Box\phi$	$[\cdot]\Box$ modality nesting	ϕ is true always during all traces of HP α
$\langle\alpha\rangle\Diamond\phi$	$\langle\cdot\rangle\Diamond$ modality nesting	ϕ is true sometimes during some trace of HP α
$[\alpha]\Diamond\phi$	$[\cdot]\Diamond$ modality nesting	ϕ is true sometimes during all traces of HP α
$\langle\alpha\rangle\Box\phi$	$\langle\cdot\rangle\Box$ modality nesting	ϕ is true always during some trace of HP α

a compositional calculus. Like CTL, dTL allows nesting with a branching time semantics [114], e.g., $[\alpha]\Box(x \geq 2 \rightarrow \langle\beta\rangle\Diamond x \leq 0)$.

For reference, the operators of differential temporal dynamic logic and typical operator nestings are summarised in Table 4.1.

Example 4.1 (Train control). Recall the simplified ETCS train control system from Sect. 2.4 with the refinements from Sect. 2.9. In Sect. 2.9, we have proven the following $d\mathcal{L}$ formula that expresses that the train control system ensures that trains stay inside their movement authority m, no matter how long the controller runs:

$$v^2 \leq 2b(m-z) \wedge b > 0 \wedge A \geq 0 \rightarrow [(ctrl;drive)^*]z \leq m. \tag{2.19*}$$

But this $d\mathcal{L}$ formula only refers to the safety of the train position at all final states of the hybrid program $(ctrl;drive)^*$, which corresponds to the rightmost state in the transition structure from Fig. 2.9a on p. 63. Formula (2.19) does not say whether $z \leq m$ is actually also ensured in all intermediate states of the transition structure in Fig. 2.9a. For hybrid system verification, it is often insufficient to prove the safety property $z \leq m$ only at the final states, because the passengers will not like to crash into another train at an intermediate state either.

Now because the dTL modality $[\alpha]$ does not need to be followed by a temporal modality $\Box$ or $\Diamond$, the $d\mathcal{L}$ formula (2.7) also is a dTL formula. In fact, all $d\mathcal{L}$ formulas are dTL formulas (those without temporal modalities). But in dTL, we can do better. We can express that the movement authority m is respected *always* throughout the system run, not just at the final states:

$$v^2 \leq 2b(m-z) \wedge b > 0 \wedge A \geq 0 \rightarrow [(ctrl;drive)^*]\Box z \leq m. \tag{4.1}$$

Unlike the $d\mathcal{L}$ formula (2.19), the dTL formula (4.1) also refers to safety at all intermediate states of the transition structure in Fig. 2.9a. In fact, for this particular system model, both formulas are equivalent, because the variables in the postcondition $z \leq m$ only change in the last step *drive* of the hybrid program . Also $z \leq m$ must have been true at all intermediate states if it holds at the final state, because z increases monotonically (the train is not allowed to drive backwards). In that sense, the final states of the hybrid program $(ctrl;drive)^*$ included all safety-critical states, because the hybrid program was written as an appropriate controller plant loop.

For other systems, where the variables of the postcondition change in multiple parts of the hybrid program, this is no longer the case, and the temporal modality $\Box$ is, in fact, crucial to express safety properly. As a simple (though somewhat contrived) example, consider the following modification, in which we allow *drive* to also drive backwards by removing the evolution domain restriction $v \geq 0$ from it:

$$v^2 \leq 2b(m-z) \wedge b > 0 \wedge A \geq 0 \rightarrow [(a:=0;drive;ctrl;drive)^*]\Box z \leq m. \quad (4.2)$$

In the hybrid program of this dTL formula, the position z changes at two places: the two occurrences of *drive*. In the first occurrence, the train always keeps its speed for some time by choosing the acceleration $a:=0$, because the sensor information is not yet available. Only in the second occurrence of *drive* do the controller actions *ctrl* actually take effect. For specifying safety adequately in (4.2), we need the temporal modality $\Box$, because we want the movement authority to be respected at all states of the system, including intermediate states. Especially we would not want the movement authority to be violated at some point during the first *drive* even if the system recovers during the second *drive*, e.g., by driving backwards.

Surprisingly, the temporal modality $\Box$ is still redundant in (4.2), because all possibilities of running its hybrid program especially include the case where the first *drive* is run for an arbitrary time $s \leq \varepsilon$ and the second *drive* is run for zero seconds. Thus all relevant safety-critical states are covered by final states and $\Box$ is not needed. But as soon as we modify *drive* to the hybrid program

$$\tau:=0;\ (z' = v, v' = a, \tau' = 1 \,\&\, \tau \leq \varepsilon);\ ?(\tau = \varepsilon),$$

the temporal modality $\Box$ becomes truly necessary. The reason is that this modification restricts the continuous evolution to take exactly ε time units (the evolution domain region restricts the evolution to $\tau \leq \varepsilon$ and the subsequent test to $?\tau = \varepsilon$) and no intermediate state is visible as a final state anymore. □

This example shows that the question about whether all relevant intermediate safety-critical states are covered by a $d\mathcal{L}$ modality can be surprisingly subtle. In larger systems, in fact, it can become quite difficult to analyse this manually. Instead, temporal dTL modality combinations like $[\alpha]\Box\phi$ can be used to ensure that all states are covered and the hybrid system α satisfies ϕ all the time for all possible executions.

Inspired by CTL* [115], syntactic and semantic extensions from dTL to dTL* are straightforward and amount to allowing propositional combinations of trace formulas. Finding appropriate proof calculi, however, is much more difficult, even for CTL* [248, 261].

4.3 Semantics of Temporal Dynamic Logic for Hybrid Systems

In standard dynamic logic [149], the logic d$\mathcal{L}$ from Chap. 2, and the logic DAL from Chap. 3, modalities only refer to the final states of system runs and the semantics is a reachability relation on states: State ω is reachable from state ν using system α if there is a run of α which terminates in ω when started in ν. For dTL, however, formulas can refer to intermediate states of runs as well. To capture this, we change the semantics of a hybrid system α to be the set of its possible *traces*, i.e., successions of states that occur during the evolution of α. The relation between the initial and the final state alone is not sufficient.

4.3.1 Trace Semantics of Hybrid Programs

States contain values of system variables during a hybrid evolution. A *state* is a map $\nu : \Sigma \to \mathbb{R}$. In addition, we distinguish a separate state Λ to denote the failure of a system run when it is *aborted* due to a test $?\chi$ that yields *false*. In particular, Λ can only occur at the end of an aborted system run and marks that no further extension is possible because of a failed test. The set of all states is denoted by $\mathrm{Sta}(\Sigma)$.

Hybrid systems evolve along piecewise continuous traces in multi-dimensional space as time passes. Continuous phases are governed by differential equations, whereas discontinuities are caused by discrete jumps in state space. Unlike in discrete cases [254, 38], traces are not just sequences of states, since hybrid systems pass through uncountably many states even in bounded time. Beyond that, continuous changes are more involved than in pure real time [6, 159], because all variables can evolve along differential equations with different slopes. Generalising the real-time traces of [159], the following definition captures hybrid behaviour by splitting the uncountable succession of states into periods σ_i that are regulated by the same control law. For discrete jumps, some of those periods are point flows of duration 0.

The (trace) semantics of hybrid programs is compositional, that is, the semantics of a complex program is defined as a simple function of the trace semantics of its parts.

Definition 4.2 (Hybrid trace). A *trace* is a (nonempty) finite or infinite sequence $\sigma = (\sigma_0, \sigma_1, \sigma_2, \dots)$ of functions $\sigma_i : [0, r_i] \to \mathrm{Sta}(\Sigma)$ with their respective durations $r_i \in \mathbb{R}$ (for $i \in \mathbb{N}$). A *position* of σ is a pair (i, ζ) with $i \in \mathbb{N}$ and ζ in the interval $[0, r_i]$; the state of σ at (i, ζ) is $\sigma_i(\zeta)$. Positions of σ are ordered lexicographically by $(i, \zeta) \prec (j, \xi)$ iff either $i < j$, or $i = j$ and $\zeta < \xi$. Further, for a state $\nu \in \mathrm{Sta}(\Sigma)$, $\hat{\nu} : 0 \mapsto \nu$ is the *point flow* at ν with duration 0. A trace *terminates* if it is a finite sequence $(\sigma_0, \sigma_1, \dots, \sigma_n)$ and $\sigma_n(r_n) \neq \Lambda$. In that case, the last state $\sigma_n(r_n)$ is denoted by $\mathrm{last}\,\sigma$. The first state $\sigma_0(0)$ is denoted by $\mathrm{first}\,\sigma$.

Unlike in [6, 159], the definition of traces also admits finite traces of bounded duration, which is necessary for compositionality of traces in $\alpha; \beta$. The semantics of hybrid programs α as the set $\tau(\alpha)$ of its possible traces depends on valuations $val(\nu, \cdot)$

of formulas and terms at intermediate states ν. The valuation of terms and interpretations of function and predicate symbols are as for real arithmetic (Chap. 2). The valuation of formulas will be defined in Definition 4.4. Again, we use $\nu[x \mapsto d]$ to denote the *modification* that agrees with state ν on all variables except for the symbol x, which is changed to $d \in \mathbb{R}$.

Definition 4.3 (Trace semantics of hybrid programs). The *trace semantics*, $\tau(\alpha)$, *of a hybrid program* α, is the set of all its possible hybrid traces and is defined inductively as follows:

1. $\tau(x_1 := \theta_1, .., x_n := \theta_n) = \{(\hat{\nu}, \hat{\omega}) \,:\, \omega = \nu[x_1 \mapsto val(\nu, \theta_1)]..[x_n \mapsto val(\nu, \theta_n)]$ for $\nu \in \mathrm{Sta}(\Sigma)\}$
2. $\tau(x_1' = \theta_1, \dots, x_n' = \theta_n \,\&\, \chi) = \{(\varphi) \,:\, \varphi$ is a state flow of order 1 and some duration $r \geq 0$ such that $\varphi \models x_1' = \theta_1 \wedge \dots \wedge x_n' = \theta_n \wedge \chi$; see Definition 3.9$\}$
3. $\tau(?\chi) = \{(\hat{\nu}) \,:\, val(\nu, \chi) = true\} \cup \{(\hat{\nu}, \hat{\Lambda}) \,:\, val(\nu, \chi) = false\}$
4. $\tau(\alpha \cup \beta) = \tau(\alpha) \cup \tau(\beta)$
5. $\tau(\alpha; \beta) = \{\sigma \circ \varsigma \,:\, \sigma \in \tau(\alpha),\, \varsigma \in \tau(\beta)$ when $\sigma \circ \varsigma$ is defined$\}$; the composition of $\sigma = (\sigma_0, \sigma_1, \sigma_2, \dots)$ and $\varsigma = (\varsigma_0, \varsigma_1, \varsigma_2, \dots)$ is

$$\sigma \circ \varsigma := \begin{cases} (\sigma_0, \dots, \sigma_n, \varsigma_0, \varsigma_1, \dots) & \text{if } \sigma \text{ terminates at } \sigma_n \text{ and } \mathrm{last}\,\sigma = \mathrm{first}\,\varsigma \\ \sigma & \text{if } \sigma \text{ does not terminate} \\ \text{not defined} & \text{otherwise} \end{cases}$$

6. $\tau(\alpha^*) = \bigcup_{n \in \mathbb{N}} \tau(\alpha^n)$, where $\alpha^{n+1} := (\alpha^n; \alpha)$ for $n \geq 1$, as well as $\alpha^1 := \alpha$ and $\alpha^0 := (?true)$.

Time passes differently during discrete and continuous change. During continuous evolution, the discrete step index i of positions (i, ζ) remains constant, whereas the continuous duration ζ remains 0 during discrete point flows. This permits multiple discrete state changes to happen at the same (super-dense) continuous time, unlike in other approaches [6].

Example 4.2. For comparing the transition semantics of hybrid programs for d$\mathcal{L}$ from Definition 2.7 and the trace semantics of hybrid programs for dTL from Definition 4.3, consider the following simple hybrid program α:

$$a := -2a;\; a := a^2.$$

The transition semantics is just the relation between initial and final states:

$$\rho(\alpha) \equiv \{(\nu, \omega) \,:\, \omega \text{ is like } \nu \text{ except that } \omega(a) = 4\nu(a)^2\}.$$

In particular, the d$\mathcal{L}$ formula $[\alpha]a \geq 0$ is valid, because all final states have a square as the value of a. In contrast, the trace semantics of α retains all intermediate states:

$$\tau(\alpha) \equiv \{(\hat{\nu}, \hat{s}, \hat{\omega}) \,:\, s \text{ is like } \nu \text{ except } s(a) = -2\nu(a) \\ \text{and } \omega \text{ is like } s \text{ except } \omega(a) = s(a)^2 = 4\nu(a)^2\}.$$

During these traces, $a \geq 0$ does not hold at all states. If the trace starts with a positive value ($\nu \models a > 0$), then it will become negative at the point flow s (where $s \models a < 0$), yet recover to a positive value ($\omega \models a > 0$) at the end. □

Example 4.3. The previous example only had discrete jumps, and, thus, the traces only involved point flows. Now consider the hybrid program β from the train context:

$$a := -b;\ z' = v, v' = a;\ ?v \geq 0;\ a := A;\ z' = v, v' = a.$$

The transition semantics of this program only considers successful runs to completion. In particular, if $A > 0$, the velocity v will always be nonnegative at the end (otherwise the test $?v \geq 0$ in the middle fails and the program aborts), because the last differential equation will accelerate and increase the velocity again. Thus, the position z at the end of the program run will never be smaller than at the beginning.

If, instead, we consider the trace semantics of β, all intermediate states are in the set of traces:

$$\begin{aligned}
\tau(\beta) \equiv \{(\hat{\mu_0}, \hat{\mu_1}, \varphi_1, \hat{\mu_2}, \hat{\mu_3}, \varphi_2) \ :\ & \mu_1 = \mu_0[a \mapsto -\mu_0(b)] \text{ and} \\
& \varphi_1 \text{ is a state flow of some duration } r_1 \geq 0 \text{ with } \varphi_1 \models z' = v \wedge v' = a \\
& \text{starting in } \varphi_1(0) = \mu_1 \text{ and ending in a state with } \varphi_1(r_1)(v) \geq 0 \\
& \text{and } \mu_2 = \varphi_1(r_1), \mu_3 = \varphi_1(r_1)[a \mapsto \varphi_1(r_1)(A)] \text{ and} \\
& \varphi_2 \text{ is a state flow of some duration } r_2 \geq 0 \text{ with } \varphi_2 \models z' = v \wedge v' = a \\
& \text{starting in } \varphi_2(0) = \mu_3 \text{ and ending in state } \varphi_2(r_2)\} \\
\cup \{(\hat{\mu_0}, \hat{\mu_1}, \varphi_1, \hat{\mu_2}, \hat{\Lambda}) \ :\ & \mu_1 = \mu_0[a \mapsto -\mu_0(b)] \text{ and} \\
& \varphi_1 \text{ is a state flow of some duration } r \geq 0 \text{ with } \varphi_1 \models z' = v \wedge v' = a \\
& \text{starting in } \varphi_1(0) = \mu_1 \text{ and ending in a state with } \varphi_1(r)(v) < 0 \\
& \text{further } \mu_2 = \varphi_1(r)\}.
\end{aligned}$$

The first set is the set of traces where the test $?v \geq 0$ in the middle succeeds and the system continues. The second set (after the union) is the set of traces that are aborted with $\hat{\Lambda}$ during their execution, because the middle test fails. Note that the traces in the first set have two continuous flows φ_1, φ_2 and four point flows $\hat{\mu_0}, \hat{\mu_1}, \hat{\mu_2}, \hat{\mu_3}$ in each trace. The traces in the second set have only one continuous flow φ_1 and three point flows $\hat{\mu_0}, \hat{\mu_1}, \hat{\mu_2}$, because the subsequent aborting point flow $\hat{\Lambda}$ does not terminate and aborts all further execution. In the trace semantics, $v < 0$ is possible in the middle of some traces, which is a fact that the transition semantics does not notice. Combining traces for $\alpha \cup \beta$, that is, for

$$(a := -2a;\ a := a^2) \cup (a := -b;\ z' = v, v' = a;\ ?v \geq 0;\ a := A;\ z' = v, v' = a)$$

is just the union $\tau(\alpha) \cup \tau(\beta)$ of the traces $\tau(\alpha)$ and $\tau(\beta)$ from Examples 4.2 and 4.3. Note that $a \leq 0$ will hold at least once during every trace of $\alpha \cup \beta$, either in the beginning, or after setting $a := -2a$ or $a := -b$, respectively, when we assume $b > 0$.

□

4.3.2 Valuation of State and Trace Formulas

In the semantics of dTL formulas, the dynamic modalities determine the set of traces according to the trace semantics of hybrid programs, and, independently, the temporal modalities determine at which points in time the respective postcondition needs to hold. The semantics of formulas is compositional and denotational, that is, the semantics of a complex formula is defined as a simple function of the semantics of its subformulas.

Definition 4.4 (Valuation of dTL formulas). For state formulas, the *valuation* $val(\nu,\cdot)$ with respect to state ν is defined inductively as follows:

1. $val(\nu, p(\theta_1,\ldots,\theta_n)) = p^\ell(val(\nu,\theta_1),\ldots,val(\nu,\theta_n))$, where p^ℓ is the relation associated with p by the fixed semantics of real arithmetic.
2. $val(\nu,\phi \wedge \psi) = true$ iff $val(\nu,\phi) = true$ and $val(\nu,\psi) = true$
3. $val(\nu,\phi \vee \psi) = true$ iff $val(\nu,\phi) = true$ or $val(\nu,\psi) = true$
4. $val(\nu,\neg\phi) = true$ iff $val(\nu,\phi) \neq true$
5. $val(\nu,\phi \rightarrow \psi) = true$ iff $val(\nu,\phi) \neq true$ or $val(\nu,\psi) = true$
6. $val(\nu,\forall x\,\phi) = true$ iff $val(\nu[x \mapsto d],\phi) = true$ for all $d \in \mathbb{R}$
7. $val(\nu,\exists x\,\phi) = true$ iff $val(\nu[x \mapsto d],\phi) = true$ for some $d \in \mathbb{R}$
8. $val(\nu,[\alpha]\pi) = true$ iff for each trace $\sigma \in \tau(\alpha)$ that starts in first $\sigma = \nu$, if $val(\sigma,\pi)$ is defined, then $val(\sigma,\pi) = true$.
9. $val(\nu,\langle\alpha\rangle\pi) = true$ iff there is a trace $\sigma \in \tau(\alpha)$ starting in first $\sigma = \nu$ such that $val(\sigma,\pi)$ is defined and $val(\sigma,\pi) = true$.

For trace formulas, the *valuation* $val(\sigma,\cdot)$ with respect to trace σ is defined inductively as:

1. If ϕ is a state formula, then $val(\sigma,\phi) = val(\text{last}\,\sigma,\phi)$ if σ terminates, whereas $val(\sigma,\phi)$ is *not defined* if σ does not terminate.
2. $val(\sigma,\Box\phi) = true$ iff $val(\sigma_i(\zeta),\phi) = true$ holds for all positions (i,ζ) of σ with $\sigma_i(\zeta) \neq \Lambda$.
3. $val(\sigma,\Diamond\phi) = true$ iff $val(\sigma_i(\zeta),\phi) = true$ holds for some position (i,ζ) of σ with $\sigma_i(\zeta) \neq \Lambda$.

As usual, a (state) formula is *valid* if it is true in all states. Further for (state) formula ϕ and state ν we write $\nu \models \phi$ iff $val(\nu,\phi) = true$. We write $\nu \not\models \phi$ iff $val(\nu,\phi) = false$. Likewise, for trace formula π and trace σ we write $\sigma \models \pi$ iff $val(\sigma,\pi) = true$ and $\sigma \not\models \pi$ iff $val(\sigma,\pi) = false$. In particular, we only write $\sigma \models \pi$ or $\sigma \not\models \pi$ if $val(\sigma,\pi)$ is defined, which it is not the case if π is a state formula and σ does not terminate. The points where a dTL property ϕ has to hold for the various combinations of temporal and dynamic modalities are illustrated in Fig. 4.1.

Example 4.4. Recall the hybrid programs α and β from Examples 4.2 and 4.3. For these, we see the following difference between the $d\mathcal{L}$ transition semantics and the dTL trace semantics, which gives the same difference between the dTL semantics for nontemporal formulas (which are $d\mathcal{L}$ formulas) and the dTL trace semantics

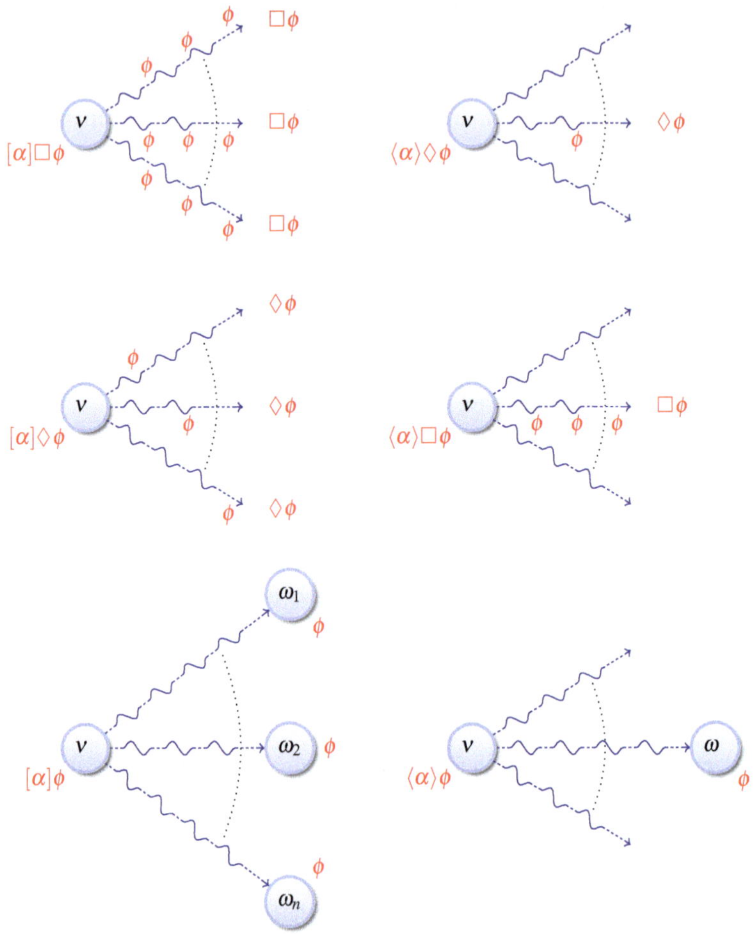

Fig. 4.1 Trace semantics of dTL formulas

for formulas with temporal modalities. The d$\mathscr{L}$ formula $[\alpha]a \geq 0$ is valid (the dTL formula $[\alpha]a \geq 0$ is also valid), because $a \geq 0$ holds at the end of all runs of α. The dTL formula $[\alpha]\Box a \geq 0$, on the other hand, is not valid, because a may be negative after the first step. Likewise, the d$\mathscr{L}$ formula $[\beta]v \geq 0$ is valid (the dTL formula $[\beta]v \geq 0$ is also valid), because $v \geq 0$ holds at the end of all runs of β. The dTL formula $[\beta]\Box v \geq 0$, instead, is not valid, because v may be negative at an intermediate state (those that later fail the middle test $?v \geq 0$ and get aborted). If we add an evolution domain restriction $v \geq 0$ to the differential equation, however, the formula $v \geq 0 \to [\beta]\Box v \geq 0$ is valid.

For the combined hybrid program $\alpha \cup \beta$ from both examples, we find that the dTL formula $[\alpha \cup \beta]\Diamond a \leq 0$ is valid, because a is nonpositive at least once during each of the traces. □

4.3.3 Conservative Temporal Extension

The following result shows that the extension of dTL by temporal operators does not change the meaning of nontemporal formulas. The trace semantics given in Definition 4.4 is equivalent to the final state reachability relation semantics given in Definition 2.6 for the sublogic $d\mathcal{L}$ of dTL.

Proposition 4.1 (Conservative temporal extension). *The logic* dTL *is a* conservative extension *of nontemporal* $d\mathcal{L}$*, i.e., the set of valid* $d\mathcal{L}$ *formulas is the same with respect to transition reachability semantics of* $d\mathcal{L}$ *(Definition 2.6) as with respect to the trace semantics of* dTL *(Definition 4.4).*

The proof of Proposition 4.1 uses the following relationship of reachability and trace semantics of dTL programs, which agree on initial and final states.

Lemma 4.1 (Trace relation). *For hybrid programs* $\alpha \in \mathrm{HP}(\Sigma)$*, we have*

$$\rho(\alpha) = \{(\mathrm{first}\,\sigma, \mathrm{last}\,\sigma) \,:\, \sigma \in \tau(\alpha) \text{ terminates}\}.$$

Proof. The proof follows an induction on the structure of α.

- The cases $x := \theta$, $x' = \theta$, and $\alpha \cup \beta$ are simple comparisons of Definitions 4.3 and 2.7.
- For $?\chi$, the reasoning splits into two directions.

"$\supseteq$" For inclusion "$\supseteq$", assume $\sigma \in \tau(?\chi)$. We distinguish between two cases. If $val(\mathrm{first}\,\sigma, \chi) = true$, then $\sigma = (\hat{v})$ has length one, $\mathrm{last}\,\sigma = \mathrm{first}\,\sigma$, and $(\mathrm{first}\,\sigma, \mathrm{first}\,\sigma) \in \rho(\alpha)$. If, however, $val(\mathrm{first}\,\sigma, \chi) = false$, then $\sigma = (\hat{v}, \hat{\Lambda})$ does not terminate; hence, there is nothing to show.

"$\subseteq$" Conversely, for inclusion "$\subseteq$", assume $(v, v) \in \rho(?\chi)$; then $val(v, \chi) = true$ and $(\hat{v}) \in \tau(\alpha)$ satisfies the conditions on σ.

- For $\alpha; \beta$, the reasoning again splits into the two directions.

"$\supseteq$" For inclusion "$\supseteq$", assume that $\sigma \circ \varsigma \in \tau(\alpha; \beta)$ terminates with $\sigma \in \tau(\alpha)$, $\varsigma \in \tau(\beta)$, and $\mathrm{last}\,\sigma = \mathrm{first}\,\varsigma$. Then, by induction hypothesis, we can assume that $(\mathrm{first}\,\sigma, \mathrm{last}\,\sigma) \in \rho(\alpha)$ and $(\mathrm{first}\,\varsigma, \mathrm{last}\,\varsigma) \in \rho(\beta)$. By the semantics of sequential composition, we have $(\mathrm{first}\,(\sigma \circ \varsigma), \mathrm{last}\,(\sigma \circ \varsigma)) \in \rho(\alpha; \beta)$.

"$\subseteq$" Conversely, for inclusion "$\subseteq$", assume that $(v, w) \in \rho(\alpha; \beta)$. That is, let $(v, z) \in \rho(\alpha)$ and $(z, w) \in \rho(\beta)$. By induction hypothesis, there is a terminating trace $\sigma \in \tau(\alpha)$ with $\mathrm{first}\,\sigma = v$ and $\mathrm{last}\,\sigma = z$. Further, by the induction hypothesis, there is a terminating $\varsigma \in \tau(\beta)$ with $\mathrm{first}\,\varsigma = z$ and $\mathrm{last}\,\varsigma = w$. Hence, $\sigma \circ \varsigma \in \tau(\alpha; \beta)$ terminates with $\mathrm{first}\,(\sigma \circ \varsigma) = v$ and $\mathrm{first}\,(\sigma \circ \varsigma) = w$.

- The case α^* is an inductive consequence of the sequential composition case. □

Proof (of Proposition 4.1). The formulas of $d\mathcal{L}$ are a subset of the dTL formulas. In the course of this proof, we use the notation $val_{d\mathcal{L}}(v, \cdot)$ to indicate that the $d\mathcal{L}$

valuation from Definition 4.4 in Sect. 2.3 is used. For $d\mathcal{L}$ formulas ψ, we show that the valuations with respect to Definitions 4.4 and 2.6 are the same for all states ν:

$$val(\nu,\psi) = val_{d\mathcal{L}}(\nu,\psi) \text{ for all } \nu.$$

We prove this by induction on the structure of ψ. The cases 1–3 of the definition of state formulas in Definition 4.1 are obvious. The other cases are proven as follows.

- If ψ has the form $[\alpha]\phi$, assume that $val(\nu,[\alpha]\phi) = false$. Then there is some terminating trace $\sigma \in \tau(\alpha)$ with first $\sigma = \nu$ such that $val(\text{last}\,\sigma,\phi) = false$. By the induction hypothesis, this implies that $val_{d\mathcal{L}}(\text{last}\,\sigma,\phi) = false$. According to Lemma 4.1, $(\nu,\text{last}\,\sigma) \in \rho(\alpha)$ holds, which implies $val_{d\mathcal{L}}(\nu,[\alpha]\phi) = false$. For the converse direction, assume that $val_{d\mathcal{L}}(\nu,[\alpha]\phi) = false$. Then there is a $(\nu,w) \in \rho(\alpha)$ with $val_{d\mathcal{L}}(w,\phi) = false$. By Lemma 4.1, there is a terminating trace $\sigma \in \tau(\alpha)$ with first $\sigma = \nu$ and last $\sigma = w$. By induction hypothesis, $val(\text{last}\,\sigma,\phi) = false$. Thus, we can conclude that both $val(\sigma,\phi) = false$ and $val(\nu,[\alpha]\phi) = false$.
- The case $\psi = \langle\alpha\rangle\phi$ is proven similarly. □

4.4 Safety Invariants in Train Control

In the European Train Control System (ETCS), trains are coordinated by decentralised Radio Block Centres (RBCs), which grant or deny movement authorities (MAs) dynamically to the individual trains by wireless communication. In emergencies, trains always have to stop within the MA issued by the RBC; see Fig. 4.2. Following

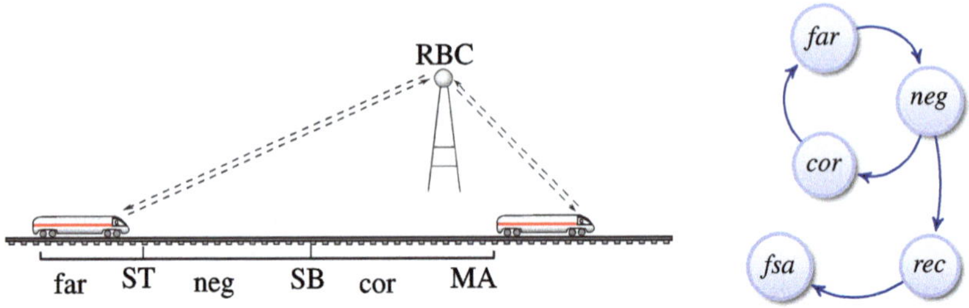

Fig. 4.2 ETCS train coordination protocol phases

the reasoning pattern for traffic agents in [89], each train negotiates with the RBC to extend its MA when approaching the end of its current MA. Since wireless communication takes time, this negotiation is initiated in due time before reaching m. To simplify the presentation, we adopt the assumption of Damm et al. [89] here that trains keep their desired speed (or at least their maximum speed limit) during negotiation. Before entering negotiation at some point ST (for start talking), the train still has sufficient distance to MA (it is in *far* mode) and can regulate its speed freely within the track limits. After the point SB (for start braking), the train has to start applying the brakes as identified in Sect. 2.9.

As a model for train movements, we use the ideal-world model from Sect. 2.4. For a safe operation of multiple traffic agents, it is crucial that the MA be respected at *every* point in time during this protocol, not only at its end. Hence, we need to consider temporal safety invariants. For instance, when the train has entered the negotiation phase at its current position z, dTL can analyse the following safety invariant of a part of the protocol cycle of the train controller:

$$\begin{aligned} \psi \;&\rightarrow [neg;cor;drive]\Box(\ell \leq L \rightarrow z < m) \qquad (4.3)\\ \text{where } neg \;&\equiv z' = v, \ell' = 1\\ cor \;&\equiv (?m - z < s; a := -b) \cup (?m - z \geq s; a := \dots)\\ drive \;&\equiv z' = v, v' = a. \end{aligned}$$

It expresses that—under a sanity condition ψ for parameters—a train will *always* remain within its MA m as long as the accumulated RBC negotiation latency ℓ is at most L. We refer to the work of Faber and Meyer [119] for details on what kind of message passing contributes to ℓ. Like in [89], we model the train to first negotiate while keeping a constant speed ($z' = v$) in *neg*. The differential equation $\ell' = 1$ defines ℓ as a clock that is never reset, but accumulates the time spent in negotiation. Thereafter, in *cor*, the train corrects its acceleration or brakes with force b (as a fail-safe recovery manoeuvre) on the basis of the remaining distance $(m - z)$. That is, if the distance of the movement authority and position is less than s (the test $?m - z < s$ succeeds), the acceleration is set to braking by $a := -b$. If, instead, the other test $?m - z \geq s$ succeeds, then acceleration is set to another value (not shown in *cor*). Finally, the train continues moving according to the differential equation system *drive* or, equivalently, $z'' = a$. Instead of manually choosing specific values for the free parameters of (4.3) as in [89, 119], we will use the techniques developed in this book to automatically synthesise constraints on the relationship of parameters that are required for a safe operation of cooperative train control.

4.5 Proof Calculus for Temporal Invariants

In this section, we introduce a sequent calculus for verifying temporal specifications of hybrid systems in differential temporal dynamic logic dTL. With the basic idea being to perform a symbolic decomposition, the calculus transforms hybrid programs successively into simpler logical formulas describing their effects. Statements about the temporal behaviour of a hybrid program are successively reduced to corresponding nontemporal statements about the intermediate states.

$$([\cup]\Box)\ \frac{[\alpha]\pi \wedge [\beta]\pi}{[\alpha \cup \beta]\pi}{}^{1} \qquad (\langle\cup\rangle\Diamond)\ \frac{\langle\alpha\rangle\pi \vee \langle\beta\rangle\pi}{\langle\alpha \cup \beta\rangle\pi}{}^{1}$$

$$([;]\Box)\ \frac{[\alpha]\Box\phi \wedge [\alpha][\beta]\Box\phi}{[\alpha;\beta]\Box\phi} \qquad (\langle;\rangle\Diamond)\ \frac{\langle\alpha\rangle\Diamond\phi \vee \langle\alpha\rangle\langle\beta\rangle\Diamond\phi}{\langle\alpha;\beta\rangle\Diamond\phi}$$

$$([?]\Box)\ \frac{\phi}{[?\chi]\Box\phi} \qquad (\langle?\rangle\Diamond)\ \frac{\phi}{\langle?\chi\rangle\Diamond\phi}$$

$$([:=]\Box)\ \frac{\phi \wedge [x:=\theta]\phi}{[x:=\theta]\Box\phi} \qquad (\langle:=\rangle\Diamond)\ \frac{\phi \vee \langle x:=\theta\rangle\phi}{\langle x:=\theta\rangle\Diamond\phi}$$

$$([']\Box)\ \frac{[x'=\theta]\phi}{[x'=\theta]\Box\phi} \qquad (\langle'\rangle\Diamond)\ \frac{\langle x'=\theta\rangle\phi}{\langle x'=\theta\rangle\Diamond\phi}$$

$$([^{*n}]\Box)\ \frac{[\alpha;\alpha^*]\Box\phi}{[\alpha^*]\Box\phi} \qquad (\langle^{*n}\rangle\Diamond)\ \frac{\langle\alpha;\alpha^*\rangle\Diamond\phi}{\langle\alpha^*\rangle\Diamond\phi}$$

$$([^*]\Box)\ \frac{[\alpha^*][\alpha]\Box\phi}{[\alpha^*]\Box\phi} \qquad (\langle^*\rangle\Diamond)\ \frac{\langle\alpha^*\rangle\langle\alpha\rangle\Diamond\phi}{\langle\alpha^*\rangle\Diamond\phi}$$

[1] π is a trace formula and—unlike the state formulas ϕ and ψ—may thus begin with a temporal modality $\Box$ or $\Diamond$.

Fig. 4.3 Rule schemata of the proof calculus for temporal differential dynamic logic

4.5.1 Proof Rules

We introduce a proof calculus for differential temporal dynamic logic dTL that inherits the proof rules of $d\mathcal{L}$ from Chap. 2 and adds new proof rules for temporal modalities. We can, in fact, get a corresponding proof calculus when we add the same proof rules for temporal modalities to the DAL proof rules from Chap. 3. In the latter case, the resulting logic is called differential-algebraic temporal dynamic logic DATL.

Inherited Nontemporal Rules

The dTL calculus is presented in Fig. 4.3 and inherits the (nontemporal) $d\mathcal{L}$ proof rules, i.e., the propositional, first-order, dynamic, and global rules from $d\mathcal{L}$. That is, it includes the propositional rules from Fig. 2.11 on p. 79 and either the free-variable quantifier rules from Fig. 2.11 or the simpler quantifier rules from Fig. 3.9 on p. 164 that are based on side deductions. The dynamic rules ($\langle;\rangle$–$[']$) and global rules ($[]gen$,$\langle\rangle gen$,ind,con) for handling nontemporal dynamic modalities are also inherited directly from Fig. 2.11. The only possible exception is that $[\cup]$,$\langle\cup\rangle$ can be generalised to apply to formulas of the form $[\alpha \cup \beta]\pi$ where π is an arbitrary trace formula, and not just a state formula as in $d\mathcal{L}$. Thus, π may begin with $\Box$ or $\Diamond$, which is why the rules are repeated in this generalised form as $[\cup]\Box$ and $\langle\cup\rangle\Diamond$ in Fig. 4.3.

$$([:=]\Box)\ \frac{\phi \wedge [x := \theta]\phi}{[x := \theta]\Box\phi}$$

$$([']\Box)\ \frac{[x' = \theta]\phi}{[x' = \theta]\Box\phi}$$

$$([;]\Box)\ \frac{[\alpha]\Box\phi \wedge [\alpha][\beta]\Box\phi}{[\alpha;\beta]\Box\phi}$$

$$([\cup]\Box)\ \frac{[\alpha]\Box\phi \wedge [\beta]\Box\phi}{[\alpha \cup \beta]\Box\phi}$$

$$([^*]\Box)\ \frac{[\alpha^*][\alpha]\Box\phi}{[\alpha^*]\Box\phi}$$

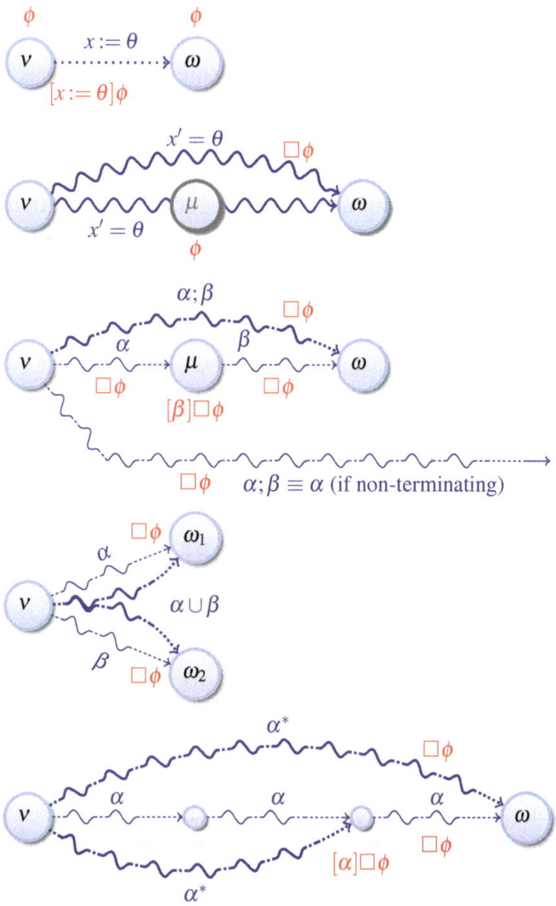

Fig. 4.4 Correspondence of temporal proof rules and trace semantics

Temporal Rules

The new temporal rules in Fig. 4.3 for the dTL calculus successively transform temporal specifications of hybrid programs into nontemporal d$\mathcal{L}$ formulas. The idea underlying this transformation is to decompose hybrid programs and recursively augment intermediate state transitions with appropriate specifications. Also see Fig. 4.4 for an illustration of the correspondence of a representative set of proof rules for temporal modalities to the trace semantics of hybrid programs (Definition 4.3).

Rule $[;]\Box$ decomposes invariants of $\alpha;\beta$ (i.e., $[\alpha;\beta]\Box\phi$ holds) into an invariant of α (i.e., $[\alpha]\Box\phi$) and an invariant of β that holds when β is started in *any* final state of α (i.e., $[\alpha]([\beta]\Box\phi)$). Its difference with the d$\mathcal{L}$ rule $[;]$ thus is that the dTL rule $[;]\Box$ also checks safety invariant ϕ at the symbolic states in between the execution of α and β, and recursively so because of the temporal modality $\Box$. Again, see Fig. 4.4 for an illustration of this proof principle.

Rule $[:=]\Box$ expresses that invariants of assignments need to hold before and after the discrete change (similarly for $[?]\Box$, except that tests do not lead to a state change, so ϕ holding before the test is all there is to it). Rule $[']\Box$ can directly reduce invariants of continuous evolutions to nontemporal formulas as restrictions of solutions of differential equations are themselves solutions of different duration and thus already included in the evolutions of $x' = \theta$. In particular, observe that the handling of differential equations within hybrid systems is fully encapsulated within the fragment of dynamic rules from Fig. 2.11. The rules $[']\Box$, $\langle'\rangle\Diamond$, $[:=]\Box$, and $\langle:=\rangle\Diamond$ directly generalise to discrete jump sets and systems of differential equations or even to DA-constraints from Chap. 3.

The (optional) iteration rule $[^{*n}]\Box$ can partially unwind loops. It relies on rule $[;]\Box$ and is simpler than $d\mathcal{L}$ rule $[^{*n}]$, because the other rules will inductively produce a premise that ϕ holds in the current state, because of the temporal modality $\Box\phi$. The dual rules $\langle\cup\rangle\Diamond, \langle;\rangle\Diamond, \langle?\rangle\Diamond, \langle:=\rangle\Diamond, \langle'\rangle\Diamond, \langle^{*n}\rangle\Diamond$ work similarly.

In Chaps. 2 and 3, the primary means for handling loops are the invariant induction (*ind*) and variant convergence (*con*) rules. Here, we take a different, completely modular approach for verifying temporal properties of loops based on the $d\mathcal{L}$ capabilities for verifying nontemporal properties of loops. Rules $[^*]\Box$ and $\langle^*\rangle\Diamond$ actually *define* temporal properties of loops inductively. Rule $[^*]\Box$ expresses that ϕ holds at all times during repetitions of α (i.e., $[\alpha^*]\Box\phi$) iff, *after* repeating α any number of times, ϕ holds at all times *during* one execution of α (i.e., $[\alpha^*]([\alpha]\Box\phi)$). See Fig. 4.4 for an illustration. Dually, $\langle^*\rangle\Diamond$ expresses that α holds at some time during repetitions of α (i.e., $\langle\alpha^*\rangle\Diamond\phi$) iff, after some number of repetitions of α, formula ϕ holds at some point during one execution of α (i.e., $\langle\alpha^*\rangle(\langle\alpha\rangle\Diamond\phi)$). In this context, the nontemporal modality $\langle\alpha^*\rangle$ can be thought of as skipping over to the iteration of α during which ϕ actually occurs, as expressed by the nested dTL formula $\langle\alpha\rangle\Diamond\phi$. The inductive definition rules $[^*]\Box$ and $\langle^*\rangle\Diamond$ completely reduce temporal properties of loops to dTL properties of standard nontemporal $d\mathcal{L}$-modalities such that standard induction (*ind*) or convergence rules (*con*) can be used for the outer nontemporal modality of the loop. Hence, after applying the inductive loop definition rules $[^*]\Box$ and $\langle^*\rangle\Diamond$, the standard $d\mathcal{L}$ loop invariant and variant rules can be used for verifying temporal properties of loops without change, except that the postcondition contains temporal modalities.

Rules for handling $[\alpha]\Diamond\phi$ and $\langle\alpha\rangle\Box\phi$ are discussed in Sect. 4.9. Finally, provability in the dTL calculus is denoted by $\Phi \vdash_{\mathrm{dTL}} \psi$, and defined according to Definition 2.11 or Definition 3.16. The notion of a dTL proof directly follows from the previous definitions in Chaps. 2 and 3.

4.5.2 Verification Example

Consider the bouncing ball example from Sect. 2.5.4. The proof in Fig. 2.20 can be generalised easily to a proof of the temporal property

$$v^2 \leq 2g(H-h) \land h \geq 0 \land g > 0 \land H \geq 0 \land 1 > c \geq 0 \\ \rightarrow [(h'' = -g \,\&\, h \geq 0; (?h > 0 \cup (?h = 0; v := -cv)))^*]\Box(0 \leq h \leq H). \quad (4.4)$$

The only aspect of the proof that changes is that the temporal proof rules in Fig. 4.3 are used instead of the dynamic proof rules from Fig. 2.11, and that the resulting extra proof goals for the invariance property at intermediate steps have to be proven.

In contrast, the proof in Fig. 2.18 for the simplified dynamics without evolution domain restriction $h \geq 0$ cannot be generalised to a proof of the temporal property

$$v^2 \leq 2g(H-h) \land h \geq 0 \land g > 0 \land H \geq 0 \land 1 > c \geq 0 \\ \rightarrow [(h'' = -g; (?h > 0 \cup (?h = 0; v := -cv)))^*]\Box(0 \leq h \leq H). \quad (4.5)$$

This difference in provability is for good reasons. The property in (4.4) is valid, but the property in (4.5) is not! While there was no noticeable semantical difference between the nontemporal properties proven in Figs. 2.18 and 2.20, there is a decisive difference between the corresponding temporal properties (4.5) and (4.4). Because there is no evolution domain restriction in (4.5), its hybrid program does not prevent continuous evolution to a negative height under the floor ($h < 0$), for which $0 \leq h \leq H$ does not hold.

The reason for this discrepancy of the temporal version compared to the nontemporal versions thus is that the nontemporal modalities do not "see" the temporary violation of $0 \leq h \leq H$. Such a temporary violation of $0 \leq h$ during the continuous evolution does not produce a successful run of the hybrid program, because it is blocked by the subsequent tests $?h = 0$ and $?h > 0$. A state with negative height fails both tests. While this behaviour does not give a successful program transition of $(\nu, \omega) \in \rho(ball)$ by Definition 2.7 so that the proof in Fig. 2.18 is correct, the behaviour still gives a valid trace $\sigma \in \tau(ball)$ by Definition 4.3. This trace σ is a partial trace, because it ends in a failure state Λ, but it is still one of the traces that $[ball]\Box(0 \leq h \leq H)$ quantifies over (quite unlike $[ball](0 \leq h \leq H)$, which only considers final states of successful traces).

4.6 Soundness

The following result shows that verification with the dTL calculus always produces correct results about the safety of hybrid systems, i.e., the dTL calculus is sound.

Theorem 4.1 (Soundness of dTL). *The* dTL *calculus is* sound, *i.e., derivable (state) formulas are valid.*

Proof. We show that all rules of the dTL calculus are *locally sound*, i.e., for all states ν, the conclusion of a rule is true in state ν when all premises are true in ν. Let ν be any state. For each rule we have to show that the conclusion is true in ν assuming the premises are true in ν. The propositional rules are locally sound by Theorem 2.1. Inductively, the soundness of the dynamic rules follows from Proposition 4.1 and local soundness of the corresponding rules in $d\mathcal{L}$. The proof for the generalisation in $[\cup]$ and $\langle\cup\rangle$ to path formulas π is a straightforward extension. The quantifier rules are sound by Theorems 2.1 and 3.1, respectively.

$[;]\Box$ Assume $\nu \models [\alpha]\Box\phi$ and $\nu \models [\alpha][\beta]\Box\phi$. Let $\sigma \in \tau(\alpha;\beta)$, i.e., $\sigma = \rho \circ \varsigma$ with $\text{first}\,\sigma = \nu$, $\rho \in \tau(\alpha)$, and $\varsigma \in \tau(\beta)$. If ρ does not terminate, then $\sigma = \rho \in \tau(\alpha)$ and $\sigma \models \Box\phi$ by premise. If, instead, ρ terminates with $\text{last}\,\rho = \text{first}\,\varsigma$, then $\rho \models \Box\phi$ by premise. Further, we know $\nu \models [\alpha][\beta]\Box\phi$. In particular for trace $\rho \in \tau(\alpha)$, we have $\text{last}\,\rho \models [\beta]\Box\phi$. Thus, $\varsigma \models \Box\phi$ because $\varsigma \in \tau(\beta)$ starts at $\text{first}\,\varsigma = \text{last}\,\rho$. By composition, $\rho \circ \varsigma \models \Box\phi$. As $\sigma = \rho \circ \varsigma$ was arbitrary, we can conclude $\nu \models [\alpha;\beta]\Box\phi$. The converse direction holds, as all traces of α are prefixes of traces of $\alpha;\beta$. Hence, the assumption $\nu \models [\alpha;\beta]\Box\phi$ directly implies $\nu \models [\alpha]\Box\phi$. Further, all traces of β that begin at a state reachable from ν by α are suffixes of traces of $\alpha;\beta$ starting in ν. Hence, $\nu \models [\alpha][\beta]\Box\phi$ is implied as well.

$[?]\Box$ Soundness of $[?]\Box$ is obvious, since, by premise, we can assume $\nu \models \phi$, and there is nothing to show for Λ states according to Definition 4.4. Conversely, $\hat{\nu}$ is a prefix of all traces in $\tau(?\chi)$ that start in ν.

$[:=]\Box$ Assuming $\nu \models \phi$ and $\nu \models [x := \theta]\phi$, we have to show that $\nu \models [x := \theta]\Box\phi$. Let $\sigma \in \tau(x := \theta)$ be any trace with $\text{first}\,\sigma = \nu$, i.e., $\sigma = (\hat{\nu}, \hat{\omega})$ by Definition 4.3. Hence, the only two states we need to consider are $\sigma_0(0) = \nu$ and $\sigma_1(0) = \omega$. By premise, $\sigma_0(0) = \nu$ yields $\sigma_0(0) \models \phi$. Similarly, for the state $\sigma_1(0) = \text{last}\,\sigma = \omega$, the premise gives $\sigma_1(0) \models \phi$. The converse direction is similar.

$[']\Box$ We prove that $[']\Box$ is locally sound by contraposition. For this, assume that $\nu \not\models [x' = \theta]\Box\phi$; then there is a trace $\sigma = (\varphi) \in \tau(x' = \theta)$ starting in $\text{first}\,\sigma = \nu$ and $\sigma \not\models \Box\phi$. Hence, there is a position $(0,\zeta)$ of σ with $\sigma_0(\zeta) \not\models \phi$. Now φ restricted to the interval $[0,\zeta]$ also solves differential equation $x' = \theta$. Thus, $(\varphi|_{[0,\zeta]}) \not\models \phi$ as $\varphi(\zeta) \not\models \phi$, since the last state is $\varphi(\zeta)$. By consequence, this gives $\nu \not\models [x' = \theta]\phi$. The converse direction is obvious as $\text{last}\,\sigma$ always is a state occurring during σ. Hence $\nu \not\models [x' = \theta]\phi$ immediately implies $\nu \not\models [x' = \theta]\Box\phi$.

$[^{*n}]\Box$ By contraposition, assume that $\nu \not\models [\alpha^*]\Box\phi$. Then there is an $n \in \mathbb{N}$ and a trace $\sigma \in \tau(\alpha^n)$ with $\text{first}\,\sigma = \nu$ such that $\sigma \not\models \Box\phi$. There are two cases. If $n > 0$ then $\sigma \in \tau(\alpha;\alpha^*)$, and thus $\nu \not\models [\alpha;\alpha^*]\Box\phi$. If, however, $n = 0$, then $\sigma = (\hat{\nu})$ and $\nu \not\models \phi$. Hence, all traces $\varsigma \in \tau(\alpha;\alpha^*)$ with $\text{first}\,\varsigma = \nu$ satisfy $\varsigma \not\models \Box\phi$. Finally, it is easy to see that all programs have at least one such trace ς (when V is nonempty) that witnesses $\nu \not\models [\alpha;\alpha^*]\Box\phi$. The converse direction is easy as all behaviour of $\alpha;\alpha^*$ is subsumed by α^*, i.e., $\tau(\alpha;\alpha^*) \subseteq \tau(\alpha^*)$.

$[^*]\Box$ Clearly, using the fact that $\tau(\alpha^*) \supseteq \tau(\alpha^*;\alpha)$, the set of states along the traces of α^* at which ϕ needs to be true for the premise is a subset of the corresponding set for the conclusion. Hence, the conclusion entails the premise. Conversely, all states during traces of α^* are also reachable by iterating α sufficiently often to completion and then following a single trace of α. In detail: If $\nu \not\models [\alpha^*]\Box\phi$, then there is a trace $\sigma \in \tau(\alpha^*)$ on which $\neg\phi$ holds true at some state, say, at $\sigma_i(\zeta) \neq \Lambda$. Let $n \geq 0$ be the (maximum) number of complete repetitions of α along σ before discrete step index i. That is, there is some discrete step index $i_n < i$ such that the prefix $\rho = (\sigma_0, \dots, \sigma_{i_n}) \in \tau(\alpha^n)$ of σ consists of n complete repetitions of α and the suffix $\varsigma = (\sigma_{i_n+1}, \sigma_{i_n+2}, \dots) \in \tau(\alpha^*)$ starts with a trace of α during which $\neg\phi$ occurs at point $\sigma_i(\zeta)$, namely at relative position $(i-(i_n+1), \zeta)$. Let $\acute{\varsigma} \in \tau(\alpha)$ be this prefix of ς. Consequently, $\acute{\varsigma} \models \langle\alpha\rangle\Diamond\neg\phi$ and the trace $\rho \circ \acute{\varsigma}$ is a witness for $\nu \models \langle\alpha^*\rangle\langle\alpha\rangle\Diamond\neg\phi$.

The proofs for $\langle;\rangle\Diamond$–$\langle^*\rangle\Diamond$ are dual, since $\langle\alpha\rangle\Diamond\phi$ is equivalent to $\neg[\alpha]\Box\neg\phi$ by duality. $\Box$

4.7 Completeness

In this section, we show that the strictly modular dTL calculus enables us to lift the Relative Completeness Theorem 2.3 for $d\mathcal{L}$ to dTL.

4.7.1 Incompleteness

The Incompleteness Theorem 2.2 directly generalises to temporal and nontemporal properties of dTL.

Theorem 4.2 (Incompleteness of dTL). *The discrete and continuous fragments of* dTL *are non-axiomatisable for temporal safety ($[\alpha]\Box\phi$) and nontemporal ($[\alpha]\phi$) fragments of* dTL. *Hence, valid* dTL *formulas are not always derivable.*

Proof. We show that the discrete and continuous fragments of the following purely temporal and nontemporal fragments of dTL are non-axiomatisable:

1. the fragment that only contains modalities of the form $[\alpha]\Box\phi$ and $\langle\alpha\rangle\Diamond\phi$
2. the fragment that only contains $[\alpha]\phi$ and $\langle\alpha\rangle\phi$ ($d\mathcal{L}$ fragment).

Case 2 is a consequence of the corresponding incompleteness result (Theorem 2.2) for fragments of the sublogic $d\mathcal{L}$, which carries over to the extension dTL by Proposition 4.1.

For Case 1, we prove that natural numbers are definable amongst the real number domain in both fragments, quite similarly to the proof of Theorem 2.2. Then these fragments extend first-order *integer* arithmetic such that the incompleteness theorem of Gödel applies.

- Natural numbers are definable in the discrete fragment without continuous evolutions $x' = \theta$ using repetitive additions:

$$nat(n) \leftrightarrow \langle x := 0; (x := x+1)^* \rangle \Diamond x = n.$$

- In the continuous fragment, natural numbers are definable as:

$$nat(n) \leftrightarrow \exists s \exists c \, (s = 0 \land c = 1 \land \langle s' = c, c' = -s, \tau' = 1 \rangle \Diamond (s = 0 \land \tau = n)).$$

  These ODEs have *sin* and *cos* as unique solutions for s and c, respectively. Their zeros characterise an isomorphic copy of natural numbers, scaled by π. □

4.7.2 Relative Completeness

Due to the modular construction of the dTL calculus, we can lift the major relative completeness result Theorem 2.3 from $d\mathcal{L}$ to dTL. By proving dTL completeness relative to Theorem 2.3, we essentially show that dTL is complete relative to $d\mathcal{L}$, which directly implies that dTL is even complete relative to FOD using Theorem 2.3 by a standard argument. Again, we restrict our attention to homogeneous combinations of path and trace quantifiers like $[\alpha]\Box\phi$ or $\langle\alpha\rangle\Diamond\phi$.

Theorem 4.3 (Relative completeness of dTL). *The* dTL *calculus is* complete relative to *FOD, i.e., every valid* dTL *formula can be derived from FOD tautologies.*

Proof (Outline). The proof is a simple extension of the proof of Theorem 2.3, because the dTL calculus successively reduces temporal properties to nontemporal properties and, in particular, handles loops by inductive definition rules in terms of $d\mathcal{L}$ modalities. The temporal rules in Fig. 4.3 transform temporal formulas to simpler formulas, i.e., to where the temporal modalities occur after simpler programs ($[\cup]\Box$, $[^*]\Box$, $\langle\cup\rangle\Diamond$, $\langle^*\rangle\Diamond$) or disappear completely ($[?]\Box$,$[:=]\Box$,$[']\Box$ and $\langle ? \rangle\Diamond$,$\langle := \rangle\Diamond$,$\langle ' \rangle\Diamond$). Hence, the inductive relative completeness proof in Sect. 2.7.2 directly generalises to dTL with the following addition: After applying $[^*]\Box$ or $\langle^*\rangle\Diamond$, loops are ultimately handled by the standard $d\mathcal{L}$ rules *ind* and *con*. To show that sufficiently strong invariants and variants exist for the temporal postconditions $[\alpha]\Box\phi$ and $\langle\alpha\rangle\Diamond\phi$, we only have to show that such temporal formulas are expressible in FOD, i.e., Lemma 2.9 generalises to dTL. □

This result, which we prove formally in the remainder of Sect. 4.7.2, gives a formal justification that the dTL calculus reduces temporal properties to nontemporal $d\mathcal{L}$ properties.

4.7.3 Expressibility and Rendition of Hybrid Trace Semantics

The central step for lifting the $d\mathcal{L}$ completeness proof to dTL is the following: To show that, after applying rule $[^*]\Box$ or $\langle^*\rangle\Diamond$, sufficiently strong invariants and variants for dTL postconditions can be expressed in $d\mathcal{L}$ for *ind* or *con* to be able to prove the result, we show that the trace semantics of hybrid programs can be characterised in FOD.

Lemma 4.2 (Hybrid program trace rendition). *For every hybrid program α with variables $\vec{x} = x_1, \dots, x_k$ there is a FOD formula $\mathscr{T}_\alpha(\vec{x}, \vec{v})$ with variables among the $2k$ distinct variables $\vec{x} = x_1, \dots, x_k$ and $\vec{v} = v_1, \dots, v_k$ such that*

$$\vDash \mathscr{T}_\alpha(\vec{x}, \vec{v}) \leftrightarrow \langle\alpha\rangle\Diamond\vec{x} = \vec{v}$$

or, equivalently, for every ν, we have that $\nu \vDash \mathscr{T}_\alpha(\vec{x}, \vec{v})$ iff

$$\sigma_i(\zeta) = \nu[\vec{x} \mapsto val(\nu, \vec{v})] \text{ for a position } (i, \zeta) \text{ of some trace } \sigma \in \tau(\alpha) \text{ starting in } \nu.$$

$$\begin{aligned}
\mathscr{T}_{x_1:=\theta_1,\dots,x_k:=\theta_k}(\vec{x}, \vec{v}) &\equiv \vec{x} = \vec{v} \vee \mathscr{S}_{x_1:=\theta_1,\dots,x_k:=\theta_k}(\vec{x}, \vec{v}) \\
\mathscr{T}_{x_1'=\theta_1,\dots,x_k'=\theta_k \,\&\, \chi}(\vec{x}, \vec{v}) &\equiv \mathscr{S}_{x_1'=\theta_1,\dots,x_k'=\theta_k \,\&\, \chi}(\vec{x}, \vec{v}) \\
\mathscr{T}_{?\chi}(\vec{x}, \vec{v}) &\equiv \mathscr{S}_{?\chi}(\vec{x}, \vec{v}) \\
\mathscr{T}_{\beta\cup\gamma}(\vec{x}, \vec{v}) &\equiv \mathscr{T}_\beta(\vec{x}, \vec{v}) \vee \mathscr{T}_\gamma(\vec{x}, \vec{v}) \\
\mathscr{T}_{\beta;\gamma}(\vec{x}, \vec{v}) &\equiv \mathscr{T}_\beta(\vec{x}, \vec{v}) \vee \exists\vec{z}\,(\mathscr{S}_\beta(\vec{x}, \vec{z}) \wedge \mathscr{T}_\gamma(\vec{z}, \vec{v})) \\
\mathscr{T}_{\beta^*}(\vec{x}, \vec{v}) &\equiv \exists\vec{z}\,(\mathscr{S}_{\beta^*}(\vec{x}, \vec{z}) \wedge \mathscr{T}_\beta(\vec{z}, \vec{v}))
\end{aligned}$$

Fig. 4.5 Explicit rendition of hybrid program trace semantics in FOD

Proof. The proof is similar to that of Lemma 2.8, yet using the definition in Fig. 4.5. We resort to corresponding characterisations from Lemma 2.8, which simplifies the characterisation of $\mathscr{T}_\alpha(\vec{x}, \vec{v})$, because we only have to augment $\mathscr{S}_\alpha(\vec{x}, \vec{v})$ by with disjunctions for intermediate states, which can again be defined in terms of $\mathscr{S}_\alpha(\vec{x}, \vec{v})$, recursively.

For instance, $\mathscr{T}_{\beta^*}(\vec{x}, \vec{v})$ characterises the states reachable during traces of β^* as the states reachable during traces of β that start after running β^* to completion for some number of iterations. □

Using this program rendition to characterise temporal trace modalities, Lemma 2.9 generalises immediately to dTL as follows:

Lemma 4.3 (dTL Expressibility). *Logic* dTL *is* expressible *in FOD: for all* dTL *formulas $\phi \in \mathrm{Fml}(\Sigma)$ there is a FOD formula $\phi^\# \in \mathrm{Fml}_{FOD}(\Sigma)$ that is equivalent, i.e., $\vDash \phi \leftrightarrow \phi^\#$. The converse holds trivially.*

Proof. The proof is by a structural induction identical to that in the proof of Lemma 2.9 with the following additions:

1. The case where ϕ is of the form $\langle\alpha\rangle\Diamond\psi$ is a consequence of Lemma 4.2:

$$\vDash \langle\alpha\rangle\Diamond\psi \leftrightarrow \exists\vec{v}\,(\mathscr{T}_\alpha(\vec{x},\vec{v}) \wedge \psi^{\#}{}_{\vec{x}}^{\vec{v}}).$$

2. The case where ϕ is $[\alpha]\Box\psi$ is again a consequence of Lemma 4.2:

$$\vDash [\alpha]\Box\psi \leftrightarrow \forall\vec{v}\,(\mathscr{T}_\alpha(\vec{x},\vec{v}) \rightarrow \psi^{\#}{}_{\vec{x}}^{\vec{v}}).$$

□

4.7.4 Modular Relative Completeness Proof for the Differential Temporal Dynamic Logic Calculus

Now we assemble the proof of Theorem 4.3 from the previous results following a simplified form of the proof of Theorem 2.3, since we can apply Theorem 2.3 for d$\mathscr{L}$ formulas.

Proof (of Theorem 4.3). The proof is a simple extension of the Relative Completeness Theorem 2.3 for d$\mathscr{L}$. Unlike for the rules of the d$\mathscr{L}$ calculus, all new temporal rules are symmetric, hence perform equivalent transformations. Consequently, whenever their conclusion is valid, their premise is valid and of smaller complexity (temporal modalities occur after simpler programs), and hence derivable by induction hypothesis.

For instance, analogously to the induction step for loops in Proposition 2.1, let $\vDash F \rightarrow [\beta^*]\Box G$; then $\vDash F \rightarrow [\beta^*][\beta]\Box G$. By Lemma 4.3, there is a d$\mathscr{L}$ formula or even FOD formula $([\beta]\Box G)^{\#}$ that characterises the temporal postcondition equivalently, i.e., such that $\vDash ([\beta]\Box G)^{\#} \leftrightarrow [\beta]\Box G$. By induction hypothesis, we can derive the simpler formula $\vdash_{\mathscr{D}} ([\beta]\Box G)^{\#} \rightarrow [\beta]\Box G$. Using Lemma 2.11, we conclude $\vdash_{\mathscr{D}} \forall^{\beta}(([\beta]\Box G)^{\#} \rightarrow [\beta]\Box G)$, thus $[\beta^*]([\beta]\Box G)^{\#} \vdash_{\mathscr{D}} [\beta^*][\beta]\Box G$ is derivable by $[]gen$. Furthermore, $\vDash F \rightarrow [\beta^*]([\beta]\Box G)^{\#}$ is a valid d$\mathscr{L}$ formula and, thus, $F \vdash_{\mathscr{D}} [\beta^*]([\beta]\Box G)^{\#}$ is derivable by Theorem 2.3. Combining these derivations by a cut with $[\beta^*]([\beta]\Box G)^{\#}$, we derive $F \vdash_{\mathscr{D}} [\beta^*]\Box G$. Since the temporal rules perform a modular reduction to nontemporal dynamic rules, the case $\vDash F \rightarrow \langle\beta^*\rangle\Diamond G$ is almost identical here, because the differences between variant rule *con* and invariant rule *ind* have already been captured in the proof of Theorem 2.3. □

4.8 Verification of Train Control Safety Invariants

Continuing the ETCS study from Sect. 4.4, we consider a slightly simplified version of equation (4.3) that gives a more concise proof. By a safe abstraction (provable in dTL), we simplify *cor* to permit braking even when $m - z \geq s$, since braking remains safe with respect to $z < m$. Recall (4.3) with additional abbreviations and a simplified *cor*:

$$\begin{aligned} \psi &\to [neg;cor;drive]\Box(\ell \leq L \to z < m) && (4.3^*)\\ \text{where } \psi &\equiv z < m \land v > 0 \land \ell = 0 \land L \geq 0\\ \phi &\equiv \ell \leq L \to z < m\\ neg &\equiv z' = v, \ell' = 1\\ cor &\equiv a := -b \cup (?m - z \geq s; a := \dots)\\ drive &\equiv z' = v, v' = a. \end{aligned}$$

Within the following proof, $\langle\!|\cdot|\!\rangle$ brackets are used instead of modalities to visually identify the discrete jump set prefix (Definition 2.11). That is, we use $\langle\!| z := \ell v + z |\!\rangle\,\phi$ to mean $\langle z := \ell v + z\rangle\phi$, just to have more readable bracket grouping. The dTL proof of the safety invariant in (4.3) splits into two cases that correspond to the respective protocol phases:

$$\to\!r\;\dfrac{[;]\Box\;\dfrac{\dfrac{\dots}{\psi \vdash [neg]\Box\phi} \qquad \dfrac{\dots}{\psi \vdash [neg][cor;drive]\Box\phi}}{\psi \vdash [neg;cor;drive]\Box\phi}}{\vdash \psi \to [neg;cor;drive]\Box\phi}$$

There, the left branch proves that ϕ holds while negotiating and is as follows:

$$\begin{array}{rl} & \psi \vdash Lv + z < m\\ \cline{2-2} \forall r, i\forall & \psi \vdash \forall l{\geq}0\,(l \leq L \to lv + z < m)\\ \cline{2-2} [:=],\langle:=\rangle & \psi \vdash \forall l{\geq}0\,\langle\!| z := lv + z, \ell := l |\!\rangle\,\phi\\ \cline{2-2} ['] & \psi \vdash [neg]\phi\\ \cline{2-2} [']\Box & \psi \vdash [neg]\Box\phi \end{array}$$

The right branch shows that ϕ continues to hold after negotiation has completed when continuing with an adjusted acceleration a in *cor*; *drive*:

$$
\begin{array}{ll}
 & \psi,\ell\geq 0 \vdash v^2 < 2b(m-Lv-z) \land Lv+z<m \\ \hline
\mathrm{i}\forall & \psi,\ell\geq 0 \vdash \langle\!|z:=\ell v+z, a:=-b|\!\rangle \forall t\geq 0\,(\ell\leq L \rightarrow \frac{a}{2}t^2+vt+z<m) \\ \hline
[:=],\langle:=\rangle & \psi,\ell\geq 0 \vdash \langle\!|z:=\ell v+z, a:=-b|\!\rangle \forall t\geq 0\, \langle\!|z:=\frac{a}{2}t^2+vt+z|\!\rangle \phi \\ \hline
[']\Box,['] & \psi,\ell\geq 0 \vdash \langle\!|z:=\ell v+z, a:=-b|\!\rangle [drive]\Box\phi \qquad \triangleright \\ \hline
[\cup] & \psi,\ell\geq 0 \vdash \langle\!|z:=\ell v+z|\!\rangle [cor][drive]\Box\phi \qquad \triangleright \\ \hline
[;]\Box & \psi,\ell\geq 0 \vdash \langle\!|z:=\ell v+z|\!\rangle [cor;drive]\Box\phi \\ \hline
\rightarrow\mathrm{r} & \psi \vdash \ell\geq 0 \rightarrow \langle\!|z:=\ell v+z|\!\rangle [cor;drive]\Box\phi \\ \hline
\forall\mathrm{r} & \psi \vdash \forall\ell\geq 0\, \langle\!|z:=\ell v+z|\!\rangle [cor;drive]\Box\phi \\ \hline
['] & \psi \vdash [neg][cor;drive]\Box\phi
\end{array}
$$

The application of rule $[;]\Box$ in this latter case spawns a third case (marked with $\triangleright$) to show that ϕ holds during *cor*. However, the reasoning in this third case is subsumed by the cases above, since the changes on a in *cor* do not interfere with condition ϕ. Generally, this optimisation of $[;]\Box$ is applicable whenever the modified vocabulary is disjoint from ϕ. Here, proof rules $[']$ and i$\forall$ are implemented in Mathematica to handle evolutions.

The leaves of the proof branches above can even be used to automatically *synthesise parameter constraints* that are necessary to avoid MA violation. The parametric safety constraint obtained by combining the open conditions conjunctively is $Lv+z<m \land v^2 < 2b(m-Lv-z)$. It simplifies to $v^2 < 2b(m-Lv-z)$ because $b>0$. This yields bounds for the speed limit and negotiation latency in order to guarantee safe driving and closing of the proof. Similarly, rule $[\cup]$ leads to a branch (marked with $\triangleright$) for the case $[?m-z\geq s; a:=\dots]$, from which corresponding conditions about the safety envelope s can be derived depending on the particular speed controller, similar to what we have shown in Sect. 2.9.

4.9 Liveness by Quantifier Alternation

Liveness specifications of the form $[\alpha]\Diamond\phi$ or $\langle\alpha\rangle\Box\phi$ are sophisticated (Σ^1_1-hard because they can express infinite occurrence in Turing machines). Beckert and Schlager [38], for instance, note that they failed to find sound rules for a discrete case that corresponds to $[\alpha;\beta]\Diamond\phi$.

For *finitary liveness semantics*, we can still find proof rules. In this section, we modify the meaning of $[\alpha]\Diamond\phi$ to refer to all *terminating* traces of α. Then, the straightforward generalisation $[;]\Diamond$ in Fig. 4.6 is sound, even in the hybrid case. But $[;]\Diamond$ still leads to an incomplete axiomatisation as it does not cover the case where, in some traces, ϕ becomes true at some point during α, and in other traces, ϕ only becomes true during β. To overcome this limitation, we use a program transformation approach. We instrument the hybrid program to monitor the occurrence of ϕ during all changes: In $[\alpha]\Diamond$, $\breve{\alpha}$ results from replacing all occurrences of $x:=\theta$ with $x:=\theta;?(\phi\rightarrow t=1)$ and $x'=\theta$ with $x'=\theta\,\&\,(\phi\rightarrow t=1)$. The latter is a continuous evolution restricted to the region of the state space that satisfies $\phi\rightarrow t=1$.

The effect is that t detects whether ϕ has occurred during any change in α. In particular, t is guaranteed to be 1 after all runs if ϕ occurs at least once along all traces of α. This trick directly works for first-order conditions ϕ. Using the combination presented in [234], nominals can be used as state labels to address the same issue for general ϕ.

$$([;]\Diamond)\ \frac{\vdash [\alpha]\Diamond\phi, [\alpha][\beta]\Diamond\phi}{\vdash [\alpha;\beta]\Diamond\phi} \qquad ([\alpha]\Diamond)\ \frac{\phi \vee \forall t\, [\check{\alpha}]t = 1}{[\alpha]\Diamond\phi}$$

Fig. 4.6 Transformation rules for alternating temporal path and trace quantifiers

Proposition 4.2 (Local soundness for temporal quantifier alternation). *The rules in Fig. 4.6 are locally sound for finitary liveness semantics.*

Proof. Let ν be any state.

$[;]\Diamond$ Assuming that the premise is true, we need to consider two cases corresponding to the two formulas of its succedent. If $\nu \models [\alpha]\Diamond\phi$, then obviously $\nu \models [\alpha;\beta]\Diamond\phi$, as every trace of $\alpha;\beta$ has a trace of α as prefix, during which ϕ holds at least once. If, however, $\nu \models [\alpha][\beta]\Diamond\phi$, then ϕ occurs at least once during all traces that start in a state reachable from ν by α. Let $\rho \circ \varsigma \in \tau(\alpha;\beta)$ with $\operatorname{first}\rho = \nu$, $\rho \in \tau(\alpha)$ and $\varsigma \in \tau(\beta)$. In finitary liveness semantics, $\rho \circ \varsigma$ can be assumed to terminate (otherwise there is nothing to show). Then, $\operatorname{last}\rho$ is a state reachable from ν by α; hence $\varsigma \models \Diamond\phi$. In particular, $\rho \circ \varsigma \models \Diamond\phi$.

$[\alpha]\Diamond$ For the soundness of $[\alpha]\Diamond$, first observe that the truth of $val(\nu,\phi)$ of ϕ depends on the state ν; hence it can only be affected during state changes. Further, the only actual changes of valuations happen during discrete jumps $x := \theta$ or continuous evolutions $x' = \theta$. All other system actions only cause control flow effects but no elementary state changes. Assume the premise is true in a state ν. If $\nu \models \phi$, the conjecture is obvious. Assume $\nu \models \forall t\, [\check{\alpha}]t = 1$. Suppose $\nu \not\models [\alpha]\phi$; then there is a trace $\sigma \in \tau(\alpha)$ with $\sigma \not\models \Diamond\phi$. Then, this trace directly corresponds to a trace $\check{\sigma}$ of $\check{\alpha}$ in which all $\phi \to t = 1$ conditions are trivially satisfied as ϕ never holds. As there are no changes of the fresh variable t during $\check{\alpha}$, the value of t remains constant during $\check{\sigma}$. But then we can conclude that there is a trace, which is essentially the same as $\check{\sigma}$ except for the constant valuation of the fresh variable t on which no conditions are imposed; hence $t = 0$ is possible. As these traces terminate in finitary liveness semantics, we can conclude $\nu \not\models \forall t\, [\check{\alpha}]t = 1$, which is a contradiction. Conversely for equivalence of premise and conclusion, assume $\nu \not\models \phi \vee \forall t\, [\check{\alpha}]t = 1$. Then, the initial state ν does not satisfy ϕ and it is possible for $\check{\alpha}$ to execute along a terminating trace σ that permits t to be $\neq 1$. Suppose there was a position (i,ζ) of σ at which $\sigma_i(\zeta) \models \phi$. Without loss of generality, we can assume (i,ζ) to be the first such position. Then, the hybrid action which regulates σ_i is accompanied by the immediate condition that $\phi \to t = 1$; hence $t = 1$ holds if σ terminates. Since the fresh variable t is rigid (is never changed during $\check{\alpha}$) and σ terminates in finitary liveness semantics, we conclude $val(\text{last}\,\sigma, t) = 1$, which is a contradiction. □

4.10 Summary

For reasoning about hybrid systems, we have introduced a temporal dynamic logic, dTL, with modal path quantifiers over traces and temporal quantifiers along the traces. It combines the capabilities of dynamic logic [149] to reason about possible system behaviour with the power of temporal logic [247, 114, 115] in reasoning about the behaviour along traces. Furthermore, we have presented a proof calculus for verifying temporal safety specifications of hybrid programs in dTL.

Our sequent calculus for dTL is a completely modular combination of temporal and nontemporal reasoning. Temporal formulas are handled using rules that augment intermediate state transitions with corresponding sub-specifications. Purely nontemporal $d\mathcal{L}$ rules handle the effects of discrete and continuous evolution. The modular nature of the dTL calculus further enables us to lift the relative completeness result from $d\mathcal{L}$ to dTL. This theoretical result shows that the dTL calculus is a sound and complete axiomatisation of the temporal behaviour of hybrid systems relative to differential equations.

As an example, we demonstrate that our logic is suitable for reasoning about safety invariants in the European Train Control System. Further, we have successfully applied our calculus to automatically synthesise (nonlinear) parametric safety constraints for this system.

Future work includes extending dTL with CTL*-like [115] formulas of the form $[\alpha](\psi \wedge \Box\phi)$ to avoid splitting of the proof into two very similar subproofs for temporal parts $[\alpha]\Box\phi$ and nontemporal parts $[\alpha]\psi$ arising in rule $[;]\Box$. Our combination of temporal logic with dynamic logic is more suitable for this purpose than previous approaches for discrete systems [38], since dTL has uniform modalities and uniform semantics for temporal and nontemporal specifications. This extension will also simplify the treatment of alternating liveness quantifiers conceptually.

Part II
Automated Theorem Proving for Hybrid Systems

Overview After having developed formal specification logics and proof calculi for specifying and verifying safety-critical properties of hybrid systems in Part I, we now turn to practical and algorithmic implementation questions. In this part, we focus on the practical aspects of implementing the proof calculi from Part I. The calculi in Part I have already been designed for the needs of automated theorem proving, most notably with the free-variable and Skolemisation techniques from Chap. 2 and the compositional proof calculi from Part I. Immediate implementations of the proof calculi from Part I in automated theorem provers can prove examples of medium complexity directly. Yet, more complex case studies still require additional algorithmic techniques for achieving high-degree automation and good scalability properties. In Chap. 5, we refine the calculi from Part I to tableau procedures and present proof strategies that navigate among their nondeterminisms to help overcome the complexity issues of integrating real quantifier elimination as a decision procedure for real arithmetic.

In Chap. 6 we introduce the "differential invariants as fixed points" paradigm. We refine the differential induction techniques from Chap. 3 to a fully automatic verification algorithm for computing the required discrete and differential invariants of a hybrid system locally in a logic-based fixed-point loop.

The algorithmic refinement techniques developed in this part of the book add better automation to the proof approach from Part I. These algorithms are crucial for automating the formal verification of properties of complex hybrid systems like the ones we consider in Part III.

Chapter 5
Deduction Modulo Real Algebra and Computer Algebra

Contents

5.1	Introduction		**234**
	5.1.1	Related Work	234
	5.1.2	Structure of This Chapter	235
5.2	Tableau Procedures Modulo		**235**
5.3	Nondeterminisms in Tableau Modulo		**238**
	5.3.1	Nondeterminisms in Branch Selection	238
	5.3.2	Nondeterminisms in Formula Selection	239
	5.3.3	Nondeterminisms in Mode Selection	240
5.4	Iterative Background Closure		**243**
5.5	Iterative Inflation		**246**
5.6	Experimental Results		**248**
5.7	Summary		**251**

Synopsis We show how deductive, real algebraic, and computer algebraic methods can be combined for verifying hybrid systems in an automated theorem proving approach. In particular, we highlight the interaction of deductive and algebraic reasoning that is used for handling the joint discrete and continuous behaviour of hybrid systems. Systematically, we derive a canonical *tableau procedure modulo* from the calculus of differential dynamic logic. We delineate the nondeterminisms in the tableau procedure carefully and analyse their practical impact in the presence of computationally expensive handling of real algebraic constraints. Based on experience with larger case studies, we analyse proof strategies for dealing with the practical challenges for integrated algebraic and deductive verification of hybrid systems. To overcome the complexity pitfalls of integrating real arithmetic, we propose the *iterative background closure* and *iterative inflation order* strategies, with which we achieve substantial computational improvements.

A. Platzer, Logical Analysis of Hybrid Systems,
DOI 10.1007/978-3-642-14509-4_5,

5.1 Introduction

While the theoretical background and formal details of our logical analysis approach for hybrid systems can be found in Part I, here we discuss the practical aspects of combining deductive, real algebraic, and computer algebraic prover technologies. In particular, we highlight the principles of how these techniques interact for verifying hybrid systems. We analyse the degrees of freedom in implementing our calculus in terms of the nondeterminisms of its canonical proof procedure. We illustrate the impact that various choices of proof strategies have on the overall performance. For hybrid system verification, we observe that the nondeterminisms in the interaction between deductive and real algebraic reasoning have considerable impact on the practical feasibility. While straightforward combinations are sufficient for verifying examples like those presented in Part I, larger case studies like those that we present in Part III are beyond the capabilities of state-of-the-art decision procedures for real arithmetic. In this chapter, we analyse and explain the causes and consequences of this effect and introduce automatic proof strategies that avoid these complexity pitfalls and work well in practise.

Here we study the modular combination in the $d\mathcal{L}$ calculus (our findings generalise directly to the extensions of the DAL and dTL calculi, so we use $d\mathcal{L}$ interchangeably with DAL and dTL in this chapter). Our observations are of more general interest, though, and we conjecture that similar results hold for other prover combinations of logics with interpreted function symbols that are handled using background decision procedures for computationally expensive theories including real arithmetic, approximations of natural arithmetic, and arrays.

5.1.1 Related Work

As we have pointed out in Sect. 1.2, there are only a few other practical approaches [201, 1] that use deduction for verifying hybrid systems and actually integrate arithmetic reasoning in STeP [201] or PVS [1], respectively. They do not work with a genuine verification logic, however, but only generate flat mathematical verification conditions for hybrid automata with a given invariant. In contrast, the symbolic decompositions in our verification logic preserve the natural problem structure, which enables us to achieve good performance in practise. See Sect. 1.2 for a detailed comparison.

Several other approaches intend to combine deductive and arithmetic reasoning, e.g. [63, 28, 3]. Their focus, however, is on general mathematical reasoning in classes of higher-order logic and is not tailored to verify hybrid systems. Our work, instead, is intended to make practical verification of hybrid systems possible and we aim at automating the verification process.

5.1.2 Structure of This Chapter

In Sect. 5.2, we analyse our calculi for differential dynamic logics from a qualitative perspective and present a corresponding tableau procedure modulo background solvers for handling real algebraic and computer algebraic constraints. We delineate their nondeterminisms carefully in Sect. 5.3 and analyse their practical impact. We present proof procedures for automated theorem proving in differential dynamic logics that navigate through the complexity pitfalls of integrating decision procedures for real arithmetic in Sects. 5.4 and 5.5. In Sect. 5.6, we evaluate their performance in larger case studies.

5.2 Tableau Procedures Modulo

In this section, we derive a canonical tableau procedure modulo background provers from the $d\mathcal{L}$ calculus and analyse the remaining nondeterminisms in the remainder of this chapter.

For the purpose of this chapter, the full details of how the respective quantifier rules of Chaps. 2 and 3 lift quantifier elimination to dynamic logic are not important. What is important to note, however, is that quantifier rules and rules for handling modalities need to interact because the actual constraints on quantified symbols depend on the effect of the hybrid programs within modalities (Sect. 2.5.3). Thus, at some point, after a number of rule applications that handle the dynamic part, quantifier rules will be used to discharge (or at least simplify) proof obligations over real algebraic or semialgebraic constraints by quantifier elimination [80, 81, 288]. The remaining subgoals will be analysed further again using dynamic rules. The quantifier rules constitute the modular interface that combines deduction for handling dynamic reasoning with algebraic constraint techniques for handling continuous reasoning about the reals. Here, we discuss the consequences and principles of this combination and analyse proof strategies.

The principle behind how the $d\mathcal{L}$ calculus in Fig. 2.11 combines deduction technology with methods for handling real algebraic constraints complies with the general background reasoning principles [32, 290, 103]. From an abstract perspective, the $d\mathcal{L}$ calculus selects a set Φ of (quantified) formulas from an open branch (Φ is called the *key*) and hands it over to the quantifier elimination procedure. The resulting formula obtained by applying quantifier elimination (QE) to Φ is then returned to the main sequent prover as a result, and the main proof continues; see Fig. 5.1. Similarly, the $d\mathcal{L}$ calculus triggers symbolic, computer algebraic computations for the rules for differential equations using their solutions (rules $\langle'\rangle$,$[']$ from Fig. 2.11) or total differentials of differential invariants (rule *DI*) or differential variants (rule *DV*) from the DAL calculus in Fig. 3.9.

In this context, the propositional rules, dynamic rules, and global rules of the $d\mathcal{L}$ calculus in Fig. 2.11 or the DAL calculus in Fig. 3.9 constitute the *foreground rules* in the main prover (middle box of Fig. 5.1), except for evolution rules $\langle'\rangle$,$[']$ and

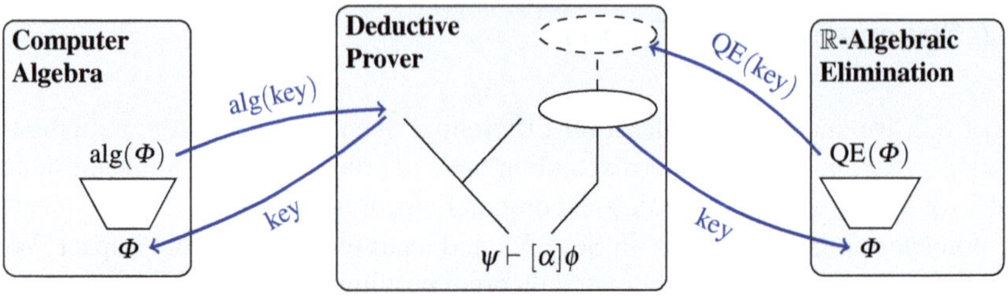

Fig. 5.1 Deductive, real algebraic, and computer algebraic prover combination

differential (in)variant rules *DI*,*DV* that represent the *computer algebraic rules* invoking a computer algebra system as a *background solver* (left box). The arithmetic quantifier rules (in particular i∀ and i∃ from Fig. 2.11 and r∀,l∀,r∃,l∃ from Fig. 3.9) form the set of rules that invoke the *background prover* (right box) for quantifier elimination. Since the primary challenges caused by the proof nondeterminisms in the $d\mathcal{L}$ calculus originate from the interaction of deductive and real algebraic rules, we simplify the presentation in the following and only distinguish between background real arithmetic rules and foreground rules, where computer algebraic rules will simply be considered as foreground rules.

The canonical tableau procedure belonging to the $d\mathcal{L}$ calculus is presented in Fig. 5.2. Observe that the tableau procedure for our $d\mathcal{L}$ calculus has a nonstandard set of nondeterministic steps (indicated by $\mathcal{B}$, $\mathcal{M}$, and $\mathcal{F}$, respectively in Fig. 5.3):

$\mathcal{B}$: *selectBranch*, i.e., select which open branch to choose for further rule applications.

$\mathcal{M}$: *selectMode*, i.e., select whether to apply foreground $d\mathcal{L}$ rules (propositional rules, dynamic rules, and global rules) or background arithmetic rules (i∀, i∃ from Fig. 2.11 or r∀,l∀,r∃,l∃ from Fig. 3.9).

$\mathcal{F}$: *selectFormula*, i.e., which formula(s) to select for rule applications from the current branch in the current mode.

Within the rule applications, there is an additional choice of whether to handle differential equations using their solution ($d\mathcal{L}$ rules $\langle'\rangle$,$[']$ from Fig. 2.11) or by differential induction (DAL rules *DI*,*DV* from Fig. 3.9). We do not follow up on this nondeterminism here but simply choose to use solutions, whenever they are first-order expressible, and fall back to differential induction (Sect. 3.5.6) when no such solution can be found. The computational cost of differential induction is generally less than when working with solutions, but the corresponding differential (in)variants have to be found first, which we handle in Chap. 6. There is a further minor nondeterminism of whether to expand loops using rules $\langle{*}^n\rangle$,$[{*}^n]$ or to go for an induction by rules *ind* and *con*. Yet, as unrolling ($\langle{*}^n\rangle$,$[{*}^n]$) only handles fixed-length reachability prefixes or bounded loops, our proof strategies prefer induction (*ind* and *con*) instead. The other $d\mathcal{L}$ rules do not produce any conflicts once a formula has been selected as they apply to formulas of distinct syntactic structures: The top-level operators uniquely determine which calculus rule to use once a formula has been selected.

```
while tableau T has open branches do
  B := selectBranch(T)           (* 𝓑 nondeterminism *)
  M := selectMode(B)             (* 𝓜 nondeterminism *)
  F := selectFormulas(B,M)       (* 𝓕 nondeterminism *)
  if M = foreground then
    R := result of applying a propositional or
           dynamic or global rule to F in B
    replace branch B by R in tableau T
  else
    send key F to background decision procedure QE
    receive result R from QE
    apply a quantifier rule to T with QE-result R
  end if
end while
```

Fig. 5.2 Tableau procedure for differential dynamic logics

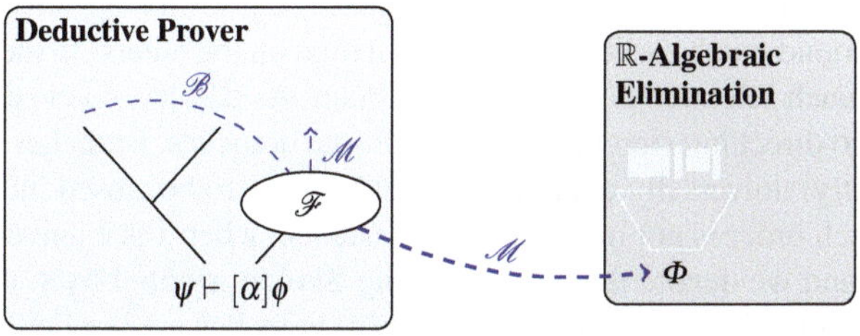

Fig. 5.3 Nondeterminisms in the tableau procedure for differential dynamic logics

At this point, notice that, unlike in the classical tableau procedure [122], we have three rather than four points of nondeterminism, since d$\mathcal{L}$ does not need closing substitutions. The reason for this is that d$\mathcal{L}$ has an interpreted domain. Rather than having to try out instantiations that have been determined by unification or heuristics as in uninterpreted first-order logic [122], we can make use of the structure in the interpreted case of first-order logic over the reals. In particular, arithmetic formulas can be reduced equivalently by QE to simpler formulas in the sense that the quantified symbols no longer occur. As this transformation is an equivalence, there is no loss of information and we do not need to backtrack [122] or simultaneously keep track of multiple local closing instantiations [134].

Despite this seemingly simpler situation, the influence of nondeterminism on the practical prover performance is quite remarkable. Even though the first-order theory of real arithmetic is decidable by quantifier elimination [81], its complexity is *doubly exponential* in the number of quantifier alternations [94]. While more efficient algorithms exist for some linear fragments [198], the practical performance is an issue in nonlinear cases. Hence, the computational cost of individual rule applications is quite different from the linear complexity of applying closing substitutions in uninterpreted tableaux.

5.3 Nondeterminisms in Tableau Modulo

In principle, exhaustive fair application of background rules by the nondeterminisms $\mathscr{M}$ and $\mathscr{F}$ remains complete for appropriate fragments of d$\mathscr{L}$. In practise, however, the complexity of real arithmetic quickly makes this naïve approach infeasible for larger case studies. In the remainder of this chapter, we discuss the consequences of the nondeterminisms and develop proof strategies to tackle the combination and integration challenges.

5.3.1 Nondeterminisms in Branch Selection

In classical uninterpreted tableaux, branch selection has no impact on completeness, and can only have impact on the proving duration, because closing substitutions can sometimes be found much earlier on one branch than on the others. In the interpreted case of d$\mathscr{L}$, branch selection is even less important. As d$\mathscr{L}$ has no closing substitutions, there is no direct interference among multiple branches. Branches with (explicitly or implicitly) universally quantified variables have to be closed independently. Hence the branch order is not important. For instance, when x is a universally quantified variable and we denote the corresponding Skolem symbol by s, the branches in the following proof can be handled separately (branches are implicitly combined by conjunction and universal quantifiers distribute over conjunctions):

$$\dfrac{\dfrac{\dfrac{\mathrm{QE}(\forall x\,(\dots bx^2 \geq 0))}{\Gamma, b>0 \vdash bs^2 \geq 0}\,\mathrm{i}\forall \qquad \dfrac{\mathrm{QE}(\forall x\,(\dots bx^4 + x^2 \geq 0))}{\Gamma, b>0 \vdash bs^4 + s^2 \geq 0}\,\mathrm{i}\forall}{\Gamma, b>0 \vdash bs^2 \geq 0 \land bs^4 + s^2 \geq 0}\,\land\mathrm{r}}{\Gamma, b>0 \vdash \forall x\,(bx^2 \geq 0 \land bx^4 + x^2 \geq 0)}\,\forall\mathrm{r}$$

For existentially quantified variables, the situation is a bit more subtle as multiple branches interfere indirectly in the sense that a simultaneous solution needs to be found for all branches at once. In $\exists v\,(v>0 \land v<0)$, for instance, the two branches resulting from the cases $v>0$ and $v<0$ cannot be handled separately, as the existential quantifier claims the existence of a simultaneous solution for $v>0$ and $v<0$, not two different solutions. Thus, when v is an existentially quantified variable and V its corresponding free existential variable, the branches in the following proof need to synchronise before quantifier elimination is applied:

$$\dfrac{\dfrac{\dfrac{\mathrm{QE}(\exists v\,((b>2 \to b(V-1)>0) \land (b>2 \to (V+1)^2 + b\varepsilon(V+1)>0)))}{\dfrac{b>2 \vdash b(V-1)>0}{b>2 \vdash [v:=V-1]bv>0}\,\mathrm{i}\forall \qquad \dfrac{b>2 \vdash (V+1)^2 + b\varepsilon(V+1)>0}{b>2 \vdash [v:=V+1]v^2 + b\varepsilon v>0}}}{b>2 \vdash [v:=V-1]bv>0 \land [v:=V+1]v^2 + b\varepsilon v>0}\,\land\mathrm{r}}{b>2 \vdash \exists v\,([v:=v-1]bv>0 \land [v:=v+1]v^2 + b\varepsilon v>0)}\,\forall\mathrm{r}$$

The order in which the intermediate steps at the two branches are handled, however, has no impact on the proof. Branches like these *synchronise* on an existential free variable V in the sense that all occurrences of V need to be first-order on all branches for quantifier elimination to be applicable. Consequently, the only fairness assumption for $\mathscr{B}$ is that whenever a formula of a branch is selected that is waiting for synchronisation with another branch to become first-order, then it transfers its branch choice to the other branch instead. In the above case the left branch synchronises with the right branch on V. Hence, rule i$\exists$ can only be applied to $b(V-1)>0$ on the left branch after rule $\langle:=\rangle$ has been applied on the right branch to produce first-order occurrences of V.

Thus, the primary remaining impact of the branch nondeterminism is that closing branches by universally quantified variable reasoning simplifies all subsequent existential variable handling, because fewer branches remain that need to be considered simultaneously. Even for existential variables, universal quantification is sound and can be very helpful; it is just not complete.

5.3.2 Nondeterminisms in Formula Selection

In background proving mode, it turns out that nondeterminism $\mathscr{F}$ is important for practical performance. In practise, when a branch closes or, at least, can be simplified significantly by a quantifier elimination call, then the running time of a single decision procedure call depends strongly on the number of irrelevant formulas that are selected in addition to the relevant ones by $\mathscr{F}$.

Clearly, when Φ is a set of formulas from a sequent that yields a tautology such that applying QE closes a branch, then selecting any superset $\Psi \supseteq \Phi$ from a branch yields the same answer in the end (a sequent forms a disjunction of its formulas; hence it can be closed to true when any subset closes). However, the running time until this result will be found in the larger Ψ can be disturbed strongly by the presence of complicated additional but irrelevant formulas. From our experience with Mathematica [303], decision procedures for full real arithmetic seem to be distracted considerably by such irrelevant additional information.

Example 5.1 (Computational distraction in quantifier elimination). One sequent from the proof of the ETCS kernel in Sects. 2.4 and 2.9 is depicted in Fig. 5.4. It results from the right branch of the proof in Sect. 2.9 by exhaustive splitting. Quantifier elimination, as performed by function Reduce in Mathematica, runs more than 24 hours without producing a result on the formula in Fig. 5.4, which has nine symbolic variables and polynomial degree 2.

The marked constraint in Fig. 5.4 corresponds to the initial state of the system, because it refers to the initial train position z and not to the current train position z_2 in the current induction step. In fact, the induction step does not depend on this part of the initial state information (it *does* depend on the initial $b>0$, though). When we remove the superfluous constraint on the initial state, the formula in Fig. 5.4 suddenly becomes provable in less than one second, compared to more than a day! The

Fig. 5.4 Computational distraction in quantifier elimination

$$
\begin{aligned}
&t_2 > 0,\ \varepsilon \geq t_2,\ v_2 \geq 0,\ A + 1/t_2 \cdot v_2 \geq 0,\ t_2 \geq 0,\\
&m - z_2 \geq v_2^2/(2b) + (A/b+1)(A/2\varepsilon^2 + \varepsilon v_2),\\
&2b(m - z_2) \geq v_2^2\\
&2b(m - z) \geq v^2, \qquad \text{/* initial state */}\\
&b > 0,\ A \geq 0\\
\vdash\ &(At_2 + v_2)^2 \leq 2b(m - 1/2(At_2^2 + 2t_2v_2 + 2z_2))
\end{aligned}
$$

d$\mathscr{L}$ calculus presented in Chap. 2 avoids this problem, because of the additional focusing quantifiers introduced by the universal closure $\forall^\alpha$ in global rules and because, later, rule i$\forall$ applies quantifier elimination after reintroducing the quantifier structure. □

Yet, such additional information accumulates in tableaux procedures quite naturally, because the purpose of a proof branch in d$\mathscr{L}$ is to keep track of all that is known about a particular (symbolic) case of the system behaviour. Generally, not all of this knowledge finally turns out to be relevant for that case and only plays a role in other branches. Nevertheless, discarding part of this knowledge arbitrarily would, of course, endanger completeness.

For instance, the train safety statement (2.7) from p. 62 in Sect. 2.4 depends on a constraint on the safety envelope s that regulates braking versus acceleration by the condition $m - z \geq s$ in *ctrl*. A maximal acceleration of A is permitted in case $m - z \geq s$, when adaptively choosing s depending on the current speed v, maximum braking force b, and maximum controller response time ε in accordance with constraint (2.18) as discovered in Sect. 2.9. This constraint is necessary for some but not all cases of the symbolic safety analysis, though. In the case where the braking behaviour of ETCS is analysed, for instance, the constraint on s is irrelevant, because braking is the safest operation that a train can do to avoid crashing into preceding trains. The unnecessary presence of several quite complicated constraints such as (2.18), however, can distract quantifier elimination procedures considerably.

5.3.3 Nondeterminisms in Mode Selection

In its own right, nondeterminism $\mathscr{M}$ has less impact on the prover performance than $\mathscr{F}$. Every part of a branch could be responsible for closing it. The foreground closing rule *ax* of the main prover can only close branches for comparably trivial reasons like $b > 0, \varepsilon > 0 \vdash \varepsilon > 0$. Hence, mode selection has to give a chance to the background procedure every once in a while, following some fair selection strategy. From the observation that some decision procedure calls can run for hours without terminating, however, we can see that realisations of nondeterminism $\mathscr{M}$ need to be devised with considerable care.

As the reason for closing a branch can be hidden in any part of the sequent, some expensive decision procedure calls are superfluous if the branch can be closed by continuing $\mathsf{d}\mathscr{L}$ reasoning on the other parts. For instance, if A is some complicated algebraic constraint, decision procedure calls triggered by nondeterminism $\mathscr{M}$ can lead to nontermination within any feasible time for

$$\ldots,\varepsilon > 0, m-z \geq s \vdash A, [drive]\varepsilon > 0, \ldots$$

If $\mathscr{M}$ chooses foreground rules, then an analysis of $[drive]\varepsilon > 0$ by $\mathsf{d}\mathscr{L}$ rules will quickly discover that the maximum reaction-time ε remains constant while driving. Then, this part of a proof closes without the need to consider constraint A at all. For this reason, proof strategies that *eagerly check* for closing branches by background procedure calls as soon as the formulas are first-order are not successful in practise; see Fig. 5.5. Similarly, for verifying the ETCS case study that we detail in Chap. 7

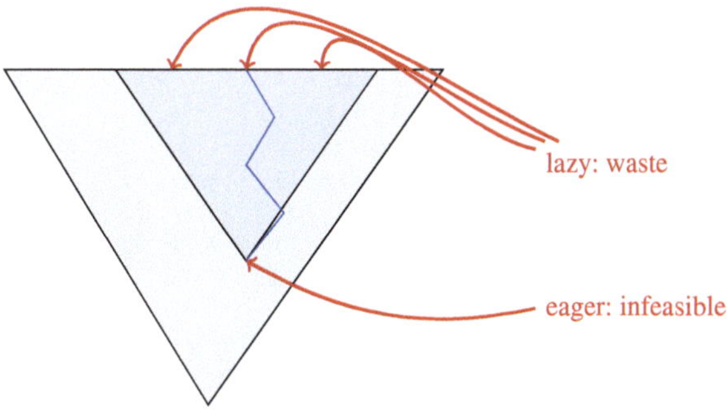

Fig. 5.5 Eager and lazy quantifier elimination in proof search space

we need to prove fairly large subgoals of first-order real arithmetic, which cannot be proven by current quantifier elimination procedures within any feasible amount of time; see Fig. 5.6 for a typical example.

Unfortunately, converse strategies with lazy checks that strongly favour foreground $\mathsf{d}\mathscr{L}$ rule applications in $\mathscr{M}$ are not appropriate either; see Fig. 5.5. The *lazy proof* strategy only uses background solvers if no more foreground proof rules can be used. There, splitting rules like $\wedge$r and $\vee$l can eagerly split the problems into multiple branches without necessarily making them any easier to solve. If this happens, slightly different but similar arithmetic problems of about the same complexity need to be solved repeatedly on multiple branches rather than just one branch, resulting in runtime blowup. The reason this can happen is a substantial syntactic redundancy in the sequent encoding of formulas. For instance, the sets of sequents before and after the following rule application are equivalent:

$$\wedge\text{r},\wedge\text{r}\ \frac{\psi \vdash v^2 \leq 2b(m-z) \quad \psi \vdash \varepsilon > 0 \quad \psi \vdash (z \geq 0 \rightarrow v \leq 0)}{\psi \vdash v^2 \leq 2b(m-z) \wedge \varepsilon > 0 \wedge (z \geq 0 \rightarrow v \leq 0)}$$

Yet, closing the three sequents above the bar by quantifier elimination is not necessarily easier than closing the single sequent below (neither conversely), because

```
  state = 0,
  2 * b * (m - z) >= v ^ 2 - d ^ 2,
  v >= 0, d >= 0, v >= 0, ep > 0, b > 0, A > 0, d >= 0
==>
  v <= vdes
-> \forall R a_3;
     (   a_3 >= 0 & a_3 <= A
      -> (        m - z
               <= (A / b + 1) * ep * v
                + (v ^ 2 - d ^ 2) / (2 * b)
                + (A / b + 1) * A * ep ^ 2 / 2
           -> \forall R t0;
                (   t0 >= 0
                 -> \forall R ts0;
                      (0 <= ts0 & ts0 <= t0
                          -> -b * ts0 + v >= 0 & ts0 + 0 <= ep)
                 ->     2 * b * (m - 1 / 2 *
                                  (-b * t0 ^ 2 + 2 * t0 * v + 2 * z))
                     >= (-b * t0 + v) ^ 2
                      - d ^ 2
                    & -b * t0 + v >= 0
                    & d >= 0))
        & (       m - z
              >  (A / b + 1) * ep * v
               + (v ^ 2 - d ^ 2) / (2 * b)
               + (A / b + 1) * A * ep ^ 2 / 2
            -> \forall R t2;
                (   t2 >= 0
                 -> \forall R ts2;
                      (0 <= ts2 & ts2 <= t2
                          -> a_3 * ts2 + v >= 0 & ts2 + 0 <= ep)
                 ->     2 * b * (m - 1 / 2 *
                                   (a_3 * t2 ^ 2 + 2 * t2 * v + 2 * z))
                     >= (a_3 * t2 + v) ^ 2
                      - d ^ 2
                    & a_3 * t2 + v >= 0
                    & d >= 0)))
```

Fig. 5.6 A large subgoal of first-order real arithmetic during ETCS verification

the working principle of arithmetic decision procedures is algebraic and not deductive. Even worse, if the sequents close by applying proof rules to ψ, then similar reasoning has to be repeated for three branches. This threefold reasoning may not even be detected as identical when ψ is again split differently on the three resulting branches.

Further, the representational equivalence in sequents is purely syntactic, i.e., up to permutation, the representations share the same disjunctive normal form. In the uninterpreted case of first-order logic, this syntactic redundancy is exploited by the propositional rules in order to transform sequents into a canonical form with atomic formulas, where partial closing situations are more readily identifiable. In the presence of a background decision procedure, however, reduction to sequents with atomic formulas is no longer necessary as it will be undone when handing the formulas over to the background decision procedure.

Moreover, the logical splitting along the propositional structure does not always help quantifier elimination procedures, because their working principle is not a deductive case analysis but (partial) cylindrical algebraic decomposition [80, 81, 83] based on cell decompositions of polynomials. See App. D.2 for explanations of the basic principle behind quantifier elimination in the background prover, which is quite different from that of symbolic logical case distinctions in the foreground prover. In Sect. 5.4, we will see that the combination of deductive analysis with algebraic arithmetic proving is an extremely important factor for accelerating quantifier elimination, but both have to be applied with care.

Finally, algebraic constraint handling techniques as in the Mathematica function Reduce can come up with a result that is only a restated version of the input if a selected (open) formula cannot be simplified or closed. For instance, the sequent $z < m \vdash v^2 \leq 2b(m-z)$ "reduces" to $\vdash b \geq v^2/(2m-2z) \vee m \leq z$ without any progress. Such arithmetical reformulation cannot even be detected by simple syntactical means and easily leads to infinite proof loops without progress when the outcome is split by $\vee$r and again handled by the background procedures.

5.4 Iterative Background Closure

In this and the following section, we propose strategies to solve the previously addressed computational issues caused by the nondeterminisms of the $d\mathcal{L}$ tableau procedure and its integration with computationally expensive background provers.

Priority-Based Strategies

We propose the priorities for rule applications in Fig. 5.7a (with rules at the top taking precedence over rules at the bottom). In this strategy, algebraic constraints are generally left intact as opposed to being split among multiple branches, because arithmetic rules have a higher priority than propositional splitting rules on first-order constraints. Further, we only accept the result of the background procedure when the number of different variable symbols has decreased to avoid infinite proof loops. We use arithmetic background rules *either* with priority 2 *or* with priority 5.

The effect of using priority 2 is that branches are checked eagerly for closing conditions or variable reductions; see Fig. 5.5. If reasoning about algebraic constraints does not yield any progress (no variables can be eliminated), then $d\mathcal{L}$ rules further analyse the system. For this choice, it is important to work with timeouts to prevent lengthy background decision procedure calls from blocking $d\mathcal{L}$ proof progress.

This problem is reduced significantly when using priority 5 for arithmetic rules instead. The effect of priority 5 is that formulas containing modalities are analysed

1. non-splitting propositional rules ¬r–∨r, ∧l–→r, *ax*
2. *arithmetic rules* if variable eliminated
3. dynamic rules ⟨?⟩–[:=]
4. splitting rules ∧r, ∨l, →l on modalities
5. *arithmetic rules* if variable eliminated
6. (in)variant global rules *ind*, *con*, *DI*, and *DV*
7. splitting rules ∧r, ∨l, →l on first-order formulas

Fig. 5.7a Proof strategy priorities

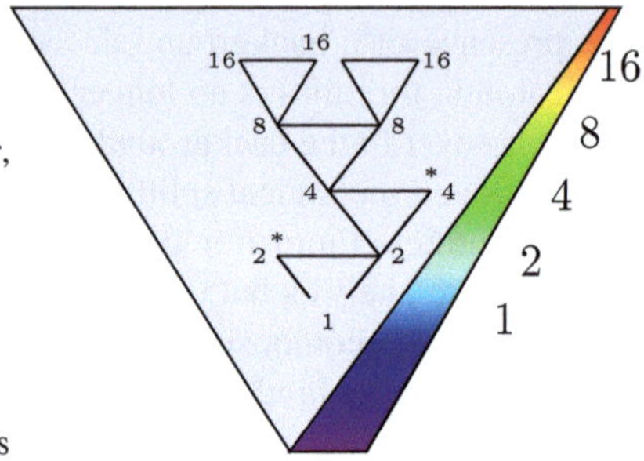

Fig. 5.7b Iterative background closure (IBC) proof strategy

and decomposed as much as possible before arithmetic reasoning is applied to algebraic constraint formulas. Then, however, the prover might consume too much time analysing the effects of programs on branches which would easily close due to simple arithmetic facts like in $\varepsilon > 0, \varepsilon < 0 \vdash [\alpha]\phi$.

A simple compromise is to use a combination of background rules with priority 2 for quick linear arithmetic [198] and expensive quantifier elimination calls for nonlinear arithmetic with priority 5. Another option is to normalise arithmetic formulas in sequents and use polynomial arithmetic to detect simple reasons for closing branches early [273], which has also been implemented in KeYmaera.

Iterative Background Closure

As a more sophisticated control strategy on top of the static priorities in Fig. 5.7a, we have introduced *iterative background closure* (IBC) [230]. There, the idea is to periodically apply arithmetic rules with a timeout T that increases by a factor of 2 after the background decision procedure runs have timed out; see Fig. 5.7b. Thus, background rules interleave with other rule applications (triangles in Fig. 5.7b), and the timeout for the subgoals increases as indicated, until the background procedure successfully eliminates variables on a branch (marked by $*$). The effect is that the prover avoids splitting in the average case but is still able to split cases when combined handling turns out to be prohibitively expensive. As an optimisation, timed-out QE will only be invoked again after the branch has been split (or after a modal formula has disappeared from the sequent) so that QE will work on a different input.

Figure 5.8 shows a proof procedure that implements IBC. Unless the QE call terminated successfully within the current timeout (line 3), IBC applies foreground rules (line 7) until the proof has split. Then, the timeout increases (line 10) and IBC handles the resulting branches recursively (line 11). Our experimental results in Sect. 5.6 show that IBC is surprisingly relevant for handling larger case studies.

```
/* prove validity of the sequent Φ ⊢ Ψ */
function IBC(Φ ⊢ Ψ, timeout):
  if QE(Φ ⊢ Ψ) succeeds within timeout then
    return QE(Φ ⊢ Ψ)
  else
    while foreground rule applicable and proof not split do
      apply foreground rule to current sequent
    end while
    let Φ_1 ⊢ Ψ_1,...,Φ_n ⊢ Ψ_n be the resulting branches
    timeout := 2*timeout
    return IBC(Φ_1 ⊢ Ψ_1 ,timeout) and ... and IBC(Φ_n ⊢ Ψ_n ,timeout)
```

Fig. 5.8 Iterative background closure (IBC) algorithm schema

And/Or Branching

More generally, the $d\mathcal{L}$ calculus gives rise to theorem proving structures with combined and/or-branching. While the branches resulting from one rule application are *and-branches* (all of them have to close for the proof to succeed), the rule alternatives—especially as caused by the tableau procedure nondeterminisms—are *or-branches* (only one of the proof search alternatives has to be successful for the proof); see Fig. 5.9. Most notably, induction rules give rise to or-branches, as there

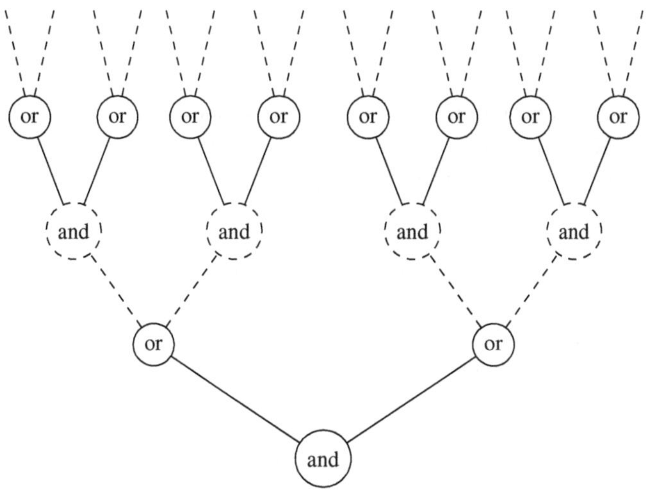

Fig. 5.9 General and/or-branching in proof strategies for differential dynamic logics

are several possible formulas (infinitely many) that could be used as (differential) (in)variants for rules *ind*,*con*,*DI*,*DV*, but one single successful (in)variant is enough for closing the proof. For these or-branch alternatives, any fair parallel exploration scheme or any fair sequential exploration scheme with time interleaving can be used. Iterative background closure is one possible sequential interleaving choice that works well in practise. On parallel computers, distributed clusters, or even multi-core processors, truly parallel exploration schemes would be faster, as they exploit the natural proving parallelism in our calculi from Part I. We will make use of and/or-branching exploration based on iterative background closure in Chap. 6.

5.5 Iterative Inflation

The iterative background closure strategy already has a decisive impact on the feasibility of verifying larger case studies. Yet, there are still cases where the $\mathscr{F}$ nondeterminism has a significant computational impact that limits scalability. These difficulties are primarily caused by superfluous constraints on previous or initial states that accumulate in the sequent; see, e.g., Example 5.1.

A general possible solution for this issue is to iteratively consider more formulas of the sequent and attempt decision procedure calls with fair time interleaving until the respective branch closes. Then, only those additional formulas need to be considered that share variables with any of the other selected formulas. Further, timeouts can be used to discontinue lengthy decision procedure calls and continue along other choices of the nondeterminisms in Fig. 5.2. For complicated cases with a prohibitive complexity, this heuristic process worked well on our examples.

Inflation Order

As a general pattern for building iterative inflation optimisations, we propose an algorithm that selects additional formulas of a sequent successively while a counterexample can be found in bounded time (e.g., using the Mathematica function *FindInstance*) and terminates when a quantifier elimination call yields "true" within the current time bound (increasing timeouts as in IBC).

The *iterative inflation* algorithm schema in Fig. 5.10 proves validity of a disjunction of formulas of first-order real arithmetic that is given as a set S of disjuncts. Validity of a sequent $\Phi \vdash \Psi$ can be proven by $IIO(\{\neg\phi \,:\, \phi \in \Phi\} \cup \{\psi \,:\, \psi \in \Psi\})$. Following the general scheme in Fig. 5.9, the IIO algorithm explores subset candidates of S in parallel by fair time interleaving. Starting from the empty subset of formulas (line 5), the algorithm will try to prove (line 8) or disprove (line 11) each candidate subset ϕ of S. The first successful QE call yielding true within the current timeout shows validity of S (line 9), because a disjunction is valid if any subset is. When a counterexample is found (line 11) or quantifier elimination produced another result than "true" within the time bounds (line 10), candidate ϕ is dropped and any remaining formulas of S are added to ϕ to form new candidates. If no candidate could be proven (line 8) or disproven (lines 10–11) within the current time bounds, the timeout increases (line 14). The algorithm returns that S is not valid if all can-

```
1  /* prove validity of the disjunctive set S */
2  /* of formulas of first-order real arithmetic */
3  function IIO(S):
4    timeout := 1
5    𝒞 := {false}
6    while 𝒞 ≠ ∅ do
7      for ϕ ∈ 𝒞 do
8        if (QE(ϕ) = true) within timeout then
9          return valid
10       else if (QE(ϕ) ≠ true) within timeout
11                or FindInstance(¬ϕ) within timeout then
12         𝒞 := (𝒞 ∪ {ϕ ∨ A : A ∈ S, A ∉ ϕ}) \ {ϕ}
13     end for
14     timeout := 2*timeout
15   end while
16   return not valid
```

Fig. 5.10 Iterative inflation order (IIO) algorithm schema

didates have counterexamples (or QE yields results other than "true") and no new formulas of S can be added to produce new candidates (line 12).

The IIO algorithm is a schematic algorithm. It allows for several refinements, optimisations, and caching improvements. Most importantly, line 12 can be refined to limit the number of candidates explored in parallel by adding only candidates with computationally promising properties. In general, a *reinclusion ordering* can be used to determine in what order new candidates are added and explored.

Canonical Reinclusion Order

One simple canonical reinclusion ordering is to order formulas by precomputing the overall resulting theoretical complexity based on various complexity results for real quantifier elimination [94, 141, 142, 263, 299, 60, 258, 259, 260], which depend on the number of variables and either the number of quantifiers (practical algorithms) or the number of quantifier alternations (theoretical complexity results). While this has a certain theoretically precise appeal, the disadvantage is that the theoretical complexity measures are rather coarse-grained and do not take into account the structure of the formula but only its asymptotic worst-case elements.

Structural Reinclusion Order

Among all possible orders for reincluding formulas, total orders share the advantage that they prevent the need for parallel exploration of multiple different possibilities of adding mutually incomparable formulas. Yet this can also turn into a disadvantage once the order starts adding computationally problematic high-degree constraints.

As a compromise of the canonical reinclusion order with structural information, we propose ordering formulas for reinclusion according to the lexicographical order of, respectively, relative variable recency, total polynomial degree, number of new variables, and maximum term depth. Favouring more recent variables follows the rationale that formulas that mention variables that have been introduced only recently into the proof are more likely to carry relevant information about the current state than those that only refer to variables from the initial proof obligation, for which more recent state variables have already been introduced during the proof. For instance, in Fig. 5.4, the train position z from the initial state may be less relevant than the current train position z_2 for the induction step. The braking force, b, however, may be as relevant as the recent z_2, because there is no updated copy of the symbolic constant b. The number of new variables that are added to the current candidate $\phi \in \mathscr{C}$ from Fig. 5.10 also has an impact on the complexity. Favouring small polynomial degrees is self-explanatory and follows the observation that the polynomial degree has a substantial impact on overall performance, as we will also see in Chap. 8. The term depth is used as an indicator for the complexity of the formula. To obtain a linear—yet arbitrary—order, this order can be extended easily by breaking ties by lexicographical comparison.

Combining this structural order with more sophisticated combinations of polynomial degree and the number of newly added variables according to the canonical orders may also be a viable alternative to improve on pure lexicographic orders.

5.6 Experimental Results

Tables 5.1–5.4 summarise verification results for our proof strategies for various case studies. Experimental results are from a 2.6 GHz AMD Opteron with 4 GB memory. The timeout for computations was 18000s = 5h and is indicated as ∞. Memory consumption of quantifier elimination is shown, excluding the front end. The dimension of the state space (column Dim) and the number of required proof steps (Steps) are indicated. To isolate effects resulting from our automatic invariant discovery that we describe in Chap. 6, we conduct experiments both with (annotated) user interactions for (differential) invariants and in automatic mode (the number of non-automatic interactions or annotations is indicated in column Int). The respective case studies are based on the examples shown in Part I, the larger case studies in Part III, and some standard examples from Part IV. For the tangential roundabout manoeuvre from Sect. 3.4, the number of participating aircraft is as indicated.

Observe that the performance of the extreme strategies of eager and lazy quantifier elimination depends on the example. For ETCS and larger aircraft roundabout manoeuvres (Tables 5.1 and 5.3), the lazy strategy performs faster, while the eager strategy is faster for the bouncing ball and water tank examples (Tables 5.2 and 5.4), where the lazy approach splits heavily (the number of required proof steps is much higher). As an intermediate strategy, IBC generally shows intermediate perform-

Table 5.1 Experimental results for proof strategies (with standalone QE) I

Case study	Int	Strategy	Time(s)	Memory(MB)	Steps	Dim
ETCS kernel	1	eager	11.1	15.1	44	9
	1	IBC	15.6	14.3	54	
	1	lazy	3.8	14.2	88	
	0	IBC	40.2	25.6	53	
	0	lazy	12.9	24.3	85	
ETCS binary safety	1	eager	9.6	11.4	144	14
	1	IBC	10.7	12.1	148	
	1	lazy	13.4	16.0	413	
	0	IBC	57.3	25.5	148	
	0	lazy	46.7	32.6	293	
ETCS safety	1	eager	∞	∞	∞	14
	1	IBC	26.8	17.6	168	
	1	lazy	25.8	29.1	423	
	0	IBC	2089.2	206.1	171	
	0	lazy	2046.5	203.3	304	
ETCS reactivity	0	eager	∞	∞	∞	14
	0	IBC	1084.1	6.1	34	
	0	lazy	∞	∞	∞	
tangential roundabout	3	eager	1.7	6.8	94	13
	3	IBC	1.5	6.8	93	
	3	lazy	1.6	6.7	139	
	0	IBC	10.3	6.9	114	
	0	lazy	10.2	6.8	197	
tangential roundabout 3	3	eager	∞	∞	∞	18
	3	IBC	75.0	24.7	165	
	3	lazy	52.3	14.9	244	
	0	IBC	1065.5	27.6	186	
	0	lazy	620.2	15.0	342	
tangential roundabout 4	3	eager	4208.2	E	E	23
	3	IBC	513.4	184.4	229	
	3	lazy	57.6	31.4	355	
	0	IBC	10998.1	184.3	256	
	0	lazy	901.7	31.4	520	
tangential roundabout 5	3	eager	7714.1	E	E	28
	3	IBC	2457.9	479.3	317	
	3	lazy	108.9	43.6	502	
	0	IBC	∞	∞	∞	
	0	lazy	3417.5	48.5	735	

Table 5.2 Experimental results for proof strategies (with standalone QE) II

Case study	Int	Strategy	Time(s)	Memory(MB)	Steps	Dim
bouncing ball	1	eager	7.7	13.6	85	7
	1	IBC	8.0	13.6	85	
	1	lazy	∞	∞	∞	
	0	IBC	66.4	19.0	43	
	0	lazy	∞	∞	∞	
water tank	1	eager	3.8	9.1	378	3
	1	IBC	3.9	9.1	375	
	1	lazy	9.2	9.5	1604	

Table 5.3 Experimental results for proof strategies (no standalone QE) I

Case study	Int	Strategy	Time(s)	Memory(MB)	Steps	Dim
ETCS kernel	1	eager	11.3	15.0	42	9
	1	IBC	6.9	14.3	47	
	1	lazy	2.8	14.2	61	
	0	IBC	47.8	25.4	46	
	0	lazy	10.5	24.2	58	
ETCS binary safety	1	eager	9.8	11.3	142	14
	1	IBC	10.4	12.2	148	
	1	lazy	7.2	15.8	235	
	0	IBC	41.0	17.4	144	
	0	lazy	18.6	12.4	204	
ETCS safety	1	eager	∞	∞	∞	14
	1	IBC	26.5	23.1	168	
	1	lazy	18.9	24.2	247	
	0	IBC	2051.3	204.1	163	
	0	lazy	2031.6	203.1	216	
ETCS reactivity	0	eager	∞	∞	∞	14
	0	IBC	104.1	61.7	49	
	0	lazy	∞	∞	∞	
tangential roundabout	3	eager	1.7	6.8	94	13
	3	IBC	1.7	6.8	93	
	3	lazy	1.9	6.7	139	
	0	IBC	10.9	6.9	114	
	0	lazy	9.9	6.8	197	
tangential roundabout 3	3	eager	∞	∞	∞	18
	3	IBC	75.3	24.7	165	
	3	lazy	52.3	15.0	244	
	0	IBC	965.8	32.8	186	
	0	lazy	636.2	15.1	342	
tangential roundabout 4	3	eager	4340.6	E	E	23
	3	IBC	513	185.7	229	
	3	lazy	57	31.4	355	
	0	IBC	3323.5	44.0	247	
	0	lazy	884.9	31.4	520	
tangential roundabout 5	3	eager	7702.3	E	E	28
	3	IBC	2439.7	533.8	317	
	3	lazy	108.9	43.6	503	
	0	IBC	16303.7	827	320	
	0	lazy	3552.6	46.9	735	

Table 5.4 Experimental results for proof strategies (no standalone QE) II

Case study	Int	Strategy	Time(s)	Memory(MB)	Steps	Dim
bouncing ball	1	eager	7.8	13.6	85	7
	1	IBC	7.8	13.6	85	
	1	lazy	20.1	13.7	166	
	0	IBC	65.0	16.9	43	
	0	lazy	78.2	26.0	92	
water tank	1	eager	3.6	9.1	387	3
	1	IBC	3.8	9.1	375	
	1	lazy	6.6	9.3	914	

ance. In our case studies, the eager strategy fails to come up with a result within the timeout fairly often. Furthermore, IBC is the only strategy that is able to prove all case studies. For the reactivity property of the ETCS case study, both lazy and eager proof strategies fail at the same time and only IBC succeeds. A result of "E" indicates that no data is available, because of out of memory errors or other limits in the KeYmaera implementation at the time of performing these experiments. For consistent performance measurements, we have used the same version of KeYmaera for all experiments.

The experiments in Tables 5.1 and 5.2 use standalone quantifier eliminations, while those in Tables 5.3 and 5.4 do not. Standalone quantifier elimination reduces quantifiers of first-order formulas immediately, instead of waiting for the whole sequent to be passed to quantifier elimination at once. By comparing Tables 5.1 and 5.2 with Tables 5.3 and 5.4, we see that the performance is generally better when we do not allow these partial reductions of standalone quantified subformulas of a sequent. The reason for this is that the result of standalone QE-reductions on subformulas (Tables 5.1 and 5.2) can produce several disjunctions which split into further subbranches, so that the redundancy effects described in Sect. 5.3 and those illustrated in Fig. 5.5 have more impact.

Finally note that, without our range of proof strategies that result from a careful analysis of the nondeterminisms in the tableau procedures for differential dynamic logics, the practical prover performance is significantly worse than all those presented in Tables 5.1–5.4. The small ETCS kernel from Chap. 2 is already provable directly when implementing the $d\mathcal{L}$ calculus naïvely. Without our range of proof-strategic improvements, however, the full ETCS system that we present in Chap. 7 requires as much as 56 user interactions to be provable [256], while our proof strategies and the algorithms that we present in Part II can, in fact, prove ETCS and other complicated systems completely automatically with zero user interaction.

5.7 Summary

From the experience of using our $d\mathcal{L}$ calculus for verifying parametric hybrid systems in traffic applications, we have investigated combinations of deductive, computer algebraic, and real algebraic reasoning from a practical perspective. We have analysed the principles of this prover combination, identified the nondeterminisms that remain in the canonical $d\mathcal{L}$ tableau procedure, and analysed their impact.

We have proposed proof strategies that navigate through these nondeterminisms, including iterative background closure and iterative inflation order strategies. Similarly to the huge importance of subsumption in resolution, background-style tableau proving requires quick techniques to rule out branches closing for simple arithmetic reasons. In our experiments with verifying cooperating traffic agents, the combinations of our proof strategies reduce the number of interactions and the overall running time significantly. In extreme cases, differences can be such that one proof strategy produces no result in a day compared to another that proves immediately, in less than a second.

Chapter 6
Computing Differential Invariants as Fixed Points

Contents

6.1	Introduction		**254**
	6.1.1	Related Work	255
	6.1.2	Structure of This Chapter	256
6.2	Inductive Verification by Combining Local Fixed Points		**256**
	6.2.1	Verification by Symbolic Decomposition	257
	6.2.2	Discrete and Differential Induction, Differential Invariants	258
	6.2.3	Flight Dynamics in Air Traffic Control	260
	6.2.4	Local Fixed-Point Computation for Differential Invariants	262
	6.2.5	Dependency-Directed Induction Candidates	263
	6.2.6	Global Fixed-Point Computation for Loop Invariants	265
	6.2.7	Interplay of Local and Global Fixed-Point Loops	268
6.3	Soundness		**269**
6.4	Optimisations		**271**
	6.4.1	Sound Interleaving with Numerical Simulation	271
	6.4.2	Optimisations for the Verification Algorithm	272
6.5	Experimental Results		**272**
6.6	Summary		**273**

Synopsis We introduce a fixed-point algorithm for verifying safety properties of hybrid systems with differential equations whose right-hand sides are polynomials in the state variables. In order to verify nontrivial systems without solving their differential equations and without numerical errors, we use differential induction as a continuous generalisation of induction, for which our algorithm computes the required *differential invariants*. As a means for combining local differential invariants into global system invariants in a sound way, our fixed-point algorithm works with differential dynamic logic as a compositional verification logic for hybrid systems. To improve the verification power, we further introduce a *saturation procedure* that refines the system dynamics successively by differential cuts with differential invariants until the property becomes provable. By complementing our symbolic verification algorithm with a robust version of numerical falsification, we obtain a fast and sound verification procedure. We verify roundabout manoeuvres in air traffic control and collision avoidance in train control.

A. Platzer, Logical Analysis of Hybrid Systems,
DOI 10.1007/978-3-642-14509-4_6, © Springer-Verlag Berlin Heidelberg 2010

6.1 Introduction

Reachability questions for systems with complex continuous dynamics are among the most challenging problems in verifying embedded systems, in particular for hybrid systems, which are models for these systems with interacting discrete and continuous transitions along differential equations. For simple systems whose differential equations have solutions that are polynomials in the state variables, quantifier elimination [81] can be used for verification as detailed in Chap. 2. Unfortunately, this symbolic approach does not scale to systems with complicated differential equations whose solutions do not support quantifier elimination (e.g., when they are transcendental functions) or cannot be given in closed form.

Numerical or approximation approaches [21, 238, 102] can deal with more general dynamics. However, numerical or approximation errors need to be handled carefully as they easily cause unsoundness, even for principal reasons [238]. More specifically, we have shown previously that even single image computations of fairly restricted classes of hybrid systems are undecidable by numerical computation [238]. Thus, numerical approaches are helpful for falsification but not (ultimately) for verification.

In this chapter, we present a verification algorithm that combines the soundness of symbolic approaches [125, 15, 231, 235] with support for nontrivial dynamics that is classically more dominant in numerical approaches [21, 238, 102]. During continuous transitions, the hybrid system follows a solution of its differential equation. But for nontrivial dynamics, these solutions are much more complicated than the original equations. Solutions quickly become transcendental even if the differential equations are linear. To overcome this, we handle continuous transitions based on their vector fields, which are described by their differential equations. We use *differential induction* (Sect. 3.5.6), a continuous generalisation of induction that works with the differential equations themselves instead of their solutions. For the induction step, we use a condition that can be checked easily based on *differential invariants* (from Sect. 3.5.6), i.e., properties whose derivative holds true in the direction of the vector field of the differential equation. The derivative is a directional derivative in the direction of (the vector field generated by) the differential equation, with derivatives generalised from functions to formulas according to the findings in Chap. 3. For this to work in practise, the most crucial steps are to find sufficiently strong local differential invariants for differential equations and compatible global invariants for the hybrid switching dynamics.

To this end, we introduce a *sound* verification algorithm for hybrid systems that computes the differential invariants and system invariants in a fixed-point loop. Based on the invariants as fixed points paradigm [73], we present the "differential invariants as fixed points" paradigm using $d\mathcal{L}$ as a verification logic that is generalised to hybrid systems accordingly. For combining multiple local differential invariants into a global invariant in a sound way, we exploit the closure properties of the underlying verification logic $d\mathcal{L}$ by forming appropriate logical combinations of multiple safety statements. In addition, we introduce a *differential saturation process* that refines the hybrid dynamics successively by differential cuts with auxiliary differential invariants until the safety statement becomes an invariant of the refined system. Finally, each fixed-point iteration of our algorithm can be combined with numerical falsification to accelerate the overall symbolic verification in a sound way. We validate our algorithm by verifying *aircraft roundabout manoeuvres* [293, 238] and train control applications automatically, continuing the examples in Part I to the full case studies in Part III.

Contributions

The major contribution in this chapter is that we introduce the "differential invariants as fixed points" paradigm and a fixed-point algorithm for computing differential invariants coupled with a differential saturation process. We turn the theoretical and conceptual proof approach from Part I into a practical verification algorithm. We show that this algorithm can verify realistic applications that were out of scope for related invariant approaches [274, 269, 252] or standard model checking approaches [156, 125, 228] based on state-space exploration, for both theoretical and scalability reasons (see discussions in Sects. 2.4 and 3.1.1).

6.1.1 Related Work

Based on successes in discrete programs [268, 267, 266], other authors [251, 274, 269, 252] argued that invariant techniques scale to more general dynamics than explicit reach-set computations or techniques that require solutions for differential equations [125, 228]. However, their techniques are focused on purely equational systems and cannot handle hybrid systems with inequalities in initial sets or switching surfaces [274, 269]. Due to tolerances, inequalities occur in most real applications. *Barrier certificates* [251, 252] only work for single inequalities, but invariants of roundabout manoeuvres require mixed equations and inequalities [237]. Prajna et al. [252] search for barrier certificates of a fixed degree by global optimisation over the set of all possible certificate combinations for the whole system at once, which is infeasible: Even with degree bound 2, this would require solving a 5,848-dimensional optimisation problem for ETCS (Chap. 7) and a 10,005-dimensional problem for roundabouts with 5 aircraft (Chap. 8).

Finally, important distinctions of our work compared to others [251, 274, 269, 252] are: (*i*) we allow arbitrary formulas as differential invariants, which provably improves verification power; (*ii*) we increase the verification power further by nesting differential invariants using differential saturation to refine the system dynamics; and (*iii*) our compositional verification logic allows local generation of differential invariants and natural local existential quantification of formal parameters for local verification subtasks.

Tomlin et al. [293] derive saddle solutions for aircraft manoeuvre games using Hamilton-Jacobi-Isaacs partial differential equations and propose roundabout manoeuvres. Their exponential state space discretisations for PDEs, however, do not scale to larger dimensions (they consider dimension 3) and can be numerically unsound [238]. Differential invariants, in contrast, work for 28-dimensional systems and are sound.

Straight-line aircraft manoeuvres have been analysed by geometrical meta-level reasoning [104, 171]. We directly verify the actual hybrid flight dynamics, including curved roundabout manoeuvres instead of straight-line manoeuvres with non-flyable instant turns. A few approaches [203, 92] have been undertaken to model check if there are *orthogonal* collisions in discretisations of roundabout manoeuvres. However, the counterexamples (see Fig. 3.6c on p. 151) found by our model checker show that *non-orthogonal* collisions can still happen in these classical manoeuvres, which is a problem that we have fixed in Chap. 3.

6.1.2 Structure of This Chapter

In Sect. 6.2, we introduce the "differential invariants as fixed points" paradigm and present an automatic verification algorithm for hybrid systems that is based on the DAL calculus and computes the differential invariants and invariants as fixed points that are required for verification. In Sect. 6.3, we show how soundness of the DAL calculus inherits in a simple and elegant way to soundness of our logic-based verification procedure. We present optimisations of the algorithm in Sect. 6.4. In Sect. 6.5, we present experimental results for the running example of roundabout manoeuvres in air traffic control and conclude in Sect. 6.6.

6.2 Inductive Verification by Combining Local Fixed Points

For verifying safety properties of hybrid systems without having to solve their differential equations, we use differential induction as a continuous form of induction based on our techniques from Chap. 3. In the induction step, we use a condition on directional derivatives in the direction of the vector field generated by the differential equation. The resulting properties are invariants of the differential equation (called *differential invariants* in Chap. 3). The crucial step for verifying discrete systems by induction is to find sufficiently strong invariants (e.g., for loops α^*). Similarly, the crucial step for verifying dynamical systems (which correspond to a single continuous mode of a hybrid system) by induction is to find sufficiently strong invariant properties of the differential equation. Consequently, for verifying hybrid systems inductively, local invariants need to be found for each differential equation and a global system invariant needs to be found that is compatible with all local invariants.

To compute the required invariants and differential invariants, we combine the invariants as in the fixed-points approach from [73] with the lifting of verification logics to hybrid systems from Chaps. 2 and 3. We introduce a verification algorithm that computes invariants of a system as fixed points of safety constraints on subsystems. In order to obtain a local algorithm that works by decomposing global properties of hybrid programs into local properties of subsystems, we exploit the fact that hybrid programs can be decomposed into subsystems and that $d\mathcal{L}$ can combine safety statements about multiple subsystems simultaneously. Note that the algorithms developed in this chapter apply for the logics $d\mathcal{L}$, DAL, and dTL from Part I, respectively, even if we mostly refer to $d\mathcal{L}$ in this chapter.

A simple *safety statement* corresponds to a $d\mathcal{L}$ formula $\psi \to [\alpha]\phi$ with a hybrid program α, a safety property ϕ about its reachable states, and an arithmetic formula ψ that characterises the set of initial states symbolically. *Validity* of formula $\psi \to [\alpha]\phi$ (i.e., truth in all states) corresponds to ϕ being true in all states reachable by hybrid program α from initial states that satisfy ψ. We describe a verification algorithm that defines the function *prove*$(\psi \to [\alpha]\phi)$ for verifying this safety statement recursively by refining the $d\mathcal{L}$ and DAL calculi to an automatic verification algorithm. The discrete base cases are discussed in Sect. 6.2.1. In Sect. 6.2.2, we contrast discrete and differential induction. Section 6.2.4 shows the fixed-point algorithm for computing differential invariants for differential equations, and Sect. 6.2.6 for computing loop invariants. In Sect. 6.2.7, we explain the interplay of local and global fixed-point loops of our verification algorithm.

6.2.1 Verification by Symbolic Decomposition

The cases of the verification procedure *prove* where $d\mathcal{L}$ can verify a property of a hybrid program directly by decomposing it into a property of its parts are shown in Fig. 6.1. They correspond to the cases where the dynamic rules of Fig. 3.9 can

```
function prove(ψ → [x:=θ]φ):
  return prove(ψ ∧ x̂ = θ → φ_x^x̂)
  where x̂ is a new auxiliary variable
function prove(ψ → [?χ]φ):
  return prove(ψ ∧ χ → φ)
function prove(ψ → [α∪β]φ):
  return prove(ψ → [α]φ) and prove(ψ → [β]φ) /* i.e. ψ → [α]φ ∧ [β]φ */
function prove(ψ → [α;β]φ):
  return prove(ψ → [α][β]φ)
function prove(ψ → [x:=*]φ):
  return prove(ψ → ∀x φ)
function prove(ψ → φ) where isFirstOrder(φ):
  return QE(ψ → φ)
function prove(ψ → Qx φ) where Qx is ∀x or ∃x:
  return QE(ψ → Qx prove(φ))
```

Fig. 6.1 $d\mathcal{L}$-based verification by symbolic decomposition

be applied without any nondeterminism, and just depending on the top-level program operator. For a concise presentation, the case in line 1 introduces an auxiliary variable $\hat{x}$ to handle discrete assignments by substituting $\hat{x}$ for x in $\phi_x^{\hat{x}}$: E.g., $x \geq 2 \rightarrow [x := x-1]x \geq 0$ is shown by proving $x \geq 2 \wedge \hat{x} = x-1 \rightarrow \hat{x} \geq 0$. Our implementation in our theorem prover KeYmaera [242] uses proof rule $\langle := \rangle$ and discrete jump sets to avoid auxiliary variables according to Chap. 2. State checks $?\chi$ are shown by assuming the test succeeds, i.e., χ holds true (line 4), nondeterministic choices split into their alternatives (line 6), sequential compositions are proven using nested modalities (line 8), and random assignments are shown by universal quantification (line 10), which is an optimisation of the handling in Chap. 3. Random assignments $[x := *]\phi$ that nondeterministically assign an arbitrary real value to x are definable by $[x' = 1 \cup x' - 1]\phi$ and are equivalent to quantifiers $\forall x\, \phi$.

The base case in line 12, where ϕ is a formula of first-order real arithmetic, can be proven by quantifier elimination (QE) in real-closed fields [81]. Despite its complexity, this can remain feasible, because the formulas resulting from our algorithm do not depend on the solutions of differential equations but only on their right-hand sides. Further, the decompositions in our verification algorithm generally lead to local properties with lower complexities. We present here only a simplified treatment of quantifiers in $\mathsf{d}\mathcal{L}$ formulas like $\forall x\, \phi$ that contain modal subformulas: The *prove* function is applied recursively to the unquantified kernel ϕ first and the resulting formula is handled by quantifier elimination (line 14), which resembles the quantifier handling by side deduction in Chap. 3. More generally, quantifier elimination can be lifted to quantifiers in $\mathsf{d}\mathcal{L}$ formulas like $\forall x\, \phi$ using the respective Skolemisation and free-variable rules from Chap. 2 . After introducing a fresh Skolem term s for variable x in ϕ, the analysis continues with the unquantified kernel ϕ_x^s. Later on when the formulas resulting from recursive application of *prove* contain no more modalities, the quantifier for s can be reintroduced (Deskolemisation) and eliminated equivalently using quantifier elimination in real-closed fields. Handling $\exists x\, \phi$ for modal formulas ϕ is similar. The crucial part is that Skolem term dependencies can be exploited to prevent unsound quantifier rearrangements; see Chap. 2.

Overall, the algorithm in Fig. 6.1 recursively reduces the safety of hybrid programs to separate properties of continuous evolutions or of repetitions, which we verify in the next sections.

6.2.2 Discrete and Differential Induction, Differential Invariants

In this section, we present algorithms for verifying loops by discrete induction (which corresponds to rule *ind* from the $\mathsf{d}\mathcal{L}$/DAL calculi) and continuous evolutions by differential induction, which is a continuous form of induction and corresponds to rule *DI* from the DAL calculus in Fig. 3.9. In either case, we prove that an invariant F holds initially (in the states characterised symbolically by ψ; thus $\psi \rightarrow F$ is valid) and finally entails the postcondition ϕ (i.e., $F \rightarrow \phi$). The cases differ in their induction step.

Definition 6.1 (Discrete induction). The formula F is a *(discrete) invariant* of $\psi \to [\alpha^*]\phi$ iff the following formulas are valid:

1. $\psi \to F$ (induction start), and
2. $F \to [\alpha]F$ (induction step).

An invariant is *sufficiently strong* if $F \to \phi$ is valid.

Definition 6.2 (Continuous invariants). Let $\mathscr{D}$ be a differential equation and χ is a first-order formula. Formula F is a *continuous invariant* of $\psi \to [\mathscr{D} \wedge \chi]\phi$ iff the following formulas are valid:

1. $\psi \wedge \chi \to F$ (induction start), and
2. $F \to [\mathscr{D} \wedge \chi]F$ (induction step).

Again, a continuous invariant is *sufficiently strong* if $F \to \phi$ is valid.

To prove that F is a continuous invariant, it is sufficient by Theorem 3.1 and proof rule *DI* from Fig. 3.9 to check a condition on the directional derivatives of all terms of the formula, which expresses that no atomic subformula of F changes its truth-value along the dynamics of the differential equation. This condition is easier to check than a reachability property ($F \to [\mathscr{D} \wedge \chi]F$) of a differential equation.

Applications like aircraft manoeuvres need invariants with mixed equations and inequalities. Thus, we use the generalisation of directional derivatives from functions to logical formulas according to the total derivation $D(\cdot)$ from Definition 3.13 on p. 155.

Definition 6.3 (Differential induction). Let the differential equation system $\mathscr{D}$ be $x_1' = \theta_1, \ldots, x_n' = \theta_n$ and χ be a first-order formula. A quantifier-free first-order formula F is a *differential invariant* of $\psi \to [\mathscr{D} \wedge \chi]\phi$ iff the following formulas are valid:

1. $\psi \wedge \chi \to F$ (induction start), and
2. $\chi \to F'_{\mathscr{D}}$ (induction step),

where $F'_{\mathscr{D}}$ is the result of substituting the differential equation $\mathscr{D}$ into the total derivation $D(F)$ of F, which corresponds to the conjunction of all directional derivatives of atomic formulas in F in the direction of the vector field of $\mathscr{D}$ (the partial derivative of b by x_i is $\frac{\partial b}{\partial x_i}$):

$$F'_{\mathscr{D}} \equiv \bigwedge_{(b \sim c) \in F} \left(\left(\sum_{i=1}^{n} \frac{\partial b}{\partial x_i} \theta_i \right) \sim \left(\sum_{i=1}^{n} \frac{\partial c}{\partial x_i} \theta_i \right) \right) \quad \text{for} \sim \in \{=, \geq, >, \leq, <\}.$$

These partial derivatives of terms are well-defined in the Euclidean space spanned by the variables and can be computed symbolically (Sect. 3.5.2).

The central property of differential invariants for verification purposes is that they can be used to replace infeasible or even impossible reachability analysis with feasible symbolic computation.

Proposition 6.1 (Principle of differential induction). *All differential invariants are continuous invariants.*

Proof. This proposition is a corollary to Theorem 3.1, which shows that the differential induction rule *DI* from the DAL calculus in Fig. 3.9 is sound. □

Example 6.1. Consider the dynamics $x' = x^2, y' = -3$. Differential invariants can show that $3x \geq 4y$ is an invariant for this dynamics without using any state-based reachability verification. We just compute symbolically:

$$(3x \geq 4y)'_{x'=x^2,y'=-3} \equiv \frac{\partial 3x}{\partial x}x^2 + \frac{\partial 3x}{\partial y}(-3) \geq \frac{\partial 4y}{\partial x}x^2 + \frac{\partial 4y}{\partial y}(-3) \equiv 3x^2 \geq -12.$$

Since the latter formula is easily found to be valid, $3x \geq 4y$ is proven to be a differential invariant and thus remains true whenever it holds true initially (case 1 of Definition 6.3). □

We have introduced this proof principle of differential induction in Chap. 3. The above notions form the basis for our development of a systematic algorithm computing the required invariants and differential invariants based on our proof rules from Chap. 3. In Sects. 6.2.4–6.2.6, we present algorithms for finding differential invariants for differential equations, and for finding global invariants for repetitions.

6.2.3 Flight Dynamics in Air Traffic Control

Aircraft collision avoidance manoeuvres resolve conflicting flight paths, e.g., by roundabout manoeuvres [293]; see Fig. 3.6 on p. 151. Their nontrivial dynamics makes safe separation of aircraft difficult to analyse, in particular as good timing and coordination of movement in space are crucial [293, 203, 104, 92, 238, 171]. Correct functioning of these manoeuvres under all circumstances is difficult to guarantee without formal verification, especially in light of the counterexample (Fig. 3.6c) that our model checker discovered [238] for the classical roundabout manoeuvre. Thus, verification is important for aircraft manoeuvres, and, at the same time, aircraft dynamics is a challenge for hybrid systems verification.

For simplicity, we consider planar movement of aircraft (if the aircraft are collision-free in a planar projection, they are collision-free in space). The parameters of two aircraft at (planar) position $x = (x_1, x_2) \in \mathbb{R}^2$ and $y = (y_1, y_2)$ with angular orientation ϑ and ς are illustrated in Fig. 6.2 (with $\vartheta = 0$). Following the work of Tomlin et al. [293], aircraft dynamics is determined by their linear speeds $v, u \in \mathbb{R}$ and angular speeds $\omega, \rho \in \mathbb{R}$, respectively:

$$x_1' = v\cos\vartheta \quad x_2' = v\sin\vartheta \quad \vartheta' = \omega \quad y_1' = u\cos\varsigma \quad y_2' = u\sin\varsigma \quad \varsigma' = \rho \tag{6.1}$$

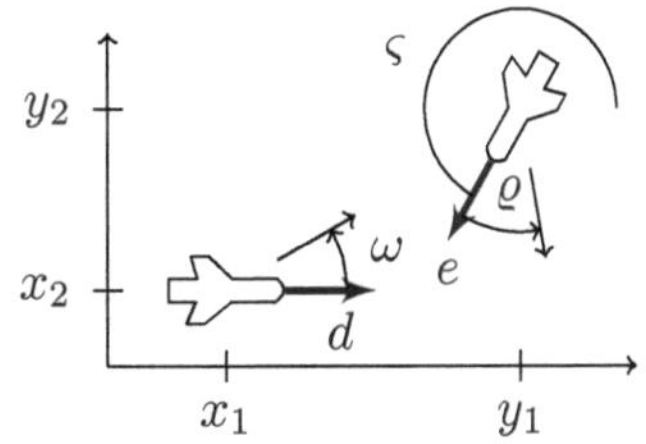

Fig. 6.2 Aircraft dynamics

That is, position x moves with speed v into the direction with angular orientation ϑ, which rotates with angular velocity ω (likewise for a second aircraft at position y with speed u, angular orientation ς, and angular velocity ρ).

In safe flight configurations, aircraft are separated by at least distance p:

$$(x_1 - y_1)^2 + (x_2 - y_2)^2 \geq p^2. \tag{6.2}$$

To handle the transcendental functions in (6.1) without facing the undecidability problems in the arithmetic of trigonometric functions, we axiomatise sin and cos by differential equations and reparametrise the system using a linear velocity vector $d = (d_1, d_2) := (v \cos \vartheta, v \sin \vartheta) \in \mathbb{R}^2$ according to Sect. 3.4.2:

$$\begin{bmatrix} x_1' = d_1 & x_2' = d_2 & d_1' = -\omega d_2 & d_2' = \omega d_1 & t' = 1 \\ y_1' = e_1 & y_2' = e_2 & e_1' = -\rho e_2 & e_2' = \rho e_1 & s' = 1 \end{bmatrix} \tag{$\mathscr{F}$}$$

Using symbolic derivations, it is easy to see that equations ($\mathscr{F}$) and (6.1) are equivalent up to reparametrisation. Here, we add clock variables t, s (with $t' = 1$ and $s' = 1$) that we need for synchronising collision avoidance manoeuvres in Chap. 8.

By a simple computation, $d_1^2 + d_2^2 \geq a^2$ is a differential invariant of ($\mathscr{F}$), thereby showing that the linear speed of aircraft does not drop below some stalling speed a during manoeuvres:

$$\begin{aligned} &(d_1^2 + d_2^2 \geq a^2)'_{\mathscr{F}} \equiv (d_1^2 + d_2^2 \geq a^2)'_{(d_1' = -\omega d_2, d_2' = \omega d_1)} \\ &\equiv \frac{\partial (d_1^2 + d_2^2)}{\partial d_1}(-\omega d_2) + \frac{\partial (d_1^2 + d_2^2)}{\partial d_2}\omega d_1 \geq \frac{\partial a^2}{\partial d_1}(-\omega d_2) + \frac{\partial a^2}{\partial d_2}\omega d_1 \\ &\equiv 2d_1(-\omega d_2) + 2d_2 \omega d_1 \geq 0. \end{aligned}$$

As a stronger statement, the equation $d_1^2 + d_2^2 = a^2$ for constant linear speed is a provable differential invariant. Similarly, conjunction $d_1^2 + d_2^2 = a^2 \wedge e_1^2 + e_2^2 \geq a^2$ can be shown to be a differential invariant of $\mathscr{F}$. The theory of differential invariants shows that this is a general phenomenon: Conjunctions of differential invariants are differential invariants but not conversely so (Sect. 3.5.6). There are examples where only the propositional combination itself is a differential invariant but none of its parts is (Sect. 3.10). Thus, allowing conjunctions in differential invariants is crucial and the verification power of differential invariants is higher than that of other approaches [251, 274, 269, 252], which do not support logical operators.

6.2.4 Local Fixed-Point Computation for Differential Invariants

As with verification with invariants, the central practical question for verification with differential invariants is how to find them. Figure 6.3 depicts our fixed-point algorithm for constructing differential invariants for each continuous evolution $\mathscr{D} \wedge \chi$ with a differential equation system $\mathscr{D}$ and evolution domain χ. The algorithm in

```
1  function prove(ψ→[𝒟∧χ]φ):
2    if prove(∀^𝒟(χ→φ)) then return true /* property proven */
3    for each F∈Candidates(ψ→[𝒟∧χ]φ, χ) do
4      if prove(ψ∧χ→F) and prove(∀^𝒟(χ→F'_𝒟)) then
5        χ := χ∧F          /* refine by differential invariant */
6        goto 2;           /* repeat fixed-point loop */
7    end for
8    return not provable using candidates
```

Fig. 6.3 Fixed-point algorithm for differential invariants (*Differential Saturation*)

Fig. 6.3 (called *Differential Saturation*) successively refines the evolution domain restriction χ by differential invariants until saturation, i.e., until χ accumulates enough information to become a strong invariant that implies postcondition ϕ (line 2). If evolution domain χ already entails ϕ, then $\psi \to [\mathscr{D} \wedge \chi]\phi$ is proven (line 2). Otherwise, the algorithm considers candidates F for augmenting χ (line 3). If F is a differential invariant (line 4), then χ can soundly be refined to $\chi \wedge F$ (line 5) without affecting the states reachable by $\mathscr{D} \wedge \chi$ (Proposition 6.2 below). Then, the fixed-point loop repeats (line 6). At each iteration of this fixed-point loop, the previous invariant χ can be used to prove the next level of refinement $\chi \wedge F$ (line 4). Hence, each differential invariant provides more information to simplify subsequent iterations. The refinement of the dynamics at line 5 is correct by the following proposition, using the fact that the conditions in line 4 imply that F is a differential invariant and, thus, a continuous invariant by Proposition 6.1.

Proposition 6.2 (Differential saturation). *If F is a continuous invariant of the formula $\psi \to [\mathscr{D} \wedge \chi]\phi$, then $\psi \to [\mathscr{D} \wedge \chi]\phi$ and $\psi \to [\mathscr{D} \wedge \chi \wedge F]\phi$ are equivalent.*

Proof. The proof is a stronger version of the soundness of the differential cut rule *DS* by Theorem 3.1: Let F be a continuous invariant, which implies that $\psi \to [\mathscr{D} \wedge \chi]F$ is valid. Let ν be a state satisfying ψ (otherwise there is nothing to show). Then, $\nu \models [\mathscr{D} \wedge \chi]F$. Since this means that F is true all along all flows φ of $\mathscr{D} \wedge \chi$ that start in ν (Definition 3.10), the latter and $\mathscr{D} \wedge \chi \wedge F$ have the same dynamics and the same reachable states from ν, i.e., $(\nu, \omega) \in \rho(\mathscr{D} \wedge \chi)$ holds if and only if $(\nu, \omega) \in \rho(\mathscr{D} \wedge \chi \wedge F)$ (Definition 2.7). Thus, we can conclude that $\psi \to [\mathscr{D} \wedge \chi]\phi$ and $\psi \to [\mathscr{D} \wedge \chi \wedge F]\phi$ are equivalent, because their semantics uses the same transition relation. □

This progressive differential saturation turns out to be crucial in practise. For instance, the aircraft separation property (6.2) cannot be proven until ($\mathscr{F}$) has been refined by invariants for d and e, because these determine x' and y'. This makes sense intuitively: Unless we have discovered some invariant about the directions d and e in which the aircraft are flying, we cannot conclude good invariants about their positions x and y, because the evolution of the positions over time depends on the directions.

Function *Candidates* determines candidates for induction (line 3) depending on transitive differential dependencies, as we will explain in Sect. 6.2.5. When these are insufficient for proving $\psi \to [\mathscr{D} \land \chi]\phi$, the algorithm fails (line 8, with improvements in subsequent sections). Again, $\forall^\alpha \phi$ denotes the *universal closure* of ϕ. It is required in lines 2 and 4, because the respective formulas need to hold in *all* states (that satisfy χ). Clearly, this set of states is overapproximated conservatively by the universal closure with respect to all variables for which there are differential equations in $\mathscr{D}$. We will reduce the number of quantifiers in Sect. 6.4.

6.2.5 Dependency-Directed Induction Candidates

In this section, we construct likely candidates for differential induction (function *Candidates* in Fig. 6.3). Later, we use the same procedure for finding global loop invariants. We construct two kinds of candidates in an order induced by differential dependencies. By following the effect of hybrid systems symbolically along their decompositions, our verification algorithm enriches precondition ψ successively with more precise information about the symbolic pre-state as obtained by the symbolic decompositions and proof steps in Figs. 6.1 and 6.3. To exploit this, we first look for invariant symbolic state information that accumulated in the respective precondition ψ and postcondition ϕ during the iterative symbolic decomposition by selecting subformulas that are not yet contained in χ. In practise, this gives particularly good candidates for highly parametric hybrid systems. Even if candidates from ψ and ϕ do not yet work in the first iteration, repetitions of outer fixed-point loops (Sect. 6.2.6) may enrich ψ and ϕ so that they become useful in later iterations.

Secondly, we generate parametric invariants. Let $V = \{x_1, \dots, x_n\}$ be a set of relevant variables. We choose fresh names $a^{(l)}_{i_1,\dots,i_n}$ for *formal parameters* of the invariant candidates and build polynomials $p_1, \dots, p_k$ of degree d with variables V using formal parameters as symbolic coefficients:

$$p_l := \sum_{i_1+\cdots+i_n \le d} a^{(l)}_{i_1,\dots,i_n} x_1^{i_1} \dots x_n^{i_n} \quad \text{for } 1 \le l \le k.$$

We define the set of *parametric candidates* (operator $\lor$ is used similarly):

$$ParaForm(k,d,V) := \left\{ \bigwedge_{l=1}^{i} p_l \ge 0 \land \bigwedge_{l=i+1}^{k} p_l = 0 \ : \ 0 \le i \le k \right\}.$$

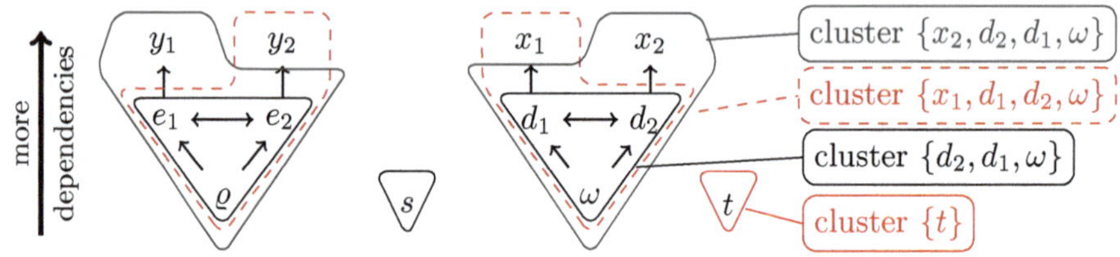

Fig. 6.4 Differential dependencies (arrows) and (triangular) variable clusters of ($\mathscr{F}$)

For instance, the parametric candidate $a_{0,0} + a_{1,0}d_1 + a_{0,1}x_2 = 0$ yields a differential invariant of ($\mathscr{F}$) for the choice $a_{0,0} = 0$, $a_{1,0} = 1$, $a_{0,1} = \omega$. By simple combinatorics, *ParaForm* contains $k+1$ candidates with $k\binom{n+d}{d}$ formal parameters $a^{(l)}_{i_1,\dots,i_n}$, which are existentially quantified. Existence of a common satisfying instantiation for these parameters can be expressed by prefixing the resulting d$\mathscr{L}$ formula that uses the parametric candidate with $\exists a^{(l)}_{i_1,\dots,i_n}$ for each of the formal parameters $a^{(l)}_{i_1,\dots,i_n}$. For this to be feasible, the number of parameters is crucial, which we minimise by respecting (differential) dependencies. We will illustrate the placement of existential quantifiers in Example 6.2 of Sect. 6.2.6 after we have presented the global fixed-point algorithm.

To accelerate the differential saturation process from Sect. 6.2.4, it is crucial to explore candidates in a promising order from simple to complex, because the algorithm in Fig. 6.3 uses successful differential invariants to refine the dynamics, thereby simplifying subsequent proofs: E.g., separation property (6.2) is only provable after the dynamics has been refined with invariants for d and e, because x' and y' depend on the direction d and e. In fact, the safety of roundabouts crucially depends on compatible directions of the aircraft, as the counterexample in Fig. 3.6c shows. We construct candidates in a natural order based on variable occurrence that is consistent with the *differential dependencies* of the differential equations. For a differential equation $\mathscr{D}$, variable x depends on variable y according to the differential equation system $\mathscr{D}$ if y occurs on the right-hand side for x' (or transitively so). The resulting set *depend*($\mathscr{D}$) of dependencies is the transitive closure of

$$\{(x,y)\ :\ (x' = \theta) \in \mathscr{D} \text{ and } y \text{ occurs in } \theta\}.$$

From the differential equation system ($\mathscr{F}$), we determine the differential dependencies indicated as arrows (pointing to the dependent variables x) in Fig. 6.4.

From these dependencies we determine an order on candidates. The idea is that, as the value of x_1 in ($\mathscr{F}$) depends on that of d_1, it makes sense to look for invariant expressions of d_1 first, because refinements with these help differential saturation in proving invariant expressions involving also x_1. Thus, we order variables by differential dependencies, which resembles the back substitution order in Gaussian elimination (if, in triangular form, x_1 depends on d_1, then equations for d_1 must be solved first, except that, in the differential case, variables may remain mutually dependent, e.g., d_1 and d_2). Now we call a set V of variables a *cluster* of the differential equation $\mathscr{D}$ iff V is closed with respect to $depend(\mathscr{D})$, i.e., variables of V only depend on variables in V:

$$x \in V \text{ and } (x,y) \in depend(\mathscr{D}) \text{ then } y \in V.$$

The resulting variable clusters for system ($\mathscr{F}$) are marked as triangular shapes in Fig. 6.4. Finally, we choose candidates from ψ and $ParaForm(k,d,V)$ starting with candidates whose variables lie in small clusters V and cover larger fractions of that cluster. Thus, the differential invariant $d_1^2 + d_2^2 \geq a^2$ of Sect. 6.2.3 within cluster $\{d_2, d_1, \omega\}$ can be discovered before invariants such as $d_1 = -\omega x_2$ that involve x_2, because x_2 depends on d_2. Similarly, $d_1 = -\omega(x_2 - c_2)$ will be discovered before $\|x-y\|^2 \geq p^2$, as the latter lies in a larger cluster with worse coverage percentage of that cluster. The successive differential saturation process along these dependencies further helps to keep the degrees in *ParaForm* small.

6.2.6 Global Fixed-Point Computation for Loop Invariants

With the uniform setup of d$\mathscr{L}$, we can adapt the algorithm in Fig. 6.3 easily to obtain a fixed-point algorithm for loops ($\psi \to [\alpha^*]\phi$) in place of continuous evolutions ($\psi \to [\mathscr{D} \wedge \chi]\phi$): In line 4 of Fig. 6.3, we replace the induction step from Definition 6.3 with the step for loops (Definition 6.1). As an optimisation, we can transfer the reuse of partial differential invariants according to Proposition 6.2 to discrete invariants for loops. Invariants χ of previous iterations can be exploited as refinements of the hybrid system dynamics, similarly to previous differential invariants that can be used in future iterations by refining the dynamics using differential saturation:

Proposition 6.3 (Loop saturation). *If χ is a discrete invariant of $\psi \to [\alpha^*]\phi$, then $\chi \wedge F$ is a discrete invariant iff $\psi \to F$ and $\chi \wedge F \to [\alpha](\chi \to F)$ are valid.*

Proof. Let χ be a discrete invariant of $\psi \to [\alpha^*]\phi$. Let, further, $\chi \wedge F$ be a discrete invariant of $\psi \to [\alpha^*]\phi$. Then $\psi \to \chi \wedge F$ and $\chi \wedge F \to [\alpha](\chi \wedge F)$ are valid by Definition 6.1. Hence, trivially, $\chi \wedge F \to [\alpha](\chi \to F)$ is valid, because all states that satisfy $\chi \wedge F$ also satisfy the weaker property $\chi \to F$. Finally, the validity of $\psi \to \chi \wedge F$ clearly entails $\psi \to F$.

Conversely, let χ be a discrete invariant. Let, further, $\chi \wedge F \to [\alpha](\chi \to F)$ and $\psi \to F$ be valid. For $\chi \wedge F$ to be a discrete invariant, we have to show that F satisfies the induction step of Definition 6.1 (the induction start $\psi \to \chi \wedge F$ is an immediate combination of the validity of $\psi \to \chi$ and $\psi \to F$). Since χ is a discrete invariant, $\chi \to [\alpha]\chi$ is valid, which entails $\chi \wedge F \to [\alpha]\chi$ as a special case. Since $\chi \wedge F \to [\alpha](\chi \to F)$ is valid and $\chi \wedge F \to [\alpha]\chi$ is valid, we conclude that $\chi \wedge F \to [\alpha](\chi \wedge F)$ is valid for the following reason. Let ν be a state satisfying the initial constraints $\chi \wedge F$. Then $\nu \models [\alpha]\chi$ and $\nu \models [\alpha](\chi \to F)$. Hence, all states ω reachable from ν by α satisfy $\omega \models \chi$ and $\omega \models \chi \to F$. Thus, they satisfy $\omega \models \chi \wedge F$, essentially by modus ponens. Consequently, we have shown that

$\chi \wedge F \rightarrow [\alpha](\chi \wedge F)$ is valid. and, hence, $\chi \wedge F$ is a discrete invariant of $\psi \rightarrow [\alpha^*]\phi$. □

The induction step from Proposition 6.3 can generally be proven faster, because it is a weaker property than that of Definition 6.1.

To adapt our approach from Sect. 6.2.5 to loops, we use discrete data-flow and control-flow dependencies of α. Dependencies can be determined immediately from the syntax of hybrid programs. There is a direct *data-flow dependency* with the value of x depending on y, if $x := \theta$ or $x' = \theta$ occurs in α with a term θ that contains y. Similarly, there is a direct *control-flow dependency* with the value of x depending on y if, for any term θ, $x := \theta$ or $x' = \theta$ occurs in α after a test $?\chi$ or evolution domain restriction $\wedge\chi$ containing y. The respective data-flow and control-flow dependencies are the transitive closures of these relations.

```
function prove(ψ→[α*]φ):
  χ := true      /* currently known invariant of ψ→[α*]φ */
  if prove(∀^α(χ→φ)) then return true /* property proven */
  for each F∈IndCandidates(ψ→[α*]φ, χ) do
    if prove(ψ∧χ→F) and prove(∀^α(χ∧F→[α](χ→F))) then
      χ := χ∧F           /* refine by discrete invariant */
      goto 3;            /* repeat fixed-point loop */
  end for
  return not provable using candidates
```

Fig. 6.5 Fixed-point algorithm for discrete loop invariants (loop saturation)

The resulting algorithm in Fig. 6.5 verifies loops. It is a direct adaption of that in Fig. 6.3, except that it uses Proposition 6.3 as an induction step for loops. The algorithm in Fig. 6.5 performs a fixed-point computation for loops and recursively combines the local differential invariants obtained by differential saturation to form a global invariant. It recursively uses *prove* for verifying its subtasks, which handle the discrete switching behaviour according to Fig. 6.1 and infer local differential invariants according to differential saturation by the fixed-point algorithm in Fig. 6.3.

Example 6.2 (Existential parameter quantification). Note that the ability of formulas in differential dynamic logic to have quantifiers in front of reachability modalities is crucial here. To illustrate, consider the simple water tank system from Fig. 6.6, where x denotes the current water level.

The second line in the hybrid program of Fig. 6.6 represents a continuous transition. It tests by $?q = on$ whether the current location q is *on*, and then fills the tank by following the differential equation $x' = 1$ restricted to the evolution domain $x \leq 9$ (i.e., the conjunction $x' = 1 \wedge x \leq 9$). The third line tests the guard $x \geq 5$ when in state *on*, then resets x by a discrete assignment $(x := x - 1)$, and then changes location q to *off*. The additional test $?x \leq 9$ in the fourth line is needed, because the hybrid automaton is only allowed to enter mode *on* when its evolution domain re-

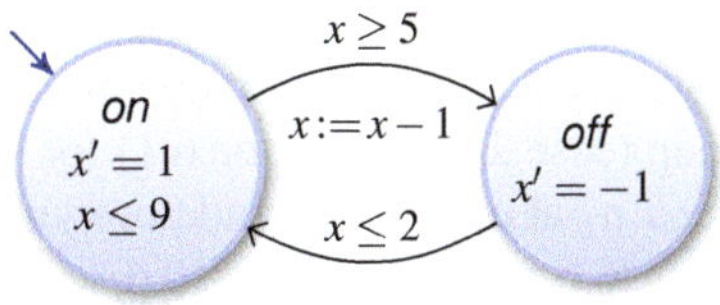

$q := on$; /* *initial location is on* */
$\big(\ (?q = on;\ x' = 1 \wedge x \le 9)$
$\cup\ (?q = on \wedge x \ge 5;\ x := x-1;\ q := off)$
$\cup\ (?q = off;\ x' = -1)$
$\cup\ (?q = off \wedge x \le 2;\ q := on;\ ?x \le 9)\big)^*$

Fig. 6.6 Hybrid program rendition of hybrid automaton for simple water tank

gion $x \le 9$ is satisfied. The * at the end of the hybrid program indicates that the transitions of a hybrid automaton repeat indefinitely.

Let *wctrl* abbreviate the body of the loop of the hybrid program for the water controller from Fig. 6.6 such that $q := on; (wctrl)^*$ corresponds to the full hybrid program from Fig. 6.6. Then the following d$\mathcal{L}$ formula states that the water level is always below 10 when it starts at level $x \le 3$:

$$x \le 3 \rightarrow [q := on; (wctrl)^*]\, x < 10. \tag{6.3}$$

During the verification run for property (6.3), we need to show a property of the following form for the loop $wctrl^*$:

$$x \le 3 \wedge q = on \rightarrow [(wctrl)^*]\, x < 10.$$

For a parametric candidate F of the form $a_1x + a_0 \ge 0$, line 5 of the algorithm in Fig. 6.5 produces subtasks for discrete induction (Definition 6.1), which will be handled recursively by the *prove* function:

$$\begin{aligned} &\text{prove}(x \le 3 \wedge q = on \wedge \chi\ \rightarrow\ a_1x + a_0 \ge 0)\\ \textbf{and}\ &\text{prove}\big(\forall^\alpha (a_1x + a_0 \ge 0 \wedge \chi\ \rightarrow\ [wctrl](\chi \rightarrow a_1x + a_0 \ge 0))\big)\end{aligned}$$

In the first iteration (where χ is still *true*), the combination of these subtasks by conjunction corresponds to proving the following overall d$\mathcal{L}$ formula:

$$\exists a_0\, \exists a_1\, \big((x \le 3 \wedge q = on \rightarrow a_1x + a_0 \ge 0) \wedge \forall x\, (a_1x + a_0 \ge 0 \rightarrow [wctrl]\, a_1x + a_0 \ge 0)\big). \tag{6.4}$$

The universal quantifier $\forall x$ in (6.4) results from the universal closure $\forall^\alpha$ with respect to all variables changed in *wctrl*, i.e., x. The existential quantifiers for a_0 and a_1 are for formal parameters in the parametric candidate (Sect. 6.2.5). Observe that the outer placement of existential quantifiers is required for the resulting instance of $a_1x + a_0 \ge 0$ to be a *common solution* implied by the precondition (left conjunct) and inductive for *wctrl* (right conjunct). In particular, the right conjunct requires finding parameter choices for which $a_1x + a_0 \ge 0$ holds true after executing one iteration *wctrl* of the loop from any initial state if it was true before.

Note that this inherently requires quantifiers around reachability properties! Yet these nestings are naturally expressible in the first-order verification logic d$\mathcal{L}$. The left conjunct expresses the fact that the parameter choices need to make $a_1x + a_0 \ge 0$ true in the beginning. The conjunction and outer placement of $\exists a_0 \exists a_1$ ensures that the required parameter choices fit together. Equation (6.4) also illustrates the need for the universal closure, because we need the same choice for the formal parameters a_0, a_1 to be able to conclude the induction step (case 2 of Definition 6.1) for all states x ($\forall x$). Different incompatible choices for a_0, a_1 at each state would not yield an inductive argument. To prove the example in Fig. 6.6, our algorithm will discover the parameter combination $a_1 = -1$ and $a_0 = 9.5$, for instance. □

For verification with parametric candidates to be feasible, we exploit the fact that we can keep existential quantifiers as local as possible in d$\mathcal{L}$, which we ensure by the symbolic decompositions in our logic. For instance, quantifiers for the formal parameters of a differential invariant will remain local to the formulas resulting from the algorithm in Fig. 6.3. Contrast this to a computationally more complicated global use of quantifiers around the whole verification problem at once. Minimising the number of parameters according to Sect. 6.2.5 further improves the computational tractability.

6.2.7 Interplay of Local and Global Fixed-Point Loops

Together, our local and global fixed-point algorithms jointly verify correctness properties of hybrid programs. Their interplay needs to be coordinated with fairness, though. If the local fixed-point algorithm in Fig. 6.3 does not converge, stronger invariants may need to be found by the global fixed-point algorithm which iteratively result in stronger preconditions ψ_i for the local fixed-point algorithm; see Fig. 6.7.

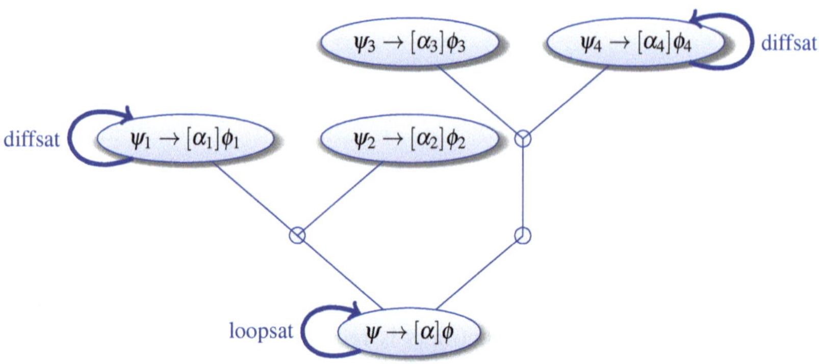

Fig. 6.7 Interplay of local (diffsat) and global (loopsat) fixed-points verification loops during symbolic decomposition

Thus, for fairness reasons, the local fixed-point algorithm should stop when it cannot prove its postcondition, either because of a counterexample or because it runs out of candidates for differential invariants. As in the work of Prajna et al. [252], the degrees of parametric invariants, therefore, need to be bounded and increased iteratively. As in [252], there is no natural measure for how these degrees should be increased. Instead, here, we exploit the fact that the candidates produced by function *Candidates* are independent and we explore them in parallel with fair time interleaving. The interleaving by iterative timeout increase is similar to iterative background closures from Sect. 5.4, except that it works for full subproofs instead of for just one local quantifier elimination call.

During fixed-point computations, wrong choices of candidates are time consuming. Thus, in practise, it is important to discover futile attempts quickly. For this, non-exhaustive sampling with numerical simulations can be used to look for counterexamples. Note that this numerical counterexample search can be performed separately for each local subtask in the decomposition. For instance, a counterexample for a candidate F for node $\psi_4 \rightarrow [\alpha_4]\phi_4$ in Fig. 6.7 will only abort the attempt to prove this particular candidate F, not attempts with other candidates. A counterexample found by local numerical simulation for the reachability property $\psi_4 \rightarrow [\alpha_4]\phi_4$ itself, on the other hand, will terminate all proof search for that subtask by propagating the need for a stronger precondition ψ_4 up the decomposition tree in Fig. 6.7. Likewise, a counterexample found for the original property $\psi \rightarrow [\alpha]\phi$ will abort the whole proof search. We discuss the implications of including numerical procedures in Sect. 6.4.

In well-designed control loop systems, global fixed points are easier to find than local fixed points of differential invariants, because the global invariant often coincides with the controllability region of the system. In these systems, the minimal control objective is to keep the system in the controllable state region. That is, if the system is in a controllable state, i.e., there is a control choice such that the system can remain safe, suitable controllers will pick one such control option such that the system again ends up in a controllable state. Inductively, this controllability region directly corresponds to a global system invariant. We make this intuition more precise in the context of Chap. 7.

6.3 Soundness

Even though the interplay of the fixed-point verification algorithms in this chapter is already quite complicated, we can exploit the fact that the algorithms are always working with formula transformations in the verification logic $\mathsf{d}\mathcal{L}$ to ensure soundness quite easily. Since all formula transformations in the algorithm are justified by sound proof rules, soundness is fully captured in the logic and the algorithm can only be incomplete but not unsound:

Theorem 6.1 (Soundness of fixed-point verification algorithm). *The verification algorithm in Sect. 6.2 is sound, i.e., whenever* `prove`$(\psi \rightarrow [\alpha]\phi)$ *returns "*`true`*", the* $\mathsf{d}\mathcal{L}$ *formula* $\psi \rightarrow [\alpha]\phi$ *is true in all states, i.e., all states reachable by* α *from states satisfying* ψ *satisfy* ϕ.

Proof. Soundness is a direct consequence of Theorem 3.1, as every statement that returns true is justified by a sound proof rule of DAL. To show this in full detail, we prove by induction on the structure of the algorithm that, for any $d\mathcal{L}$ formula ϕ, ϕ is true in every state ν where (the formula returned by) $prove(\phi)$ is true. That is, $\nu \models prove(\phi)$ implies $\nu \models \phi$. In particular, if the algorithm returns "true", the input formula is true in all states.

- In the base case (line 12 of Fig. 6.1), *prove* returns the result of quantifier elimination, which is a sound decision procedure [81]. The result of quantifier elimination is equivalent to its input. Hence one formula is true if and only if the other is.
- If α is of the form $x:=\theta$, the algorithm in line 1 of Fig. 6.1 is responsible. Consider any state ν where the formula returned by $prove(\psi \rightarrow [x:=\theta]\phi)$ is true. That is, $prove(\psi \wedge \hat{x} = \theta \rightarrow \phi_x^{\hat{x}})$ is true at ν. Hence, by the induction hypothesis, $\psi \wedge \hat{x} = \theta \rightarrow \phi_x^{\hat{x}}$ is true at ν. Now, because $\hat{x}$ was a fresh variable, the Substitution Lemma 2.2 can be used to show that $\psi \rightarrow \phi_x^{\theta}$ and $\psi \rightarrow [x:=\theta]\phi$ are true at ν.
- If α is of the form $x:=*$, the algorithm in line 10 of Fig. 6.1 is responsible. The proof is a direct consequence of the fact that ϕ being true after *all* random assignments to x is equivalent to ϕ being true for all real values of x. Hence, $\psi \rightarrow [x:=*]\phi$ is true in a state ν if and only if $\psi \rightarrow \forall x\, \phi$ is.
- For formulas of the form $\psi \rightarrow Qx\, \phi$ where Q is a quantifier, line 14 of Fig. 6.1 is responsible. If the formula returned by $prove(\psi \rightarrow Qx\, \phi)$ is true in state ν, then $\mathrm{QE}(\psi \rightarrow Qx\, prove(\phi))$ is true in ν. As quantifier elimination yields an equivalent formula, this implies that $\psi \rightarrow Qx\, prove(\phi)$ itself is true at ν. Assume that ψ holds at ν as there is nothing to show otherwise. Then $Qx\, prove(\phi)$ is true at ν. Thus for all (or some if Q is $\exists$) states ω that agree with ν except for the value of x, we know that the formula returned by $prove(\phi)$ is true at ω. The formula ϕ is structurally simpler. Hence, by induction hypothesis, ϕ is true at ω. Consequently, both $Qx\, \phi$ and $\psi \rightarrow Qx\, \phi$ are true at ν, as ω was arbitrary.
- The other cases of Fig. 6.1 are similar consequences of the soundness of the DAL rules in Fig. 3.9 by Theorem 3.1.
- If α is of the form $\mathcal{D} \wedge \chi$ for a differential equation system $\mathcal{D}$, the algorithm in Fig. 6.3 is responsible. If it returns "true" in line 2 in the first place, then the calls to *prove* in line 2 must have resulted in "true". Hence, by induction hypothesis, χ entails ϕ. Thus, soundness of rule $[DR]$ justifies soundness as follows. Postcondition ϕ is true in a subregion of the evolution domain χ. Thus $\psi \rightarrow [\mathcal{D} \wedge \chi]\phi$ is valid, trivially, because all evolutions along $\mathcal{D} \wedge \chi$ always satisfy χ and, hence, ϕ. If, however, χ was changed in line 5 during the fixed-point computation, then the calls to *prove* for the properties in line 4 must have returned "true". Thus, by the induction hypothesis, the $d\mathcal{L}$ formulas $\psi \wedge \chi \rightarrow F$ and $\forall^{\mathcal{D}}(\chi \rightarrow F'_{\mathcal{D}})$ are valid. Hence *DS* justifies soundness as follows. Formula F is a differential invariant of $\psi \rightarrow [\mathcal{D} \wedge \chi]\phi$ by Definition 6.3. Consequently, by Proposition 6.1, F also is a continuous invariant (Definition 6.2). Thus, by Proposition 6.2, the $d\mathcal{L}$ formulas $\psi \rightarrow [\mathcal{D} \wedge \chi]\phi$ and $\psi \rightarrow [\mathcal{D} \wedge \chi \wedge F]\phi$ are equivalent, and we can (soundly) verify the former by proving the latter. Consequently, the modification of the

evolution domain χ to $\chi \wedge F$ in line 5 is sound, because the algorithm will continue proving a refined but equivalent formula for a refined but equivalent system.

- If α is a loop of the form β^*, the proof is similar to the case for differential equations, except that it uses Proposition 6.3 instead of Proposition 6.1. □

Since reachability of hybrid systems is undecidable, our algorithm must be incomplete. It can fail to converge when the required invariants are not expressible in first-order logic (yet, the required invariants are always expressible in d$\mathcal{L}$ by Theorem 2.3). Consequently, a fixed point in d$\mathcal{L}$ always exists but our algorithm may fail to find appropriate (differential) invariants in first-order logic.

6.4 Optimisations

In this section, we consider some optimisations of the fixed-point verification algorithm.

6.4.1 Sound Interleaving with Numerical Simulation

During fixed-point computations, wrong choices of candidates are time consuming. Thus, it is important to discover futile proof attempts quickly. For this, we use non-exhaustive numerical simulation to look for a counterexample for each candidate. To prevent rejecting good candidates due to numerical errors, we discard fragile counterexamples. We consider counterexamples with distance $<\varepsilon$ to safe states as *fragile*, because small numerical perturbations could make it safe (the right x in Fig. 6.8 marks fragile examples or counterexamples). The left mark in

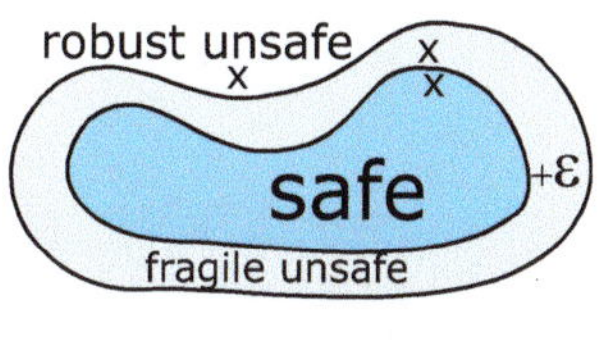

Fig. 6.8 Robustness in counterexamples

Fig. 6.8, instead, is a *robust* counterexample, i.e., only large ($\geq \varepsilon$) perturbations could make it safe. Robust counterexamples can be ensured by replacing, e.g., $a \geq b$ with $a \geq b + \varepsilon$ in the formulas for numerical reachability simulation for some estimate $\varepsilon \geq 0$ of the numerical error. Unlike in other approaches [21, 192, 228, 252, 238], numerical errors are *not* critical for soundness here, because safety is exclusively established by sound symbolic verification. Numerical computations are only used to stop futile proof attempts.

We can further exploit the symbolic decomposition performed by our algorithm in Sect. 6.2 and prefix recursive calls to $prove(\psi \rightarrow [\alpha]\phi)$ with a partial simulation of α. Using approximate cylindrical algebraic decomposition [81] in the *FindInstance* function of Mathematica, we often find good samples of states satisfying ψ to start the simulation of α.

6.4.2 Optimisations for the Verification Algorithm

Formulas with variables that do not change in a fragment of a hybrid program are trivial invariants, as their truth-value is unaffected. For instance, $\omega = \rho$ is a trivial invariant of system ($\mathscr{F}$). Hence, it can be used as an invariant without proof. A formula like $\omega^2(d_1^2 + d_2^2) > r^2$ in ψ, in contrast, is not trivially invariant, because d_i changes during ($\mathscr{F}$). Still, it has invariant consequences such as $\omega \neq 0$. To make use of these direct and indirect trivial invariants from ψ , we (soundly) weaken all universal closures of the form $\forall^\alpha \phi$ in lines 2 and 4 of Fig. 6.3 by $\psi \to \forall^\alpha \phi$, which directly reflects the rule context instantiations from Definition 2.10.

6.5 Experimental Results

We analyse aircraft roundabout manoeuvres [293] as a system with nontrivial dynamics. Curved flight as in roundabouts is challenging for verification, because of its transcendental solutions. The manoeuvre from [293] in Fig. 3.6a on p. 151 and the manoeuvre in Fig. 3.6d from [238, 237] are not flyable, because they still involve a few instant turns. A flyable roundabout manoeuvre without instant turns is depicted in Fig. 6.9. We verify safety properties for several (but not yet all) phases

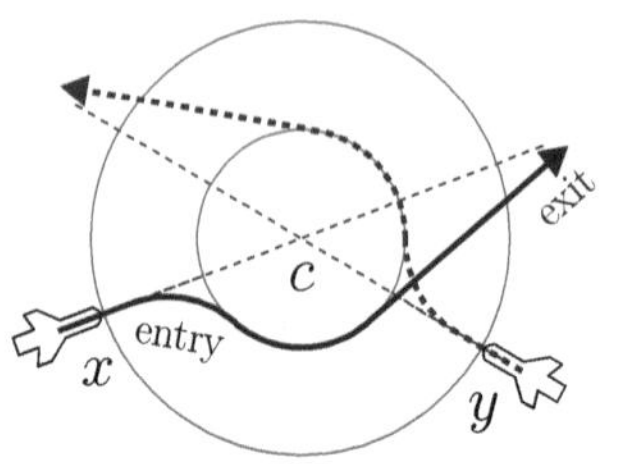

Fig. 6.9 Flyable aircraft roundabout

of Fig. 6.9 fully automatically and provide verification results in Table 6.1. Details on the case studies are presented in Part III. Finally, note that the required invariants found by our approach for the roundabout manoeuvre cannot even be found from Differential Gröbner Bases [202].

Verification results for roundabout aircraft manoeuvres [293, 92, 238, 237] and the European Train Control System (ETCS) are in Table 6.1. Results are from a 2.6 GHz AMD Opteron with 4 GB memory. Memory consumption of quantifier elimination is shown in Table 6.1, excluding the front end. The results are only slightly

Table 6.1 Experimental results for differential invariants as fixed points

Case study	Time(s)	Memory(MB)	Proof steps	Dim
tangential roundabout (2 aircraft)	14	8	117	13
tangential roundabout (3 aircraft)	387	42	182	18
tangential roundabout (4 aircraft)	730	39	234	23
tangential roundabout (5 aircraft)	1964	88	317	28
ETCS kernel safety	41	28	53	9
ETCS binary safety	56	27	147	14
ETCS safety	183	87	169	14

worse on a 1.7 GHz Pentium M laptop with 1 GB memory. The dimension of the continuous state space is indicated (column Dim). Notice that we handle *all* the variables symbolically leading to state spaces up to $\mathbb{R}^{28}$. The experimental results are encouraging, in particular as the memory consumption is fairly moderate. High memory consumption would limit scalability much more than time consumption. For instance, numerical techniques on a rather coarse numerical mesh with only 10 samples per variable would need to consider reachability analysis from 10^{28} initial states for this case. The results in Table 6.1 indicate that scalability of our approach is substantially less limited by the number of variables than in other approaches. Instead, the complexity of the constraints and dependencies among the variables have more impact. In parallel systems, there is usually a smaller fraction of dependencies among continuous variables. Multiple aircraft, e.g., are not coupled physically (see Fig. 6.4) but interact only by discrete dynamics originating from sporadic communication.

6.6 Summary

We have introduced the "differential invariants as fixed points" paradigm and we have presented a *sound* algorithm for verifying hybrid systems with nontrivial dynamics. It handles differential equations using differential invariants instead of requiring solutions of the differential equations, because the latter quickly yield undecidable arithmetic. We compute differential invariants as fixed points using $\mathsf{d}\mathcal{L}$, DAL, or dTL as verification logics for hybrid systems. In the logics we can decompose the system for computing local invariants and we obtain sound recombinations into global invariants. Moreover, we introduce a differential saturation procedure that verifies more complicated properties by refining the system dynamics successively in a sound way. We validate our algorithm on challenging *roundabout collision avoidance manoeuvres* for aircraft and on collision avoidance protocols for trains. We give verification results with a sound algorithm for challenging dynamics of curved flight in up to 28-dimensional continuous state spaces.

Our algorithm works particularly well for highly parametric hybrid systems, because their parameter constraints can often be combined faster to find invariants, than for systems with a single initial state, where simulation is more appropriate.

Our decompositional approach exploits locality in system designs. In well-designed systems, subsystems do not generally depend on all other parts of the system but are built according to modularity and locality principles in engineering. In these cases, the $d\mathcal{L}$ calculus achieves good decompositions and the required invariants for one part of the system only need very little information about other parts of the systems, so the fixed-point algorithm terminates faster. Our algorithm probably performs worse for systems that violate locality principles.

Scalability to applications from other domains like chemical process control or biomedical devices remains an interesting topic for future research. An independent question for future research, underlying virtually all hybrid systems' verification approaches, is how to scale real arithmetic handling. Semidefinite programming relaxations [224] may be an interesting direction for future research (see App. D). Differential induction and the logic $d\mathcal{L}$ generalise to liveness properties and to systems with disturbances (Chap. 3). In future work, we want to generalise the synthesis of corresponding differential (in)variants for these cases.

Part III
Case Studies and Applications in Hybrid Systems Verification

Overview In Part I we have developed differential dynamic logics as specification and verification languages for both hybrid programs and DA-programs as operational system models for hybrid systems. We have also developed proof calculi as formal verification techniques for hybrid systems in differential dynamic logic. In Part II, we have studied algorithmic refinements, automation techniques, and invariant generation procedures for the respective proof calculi.

In this part of the book, we now shift our attention to application scenarios for our logical analysis approach for hybrid systems. Extending smaller hybrid systems which have served as running examples throughout this book, we show full case studies of the European Train Control System in Chap. 7 and for aircraft collision avoidance manoeuvres in Chap. 8. We show how complex physical systems can be modeled as hybrid programs (or DA-programs) and how safety-critical properties can be formalised in differential dynamic logics and proven in their proof calculi.

Chapter 7
European Train Control System

Contents

7.1 Introduction **278**
7.1.1 Related Work 280
7.1.2 Structure of This Chapter 281
7.2 Parametric European Train Control System **281**
7.2.1 Overview of the ETCS Cooperation Protocol 281
7.2.2 Formal Model of Fully Parametric ETCS 284
7.3 Parametric Verification of Train Control **286**
7.3.1 Controllability Discovery 287
7.3.2 Iterative Control Refinement 288
7.3.3 Safety Verification 291
7.3.4 Liveness Verification 293
7.3.5 Full Correctness of ETCS 294
7.4 Disturbance and the European Train Control System **295**
7.4.1 Controllability Discovery 296
7.4.2 Iterative Control Refinement 298
7.4.3 Safety Verification 298
7.5 Experimental Results **299**
7.6 Summary **301**

Synopsis Complex physical systems have several degrees of freedom. They only work correctly when their control parameters obey corresponding constraints. Based on the informal specification of the *European Train Control System* (ETCS), we design a controller for its cooperation protocol. For the free parameters of the system, we successively identify constraints that are required to ensure collision freedom. We formally prove the parameter constraints to be sharp by characterising them equivalently in terms of reachability properties of the hybrid system dynamics. We use the calculus of our differential dynamic logic for hybrid systems and formally verify controllability, safety, liveness, and reactivity properties of the ETCS protocol that entail collision freedom. We prove that the ETCS protocol remains correct even in the presence of perturbation by disturbances in the dynamics.

A. Platzer, Logical Analysis of Hybrid Systems,
DOI 10.1007/978-3-642-14509-4_7, © Springer-Verlag Berlin Heidelberg 2010

7.1 Introduction

Complex physical control systems often contain many degrees of freedom, including how specific parameters are instantiated or adjusted [205, 91, 27]. Yet, virtually all of these systems are hybrid systems and only work correctly under certain constraints on these parameters. The *European Train Control System* (ETCS) [117] has a wide range of different possible configurations of trains, track layouts, and different driving circumstances. It is only safe for certain conditions on external parameters, e.g., as long as each train is able to avoid collisions by braking with its specific braking force on the remaining distance to the rear end of the next train. Similarly, internal control design parameters for supervisory speed control and automatic braking triggers need to be adjusted in accordance with the underlying train dynamics. Moreover, parameters must be constrained such that the system remains correct when passing from continuous models with instant reactions to sampled-data discrete-time controllers of hardware implementations. Finally, parameter choices must preserve correctness robustly in the presence of disturbances caused by unforeseen external forces (wind, friction, etc.) or internal modelling inaccuracies of ideal-world dynamics, e.g., when passing from ideal-world dynamics to *proportional-integral* (PI) controller implementations.[1] Yet, determining the range of external parameters and the choice of internal design parameters for which complex control systems like ETCS are safe, is not possible just by looking at the model, and even less so in the presence of disturbance.

Likewise, it is difficult to read off the parameter constraints that are required for correctness from a failed verification attempt of model checkers [156, 217, 126], since concrete numeric values of a counterexample trace cannot simply be translated into a generic constraint on the free parameters of the system, which would have prevented this kind of error. While approaches like counterexample-guided abstraction refinement [72, 126] are highly efficient in undoing automatic abstractions of an abstract hybrid system from spurious counterexamples, they stop when true counterexamples remain in the concrete system. For discovering constraints on free parameters, though, even concrete models will have counterexamples until all required parameter constraints have been identified. Even worse, concrete numeric values of a counterexample trace as produced by a model checker with state splitting cannot simply be translated into a uniform global constraint on the free parameters of the system, which would prevent this kind of error.

Instead, we use our techniques based on symbolic decompositions from Part I to systematically explore the design space of a hybrid system and for discovering correctness constraints on free parameters. We have already shown how some such constraints can be identified for a very simple train control model in the examples from Chap. 2. For a complex physical system, we now show step by step how a control system can be developed that meets its control design goals and desired correctness properties. Starting from a coarse skeleton of the ETCS cooperation

[1] PI is a standard control technique and also used for controlling trains [91]. We refer to joint work with Jan-David Quesel [244] for verification results of ETCS in the presence of PI control.

protocol obtained from its official specification [117], we systematically develop a safe controller and identify the parameter constraints that are required for collision freedom. Although these parameter constraints are safety-critical, they are not stated in the official specification [117]. They only originate quite indirectly from the system dynamics and the overall control objectives of ETCS. For all practical implementations, however, these constraints need to be made explicit in order to find safe parameter range choices. These constraints are also nontrivial, especially those needed to ensure a safe interplay of physics and sampled control implementations. Using the parametric constraints so discovered, we prove correctness properties of the ETCS cooperation protocol that entail collision freedom.

During the course of this case study, we prove rich properties, including safety, controllability, reactivity, and liveness, which are not uniformly expressible and verifiable in other analysis approaches. Moreover, we prove those correctness properties of the parametric ETCS case study almost fully automatically in our verification tool KeYmaera.

Contributions

We show how realistic fully parametric hybrid systems for traffic protocols can be designed and verified using our logical analysis approach. For ETCS, we identify all relevant safety constraints on free parameters, including external system parameters and internal design parameters of controllers. Safe control choices will be important for more than two million passengers in Europe per day. Similar constraints hold for other upcoming train control systems. Our first contribution is that we characterise safe parameter choices equivalently in terms of properties of the train dynamics and that we prove controllability, reactivity, safety, and liveness properties of ETCS. Other issues often arise from verification results for ideal-world dynamics that cease to hold for real-world dynamics. Our second contribution is showing how to extend formal ETCS verification to the presence of disturbances in the dynamics, which account for friction and other discrepancies of the real-world dynamics and simplistic control models. Most notably, the full ETCS model with its rich set of properties is out of scope for other approaches. Nevertheless, we have formally verified all propositions in this chapter with our verification tool KeYmaera [242]. Furthermore, we have proven ETCS to work correctly when using a proportional-integral (PI) controller for speed supervision. In contrast to their routine use in control, giving formal proofs for the correct functioning of PIs has been an essentially unsolved problem. We refer to joint work with Jan-David Quesel [245] for details on the PI verification.

ETCS further illustrates a more general phenomenon in hybrid systems: safely combining dynamics with control requires parameter constraints that are much more complicated than the original dynamics. Safe control of hybrid dynamics requires more than just an understanding of the physics underlying a system and a separate understanding of the computing processes in the system. Rather, safe choices of hybrid systems control usually depend on a joint understanding of how physics and computation interact.

7.1.1 Related Work

Model checkers for hybrid systems, for example HyTech [11] and PHAVer [126], verify by exploring the state space of the system as exhaustively as possible. In contrast to our approach they need concrete numbers for most parameters and cannot verify liveness, such as whether the train controller is guaranteed to make progress, or existential properties, e.g., whether and how a control parameter can be instantiated so that the system is always safe.

Batt et al. [27] give heuristics for splitting regions by linear constraints that can be used to determine parameter constraints. Frehse et al. [128] synthesise parameters for linear hybrid automata. However, realistic systems like ETCS require nonlinear parameter constraints, often have a nonlinear system dynamics, and are out of scope for these approaches.

Tomlin et al. [294] show a game-theoretic semidecision algorithm for hybrid controller synthesis. For systems like ETCS, which are more general than linear [127] or o-minimal hybrid automata [188], they suggest numerical approximations. We give exact results for fully parametric ETCS using symbolic techniques. We also focus on verification and parameter synthesis of parameter constraints for a given controller template, rather than for full controller synthesis. While full controller synthesis is a very interesting and tempting idea, it is less scalable than verification [75, 249]. In addition to theoretical complexity results that separate verification and synthesis in the discrete case, verification and even parameter synthesis are more modular than full synthesis. In verification, several properties can be proven separately for the same model, but in synthesis all properties need to be considered at once. A synthesised controller satisfying a safety property may not be live, e.g., it may be safe, just because it disallows movement altogether, which is perfectly useless. Another synthesised controller satisfying a liveness property, however, may be unsafe. So in synthesis all relevant properties need to be considered at once for the resulting system to make sense. In verification and also in proof-based parameter synthesis, properties can be considered one at a time, thereby gaining scalability advantages.

In independent work, Cimatti et al. [71] have recently analysed consistency of informal requirements of ETCS expressed as temporal properties, including the continuous system dynamics, using an approach based on the combination of temporal logic with regular expressions. Our work is complementary, as we focus on developing and formally verifying an actual hybrid systems controller that can be implemented later on, and not on showing the consistency of the requirement specification properties. The fact that we are able to construct a controller satisfying ETCS requirements also entails consistency of these requirements. Our goal, however, is to develop a verified controller to show how the system can be controlled safely, and not whether the requirements are incompatible.

7.1.2 Structure of This Chapter

We introduce a formal model for parametric ETCS in Sect. 7.2. Using our symbolic logical decomposition approach from Part I, we systematically derive parametric correctness constraints for ETCS and verify several correctness properties of parametric ETCS using these constraints in Sect. 7.3. In Sect. 7.4, we generalise the physical transmission model to the presence of disturbances and verify ETCS despite disturbances. We present experimental results for our verification approach in Sect. 7.5. For details on the experiment and formal KeYmaera proofs of the propositions, we refer the reader to [245].

7.2 Parametric European Train Control System

The *European Train Control System* (*ETCS*) [117, 205] is a standard to ensure safe and collision-free operation as well as high throughput of trains that currently run at speeds up to 320 km/h. Correct functioning of ETCS is highly safety-critical, because the upcoming installation of ETCS Level 3 will replace all previous track-side safety measures in order to achieve its high throughput objectives. In this section, we present a system skeleton, which corresponds to a simple representation of the train dynamics and controller reflecting the informal ETCS cooperation protocol [117]. The system obtained from the informal specification turns out to be unsafe. In Sect. 7.3, we will systematically refine this skeleton with the parameter constraints that are required for safety but not stated in the informal ETCS specification [117], and only indirectly result from the effects of the system dynamics on the safety requirements.

7.2.1 Overview of the ETCS Cooperation Protocol

ETCS Level 3 follows the *moving block principle*, i.e., neither are movement permissions known beforehand nor is the size of movement blocks fixed statically. Rather, permission to move is determined based on the current track situation by a *Radio Block Controller* (RBC). Trains are only allowed to move within their current *movement authority* (MA), which can be updated dynamically by the RBC using wireless communication. Hence the train controller needs to regulate the movement of a train locally such that it always remains within its MA, because there could be

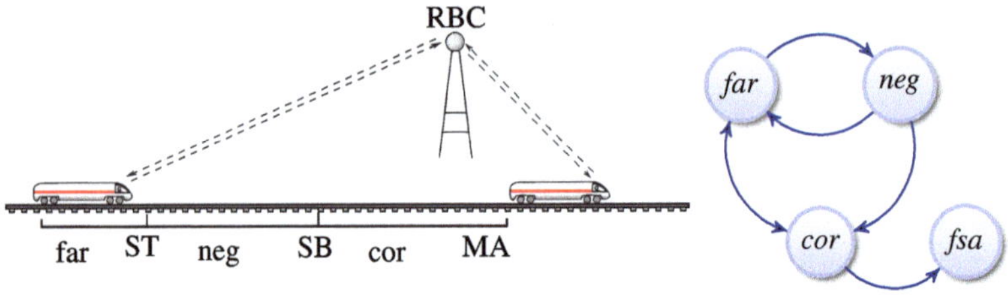

Fig. 7.1 ETCS train cooperation protocol phases and dynamic movement authorities

open gates, other trains, or speed restrictions due to tunnels beyond the end of the MA.

The automatic train protection unit (*atp*) dynamically determines a safety envelope around a train τ, within which it considers driving safe, and adjusts the train acceleration accordingly. Figure 7.1 illustrates the dynamic assignment of MA. The ETCS controller switches according to the protocol pattern in Fig. 7.1, which corresponds to a simplified version of Damm et al. [91]. When approaching the end of its MA the train switches from *far* mode (where speed can be regulated freely) to negotiation (*neg*), which, at the latest, happens at the point indicated by *ST* (for *start talking*). During negotiation the RBC grants or denies MA extensions. If the extension is not granted in time, the train starts braking in the correcting mode (*cor*) and returns to *far* after the situation has cleared. Emergency messages announced by the RBC can also put the controller into *cor* mode. If so, the train may switch to a fail-safe state (*fsa*) after the train has come to a full stop and await manual clearance by the train operator.

Lemma 7.1 (Principle of separation by movement authorities). *If each train stays within its MA and, at any time, MAs issued by the RBC form a disjoint partitioning of the track, then trains can never collide.*

Proof. To simplify notation, we assume trains are points (the proof is a simple extension when each train has some maximal length l). Consider any point in time ζ. For some $n \in \mathbb{N}$, let $z_1, \ldots, z_n$ be the positions of all the trains 1 to n at ζ. Let M_i be the MA range, i.e., the set of positions on the track for which train i has currently been issued MA. Suppose there was a collision at time ζ. Then $z_i = z_j$ at ζ for some $i, j \in \mathbb{N}$. However, by assumption, $z_i \in M_i$ and $z_j \in M_j$ at ζ, thus $M_i \cap M_j \neq \emptyset$; which contradicts the assumption of disjoint MAs. □

Lemma 7.1 effectively reduces the verification of an unbounded number of traffic agents to a finite number. We exploit MAs to decouple reasoning about global collision freedom and reduce it to local cooperation of every traffic agent with its RBC. In particular, we verify correct coordination for a train without having to consider gates or railway switches, because these only communicate via RBC mediation and can be considered as special reasons for denial of MA extensions. We only need to prove that the RBC handles all interaction between the trains by assigning or revoking MA correctly and that the trains actually respect their MA always. Yet, in order to enable the RBC to guarantee disjoint partitioning of the track, ETCS has to rely on properties like appropriate safe rear end computation of the train. Additionally, safe operation of the train plant in conjunction with its environment depends on proper functioning of the gates. As these properties have a more static nature, they are much easier to show once the actual hybrid train dynamics and movements have been proven to be controlled correctly and once the relevant safety constraints have been identified.

As trains are not allowed to drive backwards without manual clearance by track supervision personnel, the relevant part of the safety envelope is the closest distance to the end of its current MA. The point *SB*, for *start braking*, is the latest point where the train needs to start correcting its acceleration (in mode *cor*) to make sure it always stays within the bounds of its MA. In Sect. 7.3, we derive a necessary and sufficient constraint on parameter *SB* that guarantees safe driving.

We generalise the concept of MA to a vector $\mathbf{m} = (d, e, r)$, meaning that, beyond end point $\mathbf{m}.e$, the train is not permitted to have a velocity greater than the desired speed $\mathbf{m}.d$. Additionally, the train should try not to outspeed the *recommended speed* $\mathbf{m}.r$ for the current track segment, but short periods of slightly higher speed are not considered safety-critical. Figure 7.2 shows an example of possible train behaviour in conjunction with the current value of $\mathbf{m}$ that changes over time due to RBC communication.

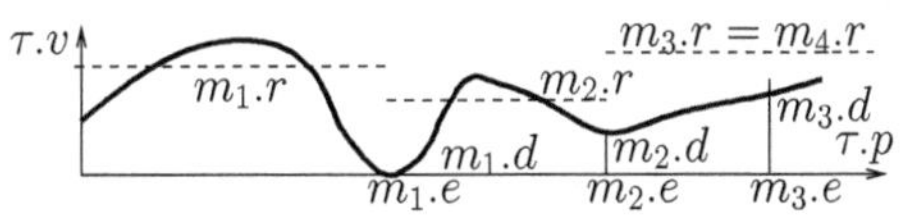

Fig. 7.2 ETCS track profile

For a train $\tau = (p, v, a)$ at position $\tau.p$ with current velocity $\tau.v$ and acceleration $\tau.a$, we want to determine sufficient conditions that ensure safety and formally verify that $\tau.v$ is always safe with respect to its current MA, satisfying:

$$\tau.p \geq \mathbf{m}.e \rightarrow \tau.v \leq \mathbf{m}.d. \qquad (\mathscr{S})$$

Formula $(\mathscr{S})$ expresses that the train velocity $\tau.v$ does not exceed the strict speed limit $\mathbf{m}.d$ after passing the point $\mathbf{m}.e$ (i.e., $\tau.p \geq \mathbf{m}.e$). Generalised MAs are a uniform composition of two safety-critical features. They are crucial aspects for ensuring collision-free operation in ETCS (Lemma 7.1) and can take into account safety-critical velocity limits due to bridges, tunnels, or passing trains. For example high-speed trains need to reduce their velocity while passing non-airtight or freight trains with a pressure-sensitive load within a tunnel. Our model captures this by reducing the speed component $\mathbf{m}.d$ of MA $\mathbf{m}$ appropriately.

$$
\begin{aligned}
ETCS_{\text{skel}} &\equiv (train \cup rbc)^* \\
train &\equiv spd;\ atp;\ drive \\
spd &\equiv \big(?(\tau.v \leq \mathbf{m}.r);\ \tau.a := *;\ ?(-b \leq \tau.a \leq A)\big) \\
&\quad \cup \big(?(\tau.v \geq \mathbf{m}.r);\ \tau.a := *;\ ?(-b \leq \tau.a \leq 0)\big) \\
atp &\equiv \text{if } (\mathbf{m}.e - \tau.p \leq SB \vee rbc.message = emergency) \text{ then } \tau.a := -b \text{ fi} \\
drive &\equiv t := 0;\ (\tau.p' = \tau.v \wedge \tau.v' = \tau.a \wedge t' = 1 \wedge \tau.v \geq 0 \wedge t \leq \varepsilon) \\
rbc &\equiv \big(rbc.message := emergency\big) \cup \big(\mathbf{m} := *;\ ?(\mathbf{m}.r > 0)\big)
\end{aligned}
$$

Fig. 7.3 Formal model of parametric ETCS cooperation protocol (skeleton)

7.2.2 Formal Model of Fully Parametric ETCS

For analysing the proper functioning of ETCS, we have developed a formal model of ETCS as a hybrid program (see Fig. 7.3) that is based on the informal ETCS specification [117]. The RBC and the train are independently distributed components running in parallel. They cooperate by message passing over wireless communication. Since the RBC is a purely digital track-side controller and has no dependent continuous dynamics, we can express parallel composition equivalently by interleaving using nondeterministic choice ($\cup$) and repetition (*): the decisions of the train controller only depend on the point in time when RBC messages arrive at the train, not on the communication latency. Thus, the nondeterministic interleaving in ETCS where either the train or ($\cup$) the RBC chooses to take action faithfully models every possible arrival time without the need for an explicit (delayed) channel model. The * at the end of $ETCS_{\text{skel}}$ indicates that the interleaving of train and RBC repeats arbitrarily often. Successive actions in each component are modelled using sequential composition ($;$). The train checks for the difference between the current speed and the recommended speed (in speed supervision *spd*) before checking if emergency braking is necessary (in automatic train protection *atp*).

Train Controller

It is quite difficult to use highly detailed models for the train and its mechanical transmission [91] directly in the verification and parameter discovery process. Hence, we first approximate it by a controller with a ranged choice for the effective acceleration $\tau.a$ between its lower bound ($-b < 0$) and upper bound ($A > 0$); we will refine the dynamics in Sect. 7.4. This nondeterministic range controller provides a model that we can use both to derive parameter constraints and to overapproximate the actual choices made by the physical train controller [91] conservatively. For Sects. 7.2 and 7.3, we model the continuous train dynamics by the differential equation system

$$\tau.p' = \tau.v \wedge \tau.v' = \tau.a \wedge t' = 1 \wedge \tau.v \geq 0 \wedge t \leq \varepsilon. \qquad (\mathscr{I})$$

It formalises the ideal-world physical laws for movement, restricted to the evolution domain $\tau.v \geq 0 \wedge t \leq \varepsilon$ in *drive*. Recall from Part I that primed variables stand for the first time derivative of the respective unprimed variable. Therefore, $\tau.p'$ gives the rate with which the position $\tau.p$ of the train τ changes, i.e., the velocity ($\tau.p' = \tau.v$). The velocity $\tau.v$ itself changes continuously according to the acceleration $\tau.a$, i.e., $\tau.v' = \tau.a$. The train speeds up when $\tau.a > 0$ and brakes when $\tau.a < 0$. In particular, for $\tau.a < 0$, the velocity would eventually become negative, which would mean the train is driving backwards. But that is prohibited without manual clearance, so we restrict the evolution domain to nonnegative speed ($\tau.v \geq 0$). Time can be measured by clocks, i.e. variables changing with constant slope 1 (along differential equation $t' = 1$). To further account conservatively for delayed effects of actuators like brakes or for delays caused by cycle times of periodic sensor polling and sampled-data discrete-time controllers, we permit the continuous movement of the train to continue for up to $\varepsilon > 0$ time units until control decisions finally take effect. This is expressed using the evolution domain restriction $t \leq \varepsilon$ on the clock t that is reset by the discrete assignment $t := 0$ before the continuous evolution starts. It is used to keep track of the progress of time advancing with constant slope 1. When the system executes the system of differential equations in *drive*, it can follow a continuous evolution respecting the constraints of ($\mathscr{I}$) all the time.

The speed supervision *spd* has two choices ($\cup$). The first option in Fig. 7.3 can be taken if the test $?(\tau.v \leq \mathbf{m}.r)$ succeeds, the second one if the test $?(\tau.v \geq \mathbf{m}.r)$ is successful. If both tests can succeed, either choice is possible by a nondeterministic choice ($\cup$). The speed supervision *spd* chooses the acceleration $\tau.a$ to keep the recommended speed $\mathbf{m}.r$ by a random assignment $\tau.a := *$, which assigns an arbitrary value to $\tau.a$. By the subsequent test $?(-b \leq \tau.a \leq 0)$ an acceleration is chosen from the interval $[-b, 0]$ if the current speed $\tau.v$ exceeds recommended speed $\mathbf{m}.r$ (otherwise the full range $[-b, A]$ is available.) Our nondeterministic controller includes controllers optimising speed and energy consumption as secondary objectives.

As a supervisory controller, the automatic train protection (*atp* in Fig. 7.3) checks whether the train has passed point *SB* ($\mathbf{m}.e - \tau.p \leq SB$, that is, the remaining distance of $\tau.p$ to $\mathbf{m}.e$ is less than or equal *SB*) or whether a message from the RBC was received notifying of a track-side emergency situation. Both events cause immediate braking with full deceleration $-b$. Thus, *atp* decisions take precedence over the recommended *spd* speed advisory. In the case where $\mathbf{m}.e - \tau.p > SB$ and no emergency message arrived, the decisions made by *spd* take effect. We will identify safety-critical constraints on parameter *SB* in Sect. 7.3.2.

Radio Block Controller

We model the RBC as a controller with two possible choices ($\cup$). It may choose to either demand immediate corrective action by sending emergency messages (*rbc.message* := *emergency*) or the RBC may update the MA by assigning arbitrary new values to its three components ($\mathbf{m} := *$). These nondeterministic changes to $\mathbf{m}$ reflect different real-world effects like extending $\mathbf{m}.e$ and $\mathbf{m}.d$ if the heading train has advanced significantly or notify of a new recommended speed $\mathbf{m}.r$ for a track segment. We will identify safety-critical constraints on MA updates in Sect. 7.3.2.

Figure 7.4 illustrates the transition structure $\rho(ETCS_{\text{skel}})$, as defined in Definition 2.7, that corresponds to the hybrid program of the ETCS skeleton in Fig. 7.3.

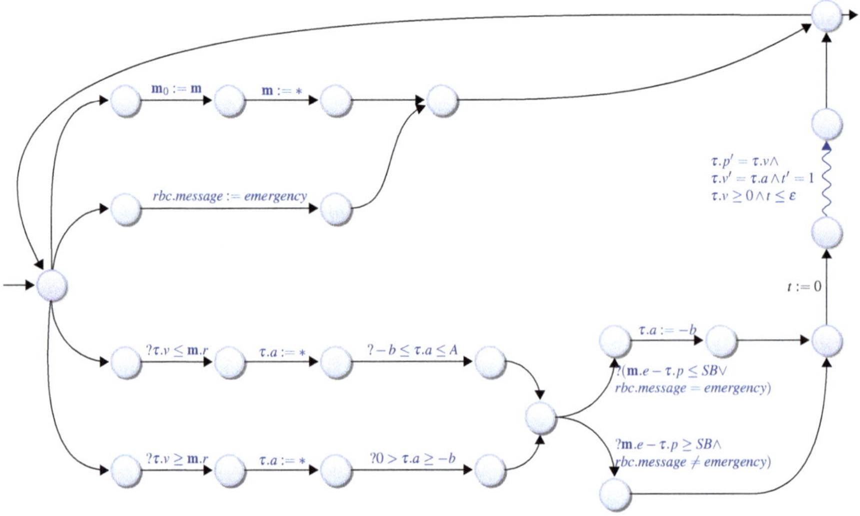

Fig. 7.4 Transition structure of ETCS skeleton

7.3 Parametric Verification of Train Control

The model in Fig. 7.3 that was read off from the informal ETCS specification [117] is unsafe, i.e., it does not always prevent collisions. To correct this we identify constraints on the free parameters of ETCS by analysing increasingly more complex correctness properties of ETCS. Using these constraints we refine the train control model iteratively into a safe model with constraints on design parameter choices and physical prerequisites on external parameters resulting from the safety requirements on ETCS and the behaviour of the train dynamics.

Iterative Refinement Process

For discovering parametric constraints required for system correctness, we follow an *iterative refinement process* that is of more general interest. It can be used to synthesise parameters with the d$\mathcal{L}$ calculus, i.e., successively identify safety-critical parameter constraints for a template controller.

1. *Controllability discovery*: Start with uncontrolled system dynamics. Use structural symbolic decomposition in $d\mathcal{L}$ until a first-order formula is obtained revealing the controllable state region, which specifies for which parameter combinations the system dynamics can actually be controlled safely by any control law.
2. *Control refinement*: Successively add partial control laws to the system while leaving its decision parameters (such as *SB* or **m**) free. Use structural symbolic decomposition to discover parametric constraints which preserve controllability under these control laws.
3. *Safety convergence*: Repeat step 2 until the resulting system is proven safe.
4. *Liveness check*: Prove that the discovered parametric constraints do not overconstrain the system inconsistently by showing that it remains live.

In practise, variants of the controllable domain as discovered by step 1 constitute good candidates for inductive invariants, and the parameter constraints discovered by step 2 ensure that the actual control choices taken by the controller never leave the controllable domain. For step 4, liveness can be verified again by structural symbolic decomposition in $d\mathcal{L}$ and no need for separate verification techniques or different system models arises.

7.3.1 Controllability Discovery in Parametric ETCS

By analysing the uncontrolled train dynamics, we obtain a controllability constraint on the external train parameters, i.e., a formula characterising the parameter combinations for which the train dynamics can be controlled safely by any control law at all. For our analysis we choose the global assumptions

$$\tau.v \geq 0 \land \mathbf{m}.d \geq 0 \land b > 0 \land A \geq 0 \qquad (\mathscr{A})$$

stating that the velocity is nonnegative ($\tau.v \geq 0$), the movement authority issued by the RBC does not force the train to drive backwards ($\mathbf{m}.d \geq 0$), and the train has some positive braking force $b > 0$ and some nonnegative maximum acceleration $A \geq 0$. The controllability constraint is now obtained by applying the $d\mathcal{L}$ proof calculus from Chap. 2 to the following $d\mathcal{L}$ (or DAL) formula:

$$(\mathscr{A} \land \tau.p \leq \mathbf{m}.e) \to [\tau.p' = \tau.v \land \tau.v' = -b \land \tau.v \geq 0]\mathscr{S}.$$

This formula expresses that—starting in some state where assumptions ($\mathscr{A}$) hold and the train has not yet passed $\mathbf{m}.e$ (i.e., $\tau.p \leq \mathbf{m}.e$)—every possible evolution of the train system that applies full brakes ($\tau.v' = -b$) is safe, i.e. does not violate ($\mathscr{S}$). By a proof similar to that in Fig. 2.14 on p. 89, we can discover that this $d\mathcal{L}$ formula only holds if $\tau.v^2 - \mathbf{m}.d^2 \leq 2b(\mathbf{m}.e - \tau.p)$ is satisfied in the initial state. We prove that the so-discovered constraint, illustrated in Fig. 7.5, characterises the set of states where the train dynamics can still respect MA by appropriate control choices (expressed by the left-hand side d$\mathcal{L}$ formula).

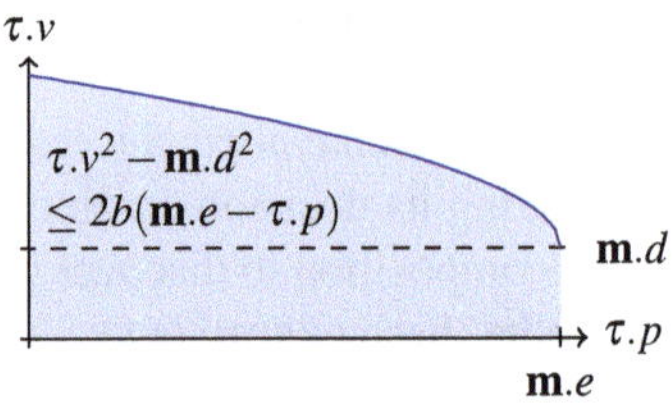

Fig. 7.5 Controllable region of ETCS

Proposition 7.1 (Controllability). *The constraint* $\tau.v^2 - \mathbf{m}.d^2 \leq 2b(\mathbf{m}.e - \tau.p)$ *is a* controllability constraint *for the train* τ *with respect to property* $(\mathcal{S})$ *on page 283, i.e., this constraint retains the ability of the train dynamics to respect the safety property. Formally, with* $\mathcal{A} \wedge \tau.p \leq \mathbf{m}.e$ *as regularity assumptions, the following equivalence is a provable* d$\mathcal{L}$ *formula:*

$$\begin{aligned}&\left[\tau.p' = \tau.v \wedge \tau.v' = -b \wedge \tau.v \geq 0\right]\left(\tau.p \geq \mathbf{m}.e \rightarrow \tau.v \leq \mathbf{m}.d\right)\\ \equiv\, &\tau.v^2 - \mathbf{m}.d^2 \leq 2b(\mathbf{m}.e - \tau.p).\end{aligned}$$

This formula expresses that *every run* of a train in braking mode always satisfies $(\mathcal{S})$ if and only if condition $\tau.v^2 - \mathbf{m}.d^2 \leq 2b(\mathbf{m}.e - \tau.p)$ holds initially. Observe how the above equivalence reduces a dynamic d$\mathcal{L}$ formula about future controllable train dynamics to a single static constraint on the current state. We use this key reduction step from dynamically safe train dynamics to controllably safe state constraints by analysing whether each part of the ETCS controller preserves train controllability.

Definition 7.1 (Controllable state). A train τ is in a *controllable state*, if the train is always able to stay within its movement authority $\mathbf{m}$ by appropriate control actions, which, by Proposition 7.1, is equivalent to

$$\tau.v^2 - \mathbf{m}.d^2 \leq 2b(\mathbf{m}.e - \tau.p) \wedge \mathcal{A}. \qquad (\mathcal{C})$$

ETCS cannot be safe unless trains start and stay in controllable states. Hence we pick $(\mathcal{C})$ as a minimal candidate for an inductive invariant. This invariant will be used to prove safety of the system by induction.

7.3.2 Iterative Control Refinement of ETCS Parameters

Starting from the constraints for controllable trains, we identify constraints for their various control decisions and refine the ETCS model correspondingly, so that the train stays controllable for every control choice.

RBC Control Constraints

For safe functioning of the ETCS it is important that trains always respect their current MA. Consequently, RBCs are not allowed to issue MAs that are physically impossible for the train to obey, such as instantaneous full stops. Rather, RBCs are only allowed to send new MAs that remain within the controllable range of the train dynamics. For technical reasons the RBC does not reliably know the train positions and velocities in its domain of responsibility to a sufficient precision, because the communication with the trains has to be performed wirelessly with possibly high communication delay and message loss. Thus, we give a fail-safe constraint for MA updates which is reliably safe even for loss of position recording communication.

Proposition 7.2 (RBC preserves train controllability). *The constraint*

$$\mathbf{m}_0.d^2 - \mathbf{m}.d^2 \leq 2b(\mathbf{m}.e - \mathbf{m}_0.e) \land \mathbf{m}_0.d \geq 0 \land \mathbf{m}.d \geq 0 \tag{$\mathcal{M}$}$$

ensures that the RBC preserves train controllability ($\mathcal{C}$) when changing the MA from $\mathbf{m}_0$ to $\mathbf{m}$. That is, the following $\mathsf{d}\mathcal{L}$ formula is provable:

$$\forall \tau \Big(\mathcal{C} \rightarrow [\mathbf{m}_0 := \mathbf{m};\ rbc]\, (\mathcal{M} \rightarrow \mathcal{C})\Big). \tag{7.1}$$

Moreover, RBC controllability is characterised by the following provable $\mathsf{d}\mathcal{L}$ formula:

$$\mathbf{m}.d \geq 0 \land b > 0 \rightarrow [\mathbf{m}_0 := \mathbf{m};\ rbc]\, \Big(\mathcal{M} \leftrightarrow \forall \tau \big((\langle \mathbf{m} := \mathbf{m}_0 \rangle \mathcal{C}) \rightarrow \mathcal{C}\big)\Big). \tag{7.2}$$

Constraint ($\mathcal{M}$) characterises that an extension is safe if it is possible to reduce the speed by braking with deceleration b from the old target speed $\mathbf{m}_0.d$ to the new target speed $\mathbf{m}.d$ within the extension range $\mathbf{m}.e - \mathbf{m}_0.e$, regardless of the current speed of train τ. It imposes constraints on feasible track profiles. Formula (7.1) expresses that, for all trains ($\forall \tau$, i.e., $\forall \tau.v \forall \tau.p$) in a controllable state ($\mathcal{C}$), every RBC change ($[rbc]$) of MA from $\mathbf{m}_0$ to $\mathbf{m}$ that complies with ($\mathcal{M}$) ensures that the train is still in a controllable state ($\mathcal{C}$). Constraint ($\mathcal{M}$) is characterised by the equivalence (7.2), expressing that for every decision of *rbc*, ($\mathcal{M}$) holds for the RBC change from $\mathbf{m}_0$ to $\mathbf{m}$ if and only if *all* trains ($\forall \tau$) that were controllable ($\mathcal{C}$) for the previous MA (set using $\langle \mathbf{m} := \mathbf{m}_0 \rangle$) still remain controllable for the new MA $\mathbf{m}$.

Train Control Constraints

Now that we have found constraints characterising when the cooperation of train and RBC is controllable, we need to find out under which circumstances the actual control choices by *spd* and *atp* retain controllability. In particular, the design parameter *SB* (start braking point relative to the end of the movement authority) needs to be chosen appropriately to preserve ($\mathcal{C}$). First we show that *there is* a safe choice of *SB*:

Proposition 7.3 (Reactivity of ETCS). *For all feasible RBC choices and all choices of speed control, there is a choice for SB that makes the train always stay within its MA, i.e., for controllable states, the following $d\mathcal{L}$ formula is provable:*

$$\mathscr{C} \rightarrow [\mathbf{m}_0 := \mathbf{m};\ rbc]\big(\mathscr{M} \rightarrow [spd]\langle SB := *\rangle[atp;\ drive]\mathscr{S}\big).$$

The formula expresses that, starting in a controllable region $\mathscr{C}$, if the RBC updates the MA from $\mathbf{m}_0$ to any $\mathbf{m}$ respecting ($\mathscr{M}$), then after arbitrary *spd* choices, the train controller is still able to find some choice for *SB* ($\langle SB := *\rangle$) such that it always respects the fresh MA when following *atp* and *drive*. Since Proposition 7.3 is provable in KeYmaera we know that *there is a safe solution for ETCS*. On the formula level the assumptions are expressed using implications such that the formula does not make any proposition if either ($\mathscr{C}$) is not initially satisfied or the RBC updates do not respect ($\mathscr{M}$). The train controller in this property is split up into the proposition that for all executions of the speed supervision ($[spd]$) there is a choice for *SB* ($\langle SB := *\rangle$) such that the automatic train protection unit (*atp*) always preserves safety during the train movement in the *drive* phase. For *atp* and *drive* we again make a statement over all possible executions of the components ($[atp; drive]$). Only the choice of *SB* is existentially quantified by an $\langle SB := *\rangle$ modality. We see that alternation of $d\mathcal{L}$ modalities is quite expressive and can capture interaction of system components.

To find a particular constraint on the choice of *SB*, we need to take the maximum reaction latency ε of the train controllers into account. With $\varepsilon > 0$, the point where the train needs to apply brakes to comply with $\mathbf{m}$ is not determined by the physical constraints ($\mathscr{C}$) alone, but needs additional safety margins to compensate for reaction delays in the controller. Therefore, we search for a constraint that characterises that for every possible end of the movement authority ($\forall \mathbf{m}.e$) and every train position ($\forall \tau.v$), train movement with an acceleration of A preserves ($\mathscr{C}$) if it started in a state where ($\mathscr{C}$) holds and the point *SB* has not been passed yet ($\mathbf{m}.e - \tau.p \geq SB \wedge \mathscr{C}$).

Proposition 7.4 (Reactivity constraint). *If the train is in a controllable state, the supervisory ETCS controller reacts appropriately in order to maintain controllability iff SB is chosen according to the following provable equivalence:*

$$\Big(\forall \mathbf{m}.e \forall \tau.p\,\big(\mathbf{m}.e - \tau.p \geq SB \wedge \mathscr{C} \rightarrow [\tau.a := A;\ drive]\mathscr{C}\big)\Big)$$
$$\equiv\ SB \geq \frac{\tau.v^2 - \mathbf{m}.d^2}{2b} + \left(\frac{A}{b} + 1\right)\left(\frac{A}{2}\varepsilon^2 + \varepsilon\ \tau.v\right). \qquad (\mathscr{B})$$

Reactivity constraint ($\mathscr{B}$) on *SB* can be derived using a projection of the train behaviour to the worst-case acceleration A in a state where *SB* has not been passed yet, similarly to the approach from Sect. 2.9. We choose this projection because the train controller needs to ensure that it can drive safely with maximum acceleration A for ε time units even right before passing *SB* in order for an acceleration choice of A to be safe. Constraint ($\mathscr{B}$) is not at all obvious from the original system model in Fig. 7.3.

$$
\begin{aligned}
ETCS_{\mathrm{r}} &\equiv (train_{\mathrm{r}} \cup rbc_{\mathrm{r}})^* \\
train_{\mathrm{r}} &\equiv spd;\ atp_{\mathrm{r}};\ drive \\
spd &\equiv \big(?(\tau.v \le \mathbf{m}.r);\ \tau.a := *;\ ?(-b \le \tau.a \le A)\big) \\
&\quad \cup \big(?(\tau.v \ge \mathbf{m}.r);\ \tau.a := *;\ ?(-b \le \tau.a \le 0)\big) \\
atp_{\mathrm{r}} &\equiv SB := \tfrac{\tau.v^2 - \mathbf{m}.d^2}{2b} + \left(\tfrac{A}{b}+1\right)\left(\tfrac{A}{2}\varepsilon^2 + \varepsilon\, \tau.v\right);\ atp \\
atp &\equiv \text{if } (\mathbf{m}.e - \tau.p \le SB \lor rbc.message = emergency) \text{ then } \tau.a := -b \text{ fi} \\
drive &\equiv t := 0;\ (\tau.p' = \tau.v \land \tau.v' = \tau.a \land t' = 1 \land \tau.v \ge 0 \land t \le \varepsilon) \\
rbc_{\mathrm{r}} &\equiv (rbc.message := emergency) \\
&\quad \cup \big(\mathbf{m}_0 := \mathbf{m};\ \mathbf{m} := *; \\
&\qquad ?(\mathbf{m}.r \ge 0 \land \mathbf{m}.d \ge 0 \land \mathbf{m}_0.d^2 - \mathbf{m}.d^2 \le 2b(\mathbf{m}.e - \mathbf{m}_0.e))\big)
\end{aligned}
$$

Fig. 7.6 ETCS cooperation protocol refined with parameter constraints

After discovering constraint ($\mathscr{B}$), however, it can be explained in retrospect: It characterises the relative braking distance required to reduce speed from $\tau.v$ to target speed $\mathbf{m}.d$ with braking deceleration b, which corresponds to controllability and is expressed by the term $\frac{\tau.v^2 - \mathbf{m}.d^2}{2b}$. In addition, the constraint involves the distance travelled during one reaction cycle of at most ε time units with acceleration A, including the additional distance needed to reduce the speed down to $\tau.v$ again after accelerating at A for ε time units, as expressed by $\left(\frac{A}{b}+1\right)\left(\frac{A}{2}\varepsilon^2 + \varepsilon\, \tau.v\right)$. This extra distance results from speed changes and depends on the relation $\frac{A}{b}$ of maximum acceleration A and maximum braking force b.

Propositions 7.1–7.4 prove $\mathsf{d}\mathcal{L}$ equivalences. Hence, counterexamples exist that demonstrate unsafety of the ETCS skeleton in Fig. 7.3 whenever the respective parameter constraints are not met. Consequently, these constraints must be obeyed for correctness of *any* model of ETCS controllers, including all technical implementation refinements. It is, thus, important to identify these safety constraints early in the overall design and verification process of ETCS. Similar parameter constraints hold for other upcoming train control systems.

7.3.3 Safety Verification of Refined ETCS

By refining the system from Fig. 7.3 with the parametric constraints obtained from Propositions 7.1–7.4, we synthesise a safe system model completing the ETCS protocol skeleton. In Fig. 7.6, we present the refined model, which bugfixes the model in Fig. 7.3 taken from the informal specification. Parts *spd*, *drive*, and the subprogram *atp* of atp_{r} stay as in Fig. 7.3. The constraints in atp_{r} and rbc_{r} are new, as is the precondition $\mathscr{C}$ on the initial states.

Proposition 7.5 (Safety of ETCS). *Assuming the train starts in a controllable state, the following global and unbounded horizon $\mathsf{d}\mathcal{L}$ safety formula about the refined ETCS system in Fig. 7.6 is provable:*

$$\mathscr{C} \to [ETCS_r](\tau.p \ge \mathbf{m}.e \to \tau.v \le \mathbf{m}.d). \tag{7.3}$$

This provable $d\mathcal{L}$ formula states that, starting in a controllable region ($\mathcal{C}$), the refined ETCS model is safe, i.e., trains always respect their movement authority even when movement authorities change dynamically based on wireless communication with the RBC.

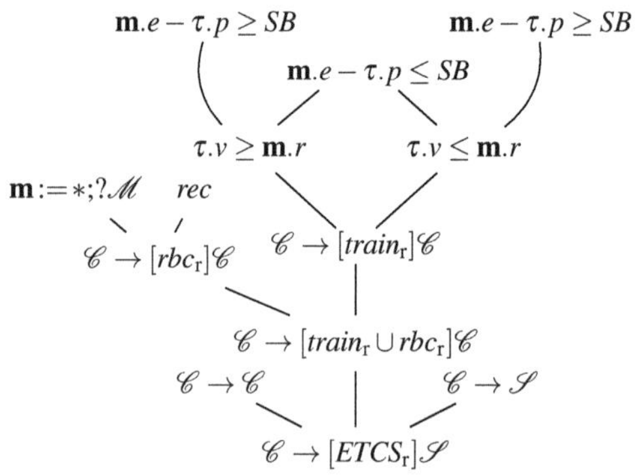

Fig. 7.7 Proof sketch for ETCS safety (Proposition 7.5)

As an example to illustrate the proof structure for the verification of Proposition 7.5, consider the sketch in Fig. 7.7. The proof starts with the conjecture at the bottom and proceeds by decomposition in the $d\mathcal{L}$ calculus to the leafs. Figure 7.7 does not give a full formal proof in the $d\mathcal{L}$ calculus but still shows a sketch of the overall proof structure for Proposition 7.5. We need to prove that the assumption that the train is in a controllable state expressed by ($\mathcal{C}$) implies $[ETCS_r]\mathcal{S}$. As the system consists of a global loop $(train_r \cup rbc_r)^*$, we prove that ($\mathcal{C}$) is an invariant of this loop and strong enough to imply ($\mathcal{S}$) by $d\mathcal{L}$ rule *ind′* from p. 86, which splits into the three lower branches indicated in Fig. 7.7. Using KeYmaera it can be shown easily that the invariant is initially valid (left branch) and implies the postcondition (right branch). As usual, proving that the invariant is preserved by the loop body is the most challenging part of the proof (middle branch), which splits into two cases by applying $d\mathcal{L}$ rule $[\cup]$, and rule $\wedge r$ from Fig. 2.11 subsequently. For the left case, we have to show that the RBC preserves the invariant, which follows from Proposition 7.2 and splits into the case where the RBC changes the MA according to constraint ($\mathcal{M}$) (in the left branch $\mathbf{m} := *$) and the case where an emergency message requests immediate recovery by emergency braking (denoted by *rec* in the right branch). For the right choice, we show that the train controller preserves the invariant. The proof splits again by rule $[\cup]$ due to the choice in the *spd* component depending on the relation of the current speed $\tau.v$ to the recommended speed $\mathbf{m}.r$. The next split with rule $[\cup]$ on both of these branches depends on the value of SB according to *atp*. If the train has passed point SB (the shared middle case), we can close this goal using Proposition 7.1, because the invariant describes a controllable state and the *spd* controller applies safety brakes. The outer branches, where the train has not yet passed SB, can be closed using Proposition 7.4, because the controllers will react to situation changes after at most ε time units (induction). Overall, the $\mathsf{d}\mathcal{L}$ formula in Proposition 7.5 can be proven automatically by our approach.

7.3.4 Liveness Verification of Refined ETCS

In order to show that the discovered parameter constraints do not over-constrain the system inconsistently, we show liveness, i.e., that an ETCS train is able to reach every track position with appropriate RBC permissions.

Proposition 7.6 (Liveness of ETCS). *The refined ETCS system is live. That is, assuming the RBC can safely grant the required MAs because preceding trains are moving on, trains are able to reach any track position P by appropriate RBC choices, which is expressed in the following provable* $\mathsf{d}\mathcal{L}$ *formula:*

$$\tau.v \geq 0 \land A > 0 \land \varepsilon > 0 \rightarrow \forall P \langle ETCS_r \rangle \, \tau.p \geq P. \tag{7.4}$$

This $\mathsf{d}\mathcal{L}$ formula expresses that, starting in a state where the velocity is non-negative and where maximum acceleration and maximum evolution time limit are positive, every point P (that is, $\forall P$) can be reached ($\tau.p \geq P$) by some execution of the ETCS model ($\langle ETCS_{\mathrm{r}} \rangle$). Here the diamond modality is used to say that not all, but *some* appropriate execution reaches a state where the postcondition ($\tau.p \geq P$) holds. The validity of formula (7.4) means that the train model is always able to reach the postcondition *for all* corresponding values of the variables. Mere satisfiability of (7.4) would correspond to a weaker property saying that the train can move only under some circumstance for appropriate positions (e.g., when $\tau.p \geq P$ trivially holds in the beginning), and not under all circumstances for all values of the free variables, as Proposition 7.6 shows.

Further note that, for proving that the ETCS model is live, a more liberal initial state is possible with regard to the controllability of the train. The only important restrictions on the initial region are those ensuring that the system of differential equations used for modelling the train movement can actually be followed for some positive amount of time $\varepsilon > 0$. As usual, the velocity of the train must be nonnegative ($\tau.v \geq 0$). If either of these assumptions is violated, the train may be unable to move, because the evolution domain ($\tau.v \geq 0 \land t \leq \varepsilon$) of the differential equation system ($\mathcal{I}$) is violated immediately and thus no continuous movement of the model would be allowed at all. Finally, the maximum acceleration A must be positive ($A > 0$) or at least the initial velocity must be $\tau.v > 0$; otherwise the train would be locked to its current position without any effective control choices.

The proof of Proposition 7.6 uses unwinding ($\mathsf{d}\mathcal{L}$ rule $\langle{}^{*n}\rangle$) and generalisation (rule $\langle\rangle gen$) to split the proof into one part to show that positive speed $\tau.v > 0$ is always reachable, and another part to show that positive speed can be maintained during sufficiently many control cycles with a non-Zeno progress towards the goal position P. The latter part of the proof can be shown using the convergence rule *con* with variant $\varphi(n) \equiv \tau.p + n\,\varepsilon\, v_0 \geq P \land \tau.v \geq v_0$, which expresses that goal P is within reach of at most n cycles of duration ε at minimum speed v_0 and that the speed $\tau.v$ does not decrease below the speed v_0 reached after the first unwinding cycle.

Contrast this simple variant with the complex variant (2.21) needed for proving the liveness property of the simple train control subsystem considered in property (2.20) on p. 121 for nonnegative acceleration $A \geq 0$. At first sight, this complexity difference sounds somewhat surprising. After all, the system considered in Chap. 2 is a lot simpler than the full ETCS protocol considered here, so the variant and proof in Chap. 2 should be simpler too. While this was perfectly true for safety properties, the situation is slightly different for liveness. In terms of liveness, the system considered in (2.20) had significantly less flexibility compared to the full ETCS system in Fig. 7.6. For safety, flexibility is harder, because we have to prove more cases and combinations to stay safe. For liveness, flexibility is good, because we have more options to choose from for satisfying the goal. The simple train controller in Chap. 2, for instance, allowed only one choice of a movement authority which then remains constant forever. Consequently, our liveness proof had to establish the existence of an appropriate MA with only one single control shot. For the ETCS system in Fig. 7.6, multiple adjustments of the MA are permitted, with which it is a lot easier to prove liveness, because inappropriate prior MAs can simply be readjusted again by RBC communication. Essentially, the liveness proof in Chap. 2 had to plan ahead and prove without any intermediate adaptation, whereas our liveness proof for Proposition 7.6 is able to use dynamic feedback to adapt to the actual movement (dynamically during the proof).

7.3.5 Full Correctness of ETCS

By collecting Propositions 7.1–7.6 together, we obtain the following main result of this chapter, which demonstrates the feasibility of d$\mathcal{L}$-based parametric discovery and formal verification. It gives important insights into the fully parametric ETCS case study and yields conclusive and fully verified choices for the free parameters in ETCS. By virtue of the parametric formulation, this result applies to all concrete instantiations of the ETCS cooperation protocol (Sect. 7.2), including controllers that further optimise speed or model refinements in hardware implementations. These constraints also apply to related train control systems.

Theorem 7.1 (Correctness of ETCS cooperation protocol). *The refined ETCS system with constraints ($\mathcal{B}$) and ($\mathcal{M}$) is correct as given in Fig. 7.6. Starting in any controllable state respecting ($\mathcal{C}$), trains remain in the controllable region at any time. They safely respect movement authorities issued by the RBC so that ETCS is collision-free. Further, trains can always react safely to all RBC decisions respecting ($\mathcal{M}$). ETCS is live: when tracks become free, trains are able to reach any track position by appropriate RBC actions. Furthermore, the refinements ($\mathcal{C}$) and ($\mathcal{B}$) are necessary and sharp: Every configuration violating ($\mathcal{C}$) or ($\mathcal{B}$), respectively, gives rise to a concrete counterexample violating safety property ($\mathcal{S}$). Finally, every RBC* *choice violating* ($\mathscr{M}$) *gives rise to a counterexample in the presence of lossy wireless communication channels.*

7.4 Disturbance and the European Train Control System

In Sects. 7.2 and 7.3, we assumed there was direct control of acceleration in ideal-world physics. In reality, actual acceleration is caused by the physical transmission of corresponding forces that depend on the electrical current in the engine and are regulated by PID controllers [91]. Yet even a full model of the physical transmission system is still somewhat incomplete, because wind and friction, slippery rail conditions, and turbulence may have some small impact on the dynamics.[2] As a conservative overapproximation of all these effects, we instead generalise the ETCS model to a model with differential inequalities, where we also take into account disturbances in the physical transmission of forces:

$$\tau.p' = \tau.v \wedge \tau.a - l \leq \tau.v' \leq \tau.a + u \wedge t' = 1 \wedge \tau.v \geq 0 \wedge t \leq \varepsilon. \qquad (\mathscr{I}_d)$$

This differential inequality tolerates a disturbance within the interval $[-l, u]$. That is, the acceleration $\tau.a$ chosen by the train controller may take effect with an error bounded by $-l$ and u. Especially, the derivative $\tau.v'$ of the velocity will not need to equal $\tau.a$ exactly in ($\mathscr{I}_d$), but $\tau.v'$ can vary arbitrarily between $\tau.a - l$ and $\tau.a + u$ over time. We thus generalise the differential equation ($\mathscr{I}$) in component *train* from Figs. 7.3 and 7.6 by replacing it with the differential inequality ($\mathscr{I}_d$) and denote the result by $train_d$. For the precise semantics of hybrid systems with differential inequalities, we refer to the semantics of DA-programs with DA-constraints for continuous evolution given in Chap. 3; see Examples 3.3 and 3.20 on pp. 137 and 176, respectively. Especially, because of the differential inequalities, the resulting programs are DA-programs and the resulting formulas are, strictly speaking, DAL formulas. See Fig. 7.8 for an illustration of how the original controllability region can be violated, because the disturbed dynamics can deviate from the ideal-world dynamics with a time-dependent, bounded disturbance.

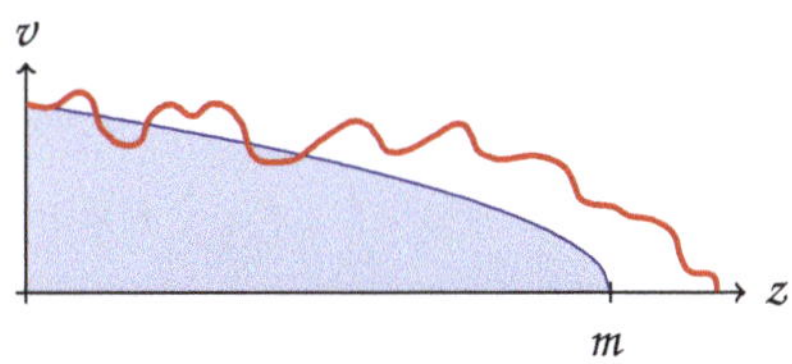

Fig. 7.8 Controllability region changes in the presence of disturbance

[2] Obtaining a fully complete model is thus de facto impossible, probably not even when resolving to the Schrödinger equation of quantum mechanics [276].

Notice that, unlike with the differential equation ($\mathscr{I}$), we cannot simply solve differential inequality ($\mathscr{I}_d$), because its actual solution depends on the precise value of the disturbance, which is a quantity that changes over time. Thus, solutions will only be relative to this disturbance function and a reachability analysis would have to consider all choices of this function, which would require higher-order logic. Instead, we use differential invariants from Chap. 3 as a first-order characterisation for the proofs behind the following propositions.

7.4.1 Controllability in ETCS with Disturbances

The controllability characterisation from Proposition 7.1 carries over to train control with disturbance when taking into account the effect that a maximum disturbance u can have on the maximum braking force b and limits the effective braking force to $(b-u)$:

Proposition 7.7 (Controllability despite disturbance). *The constraint*

$$\tau.v^2 - \mathbf{m}.d^2 \leq 2(b-u)(\mathbf{m}.e - \tau.p) \wedge \mathbf{m}.d \geq 0 \wedge b > u \geq 0 \wedge l \geq 0 \qquad (\mathscr{C}_d)$$

is a controllability constraint *with respect to property ($\mathscr{S}$) for the train τ with disturbance ($\mathscr{I}_d$), i.e., it retains the ability of the train dynamics to respect the safety property despite disturbance. Formally, with $\mathscr{A} \wedge \tau.p \leq \mathbf{m}.e \wedge b > u \geq 0 \wedge l \geq 0$ as regularity assumptions, the following equivalence is provable in the* DAL *calculus:*

$$\begin{aligned}&[\tau.p' = \tau.v \wedge -b-l \leq \tau.v' \leq -b+u \wedge t' = 1 \wedge \tau.v \geq 0 \wedge t \leq \varepsilon]\,\mathscr{S}\\ \equiv\, &\tau.v^2 - \mathbf{m}.d^2 \leq 2(b-u)(\mathbf{m}.e - \tau.p).\end{aligned}$$

Here ($\mathscr{C}_d$) results from ($\mathscr{C}$) by replacing b with $(b-u)$. In worst-case disturbance, the train cannot brake with its chosen deceleration $-b$ but might be off by a disturbance up to u. In order to guarantee that the train is still always able to stay within its MA despite the disturbance, the controller has to assume worst-case guaranteed deceleration $-(b-u)$ when making control decisions.

Since the proof of Proposition 7.7 follows interesting principles, we present a formal DAL proof for Proposition 7.7 in Fig. 7.9, using the DAL calculus from Fig. 3.9. The proof uses the following abbreviations:

$$\begin{aligned}X \approx -b &\equiv -b-l \leq X \wedge X \leq -b+u\\ \psi &\equiv \mathscr{A} \wedge \tau.p \leq \mathbf{m}.e \wedge b > u \geq 0 \wedge l \geq 0\\ \phi &\equiv \tau.v^2 - \mathbf{m}.d^2 \leq 2(b-u)(\mathbf{m}.e - \tau.p)\\ \langle \mathscr{S}_t \rangle &\equiv \langle \tau.p := \frac{-b+u}{2}t^2 + \tau.v\,t + \tau.p \wedge \tau.v := (-b+u)t + \tau.v \rangle.\end{aligned}$$

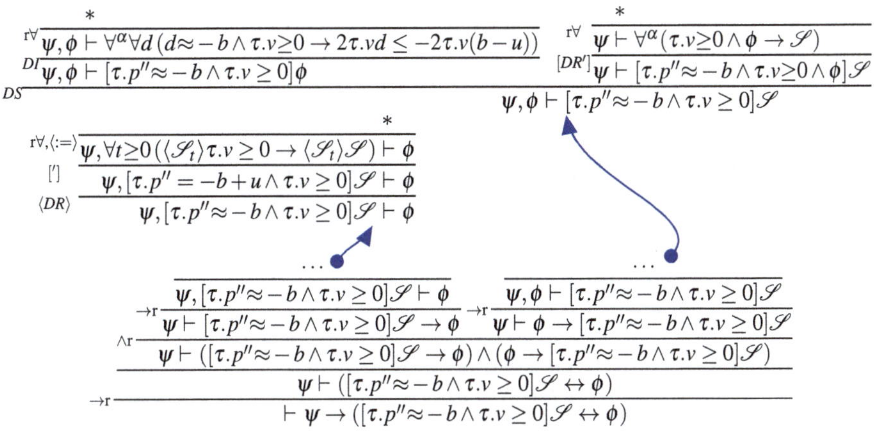

Fig. 7.9 Proof of ETCS controllability despite disturbance (Proposition 7.7)

The bottommost proof block in Fig. 7.9 splits the proof into branches for proving the two directions of the equivalence: the direction for proving necessity (middle proof block) and the direction for proving sufficiency (top proof block).

In the top block, sufficiency can be proven by differential strengthening (DAL rule *DS* from Fig. 3.9), with the controllability constraint $\mathscr{C}_d$ (or its relevant part ϕ) as an auxiliary invariant. On the left branch of the top block, validity of the differential strengthening step can be proven by differential induction rule *DI* from Fig. 3.9 to show that ϕ is actually an invariant, using the differential inequality reduction techniques from Sect. 3.5.3. On the right branch of the top block, differential weakening rule $[DR']$ from p. 175 can be used to conclude $\mathscr{S}$ from the auxiliary invariant ϕ and the evolution domain restriction $\tau.v \geq 0$. Also see Example 3.20 on p. 176 for details on this kind of reasoning.

In the middle block, necessity can be proven by differential refinement (DAL rule $\langle DR \rangle$ from Fig. 3.9). We use differential refinement to replace the differential inequality ($\mathscr{I}_d$)—which we abbreviate as $\tau.p'' \approx -b$ in Fig. 7.9—with the differential equation $\tau.p'' = -b+u$. This differential equation represents the worst case with constant maximum disturbance u in the differential inequality ($\mathscr{I}_d$). Intuitively, this replacement is possible, because the differential equation describes one possible special case of the differential inequality and we only need to find one permitted evolution that implies ϕ for proving necessity. Formally, the differential refinement is justified by proof rule $\langle DR \rangle$ and the fact that

$$\tau.p' = \tau.v \wedge \tau.v' = -b+u \wedge \tau.v \geq 0$$

entails the following differential inequality (in the sense of Lemma 3.3 from p. 158):

$$\tau.p' = \tau.v \wedge -b-l \leq \tau.v' \leq -b+u \wedge \tau.v \geq 0.$$

Note that DAL rule $\langle DR\rangle$ can be applied to the $[]$-formulas in the antecedent, because these are equivalent to $\langle\rangle$-formulas in the succedent.

7.4.2 Iterative Control Refinement of Parameters with Disturbances

When taking into account worst-case effects of disturbance on control, reactivity constraint ($\mathscr{B}$) carries over to the presence of disturbance in the train dynamics.

Proposition 7.8 (Reactivity constraint despite disturbance). *For trains in controllable state, the supervisory ETCS controller reacts appropriately despite disturbance in order to maintain controllability iff SB is chosen according to the following provable equivalence:*

$$\Big(\forall \mathbf{m}.e \forall \tau.p\,\big((\mathbf{m}.e - \tau.p \geq SB \land \tau.v^2 - \mathbf{m}.d^2 \leq 2(b-u)(\mathbf{m}.e - \tau.p)\big) \rightarrow$$
$$[\tau.a := A;\ drive_d](\tau.v^2 - \mathbf{m}.d^2 \leq 2(b-u)(\mathbf{m}.e - \tau.p))\big)\Big)$$
$$\equiv SB \geq \frac{\tau.v^2 - \mathbf{m}.d^2}{2(b-u)} + \left(\frac{A+u}{b-u} + 1\right)\left(\frac{A+u}{2}\varepsilon^2 + \varepsilon\tau.v\right). \qquad (\mathscr{B}_d)$$

For reactivity constraint ($\mathscr{B}_d$), not only the maximum deceleration but also the maximum acceleration matters—both could be affected by disturbances. Therefore, we need to substitute not only every b with $(b-u)$ but also every A with $(A+u)$, the maximum acceleration under disturbance, to get a (provable) reactivity constraint for the disturbed system.

Accordingly, we adapt the RBC constraint ($\mathscr{M}$) to account for the maximum effect of disturbance by again replacing every b with $(b-u)$:

$$\mathbf{m}_0.d^2 - \mathbf{m}.d^2 \leq 2b(\mathbf{m}.e - \mathbf{m}_0.e) \land \mathbf{m}_0.d \geq 0 \land \mathbf{m}.d \geq 0. \qquad (\mathscr{M}_d)$$

Constraint ($\mathscr{M}_d$) is fairly easy to obtain from ($\mathscr{M}$), because the RBC itself is not subject to disturbance, and only the train is.

7.4.3 Safety Verification of ETCS with Disturbances

When we refine the ETCS model with the constraints ($\mathscr{B}_d$) and ($\mathscr{M}_d$), ETCS is safe even in the presence of disturbance in the dynamics when starting in a state respecting ($\mathscr{C}_d$).

Proposition 7.9 (Safety despite disturbance). *Assuming the train starts in a controllable state satisfying ($\mathscr{C}_d$), the following global and unbounded horizon safety* *formula about the ETCS system from Fig. 7.10 with disturbance in the dynamics is provable:*

$$\mathscr{C}_d \to [ETCS_d](\tau.p \geq \mathbf{m}.e \to \tau.v \leq \mathbf{m}.d).$$

$$\begin{aligned}
ETCS_d &\equiv (train_d \cup rbc_d)^* \\
train_d &\equiv spd;\ atp_d;\ drive_d \\
spd &\equiv \left(?(\tau.v \leq \mathbf{m}.r);\ \tau.a := *;\ ?(-b \leq \tau.a \leq A)\right) \\
&\quad \cup \left(?(\tau.v \geq \mathbf{m}.r);\ \tau.a := *;\ ?(0 > \tau.a \geq -b)\right) \\
atp_d &\equiv SB := \tfrac{\tau.v^2 - \mathbf{m}.d^2}{2(b-u)} + \left(\tfrac{A+u}{b-u} + 1\right)\left(\tfrac{A+u}{2}\varepsilon^2 + \varepsilon\tau.v\right); \\
&\quad \text{if } (\mathbf{m}.e - \tau.p \leq SB \vee rbc.message = emergency) \text{ then } \tau.a := -b \text{ fi} \\
drive_d &\equiv t := 0;\ (\tau.p' = \tau.v \wedge \tau.a - l \leq \tau.v' \leq \tau.a + u \wedge t' = 1 \wedge \tau.v \geq 0 \wedge t \leq \varepsilon) \\
rbc_d &\equiv (rbc.message := emergency) \\
&\quad \cup \big(\mathbf{m}_0 := \mathbf{m}; \mathbf{m} := *; \\
&\quad\ ?(\mathbf{m}.r \geq 0 \wedge \mathbf{m}.d \geq 0 \wedge \mathbf{m}_0.d^2 - \mathbf{m}.d^2 \leq 2(b-u)(\mathbf{m}.e - \mathbf{m}_0.e))\big)
\end{aligned}$$

Fig. 7.10 Parametric ETCS cooperation protocol with disturbances

The proof sketched in Fig. 7.7 for undisturbed ETCS generalises to ETCS with disturbance when using differential induction (DAL proof rule *DI* from Fig. 3.9) and differential strengthening (*DS*) with a time-dependent version of ($\mathscr{B}_d$) as a differential invariant for the acceleration case:

$$\mathbf{m}.e - \tau.p \geq \frac{\tau.v^2 - \mathbf{m}.d^2}{2(b-u)} + \left(\frac{A+u}{b-u} + 1\right)\left(\frac{A+u}{2}(\varepsilon - t)^2 + (\varepsilon - t)\tau.v\right). \quad (7.5)$$

We also use differential strengthening by this differential invariant in the proof of Proposition 7.8. The full ETCS system with disturbance, where all abbreviations are resolved, is depicted in Fig. 7.11.

7.5 Experimental Results

Experimental results for verifying ETCS in our d$\mathscr{L}$- and DAL-based verification tool KeYmaera are presented in Table 7.1. Experimental results are from a 2.6 GHz AMD Opteron with 4 GB memory. All those correctness properties and parameter constraints of ETCS can be verified with 94% to 100% degree automation. In the universal fragment of d$\mathscr{L}$, user interactions are only needed for supplying invariants (or differential invariants), which, in turn, can be discovered using our iterative refinement process and our *differential invariants as fixed-points* algorithm from Chap. 6. For liveness properties or substantial quantifier alternations beyond the capabilities of currently available decision procedures for real arithmetic, KeYmaera still needs more user guidance. Yet, those properties can nevertheless be verified formally with KeYmaera! Further, we see that the symbolic state dimension (Dim) has more impact on the computational complexity than the number of proof steps in d$\mathscr{L}$ decompositions (Steps). Table 7.1 gives the number of user interactions in the

$$\psi \equiv \tau.v^2 - \mathbf{m}.d^2 \le 2(b-u)(\mathbf{m}.e - \tau.p) \land \mathbf{m}.d \ge 0 \land b > u \ge 0 \land l \ge 0 \to [ETCS_{\mathrm{d}}](\tau.p \ge \mathbf{m}.e \to \tau.v \le \mathbf{m}.d)$$

$$ETCS_{\mathrm{d}} \equiv \Big($$

$$\begin{aligned}
spd:\quad & \Big(((?(\tau.v \le \mathbf{m}.r);\ \tau.a := *;\ ?(-b \le \tau.a \le A)) \\
& \cup (?(\tau.v \ge \mathbf{m}.r);\ \tau.a := *;\ ?(0 > \tau.a \ge -b))); \\
atp_d:\quad & SB := \tfrac{\tau.v^2 - \mathbf{m}.d^2}{2(b-u)} + \left(\tfrac{A+u}{b-u} + 1\right)\left(\tfrac{A+u}{2}\varepsilon^2 + \varepsilon\tau.v\right); \\
& \text{if } (\mathbf{m}.e - \tau.p \le SB \lor rbc.message = emergency) \text{ then } \tau.a := -b \text{ fi}; \\
drive_d:\quad & t := 0;\ (\tau.p' = \tau.v \land \tau.a - l \le \tau.v' \le \tau.a + u \land t' = 1 \land \tau.v \ge 0 \land t \le \varepsilon)\Big) \\
& \cup \\
rbc:\quad & \Big((rbc.message := emergency) \\
& \cup \big(\mathbf{m}_0.d := \mathbf{m}.d;\ \mathbf{m}_0.e := \mathbf{m}.e;\ \mathbf{m}_0.r := \mathbf{m}.r; \\
& \quad \mathbf{m}.d := *;\ \mathbf{m}.e := *;\ \mathbf{m}.r := *; \\
& \quad ?(\mathbf{m}.r \ge 0 \land \mathbf{m}.d \ge 0 \land \mathbf{m}_0.d^2 - \mathbf{m}.d^2 \le 2(b-u)(\mathbf{m}.e - \mathbf{m}_0.e))\big)\Big)\Big)^*
\end{aligned}$$

Fig. 7.11 Parametric ETCS cooperation protocol with disturbances (full instantiation)

column Int, and, for comparison, the total number of applied proof rules in column Steps. The interactions in Propositions 7.8 and 7.9 are for the time-dependent version (7.5) of the constraint on *SB*.

The experimental results in Table 7.1 for Proposition 7.3 can be improved significantly. Most of the user interactions are only required to overcome the current limitations of the preliminary implementation of iterative inflation order proof strategies from Sect. 5.5 in KeYmaera. Finally, contrast our overall experimental results with earlier implementations [256] that required as much as 56 user interactions even for property Proposition 7.5 to be provable, which KeYmaera can prove completely automatically using our new automated theorem proving techniques for differential dynamic logics and our logic-based verification algorithms from Part II.

Table 7.1 Experimental results for the European Train Control System

Case study		Int	Time(s)	Memory(MB)	Steps	Dim
Controllability	Proposition 7.1	0	0.6	6.9	14	5
RBC Control	Proposition 7.2 property (7.1)	0	0.5	6.4	42	12
RBC Control	Proposition 7.2 property (7.2)	0	0.9	6.5	82	12
Reactivity	Proposition 7.3	13	279.1	98.3	265	14
Reactivity	Proposition 7.4	0	103.9	61.7	47	14
Safety	Proposition 7.5	0	2052.4	204.3	153	14
Liveness	Proposition 7.6 kernel	4	35.2	92.2	62	10
Liveness	Proposition 7.6 simplified	6	9.6	23.5	134	13
Controllability	Proposition 7.7 disturbance	0	2.8	8.3	26	7
Reactivity	Proposition 7.8 disturbance	1	23.7	47.6	76	15
Safety	Proposition 7.9 disturbance	1	5805.2	34	218	16

7.6 Summary

As a case study for parametric verification of hybrid systems, we have verified controllability, reactivity, safety, and liveness of the fully parametric cooperation protocol of the European Train Control System. We have demonstrated the feasibility of $\mathsf{d}\mathcal{L}$- and DAL-based verification of parametric hybrid systems and identified parametric constraints that are both sufficient and necessary for a safe collision-free operation of ETCS. We have characterised these constraints on the free parameters of ETCS equivalently in terms of corresponding reachability properties of the underlying train dynamics. We have proven that the system remains correct even when the train dynamics is subject to disturbances caused, e.g., by the physical transmission, friction, or wind. We have also verified a corresponding fully parametric proportional-integral (PI) controller for ETCS [245] using KeYmaera.

We have shown how the properties of train control can be expressed in $\mathsf{d}\mathcal{L}$. We have proven all propositions formally in our logic-based verification tool KeYmaera. Our experimental results with KeYmaera show a scalable approach by combining the power of completely automatic verification procedures with the intuition behind user guidance to tackle even highly parametric hybrid systems and properties with substantial quantifier alternation (reactivity or liveness) or disturbance; also see our report [245] for details on the experiments and KeYmaera proofs.

For future work, we want to generalise ETCS using probabilistic information about sensor and communication accuracy and availability.

Chapter 8
Air Traffic Collision Avoidance

Contents

8.1	Introduction		**304**
	8.1.1	Related Work	307
	8.1.2	Structure of This Chapter	308
8.2	Curved Flight in Roundabout Manoeuvres		**309**
	8.2.1	Flight Dynamics	309
	8.2.2	Roundabout Manoeuvre Overview	310
	8.2.3	Compositional Verification Plan	311
	8.2.4	Tangential Roundabout Manoeuvre Cycles	312
	8.2.5	Bounded Control Choices	315
	8.2.6	Flyable Entry Procedures	315
	8.2.7	Bounded Entry Duration	318
	8.2.8	Safe Entry Separation	319
8.3	Synchronisation of Roundabout Manoeuvres		**322**
	8.3.1	Successful Negotiation	322
	8.3.2	Safe Exit Separation	326
8.4	Compositional Verification		**328**
8.5	Flyable Tangential Roundabout Manoeuvre		**329**
8.6	Experimental Results		**331**
8.7	Summary		**333**

Synopsis Aircraft collision avoidance manoeuvres are important and complex applications. Curved flight exhibits nontrivial continuous behaviour. In combination with the control choices during air traffic manoeuvres, this results in hybrid systems with challenging interactions of discrete and continuous dynamics. As a case study for demonstrating the scalability of logical analysis for hybrid systems with challenging dynamics, we analyse collision freedom of roundabout manoeuvres in air traffic control, where appropriate curved flight, good timing, and compatible manoeuvring are crucial for guaranteeing safe spatial separation of aircraft throughout their flight. We show that our DAL-based proof techniques can scale to curved flight manoeuvres required in aircraft control applications. Our logical analysis approach can be used successfully to verify collision avoidance of the tangential roundabout manoeuvre automatically, even for five aircraft. Moreover, we introduce a fully flyable variant of the roundabout collision avoidance manoeuvre and verify safety properties by compositional verification in our calculus.

A. Platzer, Logical Analysis of Hybrid Systems,
DOI 10.1007/978-3-642-14509-4_8,

8.1 Introduction

In air traffic control, collision avoidance manoeuvres [293, 104, 129, 171] are used to resolve conflicting flight paths that arise during free flight. See Fig. 8.1 for a series of increasingly more realistic—yet also more complicated—aircraft collision avoidance manoeuvres. Figures 8.1a and b show successful collision avoidance manoeuvres. Figure 8.1c shows a malfunctioning collision avoidance attempt, and Fig. 8.1d shows an improved manoeuvre. Collision avoidance manoeuvres are a "last resort" for resolving air traffic conflicts that could lead to collisions. They become important whenever flight route conflicts have not been detected by the pilots during free flight or by the flight directors of the Air Route Traffic Control Centres. Consequently, complicated online trajectory prediction, manoeuvre planning, or negotiation may no longer be feasible in the short time that remains for resolving the conflict. In the tragic 2002 mid-flight collision in Überlingen [43], the aircraft collided tens of seconds after the on-board traffic alert and collision avoidance system TCAS [196] signalled a traffic alert. Thus, for safe aircraft control we need particularly reliable reactions with manoeuvres whose correctness has been established previously by a thorough offline analysis. To ensure correct functioning of aircraft collision avoidance manoeuvres under all circumstances, the temporal evolution of the aircraft in space must be analysed carefully together with the effects that manoeuvring control decisions have on their dynamics. This results in complicated superpositions of physical system dynamics with control, which can be modelled naturally as hybrid systems.

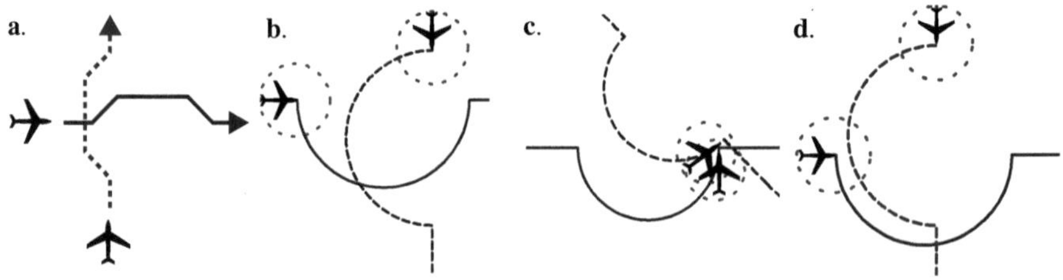

Fig. 8.1 Evolution of collision avoidance manoeuvres in air traffic control

Several numerical [293, 180, 46, 168, 171] or optimisation-based [180, 46, 167, 171] approaches have been proposed for air traffic control. It is difficult to give sound formal verification results for these approaches due to errors in numerical computations or the implicit definition of manoeuvres in terms of complicated optimisation processes. Formal verification is important to avoid collisions and prevent malfunctioning collision avoidance attempts like that in Fig. 8.1c. Formal verification results have been given by geometrical reasoning [104, 129, 295, 296] in PVS. Yet, it still remains to be proven by other techniques that the hybrid dynamics of a flight controller actually follows the geometrical shapes. In contrast, here we verify the hybrid system dynamics directly using a formally sound approach (assuming sound elementary decision procedures), consider curved flight, and achieve better automation.

Control Challenges

Because of the complicated spatio-temporal movement of aircraft, their manoeuvres are challenging for verification. Unlike in ground transportation, braking and waiting is not an option to resolve conflicts, because, in contrast to quadrotors [164], fixed-wing aircraft drop out of the sky if they do not move fast enough for sufficient lift. Consequently, aircraft manoeuvres have to be coordinated such that the aircraft always respect minimal and maximal lateral and angular speed constraints yet always remain safely separated. Further, angular velocity for curving is the primary means of control, because changes in thrust and linear speed are less efficient for aircraft.

Technical Challenges

Complexities in analysing aircraft manoeuvres manifest most prominently in difficulties with analysing hybrid systems for flight equations. General solutions of flight equations involve trigonometric functions that depend on the angular velocity ω and the orientation of the aircraft in space. For straight-line flight (angular velocity $\omega = 0$), the movement in space is just linear, so classical analysis techniques can be used [156]. These include pure straight-line manoeuvres [293, 203, 104, 129, 171]; see, e.g., Fig. 8.1a. They have to assume instant turns for heading changes of the aircraft between multiple straight line segments; see, e.g., Fig. 8.2. Instant turns, however, are impossible in mid-flight, because they are *not flyable*: Aircraft cannot suddenly change their flight direction from 0 to 45 degrees discontinuously. They need to follow a smooth curve instead, in which they slowly steer towards the desired direction by adjusting the angular velocity ω appropriately. Moreover, the

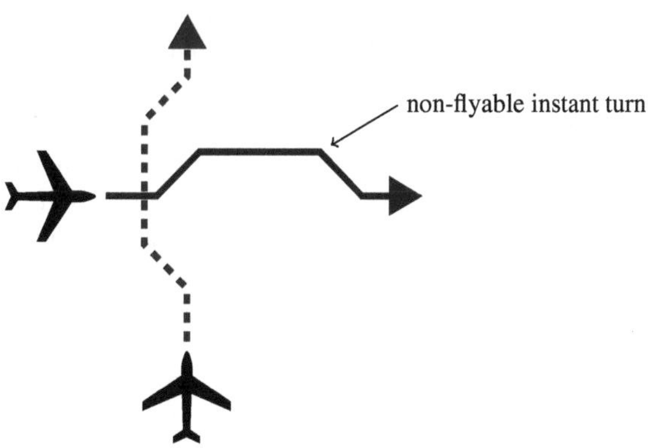

Fig. 8.2 Non-flyable straight-line manoeuvre with instant turns

area of the airspace required by manoeuvres for which instant turns could possibly be understood as adequately close approximations of properly curved flight is prohibitively large. Curved flight is thus an inherent part of real aircraft control. All the manoeuvres in Fig. 8.1 still include some instant turns and are, thus, not practical.

During curved flight, the angular velocity ω is nonzero. For $\omega \neq 0$, however, flight equations have transcendental solutions, which generally fall into undecidable classes of arithmetic. Consequently, manoeuvres with curves, as in Figs. 8.1b and d, are more realistic but also substantially more complicated for verification than straight-line manoeuvres such as that in Fig. 8.1a. In Chap. 6 we have developed a *sound* verification algorithm that works with differential invariants from Chap. 3 instead of with solutions of differential equations to address this arithmetic. Now we show how a fully curved manoeuvre can be verified, i.e., a flyable manoeuvre that only consists of curves and straight lines but *no* instant turns.

In this chapter, we introduce and verify the *fully flyable tangential roundabout manoeuvre (FTRM)*. It refines the non-flyable tangential roundabout manoeuvre (NTRM) from Fig. 8.1d, which still has discontinuities at the entry and exit points of roundabouts, to a fully flyable curved manoeuvre. Unlike most previously proposed manoeuvres [293, 46, 203, 104, 92, 129, 171], FTRM does not have any non-flyable instant turns. It is flyable and smoothly curved. Unlike the authors of other approaches emphasising the importance of flyability [180], we give formal verification results.

As a case study, we show how safety properties of collision avoidance manoeuvres in air traffic management can be verified with the DAL proof calculus from Chap. 3 using our verification algorithm from Chap. 6.

Contribution

Our main contribution is to show that reality in model design and coverage in formal verification are no longer incompatible desires even for applications as complex as aircraft manoeuvres. As a case study illustrating the use of the differential dynamic logics for hybrid systems from Part I of this book, we demonstrate how tricky and nonlinear dynamics can be verified with our verification algorithm from Chap. 6. We introduce a fully curved flight manoeuvre and verify its hybrid dynamics formally. In contrast to previous approaches, we handle curved flight and hybrid dynamics, and produce formal proofs with almost complete automation. Manual effort is still needed to simplify arithmetical complexity and modularise the proof appropriately. We further illustrate the resulting verification conditions for the respective parts of the manoeuvre. Finally, we identify the most difficult steps during the verification and present new transformations to handle the enormous computational complexity. To reduce complexity, we still use some of the simplifications assumed in related work, e.g., synchronous manoeuvring (i.e., aircraft make simultaneous, symmetric manoeuvre choices).

8.1.1 Related Work

Lafferriere et al. [189] gave important decidability results for hybrid systems with some classes of linear continuous dynamics but only random discrete resets. These results do not apply to air traffic manoeuvres, because they have non-trivial resets: the aircraft's position does not just jump randomly when switching modes but, rather, systematically according to the manoeuvre.

Tomlin et al. [293] analyse competitive aircraft manoeuvres game-theoretically using numerical approximations of partial differential equations. As a solution, they propose roundabout manoeuvres (Fig. 8.1b) and give bounded-time verification results for straight-line approximations (Fig. 8.1a). We verify actual curved roundabout manoeuvres with up to 28 variables and use a sound symbolic approach that avoids numerical approximation errors. Our symbolic techniques avoid state space discretisations that are exponential in the number of variables as required for partial differential equations. This enables our logical analysis techniques to scale better and to handle more complicated curved flight dynamics.

Flyability has been identified as one of the major challenges in Košecká et al. [180], where planning based on superposition of potential fields has been used to resolve air traffic conflicts. This planning does not guarantee flyability but, rather, defaults to classical vertical altitude changes whenever a non-flyable path is detected. The resulting manoeuvre has not yet been verified. The planning approach has been pursued further by Bicchi and Pallottino [46] with numerical simulations in two scenarios.

Numerical simulation algorithms approximating discrete-time Markov chain approximations of aircraft behaviour have been proposed by Hu et al. [168]. They approximate bounded-time probabilistic reachable sets for one initial state. We consider hybrid systems combining discrete control choices and continuous dynamics instead of uncontrolled, probabilistic continuous dynamics.

Hwang et al. [171] have presented a straight-line aircraft conflict avoidance manoeuvre that involves optimisation over complicated trigonometric computations, and have validated it using random numerical simulation and informal arguments. They also show examples where the decisions of the manoeuvre change only slightly for small perturbations.

The works of Dowek et al. [104] and Galdino et al. [129] are probably closest to ours. They consider straight-line manoeuvres and formalise geometrical proofs in PVS. As in the work of Hwang et al. [171], they do not, however, consider curved flight paths nor verify actual hybrid dynamics but work with geometrical meta-level reasoning instead.

A few attempts [203, 92] have been undertaken to model check discretisations of roundabout manoeuvres, which indicate avoidance of orthogonal collisions (Fig. 8.1b). However, counterexamples found by our model checker in previous work show that collision avoidance does not extend to other initial flight paths of the classical roundabout; see Fig. 8.1c.

Pallottino et al. [222] have presented a spatially distributed pattern for multiple roundabout circles at different positions. They reason manually about desirable properties of the system and estimate probabilistic results in [168]. Pallottino et al. thus take a view that is complementary to ours: they determine the global compatibility of multiple roundabouts while assuming correct functioning within each local roundabout. We verify that the actual hybrid dynamics of each local roundabout is collision-free. Generalising our approach to a spatial pattern of verified local roundabouts would be interesting future work.

Similarly, the work by Umeno and Lynch [296, 295] is complementary to ours. They consider real-time properties of airport protocols using Timed I/O Automata. We are interested in proving properties of the actual hybrid system dynamics.

Our approach has a very different focus than other complementary work:

- Our manoeuvre directly involves curved flight, unlike [293, 168, 104, 129, 171, 296, 295]. This makes our manoeuvre more realistic but also much more difficult to analyse.
- Unlike [180, 168, 171], we do not give results for a finite (sometimes small) number of initial flight positions in bounded-time horizons (simulation). Instead, we verify correct functioning for uncountably many initial states and give unbounded-time horizon verification results.
- Unlike [293, 180, 46, 168, 167, 171], we use symbolic instead of numerical computation so that numerical and floating-point errors cannot cause soundness problems.
- Unlike [46, 203, 168, 104, 129, 171, 296, 295], we directly analyse the hybrid system dynamics and do not use discrete or geometrical approximations that can be difficult to relate to the actual dynamics.
- Unlike [180, 293, 46, 168, 171, 203, 222] we produce formal, deductive proofs. Further unlike the formal proofs in [104, 129, 296, 295], our verification is much more automatic but also covers a different level of models.
- In [104, 129, 171, 296, 295], it remains to be proven that the hybrid dynamics and flight equations really follow geometrical intuition. Yet geometrical intuition can sometimes be quite misleading for more complicated cases such as the classical roundabout manoeuvre in Fig. 8.1b that looks promising, but fails for the situation in Fig. 8.1c. In contrast, our approach directly works for the hybrid flight dynamics. We illustrate verification results graphically to help understand them, but the figures do not prove anything.
- Unlike [216], we consider collision avoidance manoeuvres, and not just detection.
- Unlike [46, 167], we do not guarantee optimality of the resulting manoeuvre. Instead, we verify safety properties.

8.1.2 Structure of This Chapter

We verify safety-critical properties of curved flight in roundabout manoeuvres in Sect. 8.2. We analyse synchronisation properties of roundabout manoeuvres for multiple aircraft in Sect. 8.3. In Sect. 8.5, we combine the previous results to a safety theorem for fully flyable tangential roundabout manoeuvres following a compositional verification approach in our proof calculi according to Sect. 8.4. We present experimental results in Sect. 8.6 and conclude in Sect. 8.7.

8.2 Curved Flight in Roundabout Manoeuvres

In this section, we introduce and verify the fully flyable roundabout manoeuvre that is depicted in Fig. 8.3. It refines the tangential roundabout manoeuvre from Fig. 8.1d (which we considered in Chap. 3, but which still has discontinuities at the entry and exit points of roundabouts) to a fully flyable curved manoeuvre. Unlike the manoeuvres in Fig. 8.1, the resulting manoeuvre does not contain any instant turns. All of its curves are sufficiently smooth to be flyable.

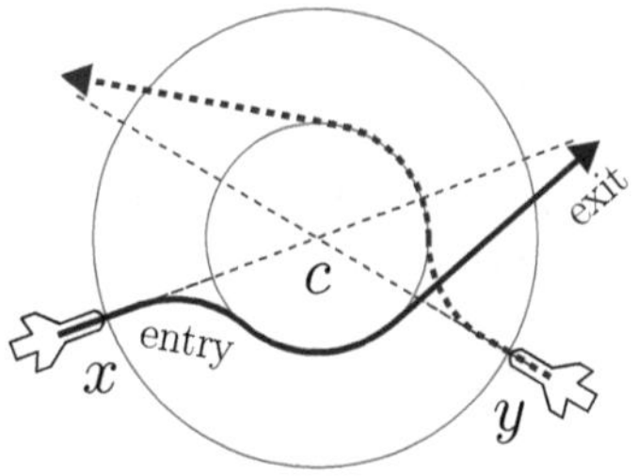

Fig. 8.3 Flyable aircraft roundabout

8.2.1 Flight Dynamics

The parameters of two aircraft at (planar) position $x = (x_1, x_2)$ and $y = (y_1, y_2)$ in $\mathbb{R}^2$ flying in directions $d = (d_1, d_2) \in \mathbb{R}^2$ and $e = (e_1, e_2)$ are illustrated in Fig. 8.4. Their dynamics is determined by their angular speeds $\omega, \varpi \in \mathbb{R}$ and linear velocity

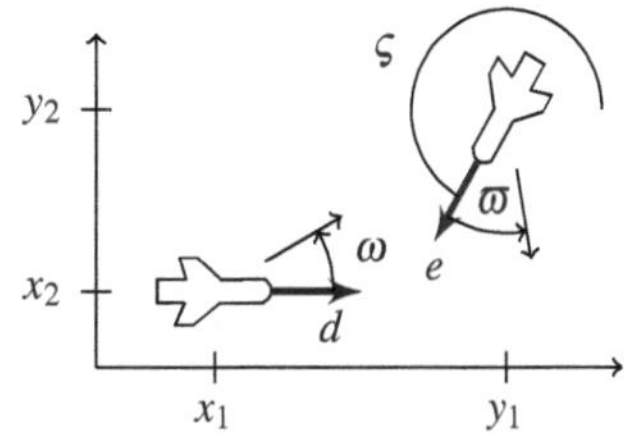

Fig. 8.4 Flight dynamics

vectors d and e, which describe both the linear velocity $\|d\| := \sqrt{d_1^2 + d_2^2}$ and the orientation of the aircraft in space. Roundabout manoeuvres are horizontal collision avoidance manoeuvres, so as in [293, 203, 167, 92, 222, 129, 171], we simplify them to planar positions. We denote the flight equations for aircraft at x with angular velocity ω by $\mathscr{F}(\omega)$ and the flight equations for aircraft at y with angular velocity ϖ by $\mathscr{G}(\varpi)$:

$$[x' = d \quad d' = \omega d^{\perp}] \tag{$\mathscr{F}(\omega)$}$$

$$[y' = e \quad e' = \varpi e^{\perp}] \tag{$\mathscr{G}(\varpi)$}$$

This differential equation system is equivalent to the system ($\mathscr{F}(\omega)$) from p. 150 that has been derived from the literature [293]. Its difference with the differential equations from Chap. 3 is that we use vectorial notation in this chapter. The vector $d^{\perp} := (-d_2, d_1)$ is the *orthogonal complement* of vector d. Differential equations $\mathscr{F}(\omega)$ express that position x is moving in direction d, which is rotating with angular velocity ω, i.e., evolves orthogonal to d. Equations $\mathscr{G}(\varpi)$ are similar for y, e and ϖ of the second aircraft. The complexity of differential equations $\mathscr{F}(\omega)$ comes from the fact that, for curved flight with $\omega \neq 0$, their solutions involve trigonometric functions that fall into undecidable classes of arithmetic. See Example B.4 on p. 362 in App. B for an explanation of the solutions of these differential equations.

In safe flight configurations, aircraft respect protected zone p. That is, they are separated by at least distance $p \geq 0$, i.e., the state satisfies formula $\mathscr{S}(p)$:

$$\mathscr{S}(p) \equiv \|x - y\|^2 \geq p^2 \equiv (x_1 - y_1)^2 + (x_2 - y_2)^2 \geq p^2 \quad \text{for } p \in \mathbb{R} \tag{8.1}$$

We treat p, like all other parameters, purely symbolically, without assuming a specific value. In practise, horizontal separation should be $\geq$5 mi, and vertical separation $\geq$1000 ft.

8.2.2 Roundabout Manoeuvre Overview

The *flyable tangential roundabout manoeuvre* (*FTRM*) consists of the phases in the protocol cycle in Fig. 8.5a, which correspond to the marked flight phases in Fig. 8.5b. During free flight, the aircraft move without restriction by repeatedly

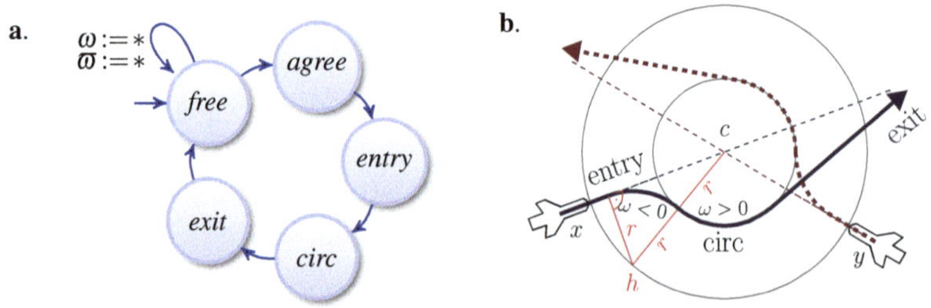

Fig. 8.5 Protocol cycle and construction of flyable roundabout manoeuvre

choosing arbitrary new angular velocities ω and ϖ respectively (as indicated by the self-loop at *free* in Fig. 8.5a). As in Chap. 3, continuous nondeterministic variation of ω and ϖ could be permitted instead of the self-loop by adding $\exists\omega\,\exists\varpi$ to the DA-constraints in *free*. The subsequent proofs carry over to this generalisation immediately.

When the aircraft come too close to one another, they agree on a roundabout manoeuvre by negotiating a compatible roundabout centre $c = (c_1, c_2)$ in coordination phase *agree* by communication. Next, the aircraft approach the roundabout circle in a right curve with $\omega < 0$ (*entry* mode) according to Fig. 8.5b, and reach a tangential position around centre c. During the *circ* mode, the aircraft follow the circular roundabout manoeuvre around the agreed centre c with a left curve of common angular velocity $\omega > 0$. Finally, the aircraft leave the roundabout in cruise mode ($\omega = 0$) in their original direction (*exit*) and enter free flight again (*free*) when they have reached sufficient distance (the protocol cycle repeats as necessary). The collision avoidance manoeuvre is symmetric when exchanging left and right curves, which corresponds to exchaing a value $\omega < 0$ for $\omega > 0$.

8.2.3 Compositional Verification Plan

For verifying safety properties and collision avoidance of FTRM, we decompose the verification problem and pursue the following overall verification plan:

- **AC1** *Tangential roundabout manoeuvre cycle*: We prove that the protected zones of aircraft are safely separated at all times during the whole manoeuvre (including repetitive collision avoidance manoeuvre initiation and multiple aircraft) with a simplified but not yet flyable entry operation $entry_n$. Subsequently, we refine this verification result to a flyable manoeuvre by verifying that we can replace $entry_n$ with its flyable variant *entry*.
- **AC2** *Bounded control choices for aircraft velocities*: We show that linear speeds remain unchanged during the whole manoeuvre (the aircraft do not stall).
- **AC3** *Flyable entry*: We prove that the simplified $entry_n$ procedure can be replaced with a flyable curve *entry* reaching the same configuration around the inner roundabout circle as $entry_n$.
- **AC4** *Bounded entry duration*: We show why the flyable *entry* procedure succeeds in bounded time, i.e., aircraft reach the roundabout circle in some bounded time $\leq T$.
- **AC5** *Safe entry separation*: Most importantly, we prove that the protected zones of aircraft are still respected during the flyable entry procedure *entry*.
- **AC6** *Successful negotiation*: We prove that the negotiation phase (*agree*) satisfies the respective requirements of multiple aircraft simultaneously and identify which constraints are needed for negotiation to succeed.
- **AC7** *Safe exit separation*: We show that, for its bounded duration, the exit procedure cannot produce collisions and that the initial *far separation* for free flight is reached again, so the FTRM cycle repeats safely.

This plan modularises the proof and allows us to identify the respective safety constraints imposed by the various manoeuvre phases successively. We present details on these verification tasks in this chapter and summarise the respective verification results in a joint safety property of FTRM in Sect. 8.5.

Finally note that to simplify notation we use square roots, vectorial notation, and norms within formulas in this chapter, because those are easily definable. For instance, for vectors $x = (x_1, x_2)$ and $y = (y_1, y_2)$, the formula $\|x - y\| \geq p$ is definable by $(x_1 - y_1)^2 + (x_2 - y_2)^2 \geq p^2$ for $p \geq 0$. Likewise, $\|x - y\| = \sqrt{3}r$ is definable by $\|x - y\|^2 = 3r^2$ for $r \geq 0$.

8.2.4 Tangential Roundabout Manoeuvre Cycles (AC1)

First, we analyse roundabouts with a simplified instant entry procedure and without an exit procedure (property AC1 from Sect. 8.2.3). This corresponds to the non-flyable NTRM depicted in Fig. 8.1d on p. 304. We refine this manoeuvre and its verification to the flyable FTRM afterwards.

Modular Correctness of Tangential Roundabout Cycles

For AC1, we have proven in Sect. 3.11 that, for arbitrary choices of the *entry* manoeuvre that satisfy the prerequisites of Theorem 3.4 from p. 199, the tangential roundabout manoeuvre safely avoids collisions, i.e., the aircraft always maintain a safe distance $\geq p$ during the curved flight in the roundabout circle. In addition, these results show that arbitrary repetitions of the protocol cycle are safe at all times for a simplified choice of the entry manoeuvre. The model and specification for this tangential roundabout, as constructed in Sects. 3.4 and 3.11, are summarised in Fig. 8.6, recalling the separation property $\mathscr{S}(p)$ from equation (8.1).

$$
\begin{aligned}
\psi &\equiv \mathscr{S}(p) \rightarrow [NTRM]\, \mathscr{S}(p) \\
\mathscr{S}(p) &\equiv \|x - y\|^2 \geq p^2 \equiv (x_1 - y_1)^2 + (x_2 - y_2)^2 \geq p^2 \\
NTRM &\equiv (\mathit{free};\ \mathit{agree};\ \mathit{entry}_n;\ \mathit{circ})^* \\
\mathit{free} &\equiv (\omega := *;\ \varpi := *;\ \mathscr{F}(\omega) \wedge \mathscr{G}(\varpi) \wedge \mathscr{S}(p))^* \\
\mathit{agree} &\equiv \omega := *;\ c := * \\
\mathit{entry}_n &\equiv d := \omega(x - c)^{\perp};\ e := \omega(y - c)^{\perp} \\
\mathit{circ} &\equiv \mathscr{F}(\omega) \wedge \mathscr{G}(\omega)
\end{aligned}
$$

Fig. 8.6 Non-flyable tangential roundabout collision avoidance manoeuvre NTRM

The simplified flight controller in Fig. 8.6 performs collision avoidance manoeuvres by tangential roundabouts and repeats these manoeuvres any number of times as needed. During each cycle of the loop of *NTRM*, the aircraft first perform arbitrary free flight (*free*) by choosing arbitrary new angular velocities ω and ϖ (repeatedly as indicated by the loop in *free*). Aircraft only fly freely while they are safely separated, which is expressed by the conjunctive constraint $\mathscr{S}(p)$ in the differential equation for *free*, which amounts to an evolution domain restriction. Then the aircraft agree on an arbitrary roundabout centre c and angular velocity ω (*agree*). We model this communication by nondeterministic assignments to the shared variables ω, c. Refinements include all negotiation processes that reach an agreement on common ω, c choices in bounded time. Next, the aircraft perform the simplified non-flyable entry procedure ($entry_n$) with instant turns (according to Fig. 8.1d). This operation identifies the goal state that entry procedures need to reach:

$$\mathscr{T} \equiv d = \omega(x-c)^{\perp} \wedge e = \omega(y-c)^{\perp}. \tag{8.2}$$

As identified in equation (3.12) on p. 198, the constraint $\mathscr{T}$ expresses that, at the positions x and y, respectively, the directions d and e of the aircraft are tangential to the roundabout circle with centre c and angular velocity ω; see Fig. 8.7. Finally,

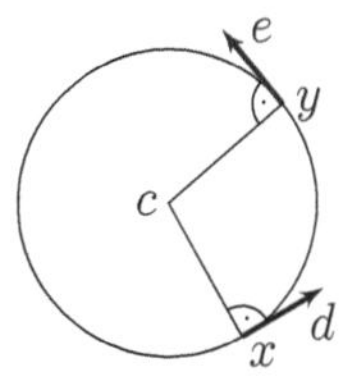

Fig. 8.7 Tangential configuration $\mathscr{T}$

the roundabout manoeuvre itself is carried out in *circ* with the common angular velocity ω that was agreed upon in *agree*. The collision avoidance roundabouts can be left again by repeating the loop and entering arbitrary free flight at any time. When further conflicts occur during free flight, the overall process repeats and the controller in Fig. 8.6 again enters roundabout conflict resolution manoeuvres. The d$\mathcal{L}$ formula ψ in Fig. 8.6 expresses the fact that aircraft following the *NTRM* protocol always stay safely separated by at least the protected zone p if they start safely separated. This formula can be proven similarly to the proof in Sect. 3.11.

Multiple Aircraft

We prove a corresponding collision avoidance property for up to five aircraft, which jointly participate in the roundabout manoeuvre. There, the safety property is mutual collision avoidance, i.e., each of the aircraft has a safe distance $\geq p$ from all the other aircraft, which yields a quadratic number of constraints. This quadratic increase in the property that actually needs to be proven for a safe roundabout of n aircraft and the increased dimension of the underlying continuous state space causes increased verification times for more aircraft in our experiments (Sect. 8.6). For instance, Fig. 8.8 illustrates the (flyable) roundabout manoeuvre with multiple air-

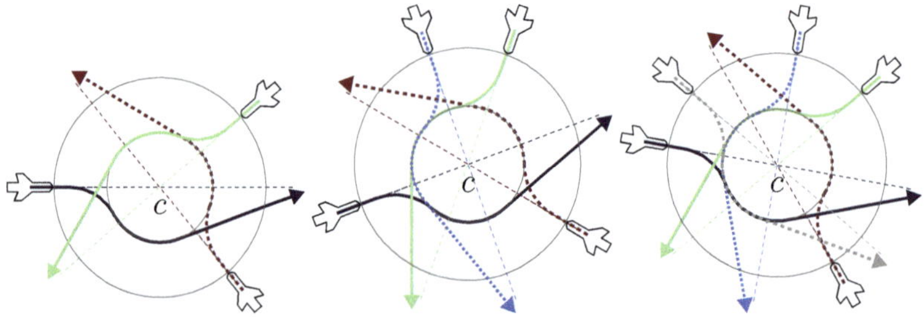

Fig. 8.8 Flyable aircraft roundabout (multiple aircraft)

craft and Fig. 8.9 contains the system and separation property specification for the four aircraft tangential roundabout manoeuvre (still with simplified entry procedure). There, $\mathsf{d}\mathcal{L}$ formula ψ expresses that the four aircraft at positions $x, y, z, u \in \mathbb{R}^2$, respectively, keep mutual distance $\geq p$ each.

$$
\begin{aligned}
\psi &\equiv \phi \rightarrow [NTRM]\phi \\
\phi &\equiv (x_1-y_1)^2+(x_2-y_2)^2 \geq p^2 \wedge (y_1-z_1)^2+(y_2-z_2)^2 \geq p^2 \\
&\quad \wedge (x_1-z_1)^2+(x_2-z_2)^2 \geq p^2 \wedge (x_1-u_1)^2+(x_2-u_2)^2 \geq p^2 \\
&\quad \wedge (y_1-u_1)^2+(y_2-u_2)^2 \geq p^2 \wedge (z_1-u_1)^2+(z_2-u_2)^2 \geq p^2 \\
NTRM &\equiv (free;\ agree;\ entry_n;\ circ)^* \\
free &\equiv (\omega_x := *;\ \omega_y := *;\ \omega_z := *;\ \omega_u := *; \\
&\quad x_1' = d_1 \wedge x_2' = d_2 \wedge d_1' = -\omega_x d_2 \wedge d_2' = \omega_x d_1 \\
&\quad \wedge y_1' = e_1 \wedge y_2' = e_2 \wedge e_1' = -\omega_y e_2 \wedge e_2' = \omega_y e_1 \\
&\quad \wedge z_1' = f_1 \wedge z_2' = f_2 \wedge f_1' = -\omega_z f_2 \wedge f_2' = \omega_z f_1 \\
&\quad \wedge u_1' = g_1 \wedge u_2' = g_2 \wedge g_1' = -\omega_u g_2 \wedge g_2' = \omega_u g_1 \wedge \phi)^* \\
agree &\equiv \omega := *;\ c := * \\
entry_n &\equiv d_1 := -\omega(x_2-c_2);\ d_2 := \omega(x_1-c_1); \\
&\quad e_1 := -\omega(y_1-c_1);\ e_2 := \omega(y_2-c_2); \\
&\quad f_1 := -\omega(z_1-c_1);\ f_2 := \omega(z_2-c_2); \\
&\quad g_1 := -\omega(u_1-c_1);\ g_2 := \omega(u_2-c_2) \\
circ &\equiv x_1' = d_1 \wedge x_2' = d_2 \wedge d_1' = -\omega_x d_2 \wedge d_2' = \omega_x d_1 \\
&\quad \wedge y_1' = e_1 \wedge y_2' = e_2 \wedge e_1' = -\omega_y e_2 \wedge e_2' = \omega_y e_1 \\
&\quad \wedge z_1' = f_1 \wedge z_2' = f_2 \wedge f_1' = -\omega_z f_2 \wedge f_2' = \omega_z f_1 \\
&\quad \wedge u_1' = g_1 \wedge u_2' = g_2 \wedge g_1' = -\omega_u g_2 \wedge g_2' = \omega_u g_1
\end{aligned}
$$

Fig. 8.9 Tangential roundabout collision avoidance manoeuvre (four aircraft)

8.2.5 Bounded Control Choices (AC2)

For property AC2 from Sect. 8.2.3, we show that bounded speed is sufficient for the choices required for the entry procedure and that the bounded speed is always maintained safely throughout the manoeuvre (no stalling).

Bounded Entry Choices

The NTRM in Fig. 8.6 maintains collision avoidance for all its choices of centre c and angular velocity ω in *agree*. Next, we show that *there always is* a choice respecting external requirements on linear speed (aircraft are safe neither at too high speeds nor when they are travailing too slowly, as they would stall), which corresponds to property AC2. The fact that all external speed requirements can be met is a consequence of the proof of property (3.14) on p. 201. Consequently, the constraints on the entry procedure are feasible with parameter choices respecting *any* bounds for velocities that are imposed by the aircraft.

Bounded Manoeuvre Speed

As a simple consequence of the proof in Example 3.16 on p. 172, it is easy to see that external requirements on the linear speed of aircraft are maintained throughout the whole roundabout manoeuvre. Example 3.16 shows that the linear speed is maintained for arbitrary repeated choices of the angular velocity ω, which, as a special case, includes the respective control choices during the roundabout manoeuvre. Thus, the aircraft do not stall.

8.2.6 Flyable Entry Procedures (AC3)

For property AC3 from Sect. 8.2.3, we generalise the verification results about NTRM with simplified entry procedures (cf. Fig. 8.1d) to FTRM (Fig. 8.5b) by replacing the non-flyable $entry_n$ procedure with flyable curves. The resulting flyable entry procedure is called *entry*. This turns the non-flyable NTRM into the flyable FTRM manoeuvre.

Flyable Entry Properties

We construct a flyable entry manoeuvre that follows the smooth entry curve from Fig. 8.5b using the anchor point h indicated in Fig. 8.10. Anchor h is positioned relative to the roundabout centre c and the x position at the start of the entry curve (i.e., with x at the right angle indicated in Fig. 8.10). This anchor h has a distance

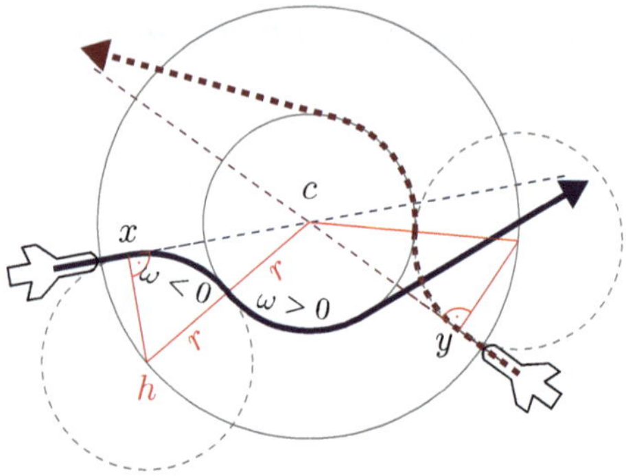

Fig. 8.10 Flyable entry characteristics

of r to x, and distance $2r$ to c, and $x-h$ is orthogonal to the aircraft direction d while taking into account the relative orientation of the roundabout as implied by angular velocity ω in the corresponding specification of the flyable entry procedure depicted in Fig. 8.10. The following property specifies that the tangential configuration of the simple choice for *agree* in Fig. 8.6 is reached by a flyable curve while waiting until aircraft x and centre c have distance r:

$$\begin{aligned}(r\omega)^2 = \|d\|^2 \wedge \|x-c\| = \sqrt{3}r \wedge \exists\lambda\geq 0\,(x+\lambda d = c)\\ \wedge \|h-c\| = 2r \wedge d = -\omega(x-h)^{\perp}\\ \rightarrow \left[\mathscr{F}(-\omega)\,\&\,\|x-c\|\geq r\right]\left(\|x-c\|\leq r \rightarrow d = \omega(x-c)^{\perp}\right) \quad (8.3)\end{aligned}$$

The assumptions in line 1 of d$\mathscr{L}$ formula (8.3) express that r is the radius corresponding to speed $\|d\|$ and angular velocity ω (which corresponds to $(r\omega)^2 = \|d\|^2$) and that *entry* starts when *agree* has finished with distance $\sqrt{3}r$ to c and is heading towards c (i.e., $\exists\lambda\geq 0\,(x+\lambda d = c)$). For the construction of the manoeuvre and positioning in space, we use the auxiliary anchor point $h\in\mathbb{R}^2$ identified in Fig. 8.10 and line 2 of (8.3). Anchor h is positioned relative to the roundabout centre c and the x position at the start of the entry curve (i.e., with x at the right angle indicated in Fig. 8.10). The entry curve around h is similar to the roundabout curve around c. It only goes the other way. Formally, h is characterised by distance r to x, and distance $2r$ to c (i.e., $\|h-c\| = 2r$) and, further, vector $x-h$ is orthogonal to d and obeys the relative orientation of the curve belonging to the right curve with angular velocity $-\omega$ (hence $d = -\omega(x-h)^{\perp}$). The formula in (8.3) specifies that the tangential goal configuration (8.2) around c is reached by a flyable curve when waiting until aircraft x and centre c have distance r, because the domain restriction of the dynamics is $\|x-c\|\geq r$ (line 3) and the postcondition assumes $\|x-c\|\leq r$, which, together, imply $\|x-c\| = r$.

The feasibility of choosing anchor point h can be shown by proving the dual existence formula: For AC3, we thus also prove that the anchor point h can always be chosen as illustrated in Fig. 8.10. That is, we show feasibility of the assumptions on h from property (8.3) by the following existence property:

$$(r\omega)^2 = \|d\|^2 \land \|x-c\| = \sqrt{3}r \land \exists \lambda \geq 0\, (x+\lambda d = c)$$
$$\rightarrow \exists h\, (d = -\omega(x-h)^{\perp} \land \|h-c\| = 2r) \tag{8.4}$$

Spatial Symmetry Reduction

The formulas (8.3) and (8.4) can be proven in a simplified version. We use a (new) *spatial symmetry reduction* to simplify these properties computationally. We exploit symmetries to reduce the spatial dimension by fixing variables and conclude safety for states that result from the fixed class of states by symmetry transformations. Without loss of generality, we can recentre the coordinate system to have centre c at position 0 (the manoeuvre is invariant under translation in space). Further, we can assume aircraft x to come from the left by changing the orientation of the coordinate system (the manoeuvre is invariant under planar rotation in space). Finally, we can assume, without loss of generality, the linear speed to be 1 (by rescaling units appropriately). Observe, however, that we cannot fix a value for both the linear speed and the angular velocity, because their units are strictly interdependent. In other words, if we fix the linear speed, we need to consider all angular velocities to verify the manoeuvre for all possible curve radii r for the roundabout manoeuvre. The x position resulting from these symmetry reductions can be determined by the Pythagoras theorem (i.e., $(2r)^2 = r^2 + x_1^2$ for the triangle enclosed by h, x, c in Fig. 8.10) or simple trigonometry, as follows:

$$x = (2r\cos\frac{\pi}{6}, 0) = (\sqrt{(2r)^2 - r^2}, 0) = (\sqrt{3}r, 0)\ . \tag{8.5}$$

To express the square root function using polynomial terms, we can easily use a random assignment for x_1 with a test condition $?(x_1^2 = (\sqrt{3}r)^2 = 3r^2)$. Consequently, we simplify formulas (8.3) and (8.4) by specialising them to the following situation:

$$\begin{aligned}
&d_1 := 1;\ d_2 := 0;\ \ c_1 := 0;\ c_2 := 0;\\
&x_2 := 0;\\
&r := *;\ \ ?(r > 0);\ \ \omega := 1/r;\\
&x_1 := *;\ \ ?(x_1^2 = 3r^2 \land x_1 \leq 0);
\end{aligned}$$

Verification results for the resulting entry procedure after symmetry reduction, and a proof of existence of a corresponding anchor point h will be shown in Sect. 8.6.

For proving the feasibility property in (8.4) within reasonable time, it is, in fact, sufficient to specialise the state by symmetry reduction to $c_1 := 0 \land c_2 := 0 \land \omega := 1$. Interestingly, this example shows another surprise of quantifier elimination complexity and the power of symmetry reduction quite impressively: Without the reduction to $\omega := 1$, QE of Mathematica runs for more than 20 days without producing a result on an Intel Xeon X5365 with 3 GHz and 16 GB memory (of which only 3 GB are used), even when adding the state space reduction $r := 1$.

8.2.7 Bounded Entry Duration (AC4)

As the first step for showing that the entry procedure finally succeeds at goal (8.2) and maintains a safe distance all the time, we show that *entry* succeeds in bounded time and cannot take arbitrarily long to succeed (property AC4 from Sect. 8.2.3).

Qualitative Analysis

By a simple consequence of the proof for formula (8.3), the entry procedure follows a circular motion around the anchor point h, which is chosen according to (8.3); see Fig. 8.10. That is, when r is the radius belonging to angular velocity ω and linear speed $\|d\|$, the property $\|x-h\| = r$ is a differential invariant (Sect. 3.5.6) of *entry*: For AC4, we prove constant distance to anchor point h, i.e., that, indeed, $\|x-h\| = r$ is an invariant of *entry*:

$$\begin{aligned}(r\omega)^2 = \|d\|^2 \wedge \|x-c\| = \sqrt{3}r \wedge \exists\lambda{\geq}0\,(x+\lambda d = c)\\ \wedge d = -\omega(x-h)^{\perp} \wedge \|h-c\| = 2r\\ \rightarrow \left[\mathscr{F}(-\omega)\,\&\,\|x-c\| \geq r\right]\left(\|x-h\| = r\right)\end{aligned} \tag{8.6}$$

Indeed, the corresponding formula in (8.6) is provable easily. For the remainder of AC4, we use informal arguments. By AC2, the speed $\|d\|$ is constant during the *entry* procedure. Thus, the aircraft proceeds with nonzero minimum progress rate $\|d\|$ around the circle with radius r. Consequently, the entry manoeuvre ends in bounded time, because the arc length of the circle of radius r around h is bounded ($2\pi r$ for radius r) and every bounded curve will be traversed in bounded time with constant linear speed and orientation.

Quantitative Analysis

To obtain a quantitative bound on the duration of the entry manoeuvre, consider the following. The flight duration for a full circle of radius r around h at constant linear speed $\|d\|$ is $\frac{2\pi r}{\|d\|}$, because its arc length is $2\pi r$. From the trigonometric identities underlying equation (8.5), we can read off that the aircraft do not even have to complete the full circle but only a $\frac{\pi}{3} = 60^\circ$ arc; see Fig. 8.10. Consequently, the maximum duration T of the *entry* procedure can be determined as:

$$T := \frac{1}{6}\cdot\frac{2\pi r}{\|d\|} = \frac{\pi r}{3\|d\|}. \tag{8.7}$$

Using the standard relation $|\omega| = \frac{v}{r}$ between the angular velocity ω, the corresponding radius r of the circle (of evasive actions), and the linear velocity $v = \|d\|$ of the aircraft, maximum duration T can also be expressed as:

$$T = \frac{1}{6} \cdot \frac{2\pi}{|\omega|}.$$

Note that the constant π in the expression (8.7) for T is not definable in the theory of real-closed fields (the transcendental number π is not even contained in the real-closed field of algebraic numbers, and thus it cannot be definable; see App. D.2). Yet, the maximum duration T only has to be some upper bound on the maximum duration of the *entry* procedure. Consequently, any overapproximation of the maximum duration computed using a rational $P \geq \pi$ will do, for instance, $\frac{3.15r}{3\|d\|}$ when using the approximation 3.15 of $\pi = 3.1415926\ldots$ in (8.7).

8.2.8 Safe Entry Separation (AC5)

In Sect. 8.2.6, we have shown that the simplified $entry_n$ procedure from NTRM can be replaced by a flyable *entry* manoeuvre that meets the requirements of approaching tangentially (according to Theorem 3.4) for each aircraft. Unlike in instant turns ($entry_n$), we have to show in addition that the flyable entry manoeuvres of multiple aircraft do not produce mutually conflicting flight paths, i.e., spatial separation of all aircraft is maintained during the entry of multiple aircraft (property AC5 from Sect. 8.2.3). See Fig. 8.8 on p. 314 for multiple aircraft FTRM where entry separation is important.

Bounded Overapproximation

We show that entry separation is a consequence of the bounded speed (AC2) and bounded duration (AC4) of the flyable entry procedure when initiating the negotiation phase *agree* with sufficient distance. We prove that, when following bounded speed for a bounded duration, aircraft only come closer by a bounded distance. This gives a fairly generic proof that generalises to other entry choices.

Let b denote the overall speed bound during FTRM according to AC2 and let T be the time bound for the duration of the entry procedure due to AC4. We overapproximate the actual behaviour during the *entry* phase by *arbitrary curved flight* (see Fig. 8.11 where angular velocities ω, ϖ are free inputs). As an overapproximation of the actual behaviour during the *entry* phase, the following strong DAL formula expresses that, when the *entry* procedure is initiated with sufficient distance $p+2bT$, the protected zone will still be respected after the two aircraft follow *any* curved flight (including the actual choices during *entry* and subsequent *circ*) with speed $\leq b$ up to T time units (see Fig. 8.11):

$$\begin{aligned}&\|d\|^2 \leq \|e\|^2 \leq b^2 \land \|x-y\|^2 \geq (p+2bT)^2 \land p \geq 0 \land b \geq 0 \land T \geq 0\\ &\quad \to \left[\tau := 0; \exists \omega\, \mathscr{F}(\omega) \land \exists \varpi\, \mathscr{G}(\varpi) \land \tau' = 1 \land \tau \leq T\right](\|x-y\|^2 \geq p^2) \quad (8.8)\end{aligned}$$

Fig. 8.11 Entry separation by bounded nondeterministic overapproximation

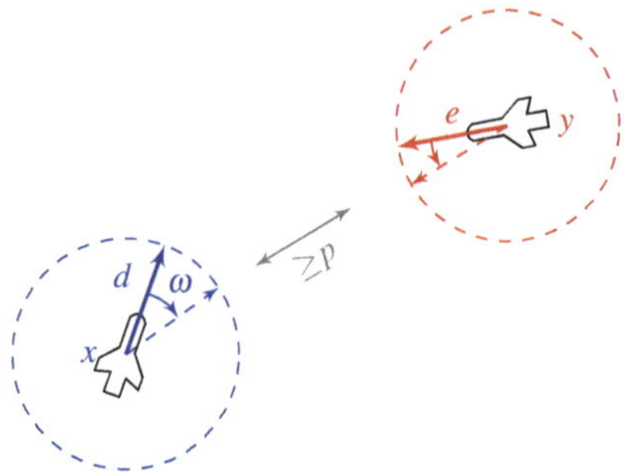

This formula is a consequence of the fact that—regardless of the actual angular velocity choices—aircraft only make limited progress in bounded time when starting from some initial point z with bounded speeds, even when changing angular velocity ω arbitrarily as a nondeterministic quantified input ($\exists\omega$ as in Chap. 3):

$$\|d\|^2 \le b^2 \land b \ge 0 \land x = z \to [\tau := 0;\ \exists\omega\, \mathscr{F}(\omega) \land \tau' = 1]\|x - z\|_\infty \le \tau b \tag{8.9}$$

The maximum distance $\|x-z\|_\infty$ from initial position z depends on clock τ and speed bound b. To reduce the polynomial degree and the computational verification complexity in the proof, we overapproximate distances from quadratic Euclidean norm $\|\cdot\|$ in terms of linearly definable supremum norm $\|\cdot\|_\infty$, which is definable as:

$$\|x\|_\infty \le c \equiv -c \le x_1 \le c \land -c \le x_2 \le c;$$
$$\|x\|_\infty \ge c \equiv (x_1 \le -c \lor c \le x_1) \lor (x_2 \le -c \lor c \le x_2).$$

The limited progress property (8.9) is provable immediately by differential induction for the postcondition (the antecedent of DI is provable as $x = z \vdash \|x-z\|_\infty = 0$):

$$\dfrac{\dfrac{\dfrac{*}{\|d\|^2 \le b^2 \land b \ge 0 \vdash \forall x, y, \tau\, \forall\omega\, (-b \le d_1 \le b \land -b \le d_2 \le b)}\ \forall r}{\|d\|^2 \le b^2 \land b \ge 0 \land x = z \vdash \langle \tau := 0\rangle [\exists\omega\, \mathscr{F}(\omega) \land \tau' = 1]\|x-z\|_\infty \le \tau b}\ DI}{\|d\|^2 \le b^2 \land b \ge 0 \land x = z \vdash [\tau := 0;\ \exists\omega\, \mathscr{F}(\omega) \land \tau' = 1]\|x-z\|_\infty \le \tau b}\ [;],[:=]$$

Cartesian Degree Reduction

Due to the complexity of QE, the proof of the original property (8.8) does not terminate quickly enough. To overcome this issue, we simplify property (8.8) and use the (linearly definable) supremum norm $\|\cdot\|_\infty$ in place of the (quadratically definable) Euclidean 2-norm $\|\cdot\|_2$, thereby yielding the following provable variant of (8.8):

$$\|d\|^2 \le \|e\|^2 \le b^2 \wedge \|x-y\|_\infty \ge (p+2bT) \wedge p \ge 0 \wedge b \ge 0 \wedge T \ge 0$$
$$\rightarrow \big[\tau := 0; \exists \omega\, \mathscr{F}(\omega) \wedge \exists \varpi\, \mathscr{G}(\varpi) \wedge \tau' = 1 \wedge \tau \le T\big](\|x-y\|_\infty \ge p) \quad (8.10)$$

It can be shown (and is provable even by QE) that the supremum norm $\|\cdot\|_\infty$ and the standard Euclidean norm $\|\cdot\|_2$ are equivalent, i.e., their values are identical up to constant factors:

$$\forall x\,(\|x\|_\infty \le \|x\|_2 \le \sqrt{n}\|x\|_\infty) \quad (8.11)$$
$$\forall x\,(\frac{1}{\sqrt{n}}\|x\|_2 \le \|x\|_\infty \le \|x\|_2) \quad (8.12)$$

where n is the dimension of the vector space, here 2. From this equivalence of norms, we can conclude that the following variant of (8.10) with 2-norms is valid:

$$\|d\|^2 \le \|e\|^2 \le b^2 \wedge \|x-y\|_2 \ge \sqrt{2}(p+2bT) \wedge p \ge 0 \wedge b \ge 0 \wedge T \ge 0$$
$$\rightarrow \big[\tau := 0; \exists \omega\, \mathscr{F}(\omega) \wedge \exists \varpi\, \mathscr{G}(\varpi) \wedge \tau' = 1 \wedge \tau \le T\big](\|x-y\|_2 \ge p) \quad (8.13)$$

The extra factor of $\sqrt{2}$ in the separation requirement results from the relaxation of the 2-norm to the ∞-norm. In the preconditions of formula (8.13), we assume $\|x-y\|_2 \ge \sqrt{2}(p+2bT)$, which implies the assumption $\|x-y\|_\infty \ge (p+2bT)$ in the precondition of (8.10) using (8.12). The postcondition $\|x-y\|_\infty \ge p$ of (8.10) in turn implies postcondition $\|x-y\|_2 \ge p$ of (8.13) using (8.11).

Now, with the bounded duration property AC4 of the entry manoeuvre, it is easy to see that *entry* is a special case of the hybrid program with quantified nondeterministic angular velocities in (8.13). Thus we conclude that the following property is valid:

$$\|x-y\| \ge \sqrt{2}(p+2bT) \wedge p \ge 0 \wedge \|d\|^2 \le \|e\|^2 \le b^2 \wedge b \ge 0 \wedge T \ge 0$$
$$\rightarrow [entry]\,(\|x-y\| \ge p) \quad (8.14)$$

Far Separation

By combining the estimation of the entry duration (8.7) at speed $\|d\| = b$ with the entry separation property (8.14), we determine the following magnitude as the *far separation* f, i.e., the initial distance guaranteeing that the *FTRM* protocol can be repeated safely in case new collision avoidance is needed:

$$f := \sqrt{2}(p+2bT) = \sqrt{2}\left(p + \frac{2}{3}\pi r\right). \quad (8.15)$$

Using f as an abbreviation for that term, we can abbreviate property (8.14) as follows:

$$\|d\|^2 \le \|e\|^2 \le b^2 \wedge \|x-y\|_2 \ge f \wedge p \ge 0 \wedge b \ge 0 \rightarrow [entry](\|x-y\|_2 \ge p) \quad (8.16)$$

8.3 Synchronisation of Roundabout Manoeuvres

In the previous sections, we have analysed collision freedom and separation properties of the various curved flight phases of FTRM. According to our verification plan in Sect. 8.2.3, we also need to show that the various actions of the respective aircraft are synchronised appropriately to ensure safety of the manoeuvre. We analyse the negotiation phase and compatible exit procedures next.

8.3.1 Successful Negotiation (AC6)

For the negotiation phase to succeed, we have to show that there is a common choice of the roundabout centre c and angular velocity ω (or radius r) so that all participating aircraft can satisfy the requirements of their respective entry procedures simultaneously. That is, we show that the assumptions of formula (8.3) can be satisfied simultaneously for all aircraft at once. We thus have to show property AC6 from Sect. 8.2.3.

Separate Success

First of all, we show that there is always a choice of centre c, radius r, and common angular velocity ω during the *agree* mode such that the antecedent of the formula (8.3) is satisfied, i.e., the *agree* mode is feasible:

$$\langle c:=*;\; r:=*;\; ?\|x-c\| = \sqrt{3}r \land r \geq 0; \omega:=*;\; ?(r\omega)^2 = d_1^2 + d_2^2\rangle \\ \left(\|x-c\| = \sqrt{3}r \land \exists \lambda \geq 0\, (x+\lambda d = c)\right) \tag{8.17}$$

The nondeterministic assignments choose any c, r, ω and the subsequent tests ensure that the *agree* mode succeeds in a state with appropriate distance $\sqrt{3}r$ to the centre c as identified in (8.3). The test $?(r\omega)^2 = d_1^2 + d_2^2$ ensures that ω is an angular velocity that fits to the linear speed $\|d\|$ and roundabout radius r.

The dual property shows that, in fact, *all* choices of *agree* that also satisfy the (hence, feasible) constraint $\exists \lambda \geq 0\, (x+\lambda d = c)$ already satisfy the *entry* requirements:

$$[c:=*; r:=*; ?\|x-c\| = \sqrt{3}r \land r \geq 0; ?\exists \lambda \geq 0\, (x+\lambda d = c); \\ \omega:=*;\; ?(r\omega)^2 = d_1^2 + d_2^2] \\ \left(\|x-c\| = \sqrt{3}r \land \exists \lambda \geq 0\, (x+\lambda d = c)\right) \tag{8.18}$$

With constraint $\exists\lambda{\geq}0\,(x+\lambda d=c)$ we chose to have the centre c on the straight line from x in its current direction d. This is one option but not the only choice for an entry construction.

Joint Negotiation Success

In addition to the entry requirements working out separately for each aircraft, we prove that all corresponding choices of *agree* satisfy the mutual requirements of multiple aircraft simultaneously. There are many possibilities for simultaneous solutions of negotiation. To simplify the computational complexity, we only consider symmetric situations with identical speed and common distance to the point of potential collision, which we can then choose as the roundabout centre c; see Fig. 8.12. This is a restriction but simplifies the computational proof complexity considerably.

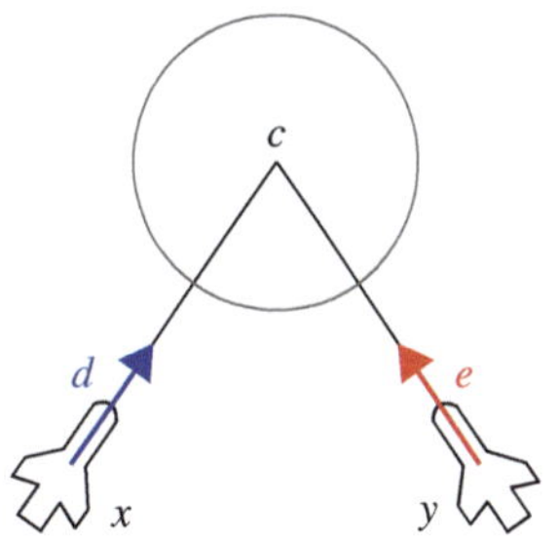

Fig. 8.12 Some mutually agreeable negotiation choices for aircraft

A very simple choice is to use the simultaneous intersection (provided there is an intersection $x+\lambda d=y+\lambda e$ after time λ) of the flight paths of the aircraft at x and y as the roundabout centre c. Then the choices for c,r,ω are compatible for multiple aircraft:

$$\begin{aligned}
&\lambda>0\land x+\lambda d=y+\lambda e\land\|d\|=\|e\|\rightarrow\\
&\big[c{:=}x+\lambda d;\ r{:=}*;\ ?\|x-c\|=\sqrt{3}r;\ ?\|y-c\|=\sqrt{3}r;\ \omega{:=}*;\ ?(r\omega)^2=\|d\|^2\big]\\
&\quad\big(\|x-c\|=\sqrt{3}r\land\lambda\geq0\land x+\lambda d=c\land\|y-c\|=\sqrt{3}r\land y+\lambda e=c\big)\qquad(8.19)
\end{aligned}$$

The tests in the dynamics again ensure that the *entry* curve starts when x, y, and c have appropriate distance $\sqrt{3}r$ (as identified in Sect. 8.2) and that r is defined to be the radius belonging to angular velocity ω and linear speed $\|d\|$.

Secondly, these choices are feasible, i.e., there always is a mutually agreeable choice, according to the following diamond formula:

$$\begin{aligned}
&\|d\|=\|e\|\land\lambda>0\land x+\lambda d=y+\lambda e\rightarrow\\
&\big\langle c{:=}x+\lambda d;\ r{:=}*;\ ?\|x-c\|=\sqrt{3}r;\ ?\|y-c\|=\sqrt{3}r;\ \omega{:=}*;\ ?(r\omega)^2=d_1^2+d_2^2\big\rangle
\end{aligned}$$

$$\left(\|x-c\| = \sqrt{3}r \wedge \lambda \geq 0 \wedge x+\lambda d = c \wedge \|y-c\| = \sqrt{3}r \wedge \lambda \geq 0 \wedge y+\lambda e = c\right) \tag{8.20}$$

This property follows from the following provable simplified variant, obtained from formula (8.20) by gluing the equations $\|x-c\| = \sqrt{3}r$ and $\|y-c\| = \sqrt{3}r$ together:

$$\|d\| = \|e\| \wedge \lambda > 0 \wedge x+\lambda d = y+\lambda e \rightarrow \langle c:=x+\lambda d;\ r:=*;\ ?\|x-c\| = \|y-c\| = \sqrt{3}r\rangle \mathit{true} \tag{8.21}$$

The fact that $\|x-c\|$ can be written in the form $\sqrt{3}r$ for some r is easy to see just by division; similarly for the choice of ω.

Far Separation

The *entry* procedure has to be initiated while the aircraft are still sufficiently far apart for safety reasons. Otherwise, there would not be enough airspace for manoeuvring and approaching the roundabout circle safely. Correspondingly, the *agree* procedure will negotiate a roundabout choice while the aircraft have far distance. Thus, the *agree* procedure will have to maintain far separation, i.e., satisfy the property

$$\|x-y\| \geq \sqrt{2}\left(p+\frac{2}{3}\pi r\right) \rightarrow [agree]\left(\|x-y\| \geq \sqrt{2}\left(p+\frac{2}{3}\pi r\right)\right) \tag{8.22}$$

This appears to be a trivial property, because *agree* models the successful completion of the negotiation, so no time elapses during *agree*; hence the positions x and y do not even change. Observe, however, that the far separation distance according to equation (8.15) depends on the protected zone p and the radius r of evasive actions. Unlike p, radius r may change during *agree*, which allows for the flexibility of changing the flight radius r adaptively when repeating the roundabout manoeuvre loop at different positions. Consequently, the far separation distance $\sqrt{2}(p+\frac{2}{3}\pi r)$ is affected when changing r.

To ensure that the new radius r is chosen such that far separation is still maintained, i.e., formula (8.22) is respected, we add a corresponding constraint to *agree*. Thus, changes of r are only accepted as long as they do not compromise far separation. We show that, when adding a corresponding constraint to property (8.19), all choices by *agree* maintain far separation of the aircraft at x and y according to (8.15):

$$\|d\| = \|e\| \wedge \lambda > 0 \wedge x+\lambda d = y+\lambda e \rightarrow$$

$$\left[c:=x+\lambda d;\ r:=*;\ ?\|x-c\| = \sqrt{3}r;\ ?\|y-c\| = \sqrt{3}r;\ ?\|x-y\| \geq \sqrt{2}\left(p+\frac{2}{3}\pi r\right);\right.$$

$$\left.\omega:=*;\ ?(r\omega)^2 = d_1^2+d_2^2\right]$$

$$\left(\|x-c\| = \sqrt{3}r \wedge \lambda \geq 0 \wedge x+\lambda d = c \wedge \|y-c\| = \sqrt{3}r \wedge \lambda \geq 0 \wedge y+\lambda e = c\right.$$

$$\wedge \|x-y\| \geq \sqrt{2}\left(p+\frac{2}{3}\pi r\right)\Big) \tag{8.23}$$

Finally, we analyse when such choices of *agree* are feasible with a diamond formula:

$$\|d\| = \|e\| \wedge \lambda > 0 \wedge x+\lambda d = y+\lambda e \rightarrow$$
$$\langle c:=x+\lambda d;\, r:=*;\, ?\|x-c\| = \|y-c\| = \sqrt{3}r\rangle \left(\|x-y\| \geq \sqrt{2}\left(p+\frac{2}{3}\pi r\right)\right) \tag{8.24}$$

The corresponding distance constraints on x, y and c for *agree*, respectively, are depicted in Fig. 8.13. Using standard trigonometric relations for each half of the triangle, we compute the resulting distance between x and y as $\|x-y\| = 2\sqrt{3}r\sin\frac{\gamma}{2}$.

Fig. 8.13 Far separation for mutually agreeable negotiation choices

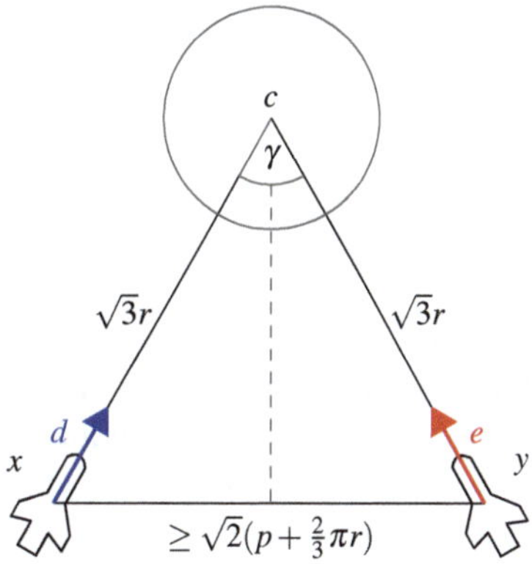

With quantifier elimination and simple evaluation for the remaining trigonometric expressions, we can determine under which circumstances d$\mathcal{L}$ formula (8.24) holds true, i.e., for all protected zones p there is a radius r satisfying the distance requirements:

$$\mathrm{QE}\left(\forall p \exists r{\geq}0\left(2\sqrt{3}r\sin\frac{\gamma}{2} \geq \sqrt{2}(p+\frac{2}{3}\pi r)\right)\right) \equiv \sin\frac{\gamma}{2} > \frac{1}{3}\sqrt{\frac{2}{3}}\pi \equiv \gamma > 117.527^\circ$$

Consequently, corresponding choices are feasible for identical speeds and all protected zones with flight paths that do not intersect with narrow collision angles. The constraint on the flight path intersection angle would relax to $\gamma > 74.4^\circ$ when removing the extra factor of $\sqrt{2}$ from (8.15), which only results from our computational simplification of Cartesian degree reduction from Sect. 8.2.8. Likewise, the constraints can be relaxed by allowing different angular velocities for *entry* and *circ*, which would be an interesting extension for future work.

Despite the presence of trigonometric expressions, the above formula is a substitution instance of first-order real arithmetic and can thus be handled by QE using

Lemma 2.5. Note that the primary difference compared to trigonometric expressions occurring in the solutions of flight equations for curved flight—which do not support quantifier elimination—is that the argument $\frac{\gamma}{2}$ of sin is not quantified over. Consequently, we can consider $\sin\frac{\gamma}{2}$ as a variable, apply QE, and then use (quantifier-free) ground evaluation.

8.3.2 *Safe Exit Separation (AC7)*

NTRM (from Fig. 8.1d) does not need an exit procedure for safety, because the manoeuvre can repeat immediately in case further air traffic conflicts arise. For FTRM, on the other hand, we need to show that the exit procedure produces safe flight paths until the aircraft are sufficiently separated: When repeating the FTRM manoeuvre, the *entry* procedure needs far separation (8.15) for manoeuvring, and not just safety distance p; see Fig. 8.5b.

Safe Separation

In order to establish the safe separation of aircraft during their *exit* procedures, we show that, for its bounded duration, the exit procedure does not conflict with other flight curves such that the initial far separation is again maintained as needed by the flyable entry procedure when re-initiating collision avoidance manoeuvres repeatedly. For showing property AC7 of Sect. 8.2.3, we have to show the following $\mathsf{d}\mathscr{L}$ formula:

$$\mathscr{T} \wedge \|x-y\|^2 \geq p^2 \rightarrow [x' = d \wedge y' = e \wedge e' = e^{\perp}; x' = d \wedge y' = e](\|x-y\|^2 \geq p^2) \tag{8.25}$$

This formula expresses that, when x already exits on a straight line while y keeps following the roundabout for a while until (after the sequential composition) both aircraft exit on straight lines, then the protected zones are respected at any point. Since, in Sect. 8.3.1, we have assumed simultaneous entry and identical speed, the aircraft can also exit simultaneously. With that simplification, property (8.25) simplifies to the following property, saying that aircraft that exit simultaneously (from tangential positions $\mathscr{T}$ of the roundabout circle) always respect their protected zones:

$$\mathscr{T} \wedge \|x-y\|^2 \geq p^2 \rightarrow [x' = d \wedge y' = e](\|x-y\|^2 \geq p^2) \tag{8.26}$$

This property expresses that safely separated aircraft that exit simultaneously along straight lines from tangential positions of a roundabout always remain safely separated. To reduce the arithmetical complexity, we overapproximate this property by showing that even the whole exit rays never cross when the aircraft exit the same roundabout tangentially (see Fig. 8.14a; the counterexample in Fig. 8.14b shows that the assumption on identical radii is required for this relaxation):

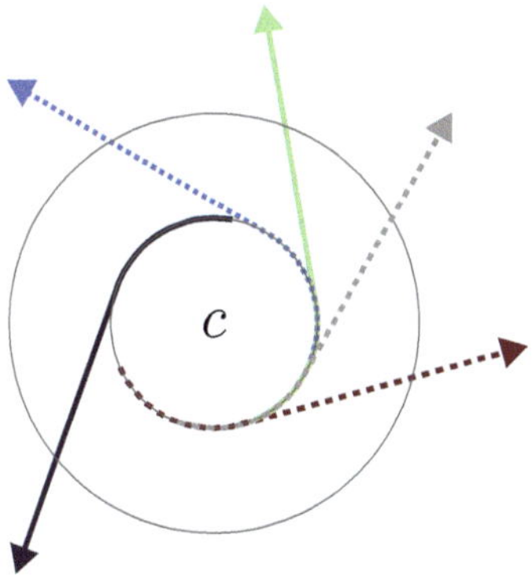

Fig. 8.14a Exit ray separation

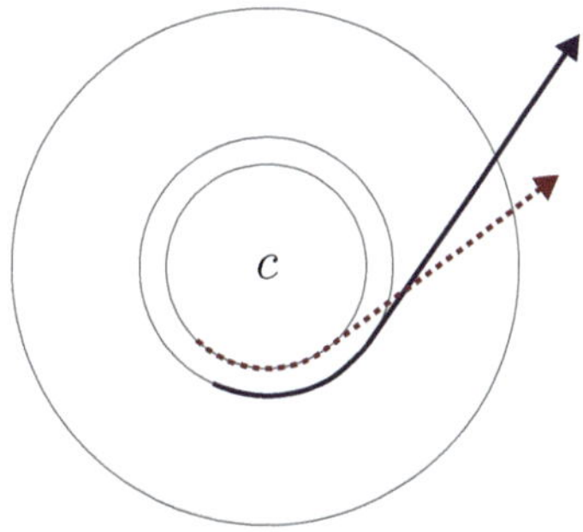

Fig. 8.14b Incompatible exit rays

$$\mathscr{T} \land \|x-c\|^2 = \|y-c\|^2 \land x \neq y \to [x' = d; y' = e]x \neq y \qquad (8.27)$$

The difference with (8.26) is that, here, the aircraft exit independently by a sequential composition of two separate differential equation systems, rather than by a single differential equation system. Again the computational complexity of proving this property can be simplified substantially by adding $c_1 := 0 \land c_2 := 0$ by spatial symmetry reduction (the manoeuvre is invariant under translation in space). From this property, the original separation property follows using the geometric fact that, for linearity reasons, rays that never cross cannot come closer than the original distance p, which can be expressed elegantly in $\mathsf{d}\mathcal{L}$ as:

$$\|x-y\|^2 \geq p^2 \land [x' = d \land y' = e]x \neq y \to [x' = d \land y' = e](\|x-y\|^2 \geq p^2) \quad (8.28)$$

Thus, by combining (8.27) with (8.28) propositionally, and by using the simple fact that the sequential independent ray evolution $x' = d; y' = e$ is an overapproximation of the synchronous evolution $x' = d \land y' = e$, we conclude that property (8.26) is valid. The rationale behind this overapproximation is that, if all combinations of two aircraft exiting independently (with a sequential composition) are safely separate, then, in particular, simultaneous exit of the aircraft along one common differential equation system is safely separate.

Far Separation

Aircraft finally reach arbitrary far separation when following the exit procedure long enough. Using the ray overapproximation principle from Fig. 8.14a, we prove that, due to different exit directions $d \neq e$, i.e., $\|d-e\| > 0$, the exit procedure will finally separate the aircraft arbitrarily far (starting from tangential configuration $\mathscr{T}$ of the roundabout):

$$\mathscr{T} \land d \neq e \to [\mathscr{F}(\omega) \land \mathscr{G}(\omega)](\|d-e\| > 0) \qquad (8.29)$$

Aircraft in a simultaneous roundabout manoeuvre can, indeed, never enter in the same direction: Either they would already have collided (when they are entering in the same direction at the same position) or would crash later (when entering in the same direction with opposing orientations at different positions of the roundabout circle), but we have already shown that the roundabout manoeuvre is collision-free (Theorem 3.4). Thus by formula (8.29), they maintain their different directions while following the roundabout. Again, we can combine (8.29) with the geometric fact that rays into different directions which never cross will be arbitrarily far apart after sufficient time (see Fig. 8.14a):

$$d \neq e \wedge [x' = d \wedge y' = e]\, x \neq y \;\rightarrow\; \forall a\, \langle x' = d \wedge y' = e \rangle (\|x-y\|^2 > a^2)$$

By combining this geometric fact with (8.29), we obtain the final separation property saying that, due to their different directions, the exit procedure will finally separate the aircraft arbitrarily far:

$$\mathscr{T} \wedge d \neq e \;\rightarrow\; \forall a\, \langle x' = d \wedge y' = e \rangle (\|x-y\|^2 > a^2) \tag{8.30}$$

8.4 Compositional Verification

Subsequently, we want to combine the previous results into a full model of the flyable tangential roundabout manoeuvre. To begin with, we show how verification results for large systems can be obtained compositionally in our proof calculi (Part I of this book) from verification lemmas about small subsystems whenever the large system has been constructed by composition from the corresponding small subsystems.

Assume we have subsystems α_i with lemmas guaranteeing $\psi_i \rightarrow [\alpha_i]\phi_i$, as is the case for the respective phases *agree*, *entry*, and *circ* of FTRM. Then, in order to obtain a verification result for their sequential composition $\alpha_1; \alpha_2; \ldots; \alpha_n$, we have to show that postcondition ϕ_{i-1} of α_{i-1} implies precondition ψ_i of α_i for $i \leq n$. In fact, this is a derived proof rule of the compositional $\mathsf{d}\mathcal{L}$ calculus from Fig. 2.11 or the DAL calculus from Fig. 3.9 and can be obtained directly from the generalisation rule $[]gen$ (with the usual cut rule application *cut* to shift the antecedent of a goal to the succedent of a new subgoal, as for rule *ind'* in Sect. 2.5.2):

$$\cfrac{\vdash \forall^\alpha(\phi_i \rightarrow \phi) \qquad \cfrac{\cfrac{\psi \vdash [\alpha_{i-1}]\psi_i \qquad \vdash \forall^\alpha(\psi_i \rightarrow [\alpha_i]\phi_i)}{\psi \vdash [\alpha_{i-1}][\alpha_i]\phi_i}\;[]gen}{\psi \vdash [\alpha_{i-1}; \alpha_i]\phi_i}\;[;]}{\psi \vdash [\alpha_{i-1}; \alpha_i]\phi}\;[]gen$$

8.5 Flyable Tangential Roundabout Manoeuvre

In this section, we combine the results of this chapter about the individual phases of flyable roundabouts into a full model of the flyable tangential roundabout manoeuvre that inherits safety in a modular way by joining the individual safety properties of the respective phases together. Essentially, we collect the various manoeuvre parts together according to the protocol cycle of Fig. 8.5 and take care to ensure that all safety prerequisites are met that we have identified for the respective flight phases previously. Formally, safety properties about the individual phases will be glued together using the generalisation rule $[]gen$ and the compositional technique from Sect. 8.4.

One possible instance of FTRM is the hybrid program in Fig. 8.15, which is composed of previously illustrated parts of the manoeuvre. The technical construction

$$\begin{aligned}
\psi &\equiv \|d\| = \|e\| \wedge r > 0 \wedge \mathscr{S}(f) \to [FTRM]\mathscr{S}(p)\\
\mathscr{C} &\equiv \|x-c\| = \sqrt{3}r \wedge \exists\lambda{\ge}0\,(x+\lambda d = c) \wedge \|y-c\| = \sqrt{3}r \wedge \exists\lambda{\ge}0\,(y+\lambda e = c)\\
FTRM &\equiv \left(free^*;\ agree;\ \Pi(entry;\ circ;\ exit)\right)^*\\
free &\equiv \omega := *;\ \varpi := *;\ \mathscr{F}(\omega) \wedge \mathscr{G}(\varpi) \wedge \mathscr{S}(f)\\
agree &\equiv c := *;\ r := *;\ ?(\mathscr{C} \wedge r > 0);\ ?\mathscr{S}(f);\\
&\qquad \omega := *;\ ?(r\omega)^2 = \|d\|^2;\ x_0 := x; d_0 := d; y_0 := y; e_0 := e\\
entry &\equiv \mathscr{F}(-\omega)\ until\ \|x-c\|^2 = r^2\\
circ &\equiv \mathscr{F}(\omega)\ until\ \exists\lambda{\ge}0\,\exists\mu{>}0\,(x+\lambda d = x_0 + \mu d_0)\\
exit &\equiv \mathscr{F}(0);\ ?\mathscr{S}(f)
\end{aligned}$$

Fig. 8.15 Flight control with flyable tangential roundabout collision avoidance

and protocol cycle of the entry procedure have already been illustrated in Fig. 8.5. The operation $\mathscr{F}(\omega)\ until\ G$ expresses that the system follows differential equation $\mathscr{F}(\omega)$ until condition G is true. It is defined in terms of evolution domain restrictions and tests to make sure the system leaves mode $\mathscr{F}(\omega)$ neither later nor earlier than G specifies. We define $\mathscr{F}(\omega)\ until\ G$ as $\mathscr{F}(\omega) \wedge \sim G;\ ?G$, using the weak negation $\sim G$ from Sect. 3.5.7 to retain the border of G in the evolution domain restriction. For instance, $\mathscr{F}(\omega)\ until\ x_1 \ge 0$ is $\mathscr{F}(\omega) \wedge x_1 \le 0; ?x_1 \ge 0$.

Finally, in FTRM, Π denotes the parallel product operator. As in the work of Hwang et al. [171], the FTRM manoeuvre is assumed to operate synchronously, i.e., all aircraft make simultaneous mode changes. Consequently, the parallel product $\Pi(entry; circ; exit)$ simplifies to the conjunction of the respective differential equations in the various modes and can be defined easily as follows (with corresponding simplifications to resolve simultaneous tests):

$$(entry_x \wedge entry_y);\ (circ_x \wedge circ_y);\ (exit_x \wedge exit_y)$$

where $entry_x$ is the entry procedure of the aircraft at position x etc.

To verify this manoeuvre, we split the proof into the modular properties that we have shown previously following the verification plan from Sect. 8.2.3. Formally, we use the generalisation rule ($[]gen$) to split the system at its sequential compositions. The corresponding subproperties are depicted in Fig. 8.16 with the mapping to previous results according to Table 8.1. Formula $\mathscr{T}$ is the characterisation of tangential configurations due to equation (8.2), and the separation formula $\mathscr{S}(p)$ is from (8.1).

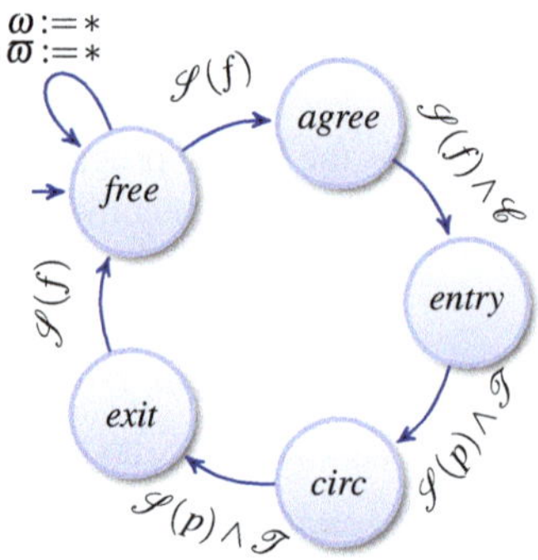

Fig. 8.16 Verification loop for flyable tangential roundabout manoeuvres

Table 8.1 Verification loop properties for flyable tangential roundabout manoeuvres

Property	Consequence of
$\mathscr{S}(f) \to [free]\mathscr{S}(f)$	Fig. 8.6
$\mathscr{S}(f) \to [agree](\mathscr{S}(f) \land \mathscr{C})$	(8.19), (8.23)
$\mathscr{C} \land \mathscr{S}(f) \to [entry]\mathscr{S}(p)$	(8.14)
$\mathscr{C} \land \mathscr{S}(f) \to [entry]\mathscr{T}$	(8.3)
$\mathscr{T} \land \mathscr{S}(p) \to [circ](\mathscr{S}(p) \land \mathscr{T})$	Fig. 8.6
$\mathscr{T} \land \mathscr{S}(p) \to [exit]\mathscr{S}(p)$	(8.26)
$\mathscr{T} \land \mathscr{S}(p) \to [exit]\mathscr{S}(f)$	(8.26), (8.30)

By combining the results about the FTRM flight phases as summarised in Fig. 8.16, we conclude that FTRM avoids collisions safely. Note that the modular proof structure in Fig. 8.16 still holds when replacing any part of the manoeuvre with a different choice that still satisfies the same specification, e.g., for different entry procedures that still succeed in tangential configuration $\mathscr{T}$ within bounded time. This includes roundabouts with *asymmetric positions*, i.e., where the initial distance to c can be different, and with *near conflicts*, where the flight paths do not intersect in one point but in a larger critical region [171]. Most notably, the separation proof in Sect. 8.2.8 is by overapproximation and tolerates asymmetric distances to c (Fig. 8.11).

Theorem 8.1 (Safety property of flyable tangential roundabouts). *FTRM is collision-free, i.e., the collision avoidance property ψ in Fig. 8.15 is valid. Furthermore, any variation of FTRM with a modified entry procedure that safely reaches tangential configuration $\mathscr{T}$ in some bounded time T is safe. That is, if the following formula holds, saying that, until time T the aircraft have safe distance p and will have reached configuration $\mathscr{T}$ at time T, where τ is a clock:*

$$\mathscr{S}(f) \rightarrow [\tau := 0; entry \wedge \tau' = 1]\big((\tau \leq T \rightarrow \mathscr{S}(p)) \wedge (\tau = T \rightarrow \mathscr{T})\big).$$

For reference, a possible instantiation of Fig. 8.15 with abbreviations resolved is shown in Fig. 8.17.

8.6 Experimental Results

Table 8.2 summarises experimental results for the air traffic control case study of flyable tangential roundabout manoeuvres. The rows marked with $*$ indicate a property where simplifications like symmetry reduction have been used to reduce the computational complexity of quantifier elimination. The results in Table 8.2 show that even aircraft manoeuvres with challenging dynamics can be verified with our logical analysis approach for hybrid systems.

Table 8.2 Experimental results for air traffic control (initial timeout = 10s)

Case study		Time(s)	Memory(MB)	Steps	Dim
tangential roundabout	(2 aircraft)	10.5	6.8	197	13
tangential roundabout	(3 aircraft)	636.1	15.1	342	18
tangential roundabout	(4 aircraft)	918.4	31.4	520	23
tangential roundabout	(5 aircraft)	3552.6	46.9	735	28
bounded speed control	(3.14)	19.6	34.4	28	12
bounded manoeuvre speed	Example 3.16	0.3	6.3	14	4
flyable roundabout entry*	(8.3)	10.2	9.6	132	8
flyable entry feasible*	(8.4)	104.4	87.9	16	10
flyable entry circular	(8.6)	2.9	7.6	81	5
limited progress	(8.9)	2	6.5	60	8
entry separation	(8.10), 29 int.	140.1	20.1	512	16
negotiation feasible	(8.17)	4.5	8.4	27	8
negotiation successful	(8.18)	2.7	21.8	34	8
mutual negotiation successful	(8.19)	0.9	6.4	60	12
mutual negotiation feasible	(8.21)	7.5	23.8	21	11
mutual far negotiation	(8.23)	2.4	8.1	67	14
simultaneous exit separation*	(8.27)	4.7	12.9	44	9
different exit directions	(8.29)	3	11.1	42	11

Experimental results are from a 2.6 GHz AMD Opteron with 4 GB memory. Memory consumption of quantifier elimination is shown in Table 8.2, excluding the front end. The dimension of the continuous state space and number of proof steps

$$\psi \equiv d_1^2+d_2^2=e_1^2+e_2^2 \wedge r>0 \wedge (x_1-y_1)^2+(x_2-y_2)^2 \geq 2\left(p+\frac{2}{3}\pi r\right)^2$$
$$\rightarrow [FTRM^*]\phi(p)$$

$$
\begin{aligned}
FTRM \equiv\ & \Big(\Big(\omega:=*;\varpi:=*;\\
\textit{free}:\ & x_1'=d_1 \wedge x_2'=d_2 \wedge d_1'=-\omega d_2 \wedge d_2'=\omega d_1\\
& \wedge y_1'=e_1 \wedge y_2'=e_2 \wedge e_1'=-\varpi e_2 \wedge e_2'=\varpi e_1\\
& \wedge (x_1-y_1)^2+(x_2-y_2)^2 \geq 2\left(p+\frac{2}{3}\pi r\right)^2\Big)^*;\\
\textit{agree}:\ & c:=*;\ r:=*;\ ?(x_1-c_1)^2+(x_2-c_2)^2=3r^2;\\
& ?\exists\lambda\geq 0\,(x_1+\lambda d_1=c_1 \wedge x_2+\lambda d_2=c_2);\\
& ?(y_1-c_1)^2+(y_2-c_2)^2=3r^2;\\
& ?\exists\lambda\geq 0\,(y_1+\lambda e_1=c_1 \wedge y_2+\lambda e_2=c_2);\\
& ?(x_1-y_1)^2+(x_2-y_2)^2 \geq 2\left(p+\frac{2}{3}\pi r\right)^2;\\
& \omega:=*;\ ?(r\omega)^2=d_1^2+d_2^2\\
& x_1^0:=x_1;\ x_2^0:=x_2;\ d_1^0:=d_1;\ d_2^0:=d_2;\\
& y_1^0:=y_1;\ y_2^0:=y_2;\ e_1^0:=e_1;\ e_2^0:=e_2;\\
\textit{entry}_x \wedge \textit{entry}_y:\ & x_1'=d_1 \wedge x_2'=d_2 \wedge d_1'=-(-\omega)d_2 \wedge d_2'=-\omega d_1\\
& \wedge y_1'=e_1 \wedge y_2'=e_2 \wedge e_1'=-(-\omega)e_2 \wedge e_2'=-\omega e_1\\
& \wedge \sim((x_1-c_1)^2+(x_2-c_2)^2=r^2);\\
& ?(x_1-c_1)^2+(x_2-c_2)^2=r^2;\\
\textit{circ}_x \wedge \textit{circ}_y:\ & x_1'=d_1 \wedge x_2'=d_2 \wedge d_1'=-\omega d_2 \wedge d_2'=\omega d_1\\
& \wedge y_1'=e_1 \wedge y_2'=e_2 \wedge e_1'=-\omega e_2 \wedge e_2'=\omega e_1\\
& \wedge \sim(\exists\lambda\geq 0\,\exists\mu>0\,(x_1+\lambda d_1=x_1^0+\mu d_1^0 \wedge x_2+\lambda d_2=x_2^0+\mu d_2^0)\\
& \qquad \wedge \exists\lambda\geq 0\,\exists\mu>0\,(y_1+\lambda e_1=y_1^0+\mu e_1^0 \wedge y_2+\lambda e_2=y_2^0+\mu e_2^0));\\
& ?(\exists\lambda\geq 0\,\exists\mu>0\,(x_1+\lambda d_1=x_1^0+\mu d_1^0 \wedge x_2+\lambda d_2=x_2^0+\mu d_2^0)\\
& \quad \wedge \exists\lambda\geq 0\,\exists\mu>0\,(y_1+\lambda e_1=y_1^0+\mu e_1^0 \wedge y_2+\lambda e_2=y_2^0+\mu e_2^0));\\
\textit{exit}_x \wedge \textit{exit}_y:\ & x_1'=d_1 \wedge x_2'=d_2 \wedge y_1'=e_1 \wedge y_2'=e_2;\\
& ?(x_1-y_1)^2+(x_2-y_2)^2 \geq 2\left(p+\frac{2}{3}\pi r\right)^2\Big)
\end{aligned}
$$

Fig. 8.17 Flight control with FTRM (synchronous instantiation)

are indicated as well. For comparison, the results reported in Table 8.2 use settings that are closer to those that we had used in Table 5.3. Using different settings for initial timeouts for differential saturation results in faster performance for larger dimensions, particularly for multiple aircraft; see Table 8.3.

Table 8.3 Experimental results for air traffic control (initial timeout = 4s)

Case study		Time(s)	Memory(MB)	Steps	Dim
tangential roundabout	(2 aircraft)	10.4	6.8	197	13
tangential roundabout	(3 aircraft)	253.6	7.2	342	18
tangential roundabout	(4 aircraft)	382.9	10.2	520	23
tangential roundabout	(5 aircraft)	1882.9	39.1	735	28
bounded speed control	(3.14)	19.5	34.4	28	12
bounded manoeuvre speed	Example 3.16	0.5	6.3	14	4
flyable roundabout entry*	(8.3)	10.1	9.6	132	8
flyable entry feasible*	(8.4)	104.5	87.9	16	10
flyable entry circular	(8.6)	3.2	7.6	81	5
limited progress	(8.9)	1.9	6.5	60	8
entry separation	(8.10), 29 int.	140.1	20.1	512	16
negotiation feasible	(8.17)	4.8	8.4	27	8
negotiation successful	(8.18)	2.6	13.1	34	8
mutual negotiation successful	(8.19)	0.8	6.4	60	12
mutual negotiation feasible	(8.21)	7.5	23.8	21	11
mutual far negotiation	(8.23)	2.4	8.1	67	14
simultaneous exit separation*	(8.27)	4.3	12.9	44	9
different exit directions	(8.29)	3.1	11.1	42	11

The experimental results in Tables 8.2 and 8.3 for property (8.10) can be improved. Currently, they still need as much as 29 simple user interactions to overcome preliminary simplifications in the implementation of our proof strategies in the KeYmaera tool. An improved implementation of iterative inflation order (Sect. 5.5) will reduce the number of required interactions to at most one interaction that specifies the postcondition of the limited progress property (8.9) as an invariant.

8.7 Summary

We have analysed complex air traffic control applications. Real aircraft can only follow sufficiently smooth flyable curves. Hence, mathematical manoeuvres that require instant turns give physically impossible conflict resolution advice. We have developed a new collision avoidance manoeuvre with smooth, fully flyable curves. Despite its complicated dynamics and manoeuvring, we have verified collision avoidance in this flyable tangential roundabout manoeuvre formally using our verification algorithm for a logic of hybrid systems. Because of the intricate spatio-temporal movement of aircraft in curved roundabouts, some of the properties require intricate arithmetic, which we handled by symmetry reduction and degree-based reductions. The proof is automatic except for the modularisation and arithmetical simplifications that are still needed to overcome the computational complexity.

While the flyable roundabout manoeuvre is a highly nontrivial and challenging study, we still use some modelling assumptions from the literature that should be relaxed in future work, e.g., synchronous, symmetric conflict resolution. Further generalisations include asymmetric entry procedures, different varying cruise speeds, disturbances, or new aircraft. The proof structure behind Theorem 8.1 is already sufficiently general, but the computational complexity is quite high. It would be interesting future work to see if the informal robustness studies of Hwang et al. [171] can be carried over to a formal verification result.

Chapter 9
Conclusion

Hybrid systems are an equally important and challenging class of systems, whose numerous occurrences in safety-critical complex physical systems call for formal verification techniques to establish their correct functioning by rigorous mathematical analysis. The characteristic feature of the modelling idea behind hybrid systems is that they admit interacting discrete and continuous dynamics to capture the superposition of physical system dynamics with control at a natural modelling level. With these superpositions, hybrid systems can model challenging system dynamics fairly easily but also require sophisticated analysis techniques.

Correct functioning of hybrid systems plays an important role in many practical applications. This includes small embedded controllers in automotive industries that regulate isolated processes like air bag inflation, and complex physical traffic systems for train or air traffic control. Similar effects occur in small biomedical devices like glucose regulators following, e.g., model-predictive control, and large-scale physical or chemical process control like injection control in combustion engines or full nuclear reactors. In other domains, hybrid effects also become increasingly important, including in robotic applications, where mobile robots have to work reliably in safety-critical environments, e.g., in driverless vehicle technology. Circuit designs also often exhibit relevant hybrid effects, where increasing clock rates lead to mixed analog and digital effects, because larger parts of the chip remain analog. That happens, for instance, when reducing the number of extra latches that stabilise values computed by analog circuits.

Logic for Hybrid Systems

As a general analysis technique for hybrid systems, we have described a systematic logical analysis approach in this book that is based on symbolic and mathematical logic, automated theorem proving, differential algebra, computer algebra, semialgebraic geometry, analysis, and results from the theory of differential equations and dynamical systems; see Fig. 9.1.

A. Platzer, Logical Analysis of Hybrid Systems,
DOI 10.1007/978-3-642-14509-4_9, © Springer-Verlag Berlin Heidelberg 2010

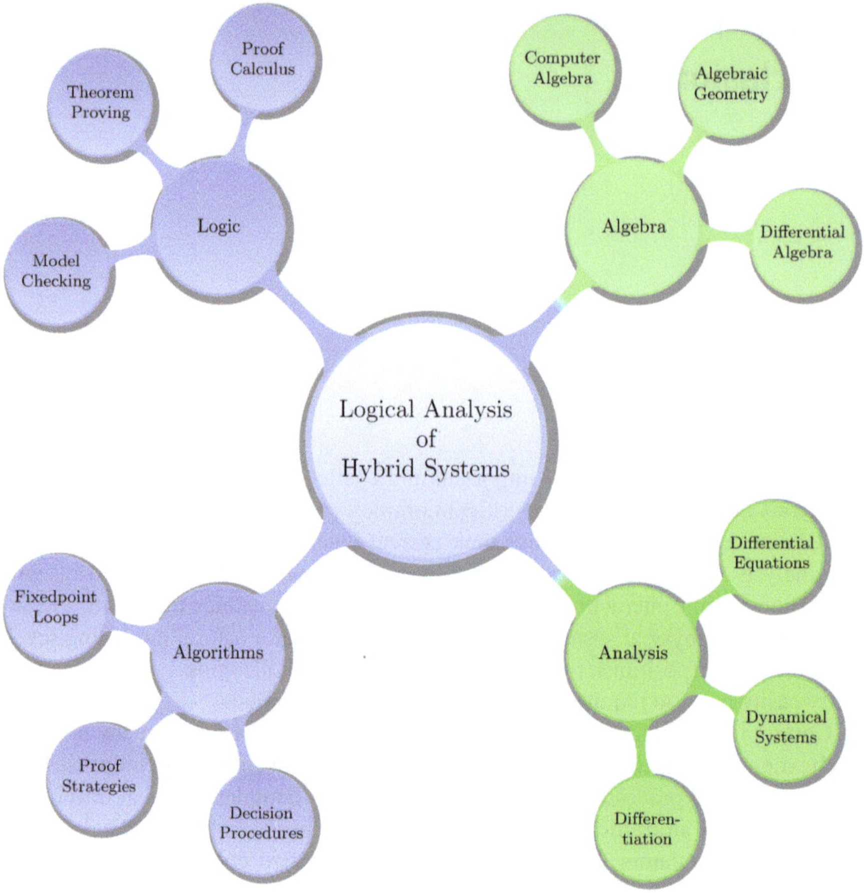

Fig. 9.1 Topics contributing to the logical analysis of hybrid systems

We introduced a series of *differential dynamic logics* ($d\mathcal{L}$, DAL, dTL) as concise languages for specifying correctness properties of hybrid systems along with concise practical proof calculi for verifying hybrid systems. The logic $d\mathcal{L}$ is a first-order dynamic logic for hybrid programs, which extend classical discrete programs to uniform operational models for hybrid systems with interacting discrete jumps and continuous evolutions along differential equations. Its first-order completion, DAL, is a first-order dynamic logic for more general differential-algebraic programs that can combine general first-order discrete jump constraints and first-order differential-algebraic constraints. In a further independent dimension, we augment $d\mathcal{L}$ and DAL to a temporal dynamic logic, dTL, with independent modal path quantifiers over traces and temporal quantifiers along traces, thereby combining the capabilities of dynamic logic to reason about possible system behaviour with the power of temporal logic in reasoning about the temporal behaviour along traces.

Proof Calculi

Our concise proof calculi for differential dynamic logics work compositionally by decomposing properties of hybrid systems symbolically into properties of their parts, which improves both traceability and scalability of results. In order to handle interacting hybrid dynamics, we lift real quantifier elimination to the deductive calculi in a new modular way that is suitable for automation, using a combination of real-valued free variables, Skolem terms, and invertible quantifier rules over the reals.

As a fundamental result aligning hybrid and continuous reasoning proof-theoretically, we have proven our calculi to axiomatise the transition behaviour of hybrid systems sound and complete relative to the handling of differential equations. This result is based on an entirely new notion of hybrid completeness and shows that the calculi are adequate for verification purposes. This is the first relative completeness result for hybrid systems and, in fact, the first notion of hybrid completeness.

Furthermore, we have complemented discrete induction with a new first-order differential induction that uses differential invariants and differential variants for proving correctness statements about first-order differential-algebraic constraints and differential inequalities purely algebraically based on the differential constraints themselves instead of their solutions. This is particularly relevant for formal verification, because solutions of differential equations quickly yield undecidable arithmetic or may not even be expressible in closed form. In combination with successive differential refinement and differential cuts for refining the system dynamics with auxiliary differential invariants, we obtain a powerful verification calculus for hybrid systems with challenging dynamics.

Finally, our sequent calculus for the temporal extension dTL of differential dynamic logic is a completely modular combination of temporal and nontemporal reasoning, where temporal formulas are handled using rules that augment intermediate state transitions with corresponding sub-specifications recursively. We have shown that this gives a sound and complete axiomatisation for temporal properties relative to the nontemporal base logic.

Logic-Based Verification Algorithms

To address practical scalability challenges for larger case studies, we have introduced proof strategies that navigate through the nondeterminisms in the proof calculi for differential dynamic logics, including iterative background closure and iterative inflation order strategies. Further, we have introduced a verification algorithm that computes invariants and differential invariants as fixed points in a proof loop for differential dynamic logic, thereby using the compositional properties of our calculi to exploit locality in system designs for formal verification. Our compositional proof calculi result in a compositional verification approach that is, essentially, a divide-and-conquer approach for hybrid systems verification.

Applications

In a series of examples and case studies from practical application domains, we have demonstrated that our techniques can be used successfully for verifying safety, controllability, and liveness properties in realistic train control applications and challenging case studies for fully curved roundabout manoeuvres in air traffic control.

In a nutshell, we have introduced and developed the theory, practise, and applications of differential dynamic logics, leading to a concise yet complete novel logical analysis approach for hybrid systems with which we can verify applications that were out of scope for other approaches.

Part IV
Appendix

Overview In the appendices, we summarise the background in logic and differential equations that we need in the main parts of this book. In App. A, we give an introduction to basic first-order logic, its syntax, semantics, and proof techniques. For reference, App. B summarises some classical results about differential equations. In App. C, we formally investigate the relationship of hybrid automata and hybrid programs by embedding hybrid automata into hybrid programs. In App. D, we briefly characterise the verification tool KeYmaera that implements the logics and automated theorem proving techniques presented in this book. We also survey various techniques that can be used to prove formulas of real arithmetic.

Appendix A
First-Order Logic and Theorem Proving

Contents

A.1	Overview		**341**
A.2	Syntax		**346**
	A.2.1	Terms	346
	A.2.2	Formulas	347
A.3	Semantics		**348**
	A.3.1	Valuation of Terms	349
	A.3.2	Valuation of Formulas	349
A.4	Proof Calculus		**350**
	A.4.1	Proof Rules	351
	A.4.2	Proof Example: Ground Proving Versus Free-Variable Proving	354
A.5	Soundness		**356**
A.6	Completeness		**356**
A.7	Computability Theory and Decidability		**357**

Synopsis In this chapter, we briefly review basic elements of first-order logic, which is the basis for the logics developed in this book. We define the formal syntax of first-order logic, i.e., the language, in which sentences and formulas of this logic can be written. We further define the semantics of first-order logic, which gives meaning to the syntactic expressions. Depending on the interpretation of their elementary symbols, some formulas then express true facts about the world while others do not. Most informative are those formulas that are valid, i.e., true under all interpretations. Finally we review sequent calculi as systematic proof procedures for proving that a formula is valid in first-order logic.

A.1 Overview

In this chapter, we review first-order logic to a level of detail that is used in this book. A comprehensive treatment of first-order logic and details about automatic proof procedures can be found in the book by Fitting [122]. Gentzen-style sequent

A. Platzer, Logical Analysis of Hybrid Systems,
DOI 10.1007/978-3-642-14509-4, © Springer-Verlag Berlin Heidelberg 2010

calculi can be found in the book by Gallier [130]. As a reference for the specifics of dynamic logic for conventional discrete programs, we recommend the book by Harel et al. [149] and the earlier work by Harel [148]. For further background on first-order logic we refer the reader to the literature [282, 42, 169, 17].

Classical logic studies how truth is preserved in reasoning and investigates valid reasoning schemes of formal proofs. In the context of hybrid systems, logic becomes useful on multiple levels: on the system level to describe the local system operations and on the reasoning level to analyse the global system behaviour. What we are most interested in for hybrid systems is what facts are true about a particular hybrid system at hand. We start with an informal introduction to first-order logic and then review its syntax, its semantics, and a proof calculus.

Logical System Modelling

On the system level, logic is very useful for describing the control logic of the system itself, for instance, for describing the conditions under which the system controller switches to a new mode. In a train control system, for instance, typical conditions in the controller are:

> **If** the train runs faster than the recommended speed, **then** the controller switches to braking mode.
> **If** the train runs slower than the recommended speed **and** the distance to the next train is large enough, **then** the controller switches to acceleration mode, **unless** the train received an emergency message.
> **If** the train received an emergency message, **then** the controller follows emergency procedures.
> **If** the controller follows emergency procedures **and** the emergency situation has **not** been cleared, **then** the controller switches to braking mode.

These switching conditions seem to imply the following consequences:

> **If** the train received an emergency message **and** the emergency situation has **not** been cleared, **or if** the train is running faster than the recommended speed, **then** the controller switches to braking mode.
> **If** the train received no emergency message **and** the distance to the next train is large enough, **then** the controller does **not** switch to braking mode **or** the train runs faster than the recommended speed.

Among many other uses of logic, logical reasoning can be used as a systematic way to analyse whether these conditions are really consequences of the above conditions. In order to find out if they are indeed consequences, we try to formalise the above conditions as logical formulas in first-order logic:

$$\begin{aligned}
&\mathit{faster}(v,r) \to \mathit{brake}(v)\\
&\mathit{slower}(v,r) \land \mathit{far}(z) \land \neg EM \to \mathit{accel}(v)\\
&EM \to EP\\
&EP \land \neg EC \to \mathit{brake}(v)
\end{aligned} \tag{A.1}$$

The first formula directly represents the first condition above using a logical implication. If the train with speed v runs faster than the recommend speed r (which we write as $faster(v,r)$), then (expressed by the implication arrow $\rightarrow$) the train brakes. By v we mean the train's velocity, by r the recommended speed. The predicate $faster(v,r)$ is meant to hold of v and r if v is more than r. By $brake(v)$ we mean that the train brakes when it has velocity v, by $accel(v)$ that it accelerates. Formula EM is meant to be true if an emergency message has been received, EP if the train follows emergency procedures. By EC we mean the condition that is true if the emergency situation has been cleared. Note that the meaning we associate with these symbols is quite arbitrary and subject to our interpretation. The logical formulas themselves are not limited to this particular interpretation of the symbols. The truth of logical formulas depends on the interpretation of symbols, which can be arbitrary. What is fixed, however, is the meaning of the logical operators as indicated in Table A.1.

Table A.1 Intuitive meaning of logical operators in first-order logic

Formula	Read	Informal meaning
$\neg\phi$	"not ϕ"	negation, holds if ϕ is false
$\phi \wedge \psi$	"ϕ and ψ"	conjunction, holds if both ϕ and ψ are true
$\phi \vee \psi$	"ϕ or ψ"	disjunction, holds if ϕ is true or ψ is true
$\phi \rightarrow \psi$	"ϕ implies ψ"	implication, holds if ϕ is false or ψ is true
$\phi \leftrightarrow \psi$	"ϕ equivalent to ψ"	equivalence, holds if ϕ and ψ are either both true or both false
$\forall x\,\phi$	"for all x: ϕ"	universal quantifier, holds if ϕ is true for all values of x
$\exists x\,\phi$	"for some x: ϕ"	existential quantifier, holds if ϕ is true for some value of x
(ϕ)		brackets, only used for grouping logical formulas

Next, we can phrase the switching conditions that we suspected to be consequences as logical formulas using the same symbols:

$$(EM \wedge \neg EC) \vee faster(v,r) \rightarrow brake(v) \tag{A.2}$$

$$\neg EM \wedge far(z) \rightarrow \neg brake(v) \vee faster(v,r) \tag{A.3}$$

Logical Consequences

Now, it makes sense to ask if the original switching conditions (A.1) actually imply the suspected consequences (A.2) and (A.3). Thus we ask if the following first-order formula, which has the conjunction of the formulas in (A.1) on the left hand side of an implication ($\rightarrow$ in the last line) and formula (A.2) on the right hand side, is true under all circumstances:

$$\begin{aligned}\Big(\big(faster(v,r) \rightarrow brake(v)\big) \wedge \big(slower(v,r) \wedge far(z) \wedge \neg EM \rightarrow accel(v)\big) \wedge \\ (EM \rightarrow EP) \wedge \big(EP \wedge \neg EC \rightarrow brake(v)\big)\Big) \\ \rightarrow \ \big((EM \wedge \neg EC) \vee faster(v,r) \rightarrow brake(v)\big)\end{aligned} \tag{A.4}$$

Using a simple proof in first-order logic, this formula turns out to be valid, i.e., true under all interpretations of the symbols. Because it is true under all interpretations, it is also true under the specific train interpretation we had in mind when we started the formalisation.

Let us try to conduct an informal proof of property (A.4). Assume all assumptions are true, because an implication does not say anything unless the assumption on the left of the implication operator ($\rightarrow$) holds. Thus we assume $(EM \land \neg EC) \lor faster(v,r)$ to be true. If this disjunction is true, then the left subformula or the right subformula needs to be true. Now if the disjunction is true because the right subformula $faster(v,r)$ is true, then we are done because, together with the first condition $faster(v,r) \rightarrow brake(v)$, this implies that $brake(v)$ must be true (if $faster(v,r)$ and $faster(v,r) \rightarrow brake(v)$ are true then $brake(v)$ is too). If, instead, the first subformula $EM \land \neg EC$ is true, then both EM and $\neg EC$ must be true, hence EC false. Thus, the combination with condition $EM \rightarrow EP$ implies that EP must be true. Since EP is true but EC is not, condition $EP \land \neg EC \rightarrow brake(v)$ can be used to conclude that $brake(v)$ must be true. Hence we have found out by logical analysis that condition (A.2) is indeed a consequence of conditions (A.1). While successful, this manual reasoning is not very scalable to large systems. Automated theorem proving studies the question how proving can be automated such that formulas such as (A.4) can be proven by a machine. In fact, formula (A.4) can be proven very easily by a computer.

Formula (A.3), however, turns out *not* to be a consequence of formulas (A.1). The reason is that—with our specific train intuition in mind—we had originally associated a special meaning with the symbols $faster, slower, accel$ and $brake$ that is not reflected in the formalisation. First-order logic in its own right, allows an arbitrary interpretation of these symbols, including one where $faster(v,r)$ and $slower(v,r)$ would be true for the same arguments v,r, or where $accel(v)$ and $brake(v)$ would both be true for all v. What we may have had in mind from the names—which are arbitrary though—is that *faster* and *slower* should be opposite notions and that *accel* and *brake* should not hold simultaneously. In order to explain these additional constraints in first-order logic, let us assume that by $faster(v,r)$ we mean exactly the opposite of $slower(v,r)$, that is $faster(v,r)$ is true if and only if $slower(v,r)$ is false. Let us further assume that an accelerating train cannot brake at the same time, and that a braking train cannot accelerate at the same time, while the train could still neither brake nor accelerate when it keeps moving with constant speed. This background knowledge corresponds to the following formulas in first-order logic:

$$\begin{aligned} &slower(v,r) \leftrightarrow \neg faster(v,r) \\ &accel(v) \rightarrow \neg brake(v) \\ &brake(v) \rightarrow \neg accel(v) \end{aligned} \tag{A.5}$$

Observe that logic can be used to show that the last two conditions are actually equivalent! The middle condition says: If the train accelerates, then it does not brake. The last condition says: If the train brakes, then it does not accelerate. But that is already entailed: if the train brakes, then it cannot accelerate for otherwise the middle condition would imply that it does not brake, contradicting the assumption that it indeed brakes. We defer logical reasoning like this and continue to work with all background assumptions, even though some of them turn out to be redundant or superfluous.

Now, with this background knowledge, (A.3) is a consequence of the conjunction of (A.1) and (A.5).

$$\begin{aligned}\Big(\big(\mathit{faster}(v,r) \to \mathit{brake}(v)\big) \land \big(\mathit{slower}(v,r) \land \mathit{far}(z) \land \neg EM \to \mathit{accel}(v)\big) \land \\ \big(EM \to EP\big) \land \big(EP \land \neg EC \to \mathit{brake}(v)\big) \land \\ \big(\mathit{slower}(v,r) \leftrightarrow \neg \mathit{faster}(v,r)\big) \land \big(\mathit{accel}(v) \to \neg \mathit{brake}(v)\big) \land \big(\mathit{brake}(v) \to \neg \mathit{accel}(v)\big)\Big) \\ \to \big(\neg EM \land \mathit{far}(z) \to \neg \mathit{brake}(v) \lor \mathit{faster}(v,r)\big) \quad \text{(A.6)}\end{aligned}$$

The interpretation of *faster* and *slower* is still arbitrary, but now only those interpretations that respect assumption $\mathit{slower}(v,r) \leftrightarrow \neg\mathit{faster}(v,r)$ matter, because the assumptions on the left side of the implication (upper three lines) are not met such that formula (A.6) is true trivially. With assumptions (A.5), we have not characterised the interpretations of *faster* and *slower* uniquely. In fact, there are still many interpretations of what *faster* means, which, indeed, is a somewhat subjective notion. At least, (A.5) characterise just enough background knowledge for our analysis above.

Limitations of First-Order Logic

Our reasoning about the switching conditions of a train in first-order logic has been successful so far. But first-order logic has its limitations that make it difficult and, in general, even impossible to prove all properties of hybrid systems in first-order logic. In reality, variables like velocity v of a train are not restricted to some arbitrary but fixed value (as first-order logic assumes), but, instead, may change its value over time as the train accelerates or brakes. Likewise, the emergency may not have been cleared yet (EC is still false), but may get cleared a minute later (EC suddenly becomes true). These changes in interpretation over time are not well reflected in first-order logic. In particular, plain first-order logic itself does not support reasoning about the differential equations and real arithmetic domains underlying hybrid systems.

Still, first-order logic is a very successful basis for hybrid systems analysis. The differential dynamic logics presented in Part I of this book are extensions of first-order logic and the proof procedures for differential dynamic logic extend first-order logic theorem proving. In this appendix, we thus review first-order logic.

A.2 Syntax of First-Order Logic

In this section, we introduce the syntax of logical formulas in *first-order logic* (FOL). First-order logic is the logic of function symbols, predicate symbols, propositional logical operators, and quantifiers over objects of the domain of discourse.

The language of first-order logic, in which sentences can be written, is defined by a formal syntax. This section presents the syntactic structure of the language of first-order logic. Given a vocabulary of symbols the language of the logic will be inductively defined by combining them using logical operators.

A.2.1 Terms

The construction of a logic starts with a *signature* Σ, which is the set of names (called *symbols*) of all entities nameable in a certain context. A signature is the vocabulary or alphabet of symbols or signs from which well-formed formulas can be built. Formulas of first-order logic are built over a set V of logical variable symbols and a signature Σ of function and predicate symbols. The difference between function and predicate symbols is that function symbols stand for functions that take the values of arguments and give back a function value. Predicate symbols, in contrast, are either true or false of their arguments. That is, they take the values of arguments and give either the truth-value "*true*" or the truth-value "*false*". No other result is permitted for predicate symbols. Function symbols are often written as f,g,h,a,b,c and predicate symbols are often written as p,q,r.

In addition to declaring which function and predicate symbols are allowed, the signature Σ further declares the arity of each function and predicate symbol, i.e., the number of arguments it expects. Formally, arity can be understood as a mapping $\Sigma \to \mathbb{N}$. Arity 0 means that the function or predicate symbol does not expect (nor admit) any arguments. A common short notation to specify the arities of function and predicate symbols is

$$\Sigma = \{a/0, c/0, f/1, g/2;\; p/1, r/2\}. \tag{A.7}$$

This notation would say that a and c are function symbols without arguments (0), f is a function symbol that expects one argument, and g a function symbol that expects two arguments. Further p would be a predicate symbol with one argument and r a predicate symbol with two arguments. In the above notation in (A.7), predicate symbols are separated from function symbols by a semicolon (;). In addition to function and predicate symbols from the signature Σ, we also allow logical variables from a set V that stand for objects. Logical variables are often written as x,y,z.

Well-formed arguments to function symbols and predicate symbols are called terms. Variables are well-formed terms, and functions applied to the appropriate number of terms as arguments are well-formed terms.

Definition A.1 (Terms). $\text{Trm}(\Sigma)$ is the set of all *terms*, which is the smallest set such that:

- If $x \in V$ is a logical variable, then $x \in \text{Trm}(\Sigma)$.
- If $f \in \Sigma$ is a function symbol of arity $n \geq 0$ and, for $1 \leq i \leq n$, $\theta_i \in \text{Trm}(\Sigma)$, then $f(\theta_1, \ldots, \theta_n) \in \text{Trm}(\Sigma)$. The case $n = 0$ is permitted.

More succinctly, we also say that the terms of first-order logic are defined by the following grammar (where $\theta_1, \ldots, \theta_n$ are terms, f a function symbol of arity n, and $x \in V$ is a logical variable):

$$\theta ::= x \mid f(\theta_1, \ldots, \theta_n).$$

Example A.1. When we assume the signature from (A.7) and the set of variables $V = \{x, y, z\}$, then the following are well-formed terms:

- $f(a)$
- $f(f(g(a, f(x))))$
- $g(f(a), f(c))$
- $f(g(g(a, f(a)), g(c, f(c))))$
- $f(x, g(g(y, a), x))$

The following, however, are no terms with respect to the chosen Σ and V:

- $f(a, c)$ — wrong number of arguments for f; exactly one argument is expected
- $g(c, f)$ — f is not a term but a function symbol that expects two arguments, not zero
- $g(g(f(a)$ — incomplete term
- $f(g(a, b))$ — b is not in V or in Σ according to (A.7). The expression is a term if we add b to V or use $\Sigma \cup \{b/0\}$ instead of Σ. □

A.2.2 First-Order Formulas

The well-formed formulas of a logic form a formal language over the alphabet $\Sigma \cup V$. The formulas consist of all words that can be built by recursively combining symbols of the signature with logical operator symbols. Formulas are defined as follows.

Definition A.2 (First-order formulas). The set $\text{Fml}_{\text{FOL}}(\Sigma)$ of *formulas of first-order logic* is the smallest set with:

- If $p \in \Sigma$ is a predicate symbol of arity $n \geq 0$ and $\theta_i \in \text{Trm}(\Sigma)$ for $1 \leq i \leq n$, then $p(\theta_1, \ldots, \theta_n) \in \text{Fml}_{\text{FOL}}(\Sigma)$.
- If $\phi, \psi \in \text{Fml}_{\text{FOL}}(\Sigma)$, then $\neg\phi, (\phi \wedge \psi), (\phi \vee \psi), (\phi \rightarrow \psi) \in \text{Fml}_{\text{FOL}}(\Sigma)$.
- If $\phi \in \text{Fml}_{\text{FOL}}(\Sigma)$ and $x \in V$, then $(\forall x\, \phi), (\exists x\, \phi) \in \text{Fml}_{\text{FOL}}(\Sigma)$.

See Table A.1 on p. 343 for an overview of operators in logical formulas. More succinctly, we also say that first-order formulas are defined by the following grammar (where ϕ, ψ are first-order formulas, θ_i are terms, p a predicate symbol of arity n, and $x \in V$ is a logical variable):

$$\phi, \psi ::= p(\theta_1, \ldots, \theta_n) \mid \neg\phi \mid \phi \wedge \psi \mid \phi \vee \psi \mid \phi \rightarrow \psi \mid \forall x \phi \mid \exists x \phi.$$

Example A.2. When we again use the signature from (A.7) and the set of variables $V = \{x, y, z\}$, then the following are well-formed formulas:

- $p(a) \rightarrow r(a, f(c))$
- $p(x) \rightarrow (r(x, f(c)) \vee \neg p(c))$
- $\forall x\, (p(x) \rightarrow r(f(x), f(c)))$
- $\big(p(a) \vee \neg(r(a,a) \wedge r(f(a), f(a)))\big) \leftrightarrow (\neg p(c) \wedge r(c,c))$
- $\forall x\, \big(p(f(x)) \rightarrow \exists y\, r(y, f(x))\big)$
- $r(a,y) \leftrightarrow \neg\forall x \exists y \forall z\, \big(r(x,y) \wedge r(y, g(x,z))\big)$

The following, however, are not formulas with respect to the chosen Σ and V:

- $p(a) \rightarrow r(g(a))$ — second argument of r missing; second argument of g missing
- $r(x,y) \wedge \rightarrow p(y)$ — formula missing between operators $\wedge$ and $\rightarrow$
- $p(a) \rightarrow \forall x$ — formula missing after quantifier
- $p(a) \rightarrow \forall r\, r(a, f(a))$ — r is not a variable symbol in V but a predicate symbol in Σ; thus r cannot be quantified over (in first-order logic). Predicate symbol r stands for a predicate, not for an individual object. The logic where quantification over functions and predicates is allowed is called higher-order logic. □

Example A.3 (Trains). The examples in App. A.1 are formulas with respect to $V = \{v, z\}$ and the following signature, where r is the only function symbol:

$$\Sigma = \{r/0;\ \mathit{EM}/0, \mathit{EP}/0, \mathit{EC}/0, \mathit{slower}/2, \mathit{faster}/2, \mathit{accel}/1, \mathit{brake}/1, \mathit{far}/1\}.$$

□

A.3 Semantics of First-Order Logic

The semantics of formulas of first-order logic depends on the interpretation of its symbols. Given such an interpretation of each of the symbols, the truth-value of the full formula is defined from the meaning of its logical operators.

A.3.1 Valuation of Terms

An *interpretation* I chooses some non-empty set D_I as the universe of objects and it assigns functions and relations over D_I to the respective function and predicate symbols in Σ. In particular, if f is a function symbol of arity n, then $I(f)$ is a function $I(f) : D_I^n \to D_I$ with n arguments. Here $D_I^n := \{(d_1,\dots,d_n) \ : \ d_i \in D_I\}$ is the nth crossproduct of D_I, i.e., the set of n-tuples of D_I. Thus $I(f)(d_1,\dots,d_n) \in D_I$ is the value of the function associated with f by I at position $(d_1,\dots,d_n) \in D_I^n$. Further, if p is a predicate symbol of arity n, then $I(p)$ would be a relation, i.e., a subset $I(p) \subseteq D_I^n$. Thus $(d_1,\dots,d_n) \in I(p)$ if and only if the predicate associated with p holds true at position $(d_1,\dots,d_n) \in D_I^n$. Interchangeably, we use a slightly different notation and consider the interpretation $I(p)$ of a predicate symbol to be the characteristic function $I(p) : D_I^n \to \{true, false\}$ of the predicate. Then p holds true at position $(d_1,\dots,d_n) \in D_I^n$ if and only if $I(p)(d_1,\dots,d_n) = true$. The difference between function symbols and predicate symbols then is that the former have values in D_I and the latter have values in $\{true, false\}$.

The meaning of a formula further depends on the interpretation of its variable symbols from V. An *assignment for logical variables* is a map $\eta : V \to D_I$ that assigns an object of the universe of I to each variable.

Given an interpretation I of function (and predicate symbols) and an assignment η of logical variables, we can evaluate terms inductively.

Definition A.3 (Valuation of terms). The *valuation* $val_{I,\eta}(\cdot)$ *of terms* with respect to interpretation I and assignment η is defined by

1. $val_{I,\eta}(x) = \eta(x)$ if $x \in V$ is a logical variable.
2. $val_{I,\eta}(f(\theta_1,\dots,\theta_n)) = I(f)\big(val_{I,\eta}(\theta_1),\dots,val_{I,\eta}(\theta_n)\big)$ when $f \in \Sigma$ is a function symbol of arity $n \geq 0$.

A.3.2 Valuation of First-Order Formulas

Given an interpretation I of function and predicate symbols and an assignment η of logical variables, we can evaluate first-order formulas. We will use $\eta[x \mapsto d]$ to denote the assignment that agrees with assignment η except on variable $x \in V$, which is assigned $d \in D_I$ instead.

Definition A.4 (Valuation of first-order formulas). The *valuation*, $val_{I,\eta}(\cdot)$, *of first-order formulas* with respect to interpretation I and assignment η is defined as

1. $val_{I,\eta}(p(\theta_1,\dots,\theta_n)) = I(p)\big(val_{I,\eta}(\theta_1),\dots,val_{I,\eta}(\theta_n)\big)$
2. $val_{I,\eta}(\phi \wedge \psi) = true$ iff $val_{I,\eta}(\phi) = true$ and $val_{I,\eta}(\psi) = true$
3. $val_{I,\eta}(\phi \vee \psi) = true$ iff $val_{I,\eta}(\phi) = true$ or $val_{I,\eta}(\psi) = true$
4. $val_{I,\eta}(\neg\phi) = true$ iff $val_{I,\eta}(\phi) \neq true$
5. $val_{I,\eta}(\phi \to \psi) = true$ iff $val_{I,\eta}(\phi) \neq true$ or $val_{I,\eta}(\psi) = true$

6. $val_{I,\eta}(\forall x\, \phi) = true$ iff $val_{I,\eta[x\mapsto d]}(\phi) = true$ for all $d \in \mathbb{R}$
7. $val_{I,\eta}(\exists x\, \phi) = true$ iff $val_{I,\eta[x\mapsto d]}(\phi) = true$ for some $d \in \mathbb{R}$

We also write $I, \eta \models \phi$ iff $val_{I,\eta}(\phi) = true$ and say that $I, \eta\, \phi$ *satisfies* ϕ. Dually, we write $I, \eta \not\models \phi$ iff $val_{I,\eta}(\phi) \neq true$. We write just $\models \phi$ iff $I, \eta \models \phi$ holds for all I, η. We say that ϕ is *valid* if $\models \phi$ i.e., ϕ is true in all interpretations and under all assignments. The study of valid formulas is an important part of logic, in particular of systematic ways to establish validity by proofs.

Example A.4 (Trains). Continuing Example A.3, consider one particular interpretation I_0 with the universe $D_{I_0} := \{0,1,2,3,\dots,100\}$ of natural numbers up to 100 and the following interpretations:

$$
\begin{aligned}
I_0(r) &= 90,\\
I_0(EM) &= \mathit{false},\\
I_0(EP) &= \mathit{false},\\
I_0(EC) &= \mathit{true},\\
I_0(\mathit{slower}) &= \{(d,e) \in \{0,1,2,\dots,100\} \ :\ d < e - 10\},\\
I_0(\mathit{faster}) &= \{(d,e) \in \{0,1,2,\dots,100\} \ :\ d > e\},\\
I_0(\mathit{accel}) &= \{d \in \{0,1,2,\dots,100\} \ :\ d > 0, d < 50\},\\
I_0(\mathit{brake}) &= \{d \in \{0,1,2,\dots,100\} \ :\ d > 80\},\\
I_0(\mathit{far}) &= \{d \in \{0,1,2,\dots,100\} \ :\ d < 20\}.
\end{aligned}
$$

Further, consider the particular assignment η_0 with $\eta_0(v) = 85$ and $\eta_0(z) = 10$. Interpretation I_0 and assignment η_0 correspond to one particular situation or system snapshot in train control. In this interpretation, formula (A.4) evaluates to true, as it does in all other interpretations, i.e., (A.4) is valid. In I_0, η_0, however, formula (A.3) evaluates to false, although all formulas in (A.1) evaluate to true. Since we argued that the similar formula (A.6) is valid under all interpretations, but the right-hand side (A.3) of its implication is false, this means that one of the assumptions from A.5 must be false in I_0, η_0. Indeed, $I_0, \eta_0 \not\models \mathit{slower}(v,r) \leftrightarrow \neg \mathit{faster}(v,r)$, because $val_{I_0,\eta_0}(\mathit{slower}(v,r)) = \mathit{false}$ but $val_{I_0,\eta_0}(\neg \mathit{faster}(v,r)) = \mathit{true}$, which, in turn, holds because $val_{I_0,\eta_0}(\mathit{faster}(v,r)) = \mathit{false}$. □

A.4 A Sequent Proof Calculus for First-Order Logic

After having defined syntax and semantics of first-order logic, we know how well-formed formulas look and how the truth-value of a formula is defined for an interpretation of its symbols. The most interesting formulas for us are those that are valid, i.e., true in all interpretations and assignments, because they will hold under all circumstances including any particular interpretation that we could have in mind.

The question is, how can these valid formulas be identified? Given a formula, how can we find out if it is actually valid?

If a formula is not valid, then it has an interpretation (and assignment) where it is false. Such an interpretation (and assignment) then is a witness for the fact that the formula is not valid. This witness is also known as a *counterexample* to the claim that the formula is valid. It can be difficult to find such a counterexample, but once we have it, it is often fairly easy to check. But if a formula is valid, how could we show that? Certainly, enumerating all interpretations and checking if the formula is true in each of these is a hopeless enterprise, because there is an infinite number of interpretations. Nevertheless, clever refinements of enumeration can give a proof procedure following the famous Herbrand theorem [160] and recent techniques based mostly on instantiation [30, 229, 29, 131].

A witness for a formula that is not valid is a counterexample, while a witness for a formula that is valid is a formal proof. In our setting, a formal proof is a derivation of the formula using a set of valid proof rules from axioms or formulas that are obviously valid for simple structural reasons, such as $\phi \rightarrow \phi$. Finding a proof can again be a difficult task, but once we have found it, the proof can be checked easily for compliance with the proof rules to establish validity. In fact, in a goal-directed sequent calculus like the one we show, the proof rules are written in a way that helps finding proofs by working in a systematic way to analyse the original formula. A full treatment of how the proving process can be automated for first-order logic is beyond the scope of this book. We refer the reader to the book by Fitting [122] for more details on automated theorem proving in first-order logic.

A.4.1 Proof Rules for First-Order Logic

Sequents are essentially a standard form for logical formulas that are convenient for proving. A *sequent* is of the form $\Gamma \vdash \Delta$, where the *antecedent* Γ and *succedent* Δ are finite sets of formulas. The semantics of sequent $\Gamma \vdash \Delta$ is that of the formula $\bigwedge_{\phi \in \Gamma} \phi \rightarrow \bigvee_{\psi \in \Delta} \psi$. The antecedent Γ can be thought of as the formulas we assume to be true, whereas the succedent Δ can be understood as formulas for which we want to show that at least one of them is true assuming all formulas of Γ are true. So for proving a sequent $\Gamma \vdash \Delta$, we assume all Γ and want to show that one of the Δ is true.

Standard propositional rules are listed in Fig. A.1. They decompose the propositional structure of formulas. Rules $\neg$r and $\neg$l use simple dualities caused by the implicative semantics of sequents. Essentially, instead of showing $\neg\phi$, we assume the contrary ϕ in rule $\neg$r. In rule $\neg$l, instead of assuming $\neg\phi$, we show the contrary ϕ. Rule $\vee$r uses the fact that formulas are combined disjunctively in succedents. Rule $\wedge$l uses the fact that they are conjunctive in antecedents. The comma between formulas in an antecedent has the same effect as a conjunction, and the comma between formulas in the succedent has the same effect as a disjunction. Rules $\vee$l and $\wedge$r split the proof into two cases, because conjuncts in the succedent can be

$$(\neg r)\frac{\Gamma,\phi \vdash \Delta}{\Gamma \vdash \neg\phi,\Delta} \qquad (\vee r)\frac{\Gamma \vdash \phi,\psi,\Delta}{\Gamma \vdash \phi \vee \psi,\Delta} \qquad (\wedge r)\frac{\Gamma \vdash \phi,\Delta \quad \Gamma \vdash \psi,\Delta}{\Gamma \vdash \phi \wedge \psi,\Delta}$$

$$(\neg l)\frac{\Gamma \vdash \phi,\Delta}{\Gamma,\neg\phi \vdash \Delta} \qquad (\vee l)\frac{\Gamma,\phi \vdash \Delta \quad \Gamma,\psi \vdash \Delta}{\Gamma,\phi \vee \psi \vdash \Delta} \qquad (\wedge l)\frac{\Gamma,\phi,\psi \vdash \Delta}{\Gamma,\phi \wedge \psi \vdash \Delta}$$

$$(\rightarrow r)\frac{\Gamma,\phi \vdash \psi,\Delta}{\Gamma \vdash (\phi \rightarrow \psi),\Delta} \qquad (ax)\frac{}{\Gamma,\phi \vdash \phi,\Delta}$$

$$(\rightarrow l)\frac{\Gamma \vdash \phi,\Delta \quad \Gamma,\psi \vdash \Delta}{\Gamma,(\phi \rightarrow \psi) \vdash \Delta} \qquad (cut)\frac{\Gamma \vdash \phi,\Delta \quad \Gamma,\phi \vdash \Delta}{\Gamma \vdash \Delta}$$

$$(\exists r)\frac{\Gamma \vdash \phi(\theta),\exists x\,\phi(x),\Delta}{\Gamma \vdash \exists x\,\phi(x),\Delta}^{1} \qquad (\forall r)\frac{\Gamma \vdash \phi(s(X_1,..,X_n)),\Delta}{\Gamma \vdash \forall x\,\phi(x),\Delta}^{2}$$

$$(\forall l)\frac{\Gamma,\phi(\theta),\forall x\,\phi(x) \vdash \Delta}{\Gamma,\forall x\,\phi(x) \vdash \Delta}^{1} \qquad (\exists l)\frac{\Gamma,\phi(s(X_1,..,X_n)) \vdash \Delta}{\Gamma,\exists x\,\phi(x) \vdash \Delta}^{2}$$

[1] θ is an arbitrary term, usually a new logical variable X.
[2] s is a new (Skolem) function and $X_1,..,X_n$ are all free logical variables of $\forall x\,\phi(x)$.

Fig. A.1 Rule schemata of the sequent calculus for first-order logic

proven separately ($\wedge$r) and, dually, disjuncts of the antecedent can be assumed separately ($\vee$l). For rule $\wedge$r we want to show conjunction $\phi \wedge \psi$, so in the left branch we proceed to show $\Gamma \vdash \phi,\Delta$, and in the right branch we show $\Gamma \vdash \psi,\Delta$. If, as in rule $\vee$l, we assume disjunction $\phi \vee \psi$ as part of the antecedent, then we do not know whether we can assume ϕ to hold or whether we can assume ψ to hold, but can only assume that one of them holds. Hence, as in a case distinction, rule $\vee$l considers both cases, the case where we assume ϕ in the antecedent, and the case where we assume ψ. If we have proven both subgoals, the subgoal assuming ϕ and the subgoal assuming ψ, then we have also justified the goal assuming only their disjunction $\phi \vee \psi$. Rules $\rightarrow$r and $\rightarrow$l can be derived from the equivalence of $\phi \rightarrow \psi$ and $\neg\phi \vee \psi$. Rule $\rightarrow$r uses that implication $\rightarrow$ has the same meaning as the turnstile arrow $\vdash$ of a sequent. Intuitively, to show implication $\phi \rightarrow \psi$, rule $\rightarrow$r assumes ϕ (in the antecedent) and shows ψ (in the succedent). Rule $\rightarrow$l assumes an implication $\phi \rightarrow \psi$ to hold in the antecedent, but we do not know if this implication holds because ϕ is false or because ψ is true, so rule $\rightarrow$l splits into those two branches.

The axiom rule *ax* closes a goal (there are no further subgoals), because assumption ϕ in the antecedent trivially entails ϕ in the succedent (sequent $\Gamma,\phi \vdash \phi,\Delta$ is a simple syntactic tautology). The *cut* rule can be used for case distinctions: The right subgoal assumes any additional formula ϕ in the antecedent that the left subgoal shows in the succedent. Dually: regardless of whether ϕ is actually true or false, both cases are covered by proof branches.

Rules $\exists$r,$\forall$l,$\forall$r,$\exists$l are standard proof rules for first-order logic. For explaining these quantifier proof rules, let us first assume for a moment there are no free variables $X_1,\ldots,X_n$ (i.e. $n = 0$) and use what is known as the ground calculus.

The quantifier proof rules work much as in mathematics. Consider $\forall$r, where we want to show a universally quantified property. When a mathematician wants to show a universally quantified property $\forall x\, \phi(x)$ to hold, he could choose a fresh symbol s (called Skolem function symbol) and prove that $\phi(s)$ holds (for s). Then the mathematician would remember that s was arbitrary and his proof did not assume anything special about the value of s. So he would conclude that $\phi(s)$ must indeed hold for all s, and that hence $\forall x\, \phi(x)$ holds true. For example, to show that the square of all numbers is nonnegative, a mathematician could start out by saying "let s be an arbitrary number", prove $s^2 \geq 0$ for s, and then conclude $\forall x\,(x^2 \geq 0)$, since s was arbitrary. Proof rule ∀r essentially makes this reasoning formal. It chooses a *new* (function) symbol s and replaces the universally quantified formula in the succedent by a formula for s (with all free logical variables $X_1, \ldots, X_n$ added as arguments, as we explain below). Notice, of course, that it is important to choose a new symbol s that has not been used (in the sequent) before. Otherwise, we would assume special properties about s that may not be justified.

Consider ∃r, where we want to show an existentially quantified property. When a mathematician proves $\exists x\, \phi(x)$, he could directly produce any witness θ for this existential property and prove that, indeed, $\phi(\theta)$, for then he would have shown $\exists x\, \phi(x)$ with this witness. For example, to show that there is a number whose cube is less than its square, a mathematician could start by saying "let me choose 0.5 and show the property for 0.5". Then he could prove $0.5^3 < 0.5^2$, because $0.125 < 0.25$, and conclude that there, thus, is such a number, i.e., $\exists x\,(x^3 < x^2)$. Proof rule ∃r does that. It allows the choice of *any* term θ for x and accepts a proof of $\phi(\theta)$ as a proof of $\exists x\, \phi(x)$. However note that the claim "θ is a witness" may turn out to be wrong, for example, the choice 2 for x would be a bad start for attempting to show $\exists x\,(x^3 < x^2)$. Consequently, proof rule ∃r keeps both options $\phi(\theta)$ and $\exists x\, \phi(x)$ in the succedent. If the proof with θ is successful, the sequent is valid and the part of the proof can be closed successfully. If the proof with θ later turns out to be unsuccessful, another attempt can be used to prove $\exists x\, \phi(x)$, e.g., by applying ∃r again with another attempt for a different witness θ_2.

Rules ∀l,∃l are dual to ∃r,∀l. Consider ∀l, where we have a universally quantified formula in the assumptions (antecedent) that we can use, and not in the succedent, which we want to show. In mathematics, when we know a universal fact, we can use this knowledge for any particular instance. If we know that all positive numbers have a square root, then we can also use the fact that 5 has a square, because 5 is a positive number. Hence from assumption $\forall x\,(x > 0 \to \mathit{hasSqrt}(x))$ in the antecedent, we can also assume instance $5 > 0 \to \mathit{hasSqrt}(5))$. Rule ∀l can produce an instance $\phi(\theta)$ for arbitrary terms θ of the assumption $\forall x\, \phi(x)$. Since we may need the universal fact $\forall x\, \phi(x)$ for multiple instantiations with $\theta_1, \theta_2, \theta_3$ during the proof, rule ∀l keeps the assumption $\forall x\, \phi(x)$ in the antecedent so that it can be used repeatedly.

Consider rule ∃l in which we can use an existentially quantified formula from the antecedent. In mathematics, if we know an existential fact, then we can give a name to the object that we then know does exist. If we know that there is a smallest integer less than 10 that is a square, we can call it s, but we cannot denote it by a different term like 5, because 5 may be (and in fact is) the wrong answer. Rule ∃l gives a fresh name s (with all logical variables $X_1, \ldots, X_n$ as arguments) to the object that exists. Since it does not make sense to give a different name for the same existing object later, $\exists x\,\phi(x)$ is removed from the antecedent when adding $\phi(s(X_1,\dots,X_n))$.

There are two ways of using the proof rules in Fig. A.1. One way is to avoid free variables X_i altogether and only choose ground terms without variables for instantiations θ in ∃r,∀l. Then the Skolem functions used in ∀r,∃l have $n = 0$ free logical variables $X_1,\dots,X_n$ as arguments. This case is called a *ground calculus*, because free variables are never used and all term instantiations are ground (no free variables).

The other way is to work with free variables and always use some fresh logical variable X for instantiation of θ every time ∃r,∀l are used. This is a free-variable calculus [147, 122, 123] where ∃r,∀l are called γ-rules and ∀r,∃l are called δ^+-rules [147], which is an improvement of what is known as the δ-rule [122, 123]. This case is called a *free-variable calculus*, because instantiations are with free variables. At the end of the proof, these free variables can be instantiated by a global substitution on the full proof to close all branches. The free variables $X_1,\dots,X_n$ in the Skolem terms keep track of the dependencies of symbols and prevent instantiations where we instantiate X_1 by a term such as $s(X_1,\dots,X_n)$ depending on X_1. The ground calculus and free-variable calculus uses of Fig. A.1 can also be mixed.

A.4.2 Proof Example: Ground Proving Versus Free-Variable Proving

Figure A.2a shows a proof in the FOL calculus from Fig. A.1 in the ground calculus version (without using free logical variables) of formula $\exists y \forall x\, c(x,y) \to \forall x \exists y\, c(x,y)$. This property says that if there is one common y for all x such that $c(x,y)$, then for each of those x there also is a y such that $c(x,y)$. Proofs are constructed bottom-up starting with the conjecture $\exists y \forall x c(x,y) \to \forall x \exists y c(x,y)$ at the bottom, and working backwards to the top by using proof rules from Fig. A.1.

$$
\dfrac{\dfrac{\dfrac{\dfrac{\dfrac{\dfrac{*}{\forall x c(x,r), c(s,r) \vdash \exists y c(s,y), c(s,t), c(s,r)}\,ax}{\forall x c(x,r), c(s,r) \vdash \exists y c(s,y), c(s,t)}\,\exists r}{\forall x c(x,r) \vdash \exists y c(s,y), c(s,t)}\,\forall l}{\exists y \forall x c(x,y) \vdash \exists y c(s,y), c(s,t)}\,\exists l}{\exists y \forall x c(x,y) \vdash \exists y c(s,y)}\,\exists r}{\exists y \forall x c(x,y) \vdash \forall x \exists y c(x,y)}\,\forall r
$$

Fig. A.2a Ground proof example

$$
\dfrac{\dfrac{\dfrac{\dfrac{\dfrac{*[X \mapsto s][Y \mapsto t]}{\forall x c(x,t), c(X,t) \vdash \exists y c(s,y), c(s,Y)}\,ax}{\forall x c(x,t) \vdash \exists y c(s,y), c(s,Y)}\,\forall l}{\exists y \forall x c(x,y) \vdash \exists y c(s,y), c(s,Y)}\,\exists l}{\exists y \forall x c(x,y) \vdash \exists y c(s,y)}\,\exists r}{\exists y \forall x c(x,y) \vdash \forall x \exists y c(x,y)}\,\forall r
$$

Fig. A.2b Free-variable proof example

One of the practical difficulties with rules ∃r,∀l is how to choose the right instantiation θ that will help close the proof. This is also visible in Fig. A.2a, where the first choice for the instantiation of ∃r with t is not successful for the proof, because the Skolem function symbol for the subsequent ∃l application must be new; hence it is called r. But $\Gamma, c(s,r) \vdash c(s,t), \Delta$ is not an instance of axiom *ax*, so the proof cannot yet be closed unless $\exists$r is used again to instantiate with r. After this, the proof closes successfully by axiom rule *ax* (the notation $*$ marks the successful end of a proof).

Thus, the primary difficulty when using ground calculi is the need to find smart instantiations that will only turn out to work or fail much later in the proof. For automated theorem provers, it is hard to predict which instantiations are likely to work or fail, although very powerful heuristics and even successful methods based entirely on instantiation [30, 229, 29, 131] have been developed already.

Figure A.2b shows a proof of the same first-order formula in the free-variable calculus version of Fig. A.1. In particular, $\exists$r,$\forall$l instantiate the formulas just with new free variables X and Y, respectively, and the actual instantiations for X and Y are chosen only at the end of the proof, once it becomes obvious which choices prove the formula. The proof closes with the choice s for free variable X and the choice t for Y, which can be detected easily because this instance of $\Gamma, c(X,t) \vdash c(s,Y), \Delta$ can be closed by *ax*. We mark this substitution for closing successfully by $*[X \mapsto s][Y \mapsto t]$ at the end of the proof. Notice also that the free-variable calculus proof Fig. A.2b here uses only one application of $\exists$r, not two applications as in Fig. A.2a, because the smart instantiation choice can easily be deferred until the end of the proof.

In contrast, the failed proof attempt in Fig. A.3 does not close, because we would

$$
\begin{array}{ll}
 & \text{not closed} \\
\hline
 & c(X,t(X)), \forall x \exists y\, c(x,y) \vdash c(s(Y),Y), \exists y \forall x\, c(x,y) \\
\hline
\exists\text{l} & \exists y\, c(X,y), \forall x \exists y\, c(x,y) \vdash c(s(Y),Y), \exists y \forall x\, c(x,y) \\
\hline
\forall\text{r} & \exists y\, c(X,y), \forall x \exists y\, c(x,y) \vdash \forall x\, c(x,Y), \exists y \forall x\, c(x,y) \\
\hline
\forall\text{l} & \forall x \exists y\, c(x,y) \vdash \forall x\, c(x,Y), \exists y \forall x\, c(x,y) \\
\hline
\exists\text{r} & \forall x \exists y\, c(x,y) \vdash \exists y \forall x\, c(x,y) \\
\hline
\rightarrow\text{r} & \vdash \forall x \exists y\, c(x,y) \rightarrow \exists y \forall x\, c(x,y)
\end{array}
$$

Fig. A.3 Wrong proof attempt in first-order logic

need to make $c(X,t(X))$ and $c(s(Y),Y)$ identical by substituting terms for X and Y. But if we try to substitute $s(Y)$ for X, then we have to apply this substitution everywhere. In particular, we have to substitute $t(X)$ for Y, which after applying the X substitution corresponds to replacing Y with $t(s(Y))$. This, however, is recursive, as Y occurs inside its own replacement (also known as occurs check). If we try to replace Y by $t(s(Y))$, then we have to apply this replacement everywhere. Thus, we no longer have to substitute $s(Y)$ for X, but now we have to substitute $s(t(s(Y)))$ for X. But then Y needs to be identical to $t(X)$; thus we need to replace Y with $t(s(t(s(Y))))$. This process never terminates! Formally, we say that $c(X,t(X))$ and $s(Y),Y)$ cannot be unified [265].

Yet the fact that we cannot prove the property in Fig. A.3 is actually good news! The formula $\forall x \exists y\, c(x,y) \rightarrow \exists y \forall x\, c(x,y)$ that we were trying to prove in the first place is not valid, so we should never be able to prove it at all. This formula expresses the false statement that, if every x has a y for which $c(x,y)$ holds, then there also is a single y such that for all x the property $c(x,y)$ holds. But being able to choose the same common y for all x is a much stronger property than just being able to choose some (possibly different) y for every x separately. Note that the attempt in Fig. A.3 also shows that it is necessary to keep the dependency on all free variables as arguments in the Skolem terms $s(X)$ and $t(Y)$. If the proof rules ∀r and ∃l would not add the free variables X and Y as arguments, then nothing would have prevented us from closing the proof in Fig. A.3 by substituting s for X and t for Y. Consequently, the Skolem dependencies, which correspond to the dependencies and orders between the quantifiers, are necessary to make ∀r and ∃l sound, i.e., ensure that they only prove valid formulas.

A.5 Soundness

The proof calculus for first-order logic in App. A.4 needs to fit the semantics of first-order logic from App. A.3. Fortunately, every first-order logic formula that can be derived in the FOL calculus from Fig. A.1 is a valid formula. This property of the calculus is called *soundness* and is crucial, because it would be disastrous if some formula would be called "proven" when it is actually not valid. For instance, being able to prove the invalid formula in Fig. A.3 would be disastrous. We certainly would not want to rely on safety-critical embedded systems that were built based on any wrong reasoning principles.

A calculus is sound iff every formula that can be derived in the calculus is also valid according to the semantics.

Theorem A.1 (Soundness of FOL). *The calculus for first-order logic is sound.*

The proof is a version of a classical soundness proof for first-order logic [147, 122, 130] and shows that every instance of the proof rules is sound, because valid premises imply that the conclusion is valid. The soundness proof for the ∀r,∃l can be found in the article by Hähnle and Schmitt [147].

A.6 Completeness

Soundness shows that all provable formulas are valid. So we know that we will never prove something that does not even hold (is not valid). The converse question is about whether all valid formulas are also provable, i.e., whether we will always be able to prove all formulas that are true (or, more precisely, valid).

Theorem A.2 (Completeness of FOL). *The first-order logic calculus is* complete, *i.e., every valid first-order logic formula can be derived in the first-order calculus.*

Completeness of a first-order logic proof calculus was first proven by Kurt Gödel [135, 136]. The proof technique was simplified considerably by Leon Henkin [155], and such Henkin completeness proofs are used very successfully today. Completeness proofs for modern first-order calculi that are such as the one in Fig. A.1 can be found in the literature [147, 122, 130].

As a corollary to the soundness and completeness results, a proof calculus for first-order logic such as that in Fig. A.1 is "perfect" in the sense that it can successfully prove exactly the valid formulas. Still, numerous improvements and refinements are needed to make first-order logic proving successful in practise.

A.7 Computability Theory and Decidability

In this section, we provide a brief, informal introduction to decidability, semidecidability, and undecidability in the context of logic. We refer the reader to the literature [280] for a formal introduction to computability theory and related notions.

First-order logic has a sound and complete calculus that we have shown in App. A.4. In particular, given any formula of first-order logic, we can use the proof calculus to check whether the formula is valid by systematically generating proofs. The primary challenge for proving formulas is to find appropriate instantiations for quantifiers, but the completeness theorem A.2 shows that appropriate instantiations exist for all valid formulas. Hence, given a first-order formula ϕ, systematically generating proofs will eventually be successful if ϕ is valid, but may go on forever if ϕ is not valid. This makes validity of first-order logic semidecidable, because there is an algorithm that always terminates and answers correctly for valid formulas but may fail to terminate for formulas that are not valid. In fact, it turns out that there is no algorithm for first-order logic that always terminates with the correct answer also for formulas that are not valid.

In general, a *decision problem* is the task to decide for each instance of the problem between "yes" and "no". For instance, in the decision problem of validity in first-order logic, the task is for each formula of first-order logic to decide between "yes, valid" and "no, not valid". A decision problem is called *decidable* iff there is an algorithm that always terminates and produces the correct answer for the decision problem. The problem is called *undecidable* otherwise, i.e., no algorithm exists that is guaranteed to terminate with the correct answer in all cases. A decision problem is called *semidecidable* iff there is an algorithm that always produces the correct answer for the decision problem and always terminates for "yes" instances of the problem, but does not need to terminate for "no" instances.

Validity in first-order logic is semidecidable, but not decidable. Validity in equational first-order logic, i.e., with equality as the only predicate symbol and no function symbols, is decidable, as has been shown by Skolem [281], based on prior work by Löwenheim [199]. Tarski [288] showed that validity in first-order real arithmetic, i.e., first-order logic interpreted over real-closed fields is decidable by quantifier elimination.

Appendix B
Differential Equations

Contents

B.1 Ordinary Differential Equations 359
B.2 Existence Theorems 363
B.3 Existence and Uniqueness Theorems 364
B.4 Linear Differential Equations with Constant Coefficients 365

Synopsis In this chapter, we give an intuition for and examples of differential equations and briefly summarise some classical results about differential equations. We state Peano's existence theorem for solutions of differential equations and the Picard-Lindelöf or Cauchy-Lipschitz existence and uniqueness theorem.

B.1 Ordinary Differential Equations

In this chapter, we briefly introduce differential equations, show examples, and summarise classical results. We refer to the book by Walter [297] for details and proofs about differential equations. For further background on differential equations, we refer you to the literature [153, 257, 116, 172]. For a presentation of differential equations that is closer to the needs in dynamical systems, we refer you to the book by Perko [227]. Differential-algebraic equations are also covered in the literature in more detail [132, 187].

In this chapter, $C^k(D,\mathbb{R}^n)$ denotes the space of k times continuously differentiable functions from domain D to $\mathbb{R}^n$.

An ordinary differential equation in explicit form is an equation $y'(t) = f(t,y)$ where $y'(t)$ is meant to be the derivative of y with respect to t. A solution is a differentiable function Y which satisfies this equation when substituted in the differential equation, i.e., when substituting $Y(t)$ for y and the derivative $Y'(t)$ of Y at t for $y'(t)$.

Definition B.1 (Ordinary differential equation). Let $f : D \to \mathbb{R}^n$ be a function on a domain $D \subseteq \mathbb{R} \times \mathbb{R}^n$. The function $Y : I \to \mathbb{R}^n$ is a *solution* on the interval $I \subseteq \mathbb{R}$

of the *initial value problem*

$$\begin{bmatrix} y'(t) = f(t,y) \\ y(t_0) = y_0 \end{bmatrix} \tag{B.1}$$

with *ordinary differential equation (ODE)* $y' = f(t,y)$, if, for all $t \in I$

1. $(t, Y(t)) \in D$,
2. $Y'(t)$ exists and $Y'(t) = f(t, Y(t))$,
3. $Y(t_0) = y_0$.

If $f : D \to \mathbb{R}^n$ is continuous, then it is easy to see that $Y : I \to \mathbb{R}^n$ is continuously differentiable. The definition is accordingly for higher-order differential equations, i.e., differential equations involving higher-order derivatives $y^{(n)}(t)$ for $n > 1$.

Let us consider the intuition for this definition. A differential equation (system) can be thought of as a vector field such as the one in Fig. B.1, where, at each point, the vector shows in which direction the solution evolves. At every point, the vector would correspond to the right-hand side of the differential equation. A solution of a differential equation adheres to this vector field at every point, i.e., the solution (e.g., the solid line in Fig. B.1) locally follows the direction indicated by the vector of the right-hand side of the differential equation. There are many solutions of the differential equation corresponding to the vector field illustrated in Fig. B.1. For the particular initial value problem, however, a solution also has to start at the position y_0 at time t_0 and then follow the differential equations or vector field from this point. In general, there could still be multiple solutions for the same initial value problem.

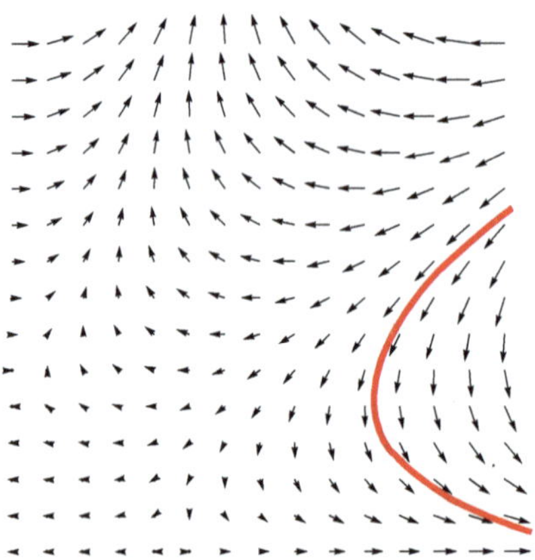

Fig. B.1 Vector field and a solution of a differential equation

Example B.1. Some differential equations are easy to solve. For instance, the initial value problem

$$\begin{bmatrix} x'(t) = 5 \\ x(0) = 2 \end{bmatrix}$$

has a solution $x(t) = 5t + 2$. This can be checked easily by inserting the solution into the differential equation and initial value equation:

$$\begin{bmatrix} (x(t))' = (5t+2)' = 5 \\ x(0) = 5 \cdot 0 + 2 = 2 \end{bmatrix}$$

□

Example B.2. Consider the initial value problem

$$\begin{bmatrix} x'(t) = -2x \\ x(1) = 3 \end{bmatrix}$$

which has a solution $x(t) = 3e^{-2(t-1)}$. The test, again, is to insert the solution into the (differential) equations of the initial value problems and check:

$$\begin{bmatrix} (3e^{-2(t-1)})' = -6e^{-2(t-1)} = -2x(t) \\ x(1) = 3e^{-2(1-1)} = 3 \end{bmatrix}$$

□

Example B.3 (Trains). For the train control examples in this book (especially in Chaps. 2 and 7), consider the differential equation system $z' = v, v' = a$ from (2.7) on p. 62 and the initial value problem

$$\begin{bmatrix} z'(t) = v(t) \\ v'(t) = a \\ z(0) = z_0 \\ v(0) = v_0 \end{bmatrix}$$

Note that this initial value problem is a *symbolic initial value problem* with symbols z_0, v_0 as initial values (not specific numbers like 5 and 2.3). Moreover, the differential equation has a constant symbol a, and not a specific number like 0.6, in the differential equation. In vectorial notation, the initial value problem with this differential equation system corresponds to a vectorial system when we denote $y(t) := (z(t), v(t))$, i.e., with dimension $n = 2$ in Definition B.1:

$$\begin{bmatrix} y'(t) = \begin{pmatrix} z \\ v \end{pmatrix}'(t) = \begin{pmatrix} v(t) \\ a \end{pmatrix} \\ y(0) = \begin{pmatrix} z \\ v \end{pmatrix}(0) = \begin{pmatrix} z_0 \\ v_0 \end{pmatrix} \end{bmatrix}$$

The solution of this initial value problem is

$$z(t) = \frac{a}{2}t^2 + v_0 t + z_0$$
$$v(t) = at + v_0$$

We can show that this is the solution by inserting the solution into the (differential) equations of the initial value problems and checking:

$$\begin{bmatrix} (\frac{a}{2}t^2 + v_0 t + z_0)' = 2\frac{a}{2}t + v_0 = v(t) \\ (at + v_0)' = a \\ z(0) = \frac{a}{2}0^2 + v_0 0 + z_0 = z_0 \\ v(0) = a0 + v_0 = v_0 \end{bmatrix}$$

□

Example B.4 (Aircraft). For the aircraft control examples in this book (especially in Chaps. 3 and 8), consider the differential equation system ($\mathscr{F}(\omega)$) from p. 150 and the corresponding initial value problem:

$$\begin{bmatrix} x_1'(t) = d_1(t) \\ x_2'(t) = d_2(t) \\ d_1'(t) = -\omega d_2(t) \\ d_2'(t) = \omega d_1(t) \\ x_1(0) = x_{1,0} \\ x_2(0) = x_{2,0} \\ d_1(0) = d_{1,0} \\ d_2(0) = d_{2,0} \end{bmatrix}$$

This initial value problem is again a symbolic initial value problem with variable symbols $x_{1,0}, x_{2,0}, d_{1,0}, d_{2,0}$ as initial values instead of specific numbers. The differential equations are also symbolic with a variable symbol ω instead of a specific number like 2.3. When we assume the particular number $\omega = 0$, which corresponds to straight-line flight (see Fig. 1.3a on p. 3), the solution is a simple linear function:

$$\begin{aligned} x_1(t) &= x_{1,0} + d_{1,0}t \\ x_2(t) &= x_{2,0} + d_{2,0}t \\ d_1(t) &= d_{1,0} \\ d_2(t) &= d_{2,0} \end{aligned}$$

For the general case with $\omega \neq 0$, which corresponds to curved flight (Fig. 1.3b–d), the solution of this initial value problem is much more complicated:

$$\begin{aligned} x_1(t) &= x_{1,0} + \frac{1}{\omega}(d_{2,0}\cos(\omega t) + d_{1,0}\sin(\omega t) - d_{2,0}) \\ x_2(t) &= x_{2,0} - \frac{1}{\omega}(d_{1,0}\cos(\omega t) - d_{2,0}\sin(\omega t) - d_{1,0}) \\ d_1(t) &= d_{1,0}\cos(\omega t) - d_{2,0}\sin(\omega t) \\ d_2(t) &= d_{2,0}\cos(\omega t) + d_{1,0}\sin(\omega t) \end{aligned}$$

We can show that this is the solution by inserting the solution into the (differential) equations of the initial value problems and checking:

$$\begin{bmatrix} x_1'(t) = \frac{1}{\omega}(-\omega d_{2,0}\sin(\omega t) + \omega d_{1,0}\cos(\omega t)) = d_1(t) \\ x_2'(t) = -\frac{1}{\omega}(-\omega d_{1,0}\sin(\omega t) - \omega d_{2,0}\cos(\omega t)) = d_2(t) \\ d_1'(t) = -\omega d_{1,0}\sin(\omega t) - \omega d_{2,0}\cos(\omega t) = -\omega d_2(t) \\ d_2'(t) = -\omega d_{2,0}\sin(\omega t) + \omega d_{1,0}\sin(\omega t) = \omega d_1(t) \\ x_1(0) = x_{1,0} + \frac{1}{\omega}(d_{2,0}\cos(0) + d_{1,0}\sin(0) - d_{2,0}) = x_{1,0} \\ x_2(0) = x_{2,0} - \frac{1}{\omega}(d_{1,0}\cos(0) - d_{2,0}\sin(0) - d_{1,0}) = x_{2,0} \\ d_1(0) = d_{1,0}\cos(0) - d_{2,0}\sin(0) = d_{1,0} \\ d_2(0) = d_{2,0}\cos(0) + d_{1,0}\sin(0) = d_{2,0} \end{bmatrix}$$

□

As a general phenomenon, observe that solutions of differential equations can be much more involved than the differential equations themselves, which is part of the representational and descriptive power of differential equations.

Often, differential equations are more difficult to solve than the above examples. In a certain sense, "most" differential equations are impossible to solve in that they have no explicit closed-form solution with elementary functions, for instance, $x''(t) = e^{t^2}$; see [304]. Likewise, differential equations like $y'(t) = \frac{2}{t^3}y$ can have non-analytic smooth solutions like:

$$y(t) = e^{-\frac{1}{t^2}}.$$

B.2 Existence Theorems

There are several classical theorems that guarantee existence and/or uniqueness of solutions if differential equations (not necessarily closed-form solutions with elementary functions, though). The existence theorem is due to Peano [225]. A proof can be found in [297, Theorem 10.IX].

Theorem B.1 (Existence theorem of Peano). *Let $f : D \to \mathbb{R}^n$ be a continuous function on an open, connected domain $D \subseteq \mathbb{R} \times \mathbb{R}^n$. Then, the initial value problem* (B.1) *with $(t_0, y_0) \in D$ has a solution. Further, every solution of* (B.1) *can be continued arbitrarily close to the border of D.*

Peano's theorem only proves that a solution exists, not for what duration it exists. Still, it shows that every solution can be *continued arbitrarily close to the border* of the domain D. That is, the closure of the graph of solution φ, when restricted to $[0,0] \times \mathbb{R}^n$, is not a compact subset of D. Especially, there is then a global solution on the interval $[0,\infty)$ if $D = \mathbb{R}^{n+1}$.

Peano's theorem shows the existence of solutions of continuous differential equations on open, connected domains, but there can still be multiple solutions.

Example B.5. The initial value problem with the following continuous differential equation

$$\begin{bmatrix} y' = \sqrt[3]{|y|} \\ y(0) = 0 \end{bmatrix}$$

has multiple solutions:

$$y(t) = 0$$

$$y(t) = \left(\frac{2}{3}t\right)^{\frac{3}{2}}$$

$$y(t) = \begin{cases} 0 & \text{for } t \leq s \\ \left(\frac{2}{3}(t-s)\right)^{\frac{3}{2}} & \text{for } t > s \end{cases}$$

where $s \geq 0$ is any nonnegative real number. □

B.3 Existence and Uniqueness Theorems

If we know that the differential equation (its right-hand side) is continuously differentiable on an open, connected domain, then the Picard-Lindelöf theorem gives a stronger result than Peano's theorem. It shows that there is a unique solution (except, of course, that the restriction of any solution to a sub-interval is again a solution). For this, recall that a function $f : D \to \mathbb{R}^n$ with $D \subseteq \mathbb{R} \times \mathbb{R}^n$ is called *Lipschitz continuous* with respect to y iff there is an $L \in \mathbb{R}$ such that for all $(t,y),(t,\bar{y}) \in D$,

$$\|f(t,y) - f(t,\bar{y})\| \leq L\|y - \bar{y}\|.$$

If, for instance, $\frac{\partial f(t,y)}{\partial y}$ exists and is bounded on D, then f is Lipschitz continuous with $\max_{(t,y)\in D} \|\frac{\partial f(t,y)}{\partial y}\|$ by mean value theorem. Similarly, f is *locally Lipschitz continuous* iff for each $(t,y) \in D$, there is a neighbourhood in which f is Lipschitz continuous. In particular, if f is continuously differentiable, i.e., $f \in C^1(D,\mathbb{R}^n)$, then f is locally Lipschitz continuous.

Most importantly, Picard-Lindelöf's theorem [195], which is also known as the Cauchy-Lipschitz theorem, guarantees existence and uniqueness of solutions. As restrictions of solutions are always solutions, we understand uniqueness up to restrictions. A proof can be found in [297, Theorem 10.VI]

Theorem B.2 (Uniqueness theorem of Picard-Lindelöf). *In addition to the assumptions of Theorem B.1, let f be locally Lipschitz continuous with respect to y (for instance, $f \in C^1(D,\mathbb{R}^n)$ is sufficient). Then, there is a unique solution of the initial value problem* (B.1).

Picard-Lindelöf's theorem does not show the duration of the solution, but shows only that the solution is unique. Under the assumptions of Picard-Lindelöf's theorem, every solution can be extended arbitrarily close to the border of D by Peano's theorem, however. The solution is unique, except that all restrictions of the solution to a sub-interval are also solutions.

The following global uniqueness theorem shows a stronger property when the domain is $[0,a] \times \mathbb{R}^n$. It is a corollary to Theorems B.1 and B.2, but used prominently in the proof of Theorem B.2, and is of independent interest. A direct proof of the following global version of the Picard-Lindelöf theorem can be found in [297, Proposition 10.VII].

Corollary B.1 (Global uniqueness theorem of Picard-Lindelöf). *Let* $f : [0,a] \times \mathbb{R}^n \to \mathbb{R}^n$ *be a continuous function that is Lipschitz continuous with respect to* y. *Then, there is a unique solution of the initial value problem* (B.1) *on* $[0,a]$.

The following result is a componentwise generalisation of [297, Proposition 6.VI] to vectorial differential equations and can be used to extend solutions.

Proposition B.1 (Continuation of solutions). *Let* $f : D \to \mathbb{R}^n$ *be a continuous function on the open, connected domain* $D \subseteq \mathbb{R} \times \mathbb{R}^n$. *If* φ *is a solution of differential equation* $y' = f(t,y)$ *on* $[0,b)$ *whose image* $\varphi([0,b))$ *lies within a compact set* $A \subseteq D$, *then* φ *can be continued to a solution on* $[0,b]$. *Furthermore, if* φ_1 *is a solution of differential equation* $y' = f(t,y)$ *on* $[0,b]$ *and* φ_2 *is a solution of* $y' = f(t,y)$ *on* $[b,c]$ *with* $\varphi_1(b) = \varphi_2(b)$, *then their concatenation*

$$\varphi(t) := \begin{cases} \varphi_1(t) & \text{for } 0 \leq t \leq b \\ \varphi_2(t) & \text{for } b < t \leq c \end{cases}$$

is a solution on $[0,c]$.

B.4 Linear Differential Equations with Constant Coefficients

For linear differential equation systems with constant coefficients there is a well-established constructive theory for obtaining closed-form solutions of initial value problems using classical techniques from linear algebra. A proof and more details can be found in [297, §18.VI].

Proposition B.2 (Linear systems with constant coefficients). *For a constant matrix* $A \in \mathbb{R}^{n\times n}$, *the initial value problem*

$$\begin{bmatrix} y'(t) = Ay(t) + b(t) \\ y(\tau) = \eta \end{bmatrix} \tag{B.2}$$

has the solution

$$y(t) = e^{A(t-\tau)}\eta + \int_{\tau}^{t} e^{A(t-s)} b(s)\, ds$$

where exponentiation of matrices is defined by the usual power series (generalized to matrices):

$$e^{At} = \sum_{n=0}^{\infty} \frac{1}{n!} A^n t^n.$$

In particular, if the matrix A is *nilpotent*, i.e., $A^n = 0$ for some $n \in \mathbb{N}$, and the terms $b(t)$ are polynomials in t, then the solution of the initial value problem is a polynomial function, because the exponential series stops at A^n and is a finite polynomial in t then:

$$e^{At} = \sum_{k=0}^{\infty} \frac{1}{k!} A^k t^k = \sum_{k=0}^{n-1} \frac{1}{k!} A^k t^k.$$

In particular, as products and sums of polynomials are polynomials (polynomials form an algebra) and polynomials in t are closed under integration, the solution identified in Proposition B.2 is a polynomial. Furthermore, this solution is unique by Theorem B.2.

Example B.6 (Trains). In the initial value problem from Example B.3, we guessed the solution of the differential equation system and then checked that it is the right solution by inserting it into the differential equations. But how do we compute the solution constructively in the first place without having to guess? The train differential equations $z' = v, v' = a$ from (2.7) on p. 62 are linear with constant coefficients. When we denote $y(t) := (z(t), v(t))$, we can rewrite (2.7) in the form of (B.2) as follows:

$$\begin{bmatrix} y'(t) = \begin{pmatrix} z'(t) \\ v'(t) \end{pmatrix} = \begin{pmatrix} 0 & 1 \\ 0 & 0 \end{pmatrix} \begin{pmatrix} z(t) \\ v(t) \end{pmatrix} + \begin{pmatrix} 0 \\ a \end{pmatrix} =: Ay(t) + b(t) \\ y(0) = \begin{pmatrix} z(0) \\ v(0) \end{pmatrix} = \begin{pmatrix} z_0 \\ v_0 \end{pmatrix} = \eta \end{bmatrix}$$

This system, as a linear differential equation system with a constant coefficient matrix, has the form required in Proposition B.2. First, we compute the exponential series for the matrix A, which terminates quickly because $A^2 = 0$:

$$\begin{aligned} e^{At} &= \sum_{n=0}^{\infty} \frac{1}{n!} A^n t^n = A^0 + At + \frac{1}{2!} \underbrace{A^2}_{0} t^2 + \underbrace{A^2}_{0} \sum_{n=3}^{\infty} \frac{1}{n!} A^{n-2} t^n \\ &= \begin{pmatrix} 1 & 0 \\ 0 & 1 \end{pmatrix} + \begin{pmatrix} 0 & 1 \\ 0 & 0 \end{pmatrix} t = \begin{pmatrix} 1 & t \\ 0 & 1 \end{pmatrix} \end{aligned}$$

Now we can use Proposition B.2 to compute a solution of this differential equation:

$$\begin{aligned} y(t) &= e^{At} \eta + \int_0^t e^{A(t-s)} b(s)\, ds \\ &= \begin{pmatrix} 1 & t \\ 0 & 1 \end{pmatrix} \begin{pmatrix} z_0 \\ v_0 \end{pmatrix} + \int_0^t \begin{pmatrix} 1 & t-s \\ 0 & 1 \end{pmatrix} \begin{pmatrix} 0 \\ a \end{pmatrix} ds \\ &= \begin{pmatrix} z_0 + v_0 t \\ v_0 \end{pmatrix} + \int_0^t \begin{pmatrix} at - as \\ a \end{pmatrix} ds \\ &= \begin{pmatrix} z_0 + v_0 t \\ v_0 \end{pmatrix} + \begin{pmatrix} \int_0^t (at - as)\, ds \\ \int_0^t a\, ds \end{pmatrix} \\ &= \begin{pmatrix} z_0 + v_0 t \\ v_0 \end{pmatrix} + \begin{pmatrix} ats - \frac{a}{2} s^2 \\ as \end{pmatrix} \Bigg|_{s=0}^{s=t} \end{aligned}$$

$$= \begin{pmatrix} z_0 + v_0 t \\ v_0 \end{pmatrix} + \begin{pmatrix} at^2 - \frac{a}{2}t^2 \\ at \end{pmatrix} - \begin{pmatrix} a0^2 - \frac{a}{2}0^2 \\ a0 \end{pmatrix}$$

$$= \begin{pmatrix} z_0 + v_0 t + \frac{a}{2}t^2 \\ v_0 + at \end{pmatrix}$$

The last equation is exactly the solution we had guessed and checked in Example B.3. Now we have computed it constructively. □

Appendix C
Hybrid Automata

Contents

C.1 Syntax and Traces of Hybrid Automata **369**
C.2 Embedding Hybrid Automata into Hybrid Programs **371**

Synopsis To formally relate the notions of hybrid programs and hybrid automata, we show that hybrid automata can be embedded canonically into hybrid programs. We further prove that reachability in hybrid automata directly corresponds to satisfying models of associated formulas in differential dynamic logic. Finally, safety corresponds to validity of those formulas in differential dynamic logic. Differential dynamic logics can express more general properties about hybrid systems than plain safety though.

C.1 Syntax and Traces of Hybrid Automata

Among several other models for hybrid systems [69, 100, 58, 270, 272, 40, 183], the model of *hybrid automata* [156, 8] is one of the more widely used notations. Even though hybrid automata are a fairly common notation for hybrid systems, there are several slightly different notions of hybrid automata or automata-based models for hybrid systems [289, 9, 218, 8, 56, 156, 11, 58, 189, 97, 228, 90]. An early mention of a formalism that is related to what has later been called hybrid automata is the work of Tavernini [289]. Even earlier inspirations for hybrid systems can be found in the work by Witsenhausen [302]. We follow the notion of hybrid automata from Henzinger [156] most closely, with corresponding care for actual definability of the relations as in other approaches [125, 189, 228, 238] and with invariants that are required to hold at all times following [8, 90, 238].

Hybrid automata are graph models with two kinds of transitions: discrete jumps in the state space caused by mode switches (edges in the graph), and continuous evolution along continuous flows within a mode (vertices in the graph).

Definition C.1 (Hybrid automata). A *hybrid automaton A* consists of

- a continuous state space $\mathbb{R}^n$;
- a finite directed graph (*control graph*) with vertices Q (as *modes*) and edges E (as *control switches*);
- flow conditions $flow_q \subseteq \mathbb{R}^n \times \mathbb{R}^n$ that determine the relationship of the continuous state $x \in \mathbb{R}^n$ and its time derivative $x' \in \mathbb{R}^n$ during continuous evolution in mode $q \in Q$;
- invariant conditions $inv_q \subseteq \mathbb{R}^n$ or evolution domain restrictions that have to be true while in mode $q \in Q$;
- jump relations $jump_e \subseteq \mathbb{R}^n \times \mathbb{R}^n$ that determine the new value of the continuous state $x \in \mathbb{R}^n$ depending on its old value when following edge $e \in E$;

We assume at least that the relations $jump_e$ and inv_q are definable in first-order real arithmetic [288], but several additional restrictions apply for $flow_q$ depending on the class of hybrid automata.

Typically, the jump relation $jump_e$ is given as a conjunction of transition guards $guard_e \subseteq \mathbb{R}^n$, which determine from which states an edge can be taken, and variable resets $reset_e \subseteq \mathbb{R}^n \times \mathbb{R}^n$, which adjust the state value to its new value. In most cases, the reset relation is specified by a list of assignments $x_1 := \theta_1, \ldots x_n := \theta_n$, which correspond to a discrete jump set of differential dynamic logic (Definition 2.3). Further, flow conditions are usually just specified by a set of differential equations $x_1' = \theta_1, \ldots, x_n' = \theta_n$. See the left part of Fig. C.1 for an example of a hybrid automaton and the right part of Fig. C.1 for a description of the same hybrid system as a hybrid program. We refer you to Sect. 1.1.1 for more examples of hybrid automata.

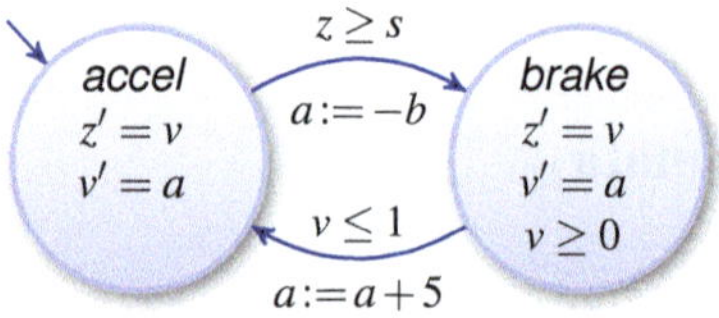

$$
\begin{array}{l}
q := accel; \qquad \text{/* initial mode is node accel */} \\
(\ \ (?q = accel;\ z' = v, v' = a) \\
\cup\ (?q = accel \land z \geq s;\ a := -b;\ q := brake;\ ?v \geq 0) \\
\cup\ (?q = brake;\ z' = v, v' = a \,\&\, v \geq 0) \\
\cup\ (?q = brake \land v \leq 1;\ a := a + 5;\ q := accel))^*
\end{array}
$$

Fig. C.1 Hybrid automaton and corresponding hybrid program

Although often neglected in the literature, definability of the constituent relations of hybrid automata, e.g., in first-order real arithmetic, is a crucial prerequisite for dealing with any state reachability question. If the parts of a hybrid automaton are not reasonably computational or constructive, then no (interesting) computable analysis can be built on top of it. For instance, if the switching conditions in the $jump_e$ relation are undecidable, then analysis procedures have no way of knowing if the system can jump along an edge. Likewise, in a hybrid automaton that uses an undecidable set like the Mandelbrot set as invariant, it would already be undecidable whether a state $x \in \mathbb{R}^n$ satisfies the invariant of the current mode. These difficulties even arise in the strong computational model of real Turing machines by Blum et al. [49].

Definition C.2 (Transition semantics of hybrid automata). The *transition system* of a hybrid automaton A is a transition relation $\curvearrowright$ defined as follows

- the state space is defined as $S := \{(q,x) \in Q \times \mathbb{R}^n \ : \ x \in inv_q\}$;
- the transition relation $\curvearrowright \subseteq S \times S$ is defined as the union $\bigcup_{e \in E} \overset{e}{\curvearrowright} \cup \bigcup_{q \in Q} \overset{q}{\curvearrowright}$ where
  1. $(q,x) \overset{e}{\curvearrowright} (\tilde{q},\tilde{x})$ iff $e \in E$ is an edge from $q \in Q$ to $\tilde{q} \in Q$ in the hybrid automaton and $(x,\tilde{x}) \in jump_e$ (*discrete transition*).
  2. $(q,x) \overset{q}{\curvearrowright} (q,\tilde{x})$ iff $q \in Q$ and there is a function $f : [0,r] \to \mathbb{R}^n$ that has a time derivative $f' : (0,r) \to \mathbb{R}^n$ such that $f(0) = x, f(r) = \tilde{x}$ and that respects $(f(\zeta), f'(\zeta)) \in flow_q$ at each $\zeta \in (0,r)$. Further, $f(\zeta) \in inv_q$ for each $\zeta \in [0,r]$ (*continuous transition*).

State $\sigma \in S$ is *reachable* from state $\sigma_0 \in S$, denoted by $\sigma_0 \curvearrowright^* \sigma$, iff, for some $n \in \mathbb{N}$, there is a sequence of states $\sigma_1, \sigma_2, \ldots, \sigma_n = \sigma \in S$ such that $\sigma_{i-1} \curvearrowright \sigma_i$ for $1 \leq i \leq n$.

Most often, the semantics of hybrid automata is further restricted to non-Zeno traces with only finitely many transitions in finite time [156, 90, 97].

C.2 Embedding Hybrid Automata into Hybrid Programs

A hybrid automaton as on the left of Fig. C.1 can be represented faithfully as the hybrid program on the right of Fig. C.1. More generally, we show that it is always possible to represent hybrid automata as hybrid programs, thus showing that hybrid automata can be embedded faithfully into differential dynamic logic.

Proposition C.1 (Hybrid automata embedding). *There is an effective mapping ι from hybrid automata to hybrid programs and DA-programs such that the following diagram commutes:*

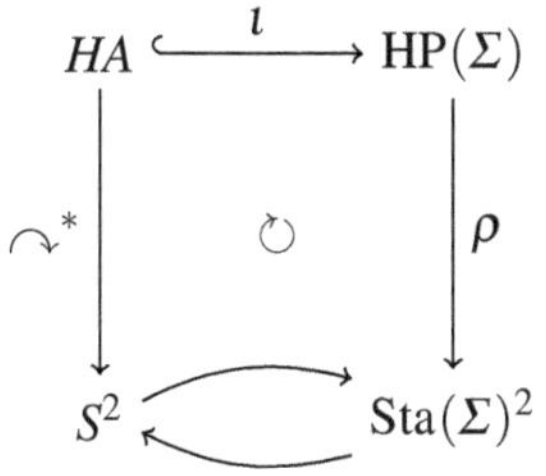

That is, the transition semantics $\rho(\iota(A))$ of the hybrid program $\iota(A)$ corresponding to a hybrid automaton A is identical to the reachability relation $\curvearrowright^*$ corresponding to hybrid automaton A when identifying states of hybrid programs in $\mathrm{Sta}(\Sigma)^2$ with states of hybrid automata in S by a canonical bijection.

Proof. Let $\Sigma = \{q, x_1, \ldots, x_n, x_1{}^+, \ldots, x_n{}^+\}$ be the signature. We use vectorial notation x for the vector $(x_1, \ldots, x_n)$ and x^+ for $(x_1{}^+, \ldots, x_n{}^+)$. We define ι as the function that maps hybrid automaton A to the following HP:

$$
\begin{aligned}
&(?q = q_i; flow_{q_i}(x,x') \,\&\, inv_{q_1} \\
&\cup\, ?q = q_i;\ (x^+ := *;\ ?jump_e(x,x^+);\ x := x^+);\ ?inv_{q_j};\ q := q_j \\
&\cup \ldots \\
&)^*
\end{aligned}
$$

The respective lines in this HP are subject to a choice for each mode q_i or each edge e from some state q_i to some state q_j. Let α^* denote this program $\iota(A)$.

States of the hybrid automaton A and states of its HP $\iota(A)$ immediately correspond to each other using the bijection $\Phi : S \to \mathrm{Sta}(\Sigma)$ that maps $(\tilde{q}, \tilde{x}) \in S$ to the state ν that is defined as $\nu(q) = \tilde{q}$ and $\nu(x_i) = \tilde{x}_i$ for $1 \le i \le n$. Observe that Φ is a bijection up to forgetful projections of internal variables x^+. In the following, we use the state identification Φ implicitly to simplify the notation. We have to show that the diagram commutes, that is, $\curvearrowright^* = \rho \circ \iota$ (up to identification of states by Φ, i.e., $\Phi \circ \curvearrowright^* = \rho \circ \iota$).

"$\subseteq$" Let $\sigma_0 \curvearrowright^* \sigma$. Hence, $n \in \mathbb{N}$ and $\sigma_1, \sigma_2, \ldots, \sigma_n = \sigma \in S$ such that $\sigma_{i-1} \curvearrowright \sigma_i$ for all $1 \le i \le n$. The proof is by induction on n.

IA If $n = 0$ then $(\sigma_0, \sigma) \in \rho(\alpha^*)$ using zero repetitions.

IS By induction hypothesis, we can assume that $(\sigma_0, \sigma_{n-1}) \in \rho(\alpha^*)$. We have to show $(\sigma_{n-1}, \sigma_n) \in \rho(\alpha)$, thereby implying $(\sigma_0, \sigma_n) \in \rho(\alpha^*)$.
Consider the case where the last transition $\sigma_{n-1} \curvearrowright \sigma_n$ is a continuous transition in mode q_i of some duration $r \ge 0$. Then, up to identification by Φ, there is a state flow $\varphi : [0, r] \to \mathrm{Sta}(\Sigma)$ with $\varphi(0) = \sigma_{n-1}, \varphi(r) = \sigma_n$ and $\varphi \models flow_{q_i} \wedge inv_{q_i}$. Thus, α can copy the transition as $(\sigma_{n-1}, \sigma_n) \in \rho(\alpha)$ using the choice $?q = q_i; flow_{q_i}(x,x') \,\&\, inv_{q_i}$. The test succeeds, because $\Phi(\sigma_{n-1})(q) = q_i$.
Consider the case where the last transition $\sigma_{n-1} \curvearrowright \sigma_n$ is a discrete transition from mode q_i to q_j along edge e. Then $(\sigma_{n-1}, \sigma) \in jump_e$. Thus, by choosing the values of σ_n for x^+, we have that $(\sigma_{n-1}, \sigma_n) \in \rho(\alpha)$ by the choice $?q = q_i;\ (x^+ := *;\ ?jump_e(x,x^+);\ x := x^+);\ ?inv_{q_j};\ q := q_j$.

"$\supseteq$" Let $(\sigma_0, \sigma_n) \in \rho(\alpha^*)$ following n repetitions of α: Let $\sigma_1, \ldots, \sigma_{n-1} \in \mathrm{Sta}(\Sigma)$ such that $(\sigma_{i-1}, \sigma_i) \in \rho(\alpha)$ for all $1 \le i \le n$. The proof is by induction on n.

IA For $n = 0$, there is nothing to show.

IS By induction hypothesis, we can assume that $\sigma_{i-1} \curvearrowright \sigma_i$ for all $1 \le i < n$. We have to show that $\sigma_{n-1} \curvearrowright \sigma_n$, thereby showing that $\sigma_0 \curvearrowright^* \sigma_n$. If $q_i := \sigma_{n-1}(q) = \sigma_n(q)$, then it is easy to see from the structure of α that $(\sigma_{n-1}, \sigma_n) \in \rho(?q = q_i; flow_{q_i}(x,x') \,\&\, inv_{q_i})$. Thus, $\sigma_{n-1} \overset{q_i}{\curvearrowright} \sigma_n$ by a continuous transition.
If, however, $q_i := \sigma_{n-1}(q)$, $q_j = \sigma_n(q)$, then it is easy to see that

$$(\sigma_{n-1}, \sigma_n) \in \rho(?q = q_i;\ (x^+ := *; ?jump_e(x, x^+); x := x^+);\ ?inv_{q_j}; q := q_j)$$

according to a line of α that originates from some edge e from q_i to q_j. Thus, $(\sigma_{n-1}, \sigma_n) \in jump_e$ and $\sigma_n \models inv_{q_j}$; hence, $\sigma_{n-1} \overset{e}{\curvearrowright} \sigma_n$ by a discrete transition. □

Corollary C.1. *There is an effective mapping from safety properties of hybrid automata to* d$\mathcal{L}$/ DAL *formulas such that the hybrid automaton A, starting in initial mode q_0, safely remains in the region $F \in \mathrm{Fml_{FOL}}(\Sigma)$ if and only if the corresponding* d$\mathcal{L}$/ DAL *formula is valid.*

Proof. Let α^* be the HP $\iota(A)$ belonging to A according to Proposition C.1. Then, when starting A in initial mode q_0, it safely remains in the region F iff the following formula is valid:

$$q = q_0 \wedge inv_{q_0} \rightarrow [\alpha^*]F.$$

□

Example C.1 (Water tank). Consider the classical simple water tank example, which regulates water level y between 1 and 12 by filling or emptying the water tank. The control in the hybrid automaton of Fig. C.2a further uses a clock variable x to model delayed reactions of pumps or valves. In node *fill*, the pump is activated and the wa-

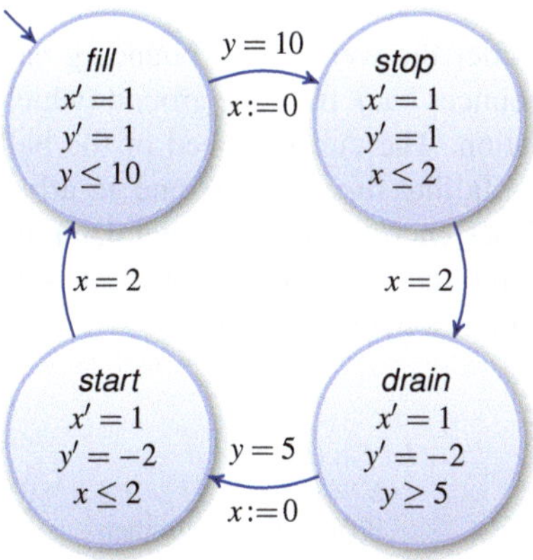

Fig. C.2a Hybrid automaton for water tank

$$\begin{aligned}
&q = fill \rightarrow [(\\
&\quad (?q = fill;\ x' = 1, y' = 1 \,\&\, y \le 10)\\
&\cup (?q = fill \wedge y = 10;\ x := 0;\ q := stop)\\
&\cup (?q = stop;\ x' = 1, y' = 1 \,\&\, x \le 2)\\
&\cup (?q = stop \wedge x = 2;\ q := drain)\\
&\cup (?q = drain;\ x' = 1, y' = -2 \,\&\, y \ge 5)\\
&\cup (?q = drain \wedge y = 5; x := 0; q := start)\\
&\cup (?q = start;\ x' = 1, y' = -2 \,\&\, x \le 2)\\
&\cup (?q = start \wedge x = 2;\ q := fill)\\
&)^*](1 \le y \wedge y \le 12)
\end{aligned}$$

Fig. C.2b Hybrid program for water tank

ter tank is filling slowly with $y' = 1$, while staying in the region $y \le 10$ all the time. At level $y = 10$, it switches to the *stop* mode, which deactivates the pump. Full deactivation of the pump takes two time units, though, such that the water is still pumped into the tank with $y' = 1$ in *stop*. When in node *drain*, water pours out with speed $y' = -2$. Again, to start the pump, the water switches to *start* first. In *start* node, the water continues to pour out with speed $y' = -2$, until, after two seconds exactly, the system switches into the *fill* node. The timing of those transitions is being taken care of deterministically by the clock variable x which has differential equation $x' = 1$ in all nodes. The fact that the evolution domain restrictions or invariant regions in the *start* and *stop* nodes are $x \leq 2$ and the outgoing edges have a guard $x = 2$ ensures that the system can only leave these nodes after two seconds exactly. The automaton cannot switch at any other time than after two seconds when $x = 2$, but it cannot remain in the *start* or *stop* nodes any longer, because the invariant region $x \leq 2$ would cease to hold then. Similarly, the switching from *fill* to *stop* is triggered by water level $y = 10$ precisely, because of the invariant region $y \leq 10$ in *fill* and the guard $y = 10$ at the outgoing edge. Finally, the switching from *drain* to *start* is at $y = 5$ precisely, for the same reasons. Consequently, the hybrid automaton in Fig. C.2b is deterministic and, thus, very easy for simulation.

Figure C.2b shows a corresponding representation of the hybrid automaton in Fig. C.2a as a hybrid program. Each line of the hybrid program corresponds to a discrete or continuous transition of the water tank hybrid automaton. The constants *fill, stop, drain, andstart* are pairwise different. The water tank is provable with the following state-dependent invariant:

$$1 \leq y \leq 12 \wedge (q = start \rightarrow y \geq 5 - 2x) \wedge (q = stop \rightarrow y \leq 10 + x).$$

□

The transformation in Proposition C.1 is a canonical embedding but hybrid programs allow for more flexible programming structures with which more natural characterisations of the system behaviour can be obtained.

Example C.2 (Parametric bouncing ball). Consider the well-known bouncing ball example [110]. A ball falls from height h and bounces back from the ground (which corresponds to $h = 0$) after an elastic deformation. The current speed of the ball is denoted by v, and t is a clock measuring the falling time. We assume an arbitrary positive gravity force g and that the ball loses energy according to a damping factor $0 \leq c < 1$. Figure C.3 depicts the hybrid automaton, an illustration of the system dynamics, and a representation of the system as a hybrid program.

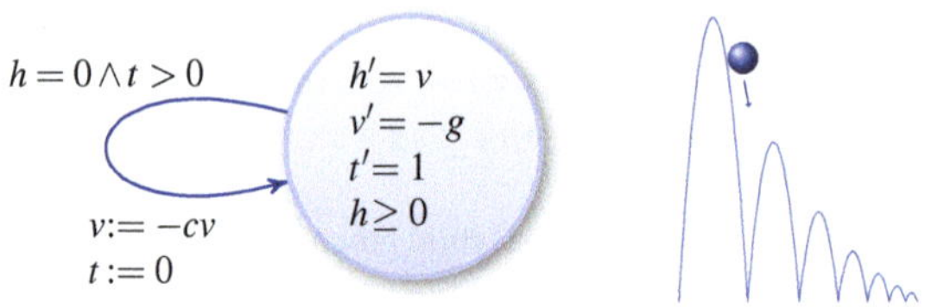

$ball \equiv ($
$\quad h' = v, v' = -g, t' = 1 \,\&\, h \geq 0;$
$\quad$ if $(h = 0 \wedge t > 0)$ then
$\quad\quad v := -cv;\ t := 0$
$\quad$ fi$)^*$

Fig. C.3 Parametric bouncing ball

The ball loses energy at every bounce. Thus the ball never bounces higher than the initial height. This can be expressed by the safety property $0 \leq h \leq H$, where H denotes the initial energy level, i.e., the initial height if $v = 0$. For instance, we can prove the following property:

$$(v^2 \leq 2g(H-h) \wedge h \geq 0 \wedge g > 0 \wedge H \geq 0 \wedge 1 > c \geq 0) \rightarrow [ball](0 \leq h \leq H).$$

This specification follows the pattern of Hoare triples. It expresses that the bouncing ball, when started in an initial state satisfying the precondition, always respects the postcondition $0 \leq h \leq H$. □

Appendix D
KeYmaera Implementation

Contents

D.1 KeYmaera: A Hybrid Theorem Prover for Hybrid Systems **377**
D.1.1 Structure of This Appendix 379
D.2 Computational Back-ends for Real Arithmetic **380**
D.2.1 Real-Closed Fields 381
D.2.2 Semialgebraic Geometry and Cylindrical Algebraic Decomposition 383
D.2.3 Nullstellensatz and Gröbner Bases 386
D.2.4 Real Nullstellensatz 392
D.2.5 Positivstellensatz and Semidefinite Programming 394
D.3 Discussion **396**
D.4 Performance Measurements **399**

Synopsis KeYmaera is a hybrid verification tool for hybrid systems that combines deductive, real algebraic, and computer algebraic prover technologies. It is an automated and interactive theorem prover for specification and verification logics for hybrid systems. KeYmaera supports *differential dynamic logic*, which is a real-valued first-order dynamic logic for hybrid systems. For automating the verification process, KeYmaera implements our generalised free-variable sequent proof calculi and automatic proof strategies that decompose the hybrid system specification symbolically. To overcome the complexity of real arithmetic, we integrate real quantifier elimination following our iterative background closure strategy. Our formal verification tool is particularly suitable for verifying parametric hybrid systems and has been used successfully for verifying collision avoidance in case studies from train and air traffic control.

D.1 KeYmaera: A Hybrid Theorem Prover for Hybrid Systems

In the interest of a comprehensive presentation, we briefly characterise the implementation of our logical analysis and verification approach from Parts I and II in our new verification tool KeYmaera [242].[1] KeYmaera [242] is a hybrid theorem prover for hybrid systems that combines the deductive, real algebraic, and computer algebraic prover technologies developed in this book. It is an automated and interactive theorem prover for differential dynamic logics for hybrid systems. KeYmaera implements the proof calculi for differential dynamic logic $d\mathcal{L}$, differential-algebraic dynamic logic DAL, and differential temporal dynamic logic dTL that we introduced in Part I.

KeYmaera has been implemented as a combination of the deductive theorem prover KeY [4, 34, 35] with computer algebraic and real algebraic techniques. The general architecture of KeYmaera is depicted in Fig. D.1. KeYmaera has built-in

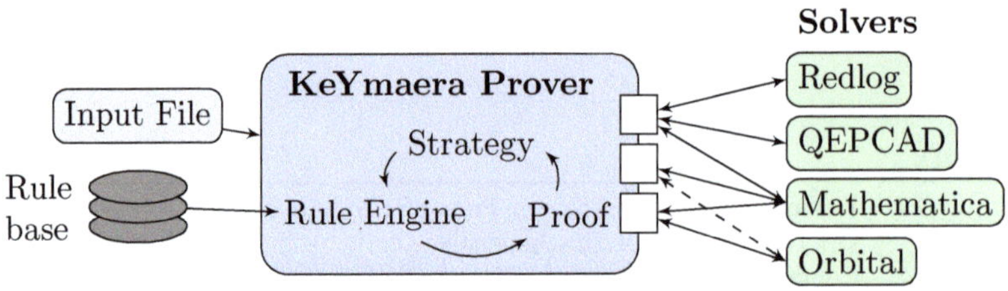

Fig. D.1 Architecture and plug-in structure of the KeYmaera prover

techniques for handling differential equations and real arithmetic, but it can also interface with other tools handling real arithmetic and/or computer algebra for improved performance. KeYmaera can interface with the computer algebra system Mathematica by Wolfram Research [303] for quantifier elimination, differential equation handling, and symbolic computation. Alternatively, it can interface with Redlog [101] or QEPCAD B [59] as quantifier elimination tools for real arithmetic. KeYmaera can also interface with computer algebraic tools and differential equation handling from the Orbital library developed by the author. KeYmaera has been developed on the basis of KeY [4, 34, 35], which is a semi-interactive theorem prover with a user-friendly graphical interface for proving correctness properties of Java programs. KeY has been developed in the group of Peter Schmitt at the University of Karlsruhe, Germany, the group of Reiner Hähnle at Chalmers University in Gothenburg, Sweden, and the group of Bernhard Beckert at the University of Koblenz-Landau, Germany. We have generalised KeY from discrete Java programs to hybrid systems by adding support for the differential dynamic logic $d\mathcal{L}$ (and DAL and dTL, respectively). Figure D.2 shows a screenshot of the graphical user interface of KeYmaera.

In the conventional KeY prover for Java programs, rule applications are comparably fast, but in KeYmaera, proof rules that use decision procedures for real arithmetic can require a substantial amount of time to produce a result. To overcome this, KeYmaera has new automatic proof strategies for the hybrid case that navigate through computationally expensive rule applications, following the strategies we

[1] KeYmaera is available from the web at `http://symbolaris.com/`

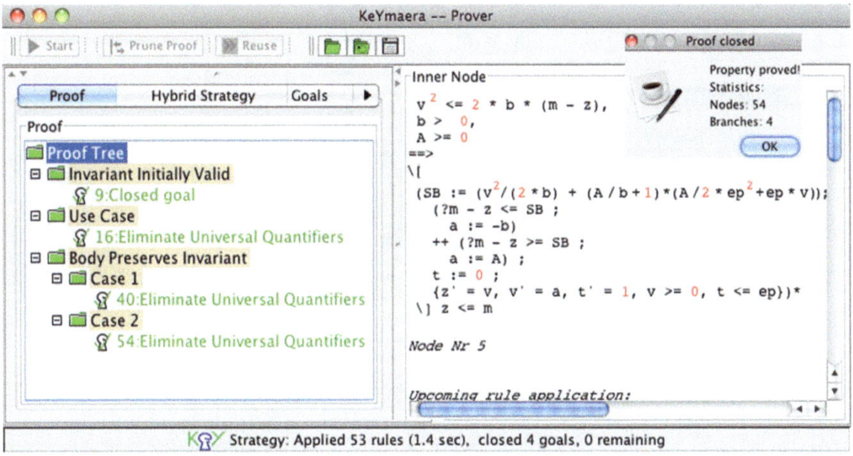

Fig. D.2 Screenshot of the KeYmaera user interface

developed in Chap. 5. KeYmaera also implements the (differential) invariant generation techniques we have developed in Chap. 6.

KeYmaera follows a plug-in architecture for integrating multiple implementations of decision procedures for real arithmetic handling, differential equation solving, and computer algebra. KeYmaera can integrate with a range of tools, some of which are shown in Fig. D.1. It does not need all of these tools to run, but can interface with any of these background solvers. KeYmaera can also run as a standalone tool based on the built-in techniques without any background solvers. Yet, the performance of KeYmaera generally improves by interfacing with additional background solvers, e.g., with Mathematica.

To overcome the complexity pitfalls of quantifier elimination and to scale to real-world application scenarios, KeYmaera implements the *iterative background closure* strategy from Sect. 5.4 that interleaves background solver calls with deductive $d\mathcal{L}$ rules. For performance reasons, the KeYmaera implementation further optimises the order of quantifiers in universal closures $\forall^\alpha \phi$ according to the modification order in the hybrid program α that produced $\forall^\alpha \phi$ (see variable dependency orders from Sects. 6.2.5 and 6.2.6). KeYmaera provides several options for adjusting the prover strategy; see Fig. D.3.

D.1.1 Structure of This Appendix

In App. D.2, we summarise techniques that can be used as back-ends for handling real arithmetic in theorem provers for differential dynamic logics. These alternatives for real arithmetic are available in KeYmaera. We discuss the consequences of the KeYmaera architecture and implementation in Sect. D.3, concerning how soundness of proof calculi is related to soundness of actual implementations. In Sect. D.4, we

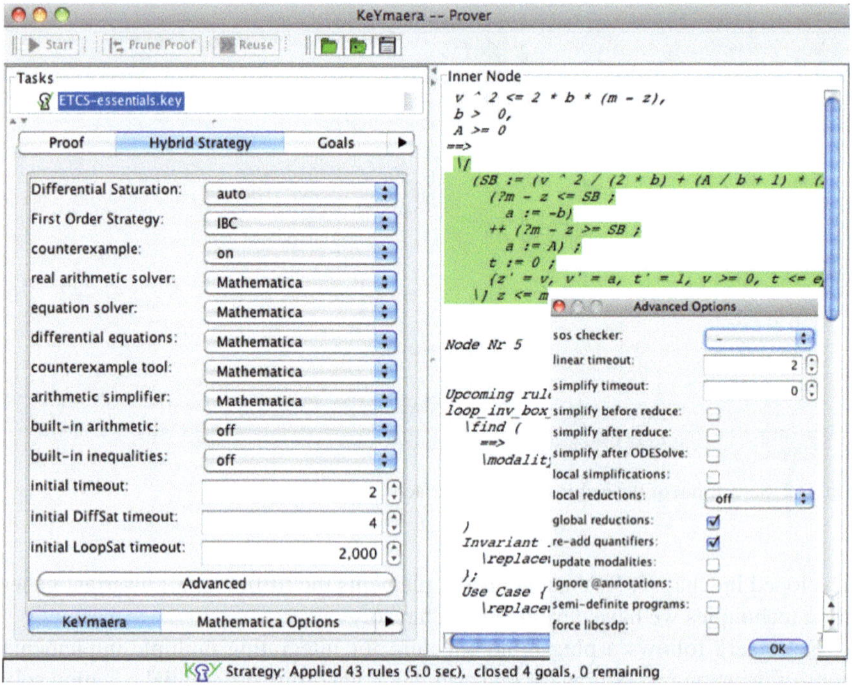

Fig. D.3 KeYmaera proof strategy options

explain the setup for our experimental evaluations and performance measurements in this book.

D.2 Computational Back-ends for Real Arithmetic

In this section, we briefly summarise a range of techniques that we use as computational back-ends for handling real arithmetic in background provers for the foreground prover KeYmaera. These techniques are various alternatives for implementing the QE quantifier elimination procedure for the quantifier rules of Fig. 2.11 or Fig. 3.9, respectively. We also refer to recent work [246], where we have introduced a new decision procedure for real arithmetic based on Gröbner bases [62], semidefinite programming [224, 151], and Stengle's real Nullstellensatz [283] that combines the techniques we present in this section.

D.2.1 Real-Closed Fields

Generalising classical techniques for finding the number of real roots for univariate polynomials due to Sturm [286], Tarski showed that validity of closed formulas in the first-order theory of real arithmetic is equivalent to the theory of real-closed fields and is decidable by quantifier elimination [287, 288]; also see Definition 2.9 on p. 77. Due to its importance, there are numerous refinements of this result that turn it into a more practical decision procedure [278, 264, 197, 79, 184, 93, 80, 19, 20, 81, 82, 84, 83, 94, 142, 299, 68, 198, 24, 285, 292, 26, 60]. A historical overview of results in this area can be found in an article by van den Dries [105]. The complexity of quantifier elimination in real-closed fields is doubly exponential in the number of quantifier alternations and has been analysed carefully [94, 141, 142, 263, 299, 60, 258, 259, 260, 259, 260, 25, 24]. Technical detail about real algebraic geometry can also be found in the book by Bochnak and Coste and Roy [50]. Algorithmic details can be found in the book by Basu, Pollack, and Roy [26].

Real numbers and real variables are omniscient in hybrid systems, because of their connection with the real physical world. Positions, velocities, accelerations, and the like are real quantities. They need to be understood properly for faithful hybrid system models. Like all other infinite structures, the reals, however, cannot be characterised uniquely up to isomorphisms in first-order logic, which is a simple corollary to the Skolem-Löwenheim theorem. Real-closed fields try to capture the properties and axioms of reals as closely as possible. Real-closed fields are an axiomatic generalisation of the field $\mathbb{R}$ of reals and share the same first-order axioms with the actual model of reals.

First we briefly recapitulate some basic notions from algebra [191, 52, 53, 112, 211]. A *ring* R is an algebraic structure that is an Abelian group with respect to addition and a (commutative) semigroup with respect to multiplication where multiplication distributes over addition, i.e., for all $a,b,c \in R$:

$$\begin{aligned} a(b+c) &= ab+ac, \\ (a+b)c &= ac+bc. \end{aligned}$$

Here R has an *Abelian group* structure with respect to addition iff, for all $a,b,c \in R$:

$$\begin{aligned} a+b &\in R, \\ (a+b)+c &= a+(b+c), \\ a+0 &= 0+a=a, \\ a+(-a) &= (-a)+a=0, \\ a+b &= b+a. \end{aligned}$$

Further, R has an *commutative semigroup* structure with respect to multiplication iff, for all $a,b,c \in R$:

$$
\begin{aligned}
a \cdot b &\in R, \\
(a \cdot b) \cdot c &= a \cdot (b \cdot c), \\
a \cdot 1 &= 1 \cdot a = a, \\
a \cdot b &= b \cdot a.
\end{aligned}
$$

A *field* K is a ring such that $1 \neq 0$ and all elements $x \in K \setminus \{0\}$ have a multiplicative inverse, i.e., a $y \in K$ such that $xy = 1$.

Now we can define formally real fields, and real-closed fields subsequently.

Definition D.1 (Formally real field). A field R is a *(formally) real field* iff any of the following (equivalent) conditions holds (see [26, Theorem 2.7]):

1. -1 is not a sum of squares in R.
2. For every $x_1, \ldots, x_n \in R$ we have that $\sum_{i=1}^{n} x_i^2 = 0$ implies $x_1 = \cdots = x_n = 0$.
3. R admits an ordering that makes R an ordered field, i.e., a total order $\leq \in R \times R$ on R that is compatible with the field structure:

   a. $x \leq y$ implies $x + z \leq y + z$ for all $x, y, z \in R$;
   b. $0 \leq x$ and $0 \leq y$ imply $0 \leq xy$ for all $x, y \in R$.

   Recall that the order $\leq$ is a total order iff the following conditions hold:

   a. Transitive: $x \leq y$ and $y \leq z$ imply $x \leq z$ for all $x, y, z \in R$;
   b. Antisymmetric: $x \leq y$ and $y \leq x$ imply $x = y$ for all $x, y \in R$;
   c. Total: for all $x, y \in R$: $x \leq y$ or $y \leq x$.

There are several equivalent characterisations of real-closed fields [26]:

Definition D.2 (Real-closed field). A field R is a *real-closed field* iff any of the following (equivalent) conditions holds (see [26, Theorem 2.11]):

1. R is an ordered field where every positive element is a square and every polynomial in $R[X]$ of odd degree has a root in R (then this order is, in fact, unique).
2. R is not algebraically closed but its field extension $R[\sqrt{-1}] = R[i]/(i^2 + 1)$ is algebraically closed.
3. R is not algebraically closed but its algebraic closure is a finite extension, i.e., finitely generated over R.
4. R has the *intermediate value property*, i.e., R is an ordered field such that for any polynomial $p \in R[X]$ with $a, b \in R, a < b$ and $p(a)p(b) < 0$, there is a ζ with $a < \zeta < b$ such that $p(\zeta) = 0$.
5. R is a real field such that no proper algebraic extension is a formally real field.

Example D.1 (Quadratic formula). The following sets are real-closed fields:

- Real numbers $\mathbb{R}$, as the prototypical but not the only example.
- Real algebraic numbers $\bar{\mathbb{Q}} \cap \mathbb{R}$, that is, real numbers in the algebraic closure of $\mathbb{Q}$, i.e., real numbers that are roots of a nonzero polynomial with rational or integer coefficients.
- Computable numbers [298], i.e., those real numbers that can be approximated by a computable function up to any desired precision.

- Definable numbers, i.e., those real numbers $a \in \mathbb{R}$ for which there is a first-order formula φ in the language of set theory with one free variable such that a is the unique real number for which φ holds true. □

According to the important Tarski-Seidenberg Principle [288, 278] or the transfer principle for real-closed fields, the first-order theory of real arithmetic is equivalent to the first-order theory of real-closed fields; see [26, Theorems 2.80 and 2.81].

Theorem D.1 (Tarski-Seidenberg principle). *The first-order theory of reals is identical to the first-order theory of real-closed fields, i.e., the set of closed first-order formulas over the signature* $+, \cdot, =, <, 0, 1$ *that are valid over the reals* $\mathbb{R}$ *is the same as the corresponding set of formulas that are valid in any real-closed field.*

Consequently, the notion of real-closed fields captures exactly the first-order logic point of view of real arithmetic. The technical device exploiting this result for deciding formulas of real arithmetic is quantifier elimination via cylindrical algebraic decomposition, which we examine next.

D.2.2 Semialgebraic Geometry and Cylindrical Algebraic Decomposition

From an algebraic or model-theoretic perspective, a quantifier-free formula of real arithmetic directly corresponds to the set of its satisfying assignments in $\mathbb{R}^n$. Formally, a *semialgebraic set* is a subset of $\mathbb{R}^n$ that is defined by a finite conjunction of polynomial equations and inequalities or any finite union of such sets, which, up to normalisation, is a quantifier-free formula of real arithmetic. From an algebraic or geometrical perspective, a purely equational quantifier-free formula of real arithmetic also corresponds directly to an algebraic variety, which is the basic object of study in algebraic geometry [150, 154, 215].

The most important part about projections in the following central theorem about semialgebraic sets is due to Tarski [288, 278, 163, 50, 106].

Theorem D.2 (Semialgebraic sets). *Semialgebraic sets are closed under finite union, finite intersection, complement, and projection (to linear subspaces).*

It is easy to see that projection of a semialgebraic set in $\mathbb{R}^n$ to a linear subspace (e.g., $\mathbb{R}^{n-1}$) corresponds to elimination of existential quantifiers. The projection of the set of points in $\mathbb{R}^n$ where a formula $\phi(x_1,\dots,x_n)$ holds true corresponds directly to the subspace $\mathbb{R}^{n-1}$ spanned by the variables $x_1,\dots,x_{n-1}$where the formula $\bar{\phi}(x_1,\dots,x_{n-1}) \equiv \exists x_n\, \phi(x_1,\dots,x_n)$ holds true. See Fig. D.4 for an illustration of how the quantifier elimination of a quantified real arithmetic formula $(\exists y\,(y \geq 0 \land 1 - x - 1.83x^2 + 1.66x^3 > y))$ corresponds to projection (to the thick region on the x axis that satisfies $0.75 < x \land x < 0.68 \lor x > 1.17$).

Simply speaking, the basic insight of Tarski [288] with subsequent simplifications by Seidenberg [278], Robinson [264], Łojasiewicz [197], Cohen [79], and

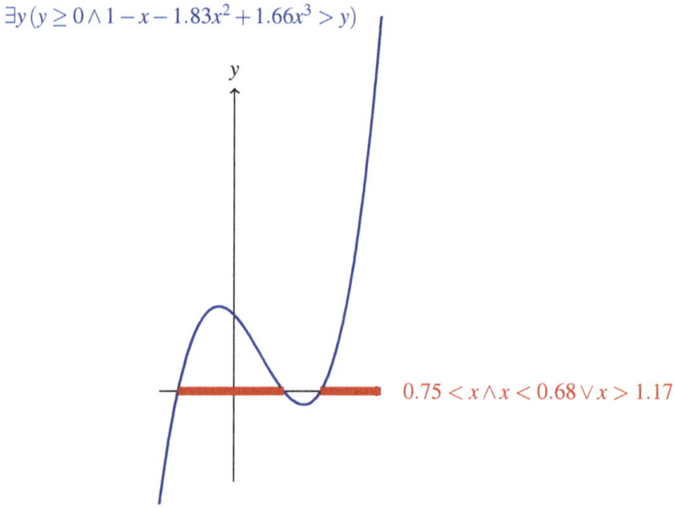

Fig. D.4 Projection of semialgebraic sets and quantifier elimination

Kreisel and Krivine [184], which led to the development of cylindrical algebraic decomposition (CAD) [80] and partial cylindrical algebraic decomposition [81] as decision procedures for real-closed fields, is that the conjunction of a set of polynomial equations and inequalities partitions the real space $\mathbb{R}^n$ into finitely many equivalence classes based on their sign combinations. The basic observation is that each polynomial $p \in K[X_1, \ldots, X_n]$ partitions the space into three equivalence classes based on its sign:

1. $\hat{+} := \{x \in \mathbb{R}^n \;:\; p(x) > 0\}$,
2. $\hat{0} := \{x \in \mathbb{R}^n \;:\; p(x) = 0\}$,
3. $\hat{-} := \{x \in \mathbb{R}^n \;:\; p(x) < 0\}$.

These classes can have multiple (but only finitely many) connected components. Further, Tarski showed that a set of polynomials partitions the real space into finitely many equivalence classes that essentially correspond to the coarsest joint refinement of all these sign relations (and possibly the relations of these polynomials on various connected components). That is, a *(natural) algebraic decomposition* of $\mathbb{R}^n$ is a partitioning of $\mathbb{R}^n$ into maximal connected regions where each of the relevant polynomials has invariant sign, which is used more systematically in the cell construction behind CAD [80]. Now, a satisfying assignment for a quantifier-free arithmetic formula exists if and only if the formula is true in one of those equivalence classes. Consequently, working from inside out, existential quantifiers can be replaced equivalently by a big disjunction, replacing the existentially quantified variable x with any representative point inside each of these finitely many equivalence classes. Let S be a (finite) representative system of the equivalence classes for all polynomials in the quantifier-free formula F; then

$$\mathrm{QE}(\exists x F) \equiv \bigvee_{x \in S} F.$$

The computationally expensive part in the quantifier elimination algorithms is the construction of the representative system S needed for a given formula $\exists x F$.

Example D.2 (Quantifier elimination by interior points and basic virtual substitution). In the following simple example, quantifier elimination can simply insert a range of representative cases into the formula and evaluate the remaining arithmetic:

$$\begin{aligned}
&\mathrm{QE}(\exists x (x > 2 \land x < \tfrac{17}{3})) && \\
&\equiv (2 > 2 \land 2 < \frac{17}{3}) && \text{border case “}x = 2\text{”} \\
&\lor (\frac{17}{3} > 2 \land \frac{17}{3} < \frac{17}{3}) && \text{border case “}x = \frac{17}{3}\text{”} \\
&\lor (\frac{2+\frac{17}{3}}{2} > 2 \land \frac{2+\frac{17}{3}}{2} < \frac{17}{3}) && \text{intermediate case “}x = \frac{2+\frac{17}{3}}{2}\text{”} \\
&\lor (-\infty > 2 \land -\infty < \frac{17}{3}) && \text{extremal case “}x = -\infty\text{”} \\
&\lor (\infty > 2 \land \infty < \frac{17}{3}) && \text{extremal case “}x = \infty\text{”} \\
&\equiv \mathit{true}.
\end{aligned}$$

The basic idea behind this is illustrated in the following real line:

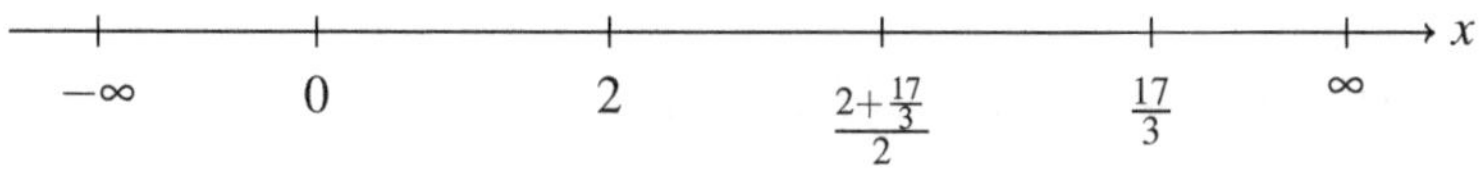

The intuition is to substitute in all the special points mentioned in the formula (2 and $\frac{17}{3}$) and to find some representative of the open interval between these points, say the middle $\frac{2+17/3}{2}$. Any other point in the open interval would be an equally good representative. The other two special cases are representatives for infinitely large (∞) or infinitely small ($-\infty$) points. Considering one disjunct above at a time, the remaining ground formulas (i.e., no variables) can be evaluated based on the numbers to obtain *true* or *false*. Here, only the intermediate case $\frac{2+17/3}{2}$ evaluates to *true*. All other disjuncts evaluate to *false*. A systematic account of quantifier elimination by a more general technique called virtual substitution can be found in the work of Weispfenning [301, 300]. □

Example D.3. A quadratic example is the following equivalence constructed by quantifier elimination:

$$\mathrm{QE}(\exists x\,(ax^2 + bx + c = 0)) \;\equiv\; a \neq 0 \land b^2 - 4ac \geq 0 \lor a = 0 \land (b = 0 \to c = 0).$$

It basically follows the standard determinant condition $b^2 \geq 4ac$. Note however, that for this to be a true equivalence, the special cases of what happens if $a = 0$ or if $b = 0$ also need to be considered by quantifier elimination procedures. □

Universal quantifiers can be handled by duality from the handling of existential quantifiers:

$$\mathrm{QE}(\forall x\, F) \equiv \neg\, \mathrm{QE}(\exists x\, \neg F).$$

For multiple quantifiers, this procedure can be used recursively by applying quantifier elimination from inside out, i.e., starting from inner quantifiers. The final result for a closed formula gives a propositional formula without variables, which is decidable by evaluating the remaining arithmetic expressions with concrete numbers. Practical quantifier elimination procedures for real arithmetic improve on this theoretical decision procedure, e.g., by minimising the number of required equivalence classes. Practical decision procedures for quantifier elimination include cylindrical algebraic decomposition [80], partial cylindrical algebraic decomposition [81], and virtual substitution [301, 300].

KeYmaera integrates Mathematica by Wolfram Research [303], Redlog [101], and QEPCAD B [59] as alternative implementations of quantifier elimination procedures. KeYmaera also includes a built-in implementation of an adaptation of John Harrison's version [152] of the Cohen-Hörmander procedure [165] that was developed by David Renshaw. Finally, KeYmaera can also interface directly with Harrison's implementation [152] or a proof-producing quantifier elimination procedure by Sean McLaughlin and John Harrison [204].

D.2.3 Nullstellensatz and Gröbner Bases

Gröbner bases [62, 31, 211, 88] can be used as a sound but incomplete procedure for proving validity of formulas in the universal fragment of equational first-order real arithmetic. This approach is, in fact, not specific to real arithmetic but also applies for other fields. Gröbner bases have been introduced by Bruno Buchberger [62] as a systematic theory and algorithm for symbolic computations in factor rings of multivariate polynomial rings. Gröbner basis algorithms can be considered as a multivariate joint generalisation of the Euclidean algorithm for computing greatest common divisors in univariate polynomial rings and of Gaussian elimination for solving linear equation systems. An analogous concept for local rings, called standard bases, was developed independently by Heisuke Hironaka in 1964.

Preliminaries

First we briefly recapitulate some basic notions from algebra [191, 52, 53, 112, 211]. We do not always give the most general definition of these notions but restrict our attention to what we actually need in our context.

For a field K, the set $K[X_1,\dots,X_n]$ of *multivariate polynomials* over K forms a ring and is defined as the free commutative and associative algebra over the indeterminates $X_1,\dots,X_n$. Products $X_1^{i_1},\dots,X_n^{i_n}$ of the indeterminates are called *monomials*. The polynomials over K are sums of monomials with coefficients in K, of the form

$$\sum_{i_1,\dots,i_n\in\mathbb{N}} a_{i_1,\dots,i_n} X_1^{i_1},\dots,X_n^{i_n},$$

for which only finitely many coefficients $a_{i_1,\dots,i_n} \in K$ are nonzero. A subset $I \subseteq R$ is an *ideal* of a ring R, denoted as $I \trianglelefteq R$, iff I is a subgroup of the additive group of R (that is $x+y \in I$ for all $x,y \in I$) and

$$rx \in I \text{ for all } x \in I, r \in R.$$

This ideal property is a "magnetic property". The elements of I are magnetic in the sense that they again land in I when multiplied with any element of the full ring R (not just when multiplied with elements in I itself).

The ideal *generated* by a set $G \subseteq R$ is the smallest ideal $I \trianglelefteq R$ containing G. In that case, G is called a *generating system* of I. In particular, if $G \subseteq K[X_1,\dots,X_n]$ is a set of polynomials that is zero at a particular point $(x_1,\dots,x_n) \in K^n$, then every element of the ideal (G) is zero at that point, because sums of zero stay zero, and products of zero with any element stay zero.

To start the development of Gröbner bases, we will need to define polynomial reductions. The notions of Gröbner bases and polynomial reductions are relative to an admissible monomial order, which also determines the leading term (largest monomial) in multivariate polynomials. An *admissible monomial order* $\prec$ is a strict well-order (well-founded total order; in particular, every non-empty set has a least element) on monomials such that $uw \prec vw$ for all monomials u,v,w with $u \prec v$. Admissible orders extend canonically to polynomials $K[X_1,\dots,X_n]$ as a multiset order [62, 99]. The monomial order determines the *leading term* in multivariate polynomials, i.e., the maximal monomial with respect to $\prec$. Here, we simply assume some fixed admissible order and refer to the literature on Gröbner bases [31, 211, 88, 62] for details on monomial orders, their admissibility conditions, and their properties.

As a multivariate generalisation of the Euclidean algorithm we need multivariate polynomial division:

Definition D.3 (Reduction). Let $f,g \in K[X_1,\dots,X_n]$ be polynomials. We say that f *reduces to* g with respect to a set $G \subset K[X_1,\dots,X_n]$ of polynomials iff for some $m \in \mathbb{N}$ there are $f_0,f_1,\dots,f_m$ in $K[X_1,\dots,X_n]$ with $f_0 = f, f_m = g$ such that, for all i,

$$f_{i+1} = f_i - h_i g_i$$

for some $h_i \in K[X_1,\dots,X_n]$, $g_i \in G$, and $f_{i+1} \prec f_i$. We write $g = \mathrm{red}_G f$ if, in addition, g cannot be reduced further, i.e., there is no $h_{m+1} \in K[X_1,\dots,X_n]$ and $g_{m+1} \in G$ with $g - h_{m+1} g_{m+1} \prec q$. Then polynomial $\mathrm{red}_G f$ is called *reduction* (or *remainder*) of f by G. Furthermore, polynomial f is called *reduced* with respect to G iff $f = \text{red}_G f$.

The most common approach to reducing a polynomial is by following a series of elementary reductions. These elementary reductions are a special case of the above definition. Let $f,g \in K[X_1,\ldots,X_n]$ be polynomials and let l be the leading term of g. If $a_{i_1,\ldots,i_n}X_1^{i_1}\cdots X_n^{i_n}$ is some (e.g., the largest) term of f which is divisible by l, then the following polynomial is called an *elementary reduction* of f by g:

$$f - \frac{a_{i_1,\ldots,i_n}X_1^{i_1}\cdots X_n^{i_n}}{l}g$$

The theory of Gröbner bases characterises exactly under which circumstances reduction produces unique remainders, which is not generally so in multivariate polynomial rings.

Definition D.4 (Gröbner basis). A finite subset G of $I \trianglelefteq K[X_1,\ldots,X_n]$, i.e., an ideal I of a polynomial ring, is called a *Gröbner basis* iff the ideal generated by G is I and any of the following (equivalent) conditions holds:

1. Reduction with respect to G gives 0 for any $p \in I$.
2. $\text{red}_G p = 0$ iff $p \in I$.
3. Reduction with respect to G gives a unique remainder.
4. The polynomials that are reduced with respect to G form representatives of the factor ring $K[X_1,\ldots,X_n]/I$.
5. The leading ideal of I, i.e., the ideal generated by leading terms of polynomials of I, is generated by the leading terms of G.

A Gröbner basis G is *reduced* if all $g \in G$ are reduced with respect to $G \setminus \{g\}$.

Because ideals are closed under multiplication by nonzero constants, we can assume Gröbner basis G to contain normalised polynomials only, i.e., where the leading coefficient of their leading terms is 1.

The most important result about Gröbner bases is that every ideal of a polynomial ring $K[X_1,\ldots,X_n]$ over a field K in finitely many variables has a unique reduced Gröbner basis (with leading coefficient 1) that can be computed effectively by the Buchberger algorithm [62] or its subsequent improvements [31, 211, 88, 120, 121]. Thus, Gröbner bases give an effective version of Hilbert's basis theorem:

Theorem D.3 (Hilbert's basis theorem). *Every ideal in the ring $K[X_1,\ldots,X_n]$ of multivariate polynomials over a field K is finitely generated. That is, every ideal $I \trianglelefteq K[X_1,\ldots,X_n]$ has a finite generating system G, i.e., the ideal generated by G is I.*

Using Gröbner Basis Eliminations

Gröbner basis reductions can be used for handling fragments of equational universal arithmetic. Assume we have a sequent of the following form (when rewriting to

(A1) $\dfrac{(f-g)z=1\vdash}{\vdash f=g}$ [1]

(A2) $\dfrac{f-g=z^2\vdash}{f\geq g\vdash}$ [1]

(A3) $\dfrac{(f-g)z^2=1\vdash}{f>g\vdash}$ [1]

(A4) $\dfrac{*}{g_1=\tilde{g}_1,\ldots,g_n=\tilde{g}_n\vdash}$ [2]

(A5) $\dfrac{*}{g_1=\tilde{g}_1,\ldots,g_n=\tilde{g}_n\vdash f=h}$ [3]

(A6) $\dfrac{\vdash 1+s_1^2+\cdots+s_n^2=0}{\vdash}$ [4]

[1] z is a fresh variable.
[2] Applicable if $\mathrm{red}_G 1=0$ where G is the Gröbner basis of the ideal $(g_1-\tilde{g}_1,\ldots,g_n-\tilde{g}_n)$.
[3] Applicable if $\mathrm{red}_G f=h$ where G is the Gröbner basis of the ideal $(g_1-\tilde{g}_1,\ldots,g_n-\tilde{g}_n)$.
[4] Polynomials $s_1,\ldots,s_n$ can be chosen arbitrarily

Fig. D.5 Rule schemata of Gröbner calculus rules

normalise equations, disequations, and inequalities as necessary):

$$\Gamma,g_1=\tilde{g}_1,\ldots,g_n=\tilde{g}_n\vdash f_1=h_1,\ldots,f_e=h_e,\Delta. \tag{D.1}$$

Let G be a Gröbner basis of the ideal generated by the $g_i-\tilde{g}_i$, i.e., of the ideal

$$(g_1-\tilde{g}_1,\ldots,g_n-\tilde{g}_n). \tag{D.2}$$

If the reduction with respect to G of some f_i equals the reduction of the corresponding h_i with respect to G, i.e., $\mathrm{red}_G f_i=\mathrm{red}_G h_i$, then the sequent (D.1) is valid and can be closed. This most naïve use of Gröbner bases for real arithmetic is summarised in the proof rules A4 and A5 of Fig. D.5. The rule A4 closes a goal (marked by $*$) if the ideal G generated by equations in the antecedent contains 1. At every point where all equations of G are zero, all equations of the ideal generated by G are also zero. If the ideal generated by G contains 1, then $1=0$ holds on the points where all polynomials of G are zero. Hence, these polynomials do not have a common zero (they are contradictory). Similarly, rule A5 closes a goal if the sides f,h of an equation in the succedent have the same remainder modulo the Gröbner basis G of the equations of the antecedent, which means $f-h\in(G)$. Thus, whenever the elements of G are zero, $f-h$ is zero. So far, rules A4 and A5 can prove some but not all properties.

The scope of the rules can be extended further by testing for *radical membership* instead of ideal membership, which can prove problems like $x^2=0\vdash x=0$ that rule A5 cannot prove. The *radical* of an ideal I is the set

$$\sqrt{I}=\bigcup_{i=1}^{\infty}\{g\in\mathbb{Q}[X_1,\ldots,X_n]\ :\ g^i\in I\}\supseteq I$$

Because the inclusion $I\subseteq\sqrt{I}$ can be strict (e.g., $\sqrt{(x^2)}=(x)$), testing for radical membership is more liberal than testing for ideal membership, while still being sound.

In practise, the rule A1, which is known as *Rabinowitch's trick*, represents a simple way of testing for radical membership. It is based on the observation that $g \in \sqrt{I}$ if and only if $1 \in (I \cup \{gz - 1\})$ (where z is a fresh variable). The latter property can be tested by first applying rule A1 and then rule A4.

Finally, inequalities can be translated to equations using proof rules A2 and A3, which exploit the fact that a real number is positive iff it is a square (rule A3 is an optimised version including Rabinowitch's trick). Rules A2, A3, and A5 are similar for inequalities in the succedent. Combined with the rules A4 and A5, this encoding of inequalities is rather weak, and not able to derive simple facts like $a \leq b \wedge b \leq c \rightarrow a \leq c$. It is, however, an important basis for extensions to full decision procedures [246], in particular, in combination with proof rule A6, which we discuss in Sect. D.2.4.

Proposition D.1 (Soundness of Gröbner basis rules). *The rules in Fig. D.5 are sound.*

Proof. The rules are locally sound. For rules A1, A2, and A3, we show satisfiability-equivalence of premise and conclusion, which implies soundness.

A1 Satisfiability-equivalence of A1 is a consequence of the Rabinowitch trick equivalence $x \neq 0 \leftrightarrow \exists z\,(xz = 1)$, using $f - g$ for x. More generally, this holds in fields where nonzero elements are exactly the elements that have some inverse z. By introducing a free new variable z, we obtain that the premise is satisfiable if and only if the conclusion is satisfiable.

A2 Satisfiability-equivalence follows from equivalence $f \geq g \leftrightarrow \exists z\,(f - g = z^2)$ in the domain of reals. More generally, this holds in real-closed fields where squares are exactly the positive numbers. By introducing a free new variable z in the rule, we see that the premise and conclusion are satisfiability-equivalent, i.e., the premise is satisfiable if and only if the conclusion is satisfiable. Using the equivalence and the soundness of $\exists$l from Theorem 2.1, soundness can also be obtained easily from the following derivation when using the state variable z, which is implicitly quantified universally in the sequent, in place of the Skolem symbol s:

$$\cfrac{\exists\text{l}\,\cfrac{f - g = s^2 \vdash}{\exists z\, f - g = z^2 \vdash}}{f \geq g \vdash}$$

A3 Satisfiability-equivalence follows from the equivalence $x > 0 \leftrightarrow \exists z\,(xz^2 = 1)$ in the reals, using $f - g$ for x.

A4 The soundness of A4 is a special case of the soundness of A5 when assuming the false formula $1 = 0$ for the succedent $f = h$.

A5 Suppose the conclusion was false in ν: $\nu \models g_1 = \tilde{g}_1 \wedge \cdots \wedge g_n = \tilde{g}_n \wedge f \neq h$. Thus, $\nu \models g = 0$ for all $g \in G$ (using the notions of algebraic geometry, this would correspond to ν being in the algebraic variety of G). Consequently, $\nu \models g = 0$ for all polynomials g in the ideal (G) of G. As a consequence of the applicability condition, we have $\text{red}_G(f - h) = 0$, which, by Definition D.4, implies that $f - h$ is in the ideal of G. In combination, we have $\nu \models f - h = 0$, and hence $\nu \models f = h$, which is a contradiction.

A6 Since a sum of squares is nonnegative over $\mathbb{R}$, the value of $1 + s_1^2 + \cdots + s_n^2$ is strictly positive and $1 + s_1^2 + \cdots + s_n^2 = 0$ is a contradiction over the reals. Consequently, if the premise is valid, then so is the conclusion. □

While rules A1, A2, A3, A4, and A5 alone without A6 are incomplete for real arithmetic, the situation is different over the complex numbers. Hilbert's Nullstellensatz shows that rules A4 and A1 yield a decision procedure for universal equational problems in $\mathbb{C}$. For this, we need the following notions. The *algebraic variety* $V(I)$ of a set of polynomials $F \trianglelefteq K[X_1, \ldots, X_n]$ is defined as the set of points where all polynomials of F are zero:

$$V(F) := \{x \in K^n \ : \ p(x) = 0 \text{ for all } p \in F\}.$$

See Fig. D.6 for examples of algebraic varieties $V(p)$ generated by a single polynomial p in two variables x and y (thus, algebraic curves in the Euclidean plain). Conversely, the *vanishing ideal* $I(V)$ of any set $V \subseteq K^n$ is defined as the set of polynomials that are zero at all points of V:

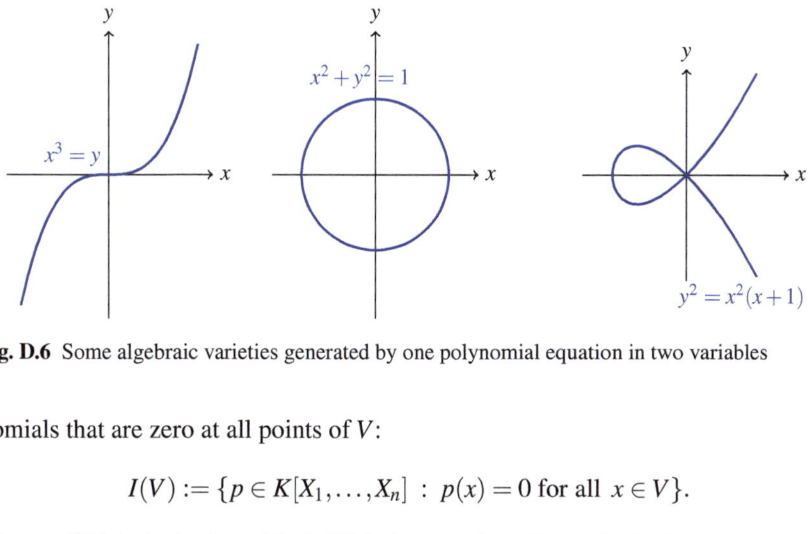

Fig. D.6 Some algebraic varieties generated by one polynomial equation in two variables

$$I(V) := \{p \in K[X_1, \ldots, X_n] \ : \ p(x) = 0 \text{ for all } x \in V\}.$$

The set $I(V)$ is, indeed, an ideal. With those notions from algebraic geometry, Hilbert's Nullstellensatz [191] can be formulated in an elegant way.

Theorem D.4 (Hilbert's Nullstellensatz). *Let K be an algebraically closed field and $I \trianglelefteq K[X_1, \ldots, X_n]$ an ideal. Then*

$$I(V(I)) = \sqrt{I}$$

In particular, $V(I) = \emptyset$ iff $\sqrt{I} = K[X_1, \ldots, X_n]$.

As a corollary, the sequent $g_1 = \tilde{g}_1, \ldots, g_n = \tilde{g}_n \vdash$ is valid iff $g_1 = \tilde{g}_1 \land \cdots \land g_n = \tilde{g}_n$ is unsatisfiable over $\mathbb{R}$. Over $\mathbb{C}$, according to Theorem D.4, the sequent is valid, i.e., the equations in the antecedent are unsatisfiable over $\mathbb{C}$ iff 1 is in the radical ideal of (D.2). With Rabinowitch's trick in rule A1, Gröbner bases with rule A4 can be used for this radical ideal membership check, giving a complete procedure for complex arithmetic.

The situation with the field of reals is more difficult. For reals, the Gröbner basis approach with rules A1, A2, A3, A4, and A5 alone gives a sound but incomplete overapproximation. To see why Gröbner bases are incomplete for real arithmetic, consider the following. Gröbner bases are a general approach for polynomial rings over fields (or even rings). They do not take into account the special properties of the reals. For instance, the sequent $x^2 = -1 \vdash$ is valid, i.e., the formula $x^2 = -1$ is unsatisfiable over $\mathbb{R}$, but the Gröbner basis of x^2+1 is $\{x^2+1\}$ and, in fact, $x^2 = -1$ is satisfiable over $\mathbb{C}$ but not over $\mathbb{R}$.

KeYmaera integrates with Mathematica by Wolfram Research [303] and the Orbital library developed by the author as implementations of Gröbner basis algorithms.

D.2.4 Real Nullstellensatz

Gröbner bases and the Nullstellensatz alone are complete for universal properties of complex arithmetic, but incomplete for the domain of real arithmetic. A corresponding completeness result for real-closed fields is Stengle's real Nullstellensatz [283], which extends previous real Nullstellensätze [186, 109, 262]:

Theorem D.5 (Real Nullstellensatz for real-closed fields). *Let R be a real-closed field (e.g., $R = \mathbb{R}$) and let $F \subset R[X_1,\dots,X_n]$ be a finite subset. Then*

$$V(F) = \emptyset$$

if and only if there are polynomials $s_1,\dots,s_m \in R[X_1,\dots,X_n]$ such that

$$1 + s_1^2 + \cdots + s_m^2 \in (F).$$

If, moreover, $F \subseteq \mathbb{Q}[X_1,\dots,X_n]$, then also the polynomials $s_1,\dots,s_m$ can be chosen among the elements of $\mathbb{Q}[X_1,\dots,X_n]$.

This theorem leads to an extremely simple, complete proof method for the universal fragment of real arithmetic: in addition to the rules A1, A2, A3, A4, and A5, we add rule A6 in Fig. D.5 for injecting the equation $1+s_1^2+\cdots+s_m^2 = 0$ into a proof goal. The proof rule A6 expresses that an equality $1+s_1^2+\cdots+s_n^2 = 0$ can always be added to the succedent at any time, because a sum of squares is nonnegative, and, if we add 1, is nonzero. Consequently, $1+s_1^2+\cdots+s_n^2 = 0$ is obviously equivalent to false. Such an equation can be very powerful when using Gröbner basis computations on equations.

Any valid formula in the universal fragment of real-closed field can then be proven as follows:

1. turn inequalities and equations in the succedent, or disequations and inequalities in the antecedent, into equations in the antecedent with rules A1, A2, and A3.
2. add the *witness* $1+s_1^2+\cdots+s_m^2$ due to the real Nullstellensatz to the succedent with rule A6; and
3. prove the resulting sequent with Gröbner Basis computations by rule A5.

Corollary D.1 (Completeness). *Along with propositional rules from Fig. 2.11, the rules in Fig. D.5 are complete for the universal fragment of real arithmetic and in real-closed fields.*

Proof. Completeness follows from Theorem D.5 using the satisfiability-equivalence properties for the transformation by A1, A2, and A3 according to Proposition D.1. □

The main difficulty with this calculus is obvious: it does not provide any guidance for choosing the witness $1+s_1^2+\cdots+s_m^2=0$. One technique to tackle the required search is semidefinite programming, following the work based on Stengle's Positivstellensatz (Sect. D.2.5) in [224, 151]. See joint work with Quesel and Rümmer [246] for details.

Example D.4. In Fig. D.7, we show a proof for the following implication (leaving out propositional reasoning):

$$x \geq y \land z \geq 0 \rightarrow xz \geq yz. \tag{D.3}$$

First, we turn the inequalities $x \geq y$ and $z \geq 0$ into equations using rule A2. Prov-

$$\begin{array}{ll}
 & * \\ \hline
\text{A5} & x-y=a^2, z=b^2, (yz-xz)c^2=1 \vdash 1+(abc)^2=0 \\ \hline
\text{A6} & x-y=a^2, z=b^2, (yz-xz)c^2=1 \vdash \\ \hline
\text{A2,A3} & x \geq y, z \geq 0, yz > xz \vdash \\ \hline
\textit{cut} & x \geq y, z \geq 0 \vdash xz \geq yz \\ \hline
\land\text{l} & x \geq y \land z \geq 0 \vdash xz \geq yz \\ \hline
\rightarrow\text{r} & \vdash x \geq y \land z \geq 0 \rightarrow xz \geq yz
\end{array}$$

Fig. D.7 Example proof using the real Nullstellensatz

ing by contradiction (or using propositional rules and the *cut* rule), the conclusion $xz \geq yz$ is considered as an assumption $yz > xz$ and subsequently eliminated with rule A3. Once this is done, we rely on an oracle to tell us the witness $1+(abc)^2$, which is introduced using rule A6. Finally, we close the proof by A5: the set $\{a^2-x+y, b^2-z, xzc^2-yzc^2+1\}$ is a Gröbner basis representing the equations in the antecedent. This Gröbner basis reduces the term $1+(abc)^2$ to 0 by subtracting multiples of the basis polynomials as indicated:

$$1+a^2b^2c^2 \overset{b^2-z}{\rightsquigarrow} 1+a^2zc^2 \overset{a^2-x+y}{\rightsquigarrow} 1+xzc^2-yzc^2 \rightsquigarrow 0.$$

Consequently, at every point where all equations in the antecedent are zero, all elements of the above Gröbner basis are zero (they are in the same ideal), and, thus, $1+(abc)^2$ is zero at the same point, because the Gröbner basis reduces this to zero. Hence, if the equations in the antecedent hold, the equation in the succedent holds too, thereby proving the property. □

D.2.5 Positivstellensatz and Semidefinite Programming

The Positivstellensatz for real-closed fields [283, 50], which has been introduced by Stengle along with the real Nullstellensatz for real-closed fields [283], can be used as a sound procedure for proving formulas in the universal fragment of first-order real arithmetic. The Positivstellensatz has recently been exploited in combination with relaxations from semidefinite programming [224, 151]. In joint work with Quesel and Rümmer, the author has recently introduced a combination of Stengle's real Nullstellensatz with Gröbner bases and semidefinite programming as a decision procedure for the universal fragment of real arithmetic [246].

A property of the ideal (G) generated by $G \subseteq R[X_1,\dots,X_n]$ is that products of an element that is zero with any element are zero and that sums of zero elements are zero. In particular, if every polynomial of G is zero at a point $x_1,\dots,x_n$, then every polynomial of (H) is zero at that point. The *multiplicative monoid* $\mathrm{mon}(H)$ generated by $H \subseteq R[X_1,\dots,X_n]$ is the set of finite products of elements of H (including the empty product 1). The intuition behind this monoid construction is that a product of nonzero elements remains nonzero. So if we know that all elements of H are nonzero at a point $x_1,\dots,x_n$, then any product in $\mathrm{mon}(H)$ is also nonzero at that point. The *cone* $\mathrm{con}(F)$ generated by a set $F \subseteq R[X_1,\dots,X_n]$ is the subsemiring of $R[X_1,\dots,X_n]$ generated by F and arbitrary squares, i.e., the smallest set containing F and squares s^2 of arbitrary polynomials $s \in R[X_1,\dots,X_n]$ that is closed under addition and multiplication. The idea behind the cone construction is that sums and products of nonnegative elements remain nonnegative and that nonnegative elements remain nonnegative if we add arbitrary squares s^2. In particular, if we assume all elements in F to be nonnegative at a point $x_1,\dots,x_n$,, then all elements of $\mathrm{con}(F)$ are nonnegative at that point as well. For more computational representations of cones and ideals, we refer you to the literature [224, 50].

With these constructions we can present Stengle's Positivstellensatz [283], which relates the feasibility of a set of equations, disequations, and inequalities to a polynomial formed from the ideal of equations, the monoid of disequations, and the cone of inequalities.

Theorem D.6 (Positivstellensatz for real-closed fields). *Let R be a real-closed field (e.g., $R=\mathbb{R}$) and F,G,H finite subsets of $R[X_1,\dots,X_n]$. Then*

$$\{x \in R^n : f(x) \geq 0 \textit{ for all } f \in F,\ g(x) = 0 \textit{ for all } g \in G,\ h(x) \neq 0 \textit{ for all } h \in H\}$$

is empty iff

$$\text{(A7)}\ \frac{*}{f_1 \geq \tilde{f}_1, \dots, f_m \geq \tilde{f}_m, g_1 = \tilde{g}_1, \dots, g_n = \tilde{g}_n, h_1 \neq \tilde{h}_1, \dots, h_l \neq \tilde{h}_l \vdash}\ ^1$$

[1] Applicable iff $s + g + m^2 = 0$ for some polynomial $s \in \text{con}(\{f_1 - \tilde{f}_1, \dots, f_m - \tilde{f}_m\})$, some polynomial $g \in (g_1 - \tilde{g}_1, \dots, g_n - \tilde{g}_n)$, and some polynomial $m \in \text{mon}(\{h_1 - \tilde{h}_1, \dots, h_l - \tilde{h}_l\})$

Fig. D.8 Rule schema of Positivstellensatz calculus rule

there are $s \in \text{con}(F), g \in (G), m \in \text{mon}(H)$ *such that* $s + g + m^2 = 0$.

If, moreover, $F, G, H \subseteq \mathbb{Q}[X_1, \dots, X_n]$, *then also the polynomials* s, g, m *can be chosen among the elements of* $\mathbb{Q}[X_1, \dots, X_n]$.

The polynomials s, g, m are polynomial infeasibility witnesses. For bounded degree, witnesses s, g, m can be searched for using numerical semidefinite programming [224] by parametrising the resulting polynomials. As (theoretical) degree bounds exist for the certificate polynomials s, g, m, the Positivstellensatz yields a theoretical decision procedure. These bounds are, however, at least triply exponential [224]. Thus, the approach advocated by Parrilo [224] is to increase the bound successively and solve the existence of bounded degree witnesses due to the Positivstellensatz by semidefinite programming [54].

As a simple corollary to Theorem D.6 we have that rule A7 in Fig. D.8 is sound.

Corollary D.2. *The rule in Fig. D.8 is sound.*

Example D.5. A proof for the implication (D.3) from p. 393 that uses the Positivstellensatz can be found in Fig. D.9. In contrast to the proof in Fig. D.7, it is now

$$\dfrac{\dfrac{\dfrac{\dfrac{\dfrac{*}{x \geq y, z \geq 0, (yz - xz)c^2 = 1 \vdash}\text{A7}}{x \geq y, z \geq 0, yz > xz \vdash}\text{A3}}{x \geq y, z \geq 0 \vdash xz \geq yz}\text{cut}}{x \geq y \wedge z \geq 0 \vdash xz \geq yz}\wedge\text{l}}{\vdash x \geq y \wedge z \geq 0 \rightarrow xz \geq yz}\rightarrow\text{r}$$

Fig. D.9 Example proof using the Positivstellensatz

unnecessary to eliminate the inequalities $x \geq y$ and $z \geq 0$ while we still use the proof rule A3 for $xz \geq yz$ (corresponding to $yz > xz$ in the antecedent). A witness for the problem is:

$$\underbrace{c^2 \cdot (x - y) \cdot z}_{s} + \underbrace{(yz - xz)c^2 - 1}_{g} + \underbrace{1}_{m^2} = 0.$$

The terms $x - y$ and z in s result from the inequalities in the sequent, while the term g is derived from the equation. □

KeYmaera supports an integration with CSDP [51] as an implementation of semidefinite programming algorithms and can interface with Harrison's implementation [151] of semidefinite programming for the Positivstellensatz in HOL Light.

D.3 Discussion

As usual, it is a nontrivial task to ensure that the soundness that has been proven for a calculus inherits to soundness of a tool implementing this calculus. Despite our simple and concise calculi from Part I, the tool KeYmaera qualifies as a nontrivial piece of code, because of its numerous features, totalling to around 33,000 SLOC[2] on top of the KeY base system with 170,000 SLOC (not all parts of the base system are used in the context of hybrid systems, though). KeYmaera can also use external background solvers as faster decision procedures. In this section, we discuss the relationship of the soundness of our calculi and the soundness of KeYmaera. We refer to [36] for a general discussion of verification for verification systems.

Deductive Kernels

One advantage of deductive verification approaches is that they can have a comparably small trusted deductive kernel. The basic rationale underlying tools like LCF [207], HOL [138], or Isabelle/HOL [220] is that programming bugs in implementations that affect completeness are less fatal than bugs that would make the tool unsound and "prove" properties that are counterfactual. Essentially, when a verification tool like KeYmaera can only prove formulas by applying proof rules to them, it is sufficient for soundness purposes to ensure the correct implementation of these calculus rules and the proof rule application mechanism. In particular, this removes all sophisticated implementations of proof strategies from the trusted computing base. See, for instance, higher-order logic proof systems like Isabelle/HOL [220] for tools that follow this approach.

In KeYmaera, however, there are some additional pitfalls. Most of the rules of the calculi for our logics from Part I can be written as schematic rules, called *taclets* [33] in the KeY system, so that the careful soundness proofs carry over from the $\mathsf{d}\mathcal{L}$/DAL/dTL calculi from Part I to the implementation KeYmaera. Yet even this requires the (long-tested) rule application mechanism of KeY [34, 35] to be correct. Other rules like i$\forall$ and i$\exists$ from Fig. 2.11, however, cannot be written as a KeY taclet but have to be implemented in Java directly. Taclet notations for some of the rules also require meta-operators that are implemented in Java, e.g., for obtaining the solution of differential equations in rules $\langle'\rangle$,$[']$. To improve the quality of these Java implementations, in addition to software engineering techniques like code reviews and testing, we work with runtime assertions to try to ensure that the results

[2] Source lines of code, not counting comments, estimated with SLOCCount from `www.dwheeler.com/sloccount/`

KeYmaera produces during the current verification run are correct. For instance, the Java rules for proof rules i$\forall$ and i$\exists$ can assert at runtime that their goal (conclusion) is actually a substitution instance of the subgoals (premises) that they constructed, thereby ensuring in retrospect that their formula transformation was justified by the quantifier elimination lifting lemma, Lemma 2.5, from p. 92; see further explanations in Sect. 2.5.3.2. Such runtime checks can eliminate most of the corresponding rule implementations in KeYmaera from the trusted basis, leaving essentially only the implementation of substitutions and quantifier rearrangements in the trusted deductive kernel.

External Black Box Procedures

The practical soundness problem is deeper, though, because, for performance reasons, the KeYmaera tool calls external decision procedures following the cooperation scheme in Chap. 5, which use the techniques described in App. D.2. Thus, soundness of KeYmaera generally requires the external decision procedures to be implemented correctly. We discuss the respective external procedures and means for removing them from the required trusted computing base in this appendix.

Differential Equation Handling

For the implementation of differential equation handling rules like $\langle ' \rangle, [']$ from the d$\mathcal{L}$ calculus in Fig. 2.11, it is easy to check at runtime by simple symbolic differentiation and symbolic computations in polynomial rings whether a solution produced by a (complicated) procedure for solving differential equations actually is a solution. These checks can eliminate the differential equation solver (Mathematica by Wolfram Research and the Orbital library by the author) from the trusted computing base completely. Similar observations hold for differential induction rules *DI*,*DV* from the DAL calculus in Fig. 3.9, which only require symbolic differentiation and symbolic polynomial computations.

Positivstellensatz and Semidefinite Programming

A pleasant property of the Positivstellensatz Theorem D.6 and the approach in Sect. D.2.5 is that it produces a witness (the polynomials f,g,h) for the validity of a formula (which corresponds to the closing of the branch by rule A7). Once the witness has been found, it is checkable by simple computations in the polynomial ring to determine whether $s+g+m^2=0$ by comparing the coefficients. Thus, complicated numerical semidefinite programming tools [54] do not need to be part of the trusted computing base concerning soundness. Similarly, the approach in Sect. D.2.4 based on the real Nullstellensatz Theorem D.5 gives checkable proof witnesses based on simple polynomial computations.

Semialgebraic Geometry and Cylindrical Algebraic Decomposition

Most quantifier elimination procedures follow a combination of cylindrical algebraic decomposition [80], partial cylindrical algebraic decomposition [81], or virtual substitution [301], which are efficient but quite intricate procedures. Unlike approaches using the Positivstellensatz, general quantifier elimination, unfortunately, does not produce simple checkable certificates.

In order to remove the implementations of quantifier elimination procedures in Mathematica and other background solvers from the trusted computing base, diversification (i.e., cross-validating results by multiple background solvers) only is a partial answer, because the solvers might still agree on the same wrong result. Further, cross-validation limits the overall performance to the worst-of-breed rather than best-of-breed running times. As a more fundamental approach, we can use verified implementations, e.g., the verified quantifier elimination procedure for linear real arithmetic by Nipkow [219] in an executable fragment of Isabelle/HOL. Another approach is to use a non-verified but proof-producing implementation of general quantifier elimination by McLaughlin and Harrison [204], which can be used in KeYmaera.

Unfortunately, the practical performance achieved by those prototype implementations is not yet sufficient for larger case studies. A good compromise is to use the paradigm of re-verification: The proof search procedures IBC and IIO generate several calls to the quantifier elimination procedure to find a proof, but only those in the final proof are soundness-critical; see Chap. 5 for details. Thus, for soundness, it is sufficient to use a fast and possibly untrusted implementation of QE during the proof search and to re-verify the final proof in a proof checker with a verified or proof-producing QE implementation [204, 219]. For this purpose, the iterative background closure strategy from Chap. 5 is especially useful, because it iteratively identifies the sweet spot for using quantifier elimination during the proof search.

Nullstellensatz and Gröbner Bases

While Gröbner basis approaches do not have as simple witnesses as Positivstellensatz approaches, their working principle is strictly based on appropriate symbolic computations in the ring of polynomials. Consequently, these polynomial reductions can be carried out from a small set of rewrite rules within a logic. Based on an implementation for integer arithmetic by Philipp Rümmer in KeY [273], KeYmaera provides a built-in real arithmetic version [246] of Gröbner bases for nonlinear equations and Fourier-Motzkin elimination for linear inequalities [214, 124]. These give fully internalised KeYmaera proofs for the arithmetic.

Simplification

As an optional procedure, KeYmaera can call simplification procedures that simplify mathematical expressions to improve readability. Yet, an external simplification procedure that simplifies x/x to 1 would be unsound for the compactified domain $\mathbb{R} \cup \{-\infty, +\infty\}$, because ∞/∞ is undefined and not identical to 1. Likewise, this simplification could be unsound depending on whether x is allowed to assume 0. In our case, the above simplification is in fact sound for the domain of reals ($\mathbb{R}$) when adopting our convention from Sects. 2.2.3 and 3.2.1 that any formula containing x/x is understood to imply that $x \neq 0$ holds. Generally, however, arbitrary mathematical simplification algorithms might perform non-equivalent transformations and are thus disabled in KeYmaera by default.

Summary

The level of formal assurance of the correct functioning of the implementation of KeYmaera is, unfortunately, not quite comparable to that of systems like LCF [207], HOL [138], or Isabelle/HOL [220]. The built-in proof rule language in KeY allows us to state most proof rules of the $d\mathcal{L}$, DAL, and dTL proof calculi primarily as schematic proof rules in the taclet language [33]. Ensuring a correct implementation of the remaining code is non-trivial. The challenge is simplified substantially by using a deductive kernel with a smaller trusted basis and by using runtime assertions that check for coherence of the current proof rule application with its formal prerequisites. Fortunately, the simplicity of the proof rules of our calculi makes it possible to come up with a correct implementation.

The most tricky part, however, is that, for performance reasons, KeYmaera relies on tuned external decision procedures to ensure scalability to our larger case studies. These procedures can be eliminated from the trusted verification base by checking witnesses when working with solutions of differential equations, the Positivstellensatz, the real Nullstellensatz, and Göbner basis rules. For full quantifier elimination over real-closed fields, verified or proof-generating implementations can be helpful.

D.4 Performance Measurements

Unless otherwise indicated, the measurements in this book have been performed on a Dual-Core AMD Opteron™ 1218 Processor (F2 stepping) running at 2.6 GHz clock speed with 64 kB instruction cache, 64 kB first level data cache, and 1 MB second-level cache per core, sharing 4 GB of DDR2-667 main memory with CAS latency 5. In these measurements, KeYmaera was used on a Java™ SE Virtual Machine, build 1.6.0_06-b02, with a HotSpot™ engine in mixed mode on a Gentoo

Linux with a 2.6.25-g SMP kernel for x86/64 bit. The measurements are further based on the Mathematica 6.0.2 kernel.

Due to several effects including Java dynamic class loading and caching effects, the timing measurements are not necessarily always statistically significant. Rather, the timing information is intended to give a feeling for differences in magnitude. The measurements have been repeated to ensure reproducibility of results to increase confidence. While the Java Virtual Machine itself already contributes to nondeterministic effects that result from dynamic class loading, garbage collection, and caching, the implementation of the KeYmaera system itself is subject to nondeterminism as well, including the proof strategies, timing effects of iterative background closures, nondeterminstic term orderings based on memory references, and so on. Finally, computational back-ends allow for variations in overall performance.

Still, qualitative performance differences can be read off from the measurements. For instance, it makes quite a remarkably dramatic practical difference whether a strategy is able to prove a case study within only 100 seconds or cannot even come up with an answer after five hours, which happened a lot when comparing our proof strategies developed in Part II to simpler procedures.

References

1. Ábrahám-Mumm, E., Steffen, M., Hannemann, U.: Verification of hybrid systems: Formalization and proof rules in PVS. In: Andler and Offutt [16], pp. 48–57. DOI 10.1109/ICECCS.2001.930163
2. Achatz, M., McCallum, S., Weispfenning, V.: Deciding polynomial-exponential problems. In: J.R. Sendra, L. González-Vega (eds.) ISSAC, pp. 215–222. ACM (2008). DOI 10.1145/1390768.1390799
3. Adams, A., Dunstan, M., Gottliebsen, H., Kelsey, T., Martin, U., Owre, S.: Computer algebra meets automated theorem proving: Integrating Maple and PVS. In: R.J. Boulton, P.B. Jackson (eds.) TPHOLs, *LNCS*, vol. 2152, pp. 27–42. Springer (2001). DOI 10.1007/3-540-44755-5_4
4. Ahrendt, W., Baar, T., Beckert, B., Bubel, R., Giese, M., Hähnle, R., Menzel, W., Mostowski, W., Roth, A., Schlager, S., Schmitt, P.H.: The KeY tool. Software and System Modeling **4**, 32–54 (2005). DOI 10.1007/s10270-004-0058-x
5. Alur, R.: Timed automata. In: N. Halbwachs, D. Peled (eds.) CAV, *LNCS*, vol. 1633, pp. 8–22. Springer (1999). DOI 10.1007/3-540-48683-6_3
6. Alur, R., Courcoubetis, C., Dill, D.L.: Model-checking for real-time systems. In: Mitchell [210], pp. 414–425
7. Alur, R., Courcoubetis, C., Dill, D.L.: Model-checking in dense real-time. Inf. Comput. **104**(1), 2–34 (1993). DOI 10.1006/inco.1993.1024
8. Alur, R., Courcoubetis, C., Halbwachs, N., Henzinger, T.A., Ho, P.H., Nicollin, X., Olivero, A., Sifakis, J., Yovine, S.: The algorithmic analysis of hybrid systems. Theor. Comput. Sci. **138**(1), 3–34 (1995). DOI 10.1016/0304-3975(94)00202-T
9. Alur, R., Courcoubetis, C., Henzinger, T.A., Ho, P.H.: Hybrid automata: An algorithmic approach to the specification and verification of hybrid systems. In: Grossman et al. [144], pp. 209–229
10. Alur, R., Dill, D.L.: A theory of timed automata. Theor. Comput. Sci. **126**(2), 183–235 (1994). DOI 10.1016/0304-3975(94)90010-8
11. Alur, R., Henzinger, T.A., Ho, P.H.: Automatic symbolic verification of embedded systems. IEEE T. Software Eng. **22**(3), 181–201 (1996)
12. Alur, R., Henzinger, T.A., Sontag, E.D. (eds.): Hybrid Systems III: Verification and Control, Proceedings of the DIMACS/SYCON Workshop, October 22–25, 1995, Rutgers University, New Brunswick, NJ, USA, *LNCS*, vol. 1066. Springer (1996)
13. Alur, R., Henzinger, T.A., Vardi, M.Y.: Parametric real-time reasoning. In: STOC, pp. 592–601 (1993). DOI 10.1145/167088.167242
14. Alur, R., Pappas, G.J. (eds.): Hybrid Systems: Computation and Control, 7th International Workshop, HSCC 2004, Philadelphia, PA, USA, March 25–27, 2004, Proceedings, *LNCS*, vol. 2993. Springer (2004). DOI 10.1007/b96398

15. Anai, H., Weispfenning, V.: Reach set computations using real quantifier elimination. In: M.D.D. Benedetto, A.L. Sangiovanni-Vincentelli (eds.) HSCC, *LNCS*, vol. 2034, pp. 63–76. Springer (2001). DOI 10.1007/3-540-45351-2_9
16. Andler, S.F., Offutt, J. (eds.): 7th International Conference on Engineering of Complex Computer Systems (ICECCS 2001), 11–13 June 2001, Skövde, Sweden. IEEE Computer Society, Los Alamitos (2001)
17. Andrews, P.B.: An Introduction to Mathematical Logic and Type Theory: To Truth Through Proof, 2 edn. Kluwer (2002)
18. Armando, A., Baumgartner, P., Dowek, G. (eds.): Automated Reasoning, Third International Joint Conference, IJCAR 2008, Sydney, Australia, Proceedings, *LNCS*, vol. 5195. Springer (2008)
19. Arnon, D.S., Collins, G.E., McCallum, S.: Cylindrical algebraic decomposition I: The basic algorithm. SIAM J. Comput. **13**(4), 865–877 (1984)
20. Arnon, D.S., Collins, G.E., McCallum, S.: Cylindrical algebraic decomposition II: An adjacency algorithm for the plane. SIAM J. Comput. **13**(4), 878–889 (1984)
21. Asarin, E., Dang, T., Girard, A.: Reachability analysis of nonlinear systems using conservative approximation. In: Maler and Pnueli [200], pp. 20–35. DOI 10.1007/3-540-36580-X_5
22. Asarin, E., Dang, T., Maler, O.: The d/dt tool for verification of hybrid systems. In: E. Brinksma, K.G. Larsen (eds.) CAV, *LNCS*, vol. 2404, pp. 365–370. Springer (2002). DOI 10.1007/3-540-45657-0_30
23. Barnett, M., Chang, B.Y.E., DeLine, R., Jacobs, B., Leino, K.R.M.: Boogie: A modular reusable verifier for object-oriented programs. In: F.S. de Boer, M.M. Bonsangue, S. Graf, W.P. de Roever (eds.) FMCO, *LNCS*, vol. 4111, pp. 364–387. Springer (2005). DOI 10.1007/11804192_17
24. Basu, S.: New results on quantifier elimination over real closed fields and applications to constraint databases. J. ACM **46**(4), 537–555 (1999). DOI 10.1145/320211.320240
25. Basu, S., Pollack, R., Roy, M.F.: On the combinatorial and algebraic complexity of quantifier elimination. J. ACM **43**(6), 1002–1045 (1996). DOI 10.1145/235809.235813
26. Basu, S., Pollack, R., Roy, M.F.: Algorithms in Real Algebraic Geometry, 2 edn. Springer (2006)
27. Batt, G., Belta, C., Weiss, R.: Model checking genetic regulatory networks with parameter uncertainty. In: Bemporad et al. [41], pp. 61–75. DOI 10.1007/978-3-540-71493-4_8
28. Bauer, A., Clarke, E.M., Zhao, X.: Analytica — an experiment in combining theorem proving and symbolic computation. J. Autom. Reasoning **21**(3), 295–325 (1998). DOI 10.1023/A: 1006079212546
29. Baumgartner, P.: FDPLL — a first order Davis-Putnam-Longeman-Loveland procedure. In: D.A. McAllester (ed.) CADE, *LNCS*, vol. 1831, pp. 200–219. Springer (2000). DOI 10. 1007/10721959_16
30. Baumgartner, P., Tinelli, C.: The model evolution calculus as a first-order DPLL method. Artif. Intell. **172**(4-5), 591–632 (2008). DOI 10.1016/j.artint.2007.09.005
31. Becker, T., Weispfenning, V.: Gröbner Bases, *Springer Graduate Texts in Mathematics*, vol. 141. Springer (1998)
32. Beckert, B.: Equality and other theories. In: M. D'Agostino, D. Gabbay, R. Hähnle, J. Posegga (eds.) Handbook of Tableau Methods. Kluwer (1999)
33. Beckert, B., Giese, M., Habermalz, E., Hähnle, R., Roth, A., Rümmer, P., Schlager, S.: Taclets: A new paradigm for constructing interactive theorem provers. Revista de la Real Academia de Ciencias Exactas, Físicas y Naturales, Serie A: Matemáticas (RACSAM) **98**(1) (2004)
34. Beckert, B., Giese, M., Hähnle, R., Klebanov, V., Rümmer, P., Schlager, S., Schmitt, P.H.: The KeY System 1.0 (deduction component). In: F. Pfenning (ed.) Proceedings, International Conference on Automated Deduction, Bremen, Germany, LNCS. Springer (2007)
35. Beckert, B., Hähnle, R., Schmitt, P.H. (eds.): Verification of Object-Oriented Software: The KeY Approach, *LNCS*, vol. 4334. Springer (2007). DOI 10.1007/978-3-540-69061-0
36. Beckert, B., Klebanov, V.: Must program verification systems and calculi be verified? In: 3rd International Verification Workshop VERIFY'06, at FLoC, Seattle, USA, pp. 34–41 (2006)

37. Beckert, B., Platzer, A.: Dynamic logic with non-rigid functions: A basis for object-oriented program verification. In: U. Furbach, N. Shankar (eds.) IJCAR, *LNCS*, vol. 4130, pp. 266–280. Springer (2006). DOI 10.1007/11814771_23
38. Beckert, B., Schlager, S.: A sequent calculus for first-order dynamic logic with trace modalities. In: Goré et al. [139], pp. 626–641. DOI 10.1007/3-540-45744-5_51
39. van Beek, D.A., Man, K.L., Reniers, M.A., Rooda, J.E., Schiffelers, R.R.H.: Deriving simulators for hybrid Chi models. Intelligent Control, 2006. IEEE International Symposium on pp. 42–49 (2006). DOI 10.1109/CACSD-CCA-ISIC.2006.4776622
40. van Beek, D.A., Man, K.L., Reniers, M.A., Rooda, J.E., Schiffelers, R.R.H.: Syntax and consistent equation semantics of hybrid Chi. J. Log. Algebr. Program. **68**(1-2), 129–210 (2006). DOI 10.1016/j.jlap.2005.10.005
41. Bemporad, A., Bicchi, A., Buttazzo, G. (eds.): Hybrid Systems: Computation and Control, 10th International Conference, HSCC 2007, Pisa, Italy, Proceedings, *LNCS*, vol. 4416. Springer (2007)
42. Ben-Ari, M.: Mathematical Logic for Computer Science, 2 edn. Springer (2003)
43. BFU: Investigation report. Tech. Rep. AX001-1-2/02, German Federal Bureau of Aircraft Accidents Investigation (2004)
44. Bianconi, R.: Nondefinability results for expansions of the field of real numbers by the exponential function and by the restricted sine functions. J. Symb. Log. **62**(4), 1173–1178 (1997)
45. Bianconi, R.: Undefinability results in o-minimal expansions of the real numbers. Ann. Pure Appl. Logic **134**(1), 43–51 (2005). DOI 10.1016/j.apal.2004.06.010
46. Bicchi, A., Pallottino, L.: On optimal cooperative conflict resolution for air traffic management systems. IEEE Trans. Intelligent Transportation Systems **1**(4), 221–231 (2000)
47. Biere, A., Cimatti, A., Clarke, E.M., Zhu, Y.: Symbolic model checking without BDDs. In: R. Cleaveland (ed.) TACAS, *LNCS*, vol. 1579, pp. 193–207. Springer (1999). DOI 10.1007/3-540-49059-0_14
48. Blackburn, P.: Internalizing labelled deduction. J. Log. Comput. **10**(1), 137–168 (2000)
49. Blum, L., Cucker, F., Shub, M., Smale, S.: Complexity and real computation. Springer, Secaucus, NJ, USA (1998)
50. Bochnak, J., Coste, M., Roy, M.F.: Real Algebraic Geometry, *Ergebnisse der Mathematik und ihrer Grenzgebiete*, vol. 36. Springer (1998)
51. Borchers, B.: CSDP, a C library for semidefinite programming. Optimization Methods and Software **11**(1-4), 613–623 (1999). DOI 10.1080/10556789908805765
52. Bourbaki, N.: Algebra I: Chapters 1–3. Elements of mathematics. Springer (1989)
53. Bourbaki, N.: Commutative Algebra. Elements of mathematics. Hermann, Paris (1989). Translated from the French
54. Boyd, S., Vandenberghe, L.: Convex Optimization. Cambridge University Press (2004)
55. Branicky, M.S.: General hybrid dynamical systems: Modeling, analysis, and control. In: Alur et al. [12], pp. 186–200. DOI 10.1007/BFb0020945
56. Branicky, M.S.: Studies in hybrid systems: Modeling, analysis, and control. Ph.D. thesis, Dept. Elec. Eng. and Computer Sci., Massachusetts Inst. Technol., Cambridge, MA (1995)
57. Branicky, M.S.: Universal computation and other capabilities of hybrid and continuous dynamical systems. Theor. Comput. Sci. **138**(1), 67–100 (1995). DOI 10.1016/0304-3975(94)00147-B
58. Branicky, M.S., Borkar, V.S., Mitter, S.K.: A unified framework for hybrid control: Model and optimal control theory. IEEE T. Automat. Contr. **43**(1), 31–45 (1998). DOI 10.1109/9.654885
59. Brown, C.W.: QEPCAD B: A program for computing with semi-algebraic sets using CADs. SIGSAM Bull. **37**(4), 97–108 (2003). DOI 10.1145/968708.968710
60. Brown, C.W., Davenport, J.H.: The complexity of quantifier elimination and cylindrical algebraic decomposition. In: D. Wang (ed.) ISSAC, pp. 54–60. ACM (2007). DOI 10.1145/1277548.1277557
61. Bruyère, V., Raskin, J.F.: Real-time model-checking: Parameters everywhere. Logical Methods in Computer Science **3**(1) (2007). DOI 10.2168/LMCS-3(1:7)2007. Online journal

62. Buchberger, B.: An algorithm for finding the basis elements of the residue class ring of a zero dimensional polynomial ideal. Ph.D. thesis, University of Innsbruck (1965)
63. Buchberger, B., Jebelean, T., Kriftner, F., Marin, M., Tomuta, E., Vasaru, D.: A survey of the Theorema project. In: ISSAC, pp. 384–391 (1997)
64. Buehler, M.: Summary of DGC 2005 results. Journal of Field Robotics **23**, 465–466 (2008). DOI 10.1002/rob.20145
65. Burch, J.R., Clarke, E.M., McMillan, K.L., Dill, D.L., Hwang, L.J.: Symbolic model checking: 10^{20} states and beyond. In: Mitchell [210], pp. 428–439
66. Burch, J.R., Clarke, E.M., McMillan, K.L., Dill, D.L., Hwang, L.J.: Symbolic model checking: 10^{20} states and beyond. Inf. Comput. **98**(2), 142–170 (1992). DOI 10.1016/0890-5401(92)90017-A
67. Cassez, F., Larsen, K.G.: The impressive power of stopwatches. In: CONCUR, pp. 138–152 (2000). DOI 10.1007/3-540-44618-4_12
68. Caviness, B.F., Johnson, J.R. (eds.): Quantifier Elimination and Cylindrical Algebraic Decomposition, Texts and Monographs in Symbolic Computation. Springer (1998)
69. Chaochen, Z., Ji, W., Ravn, A.P.: A formal description of hybrid systems. In: Alur et al. [12], pp. 511–530
70. Chutinan, A., Krogh, B.H.: Computational techniques for hybrid system verification. IEEE T. Automat. Contr. **48**(1), 64–75 (2003). DOI 10.1109/TAC.2002.806655
71. Cimatti, A., Roveri, M., Tonetta, S.: Requirements validation for hybrid systems. In: A. Bouajjani, O. Maler (eds.) CAV, *LNCS*, vol. 5643. Springer (2009). DOI 10.1007/978-3-642-02658-4_17
72. Clarke, E., Grumberg, O., Jha, S., Lu, Y., Veith, H.: Counterexample-guided abstraction refinement for symbolic model checking. J. ACM **50**(5), 752–794 (2003). DOI 10.1145/876638.876643
73. Clarke, E.M.: Program invariants as fixedpoints. Computing **21**(4), 273–294 (1979). DOI 10.1007/BF02248730
74. Clarke, E.M., Biere, A., Raimi, R., Zhu, Y.: Bounded model checking using satisfiability solving. Formal Methods in System Design **19**(1), 7–34 (2001)
75. Clarke, E.M., Emerson, E.A.: Design and synthesis of synchronization skeletons using branching-time temporal logic. In: D. Kozen (ed.) Logic of Programs, *LNCS*, vol. 131, pp. 52–71. Springer (1981)
76. Clarke, E.M., Emerson, E.A., Sifakis, J.: Model checking: algorithmic verification and debugging. Commun. ACM **52**(11), 74–84 (2009). DOI 10.1145/1592761.1592781
77. Clarke, E.M., Fehnker, A., Han, Z., Krogh, B.H., Ouaknine, J., Stursberg, O., Theobald, M.: Abstraction and counterexample-guided refinement in model checking of hybrid systems. Int. J. Found. Comput. Sci. **14**(4), 583–604 (2003). DOI 10.1142/S012905410300190X
78. Clarke, E.M., Grumberg, O., Peled, D.A.: Model Checking. MIT Press, Cambridge, MA, USA (1999)
79. Cohen, P.J.: Decision procedures for real and *p*-adic fields. Communications in Pure and Applied Mathematics **22**, 131–151 (1969)
80. Collins, G.E.: Hauptvortrag: Quantifier elimination for real closed fields by cylindrical algebraic decomposition. In: H. Barkhage (ed.) Automata Theory and Formal Languages, *LNCS*, vol. 33, pp. 134–183. Springer (1975). DOI 10.1007/3-540-07407-4_17
81. Collins, G.E., Hong, H.: Partial cylindrical algebraic decomposition for quantifier elimination. J. Symb. Comput. **12**(3), 299–328 (1991). DOI 10.1016/S0747-7171(08)80152-6
82. Collins, G.E., Johnson, J.R.: Quantifier elimination and the sign variation method for real root isolation. In: ISSAC, pp. 264–271 (1989). DOI 10.1145/74540.74574
83. Collins, G.E., Johnson, J.R., Krandick, W.: Interval arithmetic in cylindrical algebraic decomposition. J. Symb. Comput. **34**(2), 145–157 (2002). DOI 10.1006/jsco.2002.0547
84. Collins, G.E., Johnson, J.R., Küchlin, W.: Parallel real root isolation using the coefficient sign variation method. In: R. Zippel (ed.) CAP, *LNCS*, vol. 584, pp. 71–87. Springer (1990)
85. Collins, P., Lygeros, J.: Computability of finite-time reachable sets for hybrid systems. In: CDC-ECC'05, pp. 4688– 4693. IEEE (2005)

86. Comon, H., Jurski, Y.: Timed automata and the theory of real numbers. In: J.C.M. Baeten, S. Mauw (eds.) CONCUR, *LNCS*, vol. 1664, pp. 242–257. Springer (1999). DOI 10.1007/3-540-48320-9_18
87. Cook, S.A.: Soundness and completeness of an axiom system for program verification. SIAM J. Comput. **7**(1), 70–90 (1978). DOI 10.1137/0207005
88. Cox, D.A., Little, J., O'Shea, D.: Ideals, Varieties and Algorithms: An Introduction to Computational Algebraic Geometry and Commutative Algebra. Undergraduate Texts in Mathematics. Springer, New York (1992)
89. Damm, W., Hungar, H., Olderog, E.R.: On the verification of cooperating traffic agents. In: F.S. de Boer, M.M. Bonsangue, S. Graf, W.P. de Roever (eds.) FMCO, *LNCS*, vol. 3188, pp. 77–110. Springer (2003). DOI 10.1007/b100112
90. Damm, W., Hungar, H., Olderog, E.R.: Verification of cooperating traffic agents. International Journal of Control **79**(5), 395–421 (2006). DOI 10.1080/00207170600587531
91. Damm, W., Mikschl, A., Oehlerking, J., Olderog, E.R., Pang, J., Platzer, A., Segelken, M., Wirtz, B.: Automating verification of cooperation, control, and design in traffic applications. In: C.B. Jones, Z. Liu, J. Woodcock (eds.) Formal Methods and Hybrid Real-Time Systems, *LNCS*, vol. 4700, pp. 115–169. Springer (2007). DOI 10.1007/978-3-540-75221-9_6
92. Damm, W., Pinto, G., Ratschan, S.: Guaranteed termination in the verification of LTL properties of non-linear robust discrete time hybrid systems. In: Peled and Tsay [226], pp. 99–113. DOI 10.1007/11562948_10
93. Dantzig, G.B., Eaves, B.C.: Fourier-Motzkin elimination and its dual. J. Comb. Theory, Ser. A **14**(3), 288–297 (1973)
94. Davenport, J.H., Heintz, J.: Real quantifier elimination is doubly exponential. J. Symb. Comput. **5**(1/2), 29–35 (1988). DOI 10.1016/S0747-7171(88)80004-X
95. Davoren, J.M.: On hybrid systems and the modal μ-calculus. In: P.J. Antsaklis, W. Kohn, M.D. Lemmon, A. Nerode, S. Sastry (eds.) Hybrid Systems, *LNCS*, vol. 1567, pp. 38–69. Springer (1997). DOI 10.1007/3-540-49163-5_3
96. Davoren, J.M., Coulthard, V., Markey, N., Moor, T.: Non-deterministic temporal logics for general flow systems. In: Alur and Pappas [14], pp. 280–295. DOI 10.1007/b96398
97. Davoren, J.M., Nerode, A.: Logics for hybrid systems. IEEE **88**(7), 985–1010 (2000). DOI 10.1109/5.871305
98. Daws, C., Olivero, A., Tripakis, S., Yovine, S.: The tool KRONOS. In: Hybrid Systems III, *LNCS*, vol. 1066, pp. 208–219 (1996)
99. Dershowitz, N., Manna, Z.: Proving termination with multiset orderings. Commun. ACM **22**(8), 465–476 (1979). DOI 10.1145/359138.359142
100. Deshpande, A., Göllü, A., Varaiya, P.: SHIFT: A formalism and a programming language for dynamic networks of hybrid automata. In: P.J. Antsaklis, W. Kohn, A. Nerode, S. Sastry (eds.) Hybrid Systems, *LNCS*, vol. 1273, pp. 113–133. Springer (1996). DOI 10.1007/BFb0031558
101. Dolzmann, A., Sturm, T.: Redlog: Computer algebra meets computer logic. ACM SIGSAM Bull. **31**, 2–9 (1997). DOI 10.1145/261320.261324
102. Donzé, A., Maler, O.: Systematic simulation using sensitivity analysis. In: Bemporad et al. [41], pp. 174–189. DOI 10.1007/978-3-540-71493-4_16
103. Dowek, G., Hardin, T., Kirchner, C.: Theorem proving modulo. J. Autom. Reasoning **31**(1), 33–72 (2003). DOI 10.1023/A:1027357912519
104. Dowek, G., Muñoz, C., Carreño, V.A.: Provably safe coordinated strategy for distributed conflict resolution. In: Proceedings of the AIAA Guidance Navigation, and Control Conference and Exhibit 2005, AIAA-2005-6047 (2005)
105. van den Dries, L.: Alfred Tarski's elimination theory for real closed fields. J. Symb. Log. **53**(1), 7–19 (1988)
106. van den Dries, L.: Tame Topology and O-minimal Structures. Cambridge University Press (1998)
107. van den Dries, L., Miller, C.: On the real exponential field with restricted analytic functions. Israel J. Math. **85**(1-3), 19–56 (1994). DOI 10.1007/BF02758635

108. van den Dries, L., Speissegger, P.: The real field with convergent generalized power series. Trans. Amer. Math. Soc. **350**, 4377–4421 (1998). DOI 10.1090/S0002-9947-98-02105-9
109. Dubois, D.W.: A Nullstellensatz for ordered fields. Ark. Mat. **8**(2), 111–114 (1970). DOI 10.1007/BF02589551
110. Egerstedt, M., Johansson, K.H., Sastry, S., Lygeros, J.: On the regularization of Zeno hybrid automata. Systems and Control Letters **38**, 141–150 (1999)
111. Egerstedt, M., Mishra, B. (eds.): Hybrid Systems: Computation and Control, 11th International Conference, HSCC 2008, St. Louis, MO, USA, Proceedings, *LNCS*, vol. 4981. Springer (2008)
112. Eisenbud, D.: Commutative algebra with a view toward algebraic geometry, *Graduate Texts in Mathematics*, vol. 150, 3 edn. Springer, New York (1999)
113. Emerson, A.: Temporal and modal logic. In: J. van Leeuwen (ed.) Handbook of Theoretical Computer Science, Volume B: Formal Models and Sematics (B), pp. 995–1072. MIT Press (1990)
114. Emerson, E.A., Clarke, E.M.: Using branching time temporal logic to synthesize synchronization skeletons. Sci. Comput. Program. **2**(3), 241–266 (1982)
115. Emerson, E.A., Halpern, J.Y.: "Sometimes" and "Not Never" revisited: on branching versus linear time temporal logic. J. ACM **33**(1), 151–178 (1986). DOI 10.1145/4904.4999
116. Eriksson, K., Estep, D., Hansbo, P., Johnson, C.: Computational Differential Equations. Cambridge University Press (1996)
117. ERTMS User Group: ERTMS/ETCS System requirements specification. `http://www.era.europa.eu` (2002)
118. Etessami, K., Rajamani, S.K. (eds.): Computer Aided Verification, 17th International Conference, CAV 2005, Edinburgh, Scotland, UK, July 6–10, 2005, Proceedings, *LNCS*, vol. 3576. Springer (2005)
119. Faber, J., Meyer, R.: Model checking data-dependent real-time properties of the European Train Control System. In: FMCAD, pp. 76–77. IEEE Computer Society Press (2006). DOI 10.1109/FMCAD.2006.21
120. Faugére, J.C.: A new efficient algorithm for computing Gröbner bases (F4). Journal of Pure and Applied Algebra **139**(1-3), 61–88 (1999). DOI 10.1016/S0022-4049(99)00005-5
121. Faugére, J.C.: A new efficient algorithm for computing Gröbner bases without reduction to zero (F5). In: ISAAC. ACM Press (2002). DOI 10.1145/780506.780516
122. Fitting, M.: First-Order Logic and Automated Theorem Proving, 2 edn. Springer, New York (1996)
123. Fitting, M., Mendelsohn, R.L.: First-Order Modal Logic. Kluwer, Norwell, MA, USA (1999)
124. Fourier, J.B.J.: Solution d'une question particulière du calcul des inégalités. Nouveau Bulletin des Sciences par la Société Philomatique de Paris pp. 99–100 (1823)
125. Fränzle, M.: Analysis of hybrid systems: An ounce of realism can save an infinity of states. In: J. Flum, M. Rodríguez-Artalejo (eds.) CSL, *LNCS*, vol. 1683, pp. 126–140. Springer (1999)
126. Frehse, G.: PHAVer: Algorithmic verification of hybrid systems past HyTech. In: Morari and Thiele [212], pp. 258–273. DOI 10.1007/b106766
127. Frehse, G.: PHAVer: algorithmic verification of hybrid systems past HyTech. STTT **10**(3), 263–279 (2008). DOI 10.1007/s10009-007-0062-x
128. Frehse, G., Jha, S.K., Krogh, B.H.: A counterexample-guided approach to parameter synthesis for linear hybrid automata. In: Egerstedt and Mishra [111], pp. 187–200. DOI 10.1007/978-3-540-78929-1_14
129. Galdino, A.L., Muñoz, C., Ayala-Rincón, M.: Formal verification of an optimal air traffic conflict resolution and recovery algorithm. In: D. Leivant, R. de Queiroz (eds.) WoLLIC, *LNCS*, vol. 4576, pp. 177–188. Springer (2007)
130. Gallier, J.H.: Logic for Computer Science. Longman Higher Education (1986)
131. Ganzinger, H., Korovin, K.: Theory instantiation. In: M. Hermann, A. Voronkov (eds.) LPAR, *LNCS*, vol. 4246, pp. 497–511. Springer (2006). DOI 10.1007/11916277_34
132. Gear, C.W.: Differential-algebraic equations index transformations. SIAM J. Sci. Stat. Comput. **9**(1), 39–47 (1988). DOI 10.1137/0909004

133. Gentzen, G.: Untersuchungen über das logische Schließen. Math. Zeit. **39**, 405–431 (1935). DOI 10.1007/BF01201363
134. Giese, M.: Incremental closure of free variable tableaux. In: Goré et al. [139], pp. 545–560. DOI 10.1007/3-540-45744-5_46
135. Gödel, K.: Über die Vollständigkeit des Logikkalküls. Ph.D. thesis, Universität Wien (1929)
136. Gödel, K.: Die Vollständigkeit der Axiome des logischen Funktionenkalküls. Mon. hefte Math. Phys. **37**, 349–360 (1930). DOI 10.1007/BF01696781
137. Gödel, K.: Über formal unentscheidbare Sätze der Principia Mathematica und verwandter Systeme I. Mon. hefte Math. Phys. **38**, 173–198 (1931). DOI 10.1007/BF01700692
138. Gordon, M.J.C., Melham, T.F.: Introduction to HOL: A Theorem-Proving Environment for Higher-Order Logic. Cambridge Univ. Press (1993)
139. Goré, R., Leitsch, A., Nipkow, T. (eds.): Automated Reasoning, First International Joint Conference, IJCAR 2001, Siena, Italy, June 18–23, 2001, Proceedings, *LNCS*, vol. 2083. Springer (2001)
140. Graça, D.S., Campagnolo, M.L., Buescu, J.: Computability with polynomial differential equations. Advances in Applied Mathematics (2007)
141. Grigoriev, D.: Complexity of deciding Tarski algebra. J. Symb. Comput. **5**(1/2), 65–108 (1988). DOI 10.1016/S0747-7171(88)80006-3
142. Grigoriev, D., Vorobjov, N.: Solving systems of polynomial inequalities in subexponential time. J. Symb. Comput. **5**(1/2), 37–64 (1988). DOI 10.1016/S0747-7171(88)80005-1
143. Gross, J.: Schlussbericht über die Entgleisung von Güterzug 43647 der BLS AG auf der Weiche 34 (Einfahrt Lötschberg-Basisstrecke). Tech. Rep. 07101601, Unfalluntersuchungsstelle Bahnen und Schiffe (2007)
144. Grossman, R.L., Nerode, A., Ravn, A.P., Rischel, H. (eds.): Hybrid Systems, *LNCS*, vol. 736. Springer (1993)
145. Gulwani, S., Tiwari, A.: Constraint-based approach for analysis of hybrid systems. In: Gupta and Malik [146], pp. 190–203. DOI 10.1007/978-3-540-70545-1
146. Gupta, A., Malik, S. (eds.): Computer Aided Verification, CAV 2008, Princeton, NJ, USA, Proceedings, *LNCS*, vol. 5123. Springer (2008)
147. Hähnle, R., Schmitt, P.H.: The liberalized δ-rule in free variable semantic tableaux. J. Autom. Reasoning **13**(2), 211–221 (1994). DOI 10.1007/BF00881956
148. Harel, D.: First-Order Dynamic Logic. Springer, New York (1979)
149. Harel, D., Kozen, D., Tiuryn, J.: Dynamic logic. MIT Press, Cambridge (2000)
150. Harris, J.: Algebraic Geometry: A First Course. Graduate Texts in Mathematics. Springer (1995)
151. Harrison, J.: Verifying nonlinear real formulas via sums of squares. In: K. Schneider, J. Brandt (eds.) TPHOLs, *LNCS*, vol. 4732, pp. 102–118. Springer (2007). DOI 10.1007/978-3-540-74591-4_9
152. Harrison, J.: Handbook of Practical Logic and Automated Reasoning. Cambridge Univ. Press (2009)
153. Hartman, P.: Ordinary Differential Equations. John Wiley (1964)
154. Hartshorne, R.: Algebraic Geometry, *Graduate Texts in Mathematics*, vol. 52. Springer (1977)
155. Henkin, L.: The completeness of the first-order functional calculus. J. Symb. Log. **14**(3), 159–166 (1949)
156. Henzinger, T.A.: The theory of hybrid automata. In: LICS, pp. 278–292. IEEE Computer Society, Los Alamitos (1996)
157. Henzinger, T.A., Ho, P.H.: HYTECH: The Cornell HYbrid TECHnology tools. In: P.J. Antsaklis, W. Kohn, A. Nerode, S. Sastry (eds.) Hybrid Systems, *LNCS*, vol. 999, pp. 265–293. Springer (1994)
158. Henzinger, T.A., Ho, P.H., Wong-Toi, H.: HyTech: A model checker for hybrid systems. In: O. Grumberg (ed.) CAV, *LNCS*, vol. 1254, pp. 460–463. Springer (1997)
159. Henzinger, T.A., Nicollin, X., Sifakis, J., Yovine, S.: Symbolic model checking for real-time systems. In: LICS, pp. 394–406. IEEE Computer Society (1992). DOI 10.1006/inco.1994.1045

160. Herbrand, J.: Recherches sur la théorie de la démonstration. Travaux de la Société des Sciences et des Lettres de Varsovie, Class III, Sciences Mathématiques et Physiques **33**, 33–160 (1930)
161. Hoare, C.A.R.: An axiomatic basis for computer programming. Commun. ACM **12**(10), 576–580 (1969). DOI 10.1145/363235.363259
162. Hoare, C.A.R.: Communicating Sequential Processes. Prentice Hall International (1985)
163. Hodges, W.: Model Theory. Cambridge University Press (1993)
164. Hoffmann, G.M., Huang, H., Waslander, S.L., Tomlin, C.J.: Quadrotor helicopter flight dynamics and control: Theory and experiment. In: Proceedings of the AIAA Guidance, Navigation, and Control Conference. Hilton Head, SC (2007). AIAA Paper Number 2007-6461
165. Hörmander, L.: The Analysis of Linear Partial Differential Operators II, *Grundlehren der mathematischen Wissenschaften*, vol. 257. Springer (1983)
166. Hsu, A., Eskafi, F., Sachs, S., Varaiya, P.: Design of platoon maneuver protocols for IVHS. PATH Research Report UCB-ITS-PRR-91-6, Institute of Transportation Studies, University of California, Berkeley (1991)
167. Hu, J., Prandini, M., Sastry, S.: Optimal coordinated motions of multiple agents moving on a plane. SIAM Journal on Control and Optimization **42**, 637–668 (2003)
168. Hu, J., Prandini, M., Sastry, S.: Probabilistic safety analysis in three-dimensional aircraft flight. In: CDC, vol. 5, pp. 5335 – 5340 (2003). DOI 10.1109/CDC.2003.1272485
169. Huth, M., Ryan, M.: Logic in Computer Science: Modelling and Reasoning about Systems, 2 edn. Cambridge Univ. Press (2004)
170. Hutter, D., Langenstein, B., Sengler, C., Siekmann, J.H., Stephan, W., Wolpers, A.: Deduction in the verification support environment (VSE). In: M.C. Gaudel, J. Woodcock (eds.) FME, *LNCS*, vol. 1051, pp. 268–286. Springer (1996)
171. Hwang, I., Kim, J., Tomlin, C.: Protocol-based conflict resolution for air traffic control. Air Traffic Control Quarterly **15**(1), 1–34 (2007)
172. Ilyashenko, Y., Yakovenko, S.: Lectures on Analytic Differential Equations, *Graduate Studies in Mathematics*, vol. 86. AMS (2008)
173. Jhala, R., McMillan, K.L.: Interpolant-based transition relation approximation. In: Etessami and Rajamani [118], pp. 39–51. DOI 10.1007/11513988_6
174. Jhala, R., McMillan, K.L.: Interpolant-based transition relation approximation. Logical Methods in Computer Science **3**(4) (2007). DOI 10.2168/LMCS-3(4:1)2007
175. Jifeng, H.: From CSP to hybrid systems. In: A.W. Roscoe (ed.) A classical mind: essays in honour of C. A. R. Hoare, pp. 171–189. Prentice Hall, Hertfordshire, UK (1994)
176. Johansson, K.H., Sastry, S., Zhang, J., Lygeros, J.: Zeno hybrid systems. Int. J. Robust and Nonlinear Control **11**, 435–451 (2001). DOI 10.1002/rnc.592
177. Katok, A., Hasselblatt, B.: Introduction to the Modern Theory of Dynamical Systems. Cambridge University Press, New York, NY (1996)
178. Kesten, Y., Manna, Z., Pnueli, A.: Verification of clocked and hybrid systems. Acta Inf. **36**(11), 837–912 (2000). DOI 10.1007/s002360050177
179. Kolchin, E.R.: Differential Algebra and Algebraic Groups. Academic Press, New York (1972)
180. Košecká, J., Tomlin, C., Pappas, G., Sastry, S.: 2-1/2D conflict resolution maneuvers for ATMS. In: CDC, vol. 3, pp. 2650–2655. Tampa, FL, USA (1998). DOI 10.1109/CDC.1998.757853
181. Kozen, D.: Results on the propositional μ-calculus. Theor. Comput. Sci. **27**, 333–354 (1983). DOI 10.1016/0304-3975(82)90125-6
182. Kozen, D.: Kleene algebra with tests. ACM Trans. Program. Lang. Syst. **19**(3), 427–443 (1997). DOI 10.1145/256167.256195
183. Kratz, F., Sokolsky, O., Pappas, G.J., Lee, I.: R-Charon, a modeling language for reconfigurable hybrid systems. In: J.P. Hespanha, A. Tiwari (eds.) HSCC, *LNCS*, vol. 3927, pp. 392–406. Springer (2006). DOI 10.1007/11730637_30
184. Kreisel, G., Krivine, J.L.: Elements of mathematical logic: Model Theory, 2 edn. North-Holland (1971)

185. Kripke, S.A.: Semantical considerations on modal logic. Acta Philosophica Fennica **16**, 83–94 (1963)
186. Krivine, J.L.: Anneaux préordonnés. Journal d'analyse mathématique **12**, 307–326 (1964)
187. Kunkel, P., Mehrmann, V.: Differential-Algebraic Equations: Analysis and Numerical Solution. European Mathematical Society (2006)
188. Lafferriere, G., Pappas, G.J., Sastry, S.: O-minimal hybrid systems. Mathematics of Control, Signals, and Systems **13**(1), 1–21 (2000). DOI 10.1007/PL00009858
189. Lafferriere, G., Pappas, G.J., Yovine, S.: A new class of decidable hybrid systems. In: F.W. Vaandrager, J.H. van Schuppen (eds.) HSCC, *LNCS*, vol. 1569, pp. 137–151. Springer (1999). DOI 10.1007/3-540-48983-5_15
190. Lafferriere, G., Pappas, G.J., Yovine, S.: Symbolic reachability computation for families of linear vector fields. J. Symb. Comput. **32**(3), 231–253 (2001). DOI 10.1006/jsco.2001.0472
191. Lang, S.: Algebra. Addison-Wesley Publishing Company (1978)
192. Lanotte, R., Tini, S.: Taylor approximation for hybrid systems. In: Morari and Thiele [212], pp. 402–416. DOI 10.1007/b106766
193. Larsen, K.G., Pettersson, P., Yi, W.: UPPAAL in a nutshell. STTT **1**(1-2), 134–152 (1997)
194. Liberzon, D.: Switching in Systems and Control. Systems and Control: Foundations and Applications. Birkhäuser, Boston, MA (2003)
195. Lindelöf, M.E.: Sur l'application de la méthode des approximations successives aux équations différentielles ordinaires du premier ordre. Comptes rendus hebdomadaires des séances de l'Académie des sciences **114**, 454–457 (1894)
196. Livadas, C., Lygeros, J., Lynch, N.A.: High-level modeling and analysis of TCAS. Proc. IEEE – Special Issue on Hybrid Systems: Theory & Applications **88**(7), 926–947 (2000)
197. Łojasiewicz, S.: Triangulations of semi-analytic sets. Annali della Scuola Normale Superiore di Pisa **18**, 449–474 (1964)
198. Loos, R., Weispfenning, V.: Applying linear quantifier elimination. Comput. J. **36**(5), 450–462 (1993). DOI 10.1093/comjnl/36.5.450
199. Löwenheim, L.: Über Möglichkeiten im Relativkalkül. Mathematische Annalen **76**, 447–470 (1915). DOI 10.1007/BF01458217
200. Maler, O., Pnueli, A. (eds.): Hybrid Systems: Computation and Control, 6th International Workshop, HSCC 2003 Prague, Czech Republic, April 3–5, 2003, Proceedings, *LNCS*, vol. 2623. Springer (2003)
201. Manna, Z., Sipma, H.: Deductive verification of hybrid systems using STeP. In: T.A. Henzinger, S. Sastry (eds.) HSCC, *LNCS*, vol. 1386, pp. 305–318. Springer (1998). DOI 10.1007/3-540-64358-3_47
202. Mansfield, E.L.: Differential Gröbner bases. Ph.D. thesis, University of Sydney (1991)
203. Massink, M., Francesco, N.D.: Modelling free flight with collision avoidance. In: Andler and Offutt [16], pp. 270–280. DOI 10.1109/ICECCS.2001.930186
204. McLaughlin, S., Harrison, J.: A proof-producing decision procedure for real arithmetic. In: R. Nieuwenhuis (ed.) CADE, *LNCS*, vol. 3632, pp. 295–314. Springer (2005). DOI 10.1007/11532231_22
205. Meyer, R., Faber, J., Hoenicke, J., Rybalchenko, A.: Model checking duration calculus: A practical approach. Formal Aspects of Computing pp. 1–25 (2008). DOI 10.1007/s00165-008-0082-7
206. Miller, C.L.: Expansions of the real field with power functions. Ann. Pure Appl. Logic **68**(1), 79–94 (1994)
207. Milner, R.: Logic for computable functions: description of a machine implementation. Tech. rep., Stanford University, Stanford, CA, USA (1972)
208. Milner, R.: Communicating and Mobile Systems: the π-Calculus. Cambridge Univ. Press (1999)
209. Misner, C.W., Thorne, K.S., Wheeler, J.A.: Gravitation. W. H. Freeman, New York (1973)
210. Mitchell, J. (ed.): Proceedings, Fifth Annual IEEE Symposium on Logic in Computer Science, 4–7 June 1990, Philadelphia, Pennsylvania, USA. IEEE Computer Society (1990)
211. Mora, T.: Solving Polynomial Equation Systems II: Macaulay's Paradigm and Gröbner Technology. Cambridge University Press (2005)

212. Morari, M., Thiele, L. (eds.): Hybrid Systems: Computation and Control, 8th International Workshop, HSCC 2005, Zurich, Switzerland, March 9–11, 2005, Proceedings, *LNCS*, vol. 3414. Springer (2005)
213. Morayne, M.: On differentiability of Peano type functions. Colloquium Mathematicum **LIII**, 129–132 (1987)
214. Motzkin, T.S.: Beiträge zur Theorie der Linearen Ungleichungen. Ph.D. thesis, Basel, Jerusalem (1936)
215. Mumford, D.: Algebraic Geometry I: Complex Projective Varieties. Classics in Mathematics. Springer (1995)
216. Muñoz, C., Carreño, V., Dowek, G., Butler, R.W.: Formal verification of conflict detection algorithms. STTT **4**(3), 371–380 (2003). DOI 10.1007/s10009-002-0084-3
217. Mysore, V., Piazza, C., Mishra, B.: Algorithmic algebraic model checking II: Decidability of semi-algebraic model checking and its applications to systems biology. In: Peled and Tsay [226], pp. 217–233. DOI 10.1007/11562948_18
218. Nicollin, X., Olivero, A., Sifakis, J., Yovine, S.: An approach to the description and analysis of hybrid systems. In: Grossman et al. [144], pp. 149–178
219. Nipkow, T.: Linear quantifier elimination. In: Armando et al. [18]
220. Nipkow, T., Paulson, L.C., Wenzel, M.: Isabelle/HOL — A Proof Assistant for Higher-Order Logic, *LNCS*, vol. 2283. Springer (2002)
221. Olderog, E.R., Dierks, H.: Real-Time Systems: Formal Specification and Automatic Verification. Cambridge Univ. Press (2008)
222. Pallottino, L., Scordio, V.G., Frazzoli, E., Bicchi, A.: Decentralized cooperative policy for conflict resolution in multi-vehicle systems. IEEE Trans. on Robotics **23**(6), 1170–1183 (2007)
223. Parker, R.S., Doyle, F.J., Peppas, N.A.: The intravenous route to blood glucose control. IEEE Engineering in Medicine and Biology **20**(1), 65–73 (2001). DOI 10.1109/51.897829
224. Parrilo, P.A.: Semidefinite programming relaxations for semialgebraic problems. Math. Program. **96**(2), 293–320 (2003). DOI 10.1007/s10107-003-0387-5
225. Peano, G.: Demonstration de l'intégrabilité des équations différentielles ordinaires. Mathematische Annalen **37**(2), 182–228 (1890). DOI 10.1007/BF01200235
226. Peled, D., Tsay, Y.K. (eds.): Automated Technology for Verification and Analysis, Third International Symposium, ATVA 2005, Taipei, Taiwan, October 4–7, 2005, Proceedings, *LNCS*, vol. 3707. Springer (2005)
227. Perko, L.: Differential equations and dynamical systems, 3 edn. Springer, New York, NY, USA (2006)
228. Piazza, C., Antoniotti, M., Mysore, V., Policriti, A., Winkler, F., Mishra, B.: Algorithmic algebraic model checking I: Challenges from systems biology. In: Etessami and Rajamani [118], pp. 5–19. DOI 10.1007/11513988_3
229. Plaisted, D.A., Zhu, Y.: Ordered semantic hyper-linking. J. Autom. Reasoning **25**(3), 167–217 (2000)
230. Platzer, A.: Combining deduction and algebraic constraints for hybrid system analysis. In: B. Beckert (ed.) VERIFY'07 at CADE, Bremen, Germany, *CEUR Workshop Proceedings*, vol. 259, pp. 164–178. CEUR-WS.org (2007)
231. Platzer, A.: Differential dynamic logic for verifying parametric hybrid systems. In: N. Olivetti (ed.) TABLEAUX, *LNCS*, vol. 4548, pp. 216–232. Springer (2007). DOI 10.1007/978-3-540-73099-6_17
232. Platzer, A.: Differential logic for reasoning about hybrid systems. In: Bemporad et al. [41], pp. 746–749. DOI 10.1007/978-3-540-71493-4_75
233. Platzer, A.: A temporal dynamic logic for verifying hybrid system invariants. In: S.N. Artëmov, A. Nerode (eds.) LFCS, *LNCS*, vol. 4514, pp. 457–471. Springer (2007). DOI 10.1007/978-3-540-72734-7_32
234. Platzer, A.: Towards a hybrid dynamic logic for hybrid dynamic systems. In: P. Blackburn, T. Bolander, T. Braüner, V. de Paiva, J. Villadsen (eds.) LICS International Workshop on Hybrid Logic, HyLo'06, Seattle, USA, Proceedings, *ENTCS*, vol. 174, pp. 63–77 (2007). DOI 10.1016/j.entcs.2006.11.026

235. Platzer, A.: Differential dynamic logic for hybrid systems. Journal of Automated Reasoning **41**(2), 143–189 (2008). DOI 10.1007/s10817-008-9103-8
236. Platzer, A.: Differential dynamic logics: Automated theorem proving for hybrid systems. Ph.D. thesis, Department of Computing Science, University of Oldenburg (2008)
237. Platzer, A.: Differential-algebraic dynamic logic for differential-algebraic programs. Journal of Logic and Computation **20**(1), 309–352 (2010). DOI 10.1093/logcom/exn070. Advance Access published on November 18, 2008
238. Platzer, A., Clarke, E.M.: The image computation problem in hybrid systems model checking. In: Bemporad et al. [41], pp. 473–486. DOI 10.1007/978-3-540-71493-4_37
239. Platzer, A., Clarke, E.M.: Computing differential invariants of hybrid systems as fixedpoints. In: Gupta and Malik [146], pp. 176–189. DOI 10.1007/978-3-540-70545-1_17
240. Platzer, A., Clarke, E.M.: Computing differential invariants of hybrid systems as fixedpoints. Formal Methods in System Design **35**(1), 98–120 (2009). DOI 10.1007/s10703-009-0079-8
241. Platzer, A., Clarke, E.M.: Formal verification of curved flight collision avoidance maneuvers: A case study. In: A. Cavalcanti, D. Dams (eds.) FM, *LNCS*, vol. 5850, pp. 547–562. Springer (2009). DOI 10.1007/978-3-642-05089-3_35
242. Platzer, A., Quesel, J.D.: KeYmaera: A hybrid theorem prover for hybrid systems. In: Armando et al. [18], pp. 171–178. DOI 10.1007/978-3-540-71070-7_15
243. Platzer, A., Quesel, J.D.: Logical verification and systematic parametric analysis in train control. In: Egerstedt and Mishra [111], pp. 646–649. DOI 10.1007/978-3-540-78929-1_55
244. Platzer, A., Quesel, J.D.: European Train Control System: A case study in formal verification. In: K. Breitman, A. Cavalcanti (eds.) ICFEM, *LNCS*, vol. 5885, pp. 246–265. Springer (2009). DOI 10.1007/978-3-642-10373-5_13
245. Platzer, A., Quesel, J.D.: European Train Control System: A case study in formal verification. Tech. Rep. 54, Reports of SFB/TR 14 AVACS (2009). ISSN: 1860-9821, http://www.avacs.org.
246. Platzer, A., Quesel, J.D., Rümmer, P.: Real world verification. In: R.A. Schmidt (ed.) CADE, *LNCS*, vol. 5663, pp. 485–501. Springer (2009). DOI 10.1007/978-3-642-02959-2_35
247. Pnueli, A.: The temporal logic of programs. In: FOCS, pp. 46–57. IEEE (1977)
248. Pnueli, A., Kesten, Y.: A deductive proof system for CTL*. In: L. Brim, P. Jancar, M. Kretínský, A. Kucera (eds.) CONCUR, *LNCS*, vol. 2421, pp. 24–40. Springer (2002). DOI 10.1007/3-540-45694-5_2
249. Pnueli, A., Rosner, R.: On the synthesis of a reactive module. In: POPL, pp. 179–190 (1989). DOI 10.1145/75277.75293
250. Pour-El, M.B., Richards, I.: A computable ordinary differential equation which possesses no computable solution. Annals of Mathematical Logic **17**, 61–90 (1979). DOI 10.1016/0003-4843(79)90021-4
251. Prajna, S., Jadbabaie, A.: Safety verification of hybrid systems using barrier certificates. In: Alur and Pappas [14], pp. 477–492. DOI 10.1007/b96398
252. Prajna, S., Jadbabaie, A., Pappas, G.J.: A framework for worst-case and stochastic safety verification using barrier certificates. IEEE T. Automat. Contr. **52**(8), 1415–1429 (2007). DOI 10.1109/TAC.2007.902736
253. Pratt, V.R.: Semantical considerations on Floyd-Hoare logic. In: FOCS, pp. 109–121. IEEE (1976)
254. Pratt, V.R.: Process logic. In: POPL, pp. 93–100 (1979). DOI 10.1145/567752.567761
255. Queille, J.P., Sifakis, J.: Specification and verification of concurrent systems in CESAR. In: M. Dezani-Ciancaglini, U. Montanari (eds.) Symposium on Programming, *LNCS*, vol. 137, pp. 337–351. Springer (1982). DOI 10.1007/3-540-11494-7_22
256. Quesel, J.D.: A theorem prover for differential dynamic logic. Master's thesis, University of Oldenburg, Department of Computing Science. Correct System Design Group (2007)
257. Reid, W.T.: Ordinary Differential Equations. John Wiley (1971)
258. Renegar, J.: On the computational complexity and geometry of the first-order theory of the reals, part I: Introduction. Preliminaries. The geometry of semi-algebraic sets. The decision problem for the existential theory of the reals. J. Symb. Comput. **13**(3), 255–300 (1992)

259. Renegar, J.: On the computational complexity and geometry of the first-order theory of the reals, part II: The general decision problem. Preliminaries for quantifier elimination. J. Symb. Comput. **13**(3), 301–328 (1992)
260. Renegar, J.: On the computational complexity and geometry of the first-order theory of the reals, part III: Quantifier elimination. J. Symb. Comput. **13**(3), 329–352 (1992)
261. Reynolds, M.: An axiomatization of PCTL*. Inf. Comput. **201**(1), 72–119 (2005). DOI 10.1016/j.ic.2005.03.005
262. Risler, J.J.: Une caractérisation des idéaux des variétés algébraiques réelles. C.R.A.S. Paris série A **271**, 1171–1173 (1970)
263. Risler, J.J.: Some aspects of complexity in real algebraic geometry. J. Symb. Comput. **5**(1/2), 109–119 (1988). DOI 10.1016/S0747-7171(88)80007-5
264. Robinson, A.: Complete Theories. North-Holland (1956)
265. Robinson, J.A.: A machine-oriented logic based on the resolution principle. J. Assoc. Comput. Mach. **12**, 23–41 (1965). DOI 10.1145/321250.321253
266. Rodríguez-Carbonell, E., Kapur, D.: An abstract interpretation approach for automatic generation of polynomial invariants. In: R. Giacobazzi (ed.) SAS, *Lecture Notes in Computer Science*, vol. 3148, pp. 280–295. Springer (2004). DOI 10.1007/b99688
267. Rodríguez-Carbonell, E., Kapur, D.: Automatic generation of polynomial invariants of bounded degree using abstract interpretation. Sci. Comput. Program. **64**(1), 54–75 (2007). DOI 10.1016/j.scico.2006.03.003
268. Rodríguez-Carbonell, E., Kapur, D.: Generating all polynomial invariants in simple loops. J. Symb. Comput. **42**(4), 443–476 (2007). DOI 10.1016/j.jsc.2007.01.002
269. Rodríguez-Carbonell, E., Tiwari, A.: Generating polynomial invariants for hybrid systems. In: Morari and Thiele [212], pp. 590–605. DOI 10.1007/b106766
270. Rönkkö, M., Ravn, A.P., Sere, K.: Hybrid action systems. Theor. Comput. Sci. **290**(1), 937–973 (2003)
271. Rounds, W.C.: A spatial logic for the hybrid π-calculus. In: Alur and Pappas [14], pp. 508–522. DOI 10.1007/b96398
272. Rounds, W.C., Song, H.: The ϕ-calculus: A language for distributed control of reconfigurable embedded systems. In: HSCC, *LNCS*, vol. 2623, pp. 435–449 (2003). DOI 10.1007/3-540-36580-X_32
273. Rümmer, P.: A sequent calculus for integer arithmetic with counterexample generation. In: B. Beckert (ed.) VERIFY'07 at CADE, Bremen, Germany, *CEUR Workshop Proceedings*, vol. 259, pp. 179–194. CEUR-WS.org (2007)
274. Sankaranarayanan, S., Sipma, H., Manna, Z.: Constructing invariants for hybrid systems. In: Alur and Pappas [14], pp. 539–554. DOI 10.1007/b96398
275. Schobbens, P.Y., Raskin, J.F., Henzinger, T.A.: Axioms for real-time logics. Theor. Comput. Sci. **274**(1-2), 151–182 (2002). DOI 10.1016/S0304-3975(00)00308-X
276. Schrödinger, E.: An undulatory theory of the mechanics of atoms and molecules. Phys. Rev. **28**, 1049–1070 (1926). DOI 10.1103/PhysRev.28.1049
277. Scott, D., Strachey, C.: Toward a mathematical semantics for computer languages? Tech. Rep. PRG-6, Oxford Programming Research Group (1971)
278. Seidenberg, A.: A new decision method for elementary algebra. Annals of Mathematics **60**, 365–374 (1954)
279. Sibirsky, K.S.: Introduction to Topological Dynamics. Noordhoff, Leyden (1975)
280. Sipser, M.: Introduction to the Theory of Computation, 2 edn. Course Technology (2005)
281. Skolem, T.: Logisch-kombinatorische Untersuchungen über die Erfüllbarkeit oder Beweisbarkeit mathematischer Sätze nebst einem Theorem über dichte Mengen. Videnskapsselskapet Skrifter, I. Matematisk-naturvidenskabelig Klasse **6**, 1–36 (1920)
282. Smullyan, R.M.: First-Order Logic. Dover (1995)
283. Stengle, G.: A Nullstellensatz and a Positivstellensatz in semialgebraic geometry. Math. Ann. **207**(2), 87–97 (1973). DOI 10.1007/BF01362149
284. Stirling, C.: Modal and temporal logics. In: Handbook of Logic in Computer Science (vol. 2): Background: Computational Structures, pp. 477–563. Oxford University Press, Inc., New York, NY, USA (1992)

285. Strzebonski, A.W.: Cylindrical algebraic decomposition using validated numerics. J. Symb. Comput. **41**(9), 1021–1038 (2006). DOI 10.1016/j.jsc.2006.06.004
286. Sturm, C.F.: Mémoire sur la résolution des équations numériques. Mémoires des Savants Etrangers **6**, 271–318 (1835)
287. Tarski, A.: Sur les ensembles définissables de nombres réels I. Fundam. Math. **17**, 210–239 (1931)
288. Tarski, A.: A Decision Method for Elementary Algebra and Geometry, 2 edn. University of California Press, Berkeley (1951)
289. Tavernini, L.: Differential automata and their discrete simulators. Non-Linear Anal. **11**(6), 665–683 (1987). DOI 10.1016/0362-546X(87)90034-4
290. Tinelli, C.: Cooperation of background reasoners in theory reasoning by residue sharing. J. Autom. Reasoning **30**(1), 1–31 (2003). DOI 10.1023/A:1022587501759
291. Tiwari, A.: Approximate reachability for linear systems. In: Maler and Pnueli [200], pp. 514–525. DOI 10.1007/3-540-36580-X_37
292. Tiwari, A.: An algebraic approach for the unsatisfiability of nonlinear constraints. In: C.H.L. Ong (ed.) CSL, *LNCS*, vol. 3634, pp. 248–262. Springer (2005). DOI 10.1007/11538363_18
293. Tomlin, C., Pappas, G.J., Sastry, S.: Conflict resolution for air traffic management: a study in multi-agent hybrid systems. IEEE T. Automat. Contr. **43**(4), 509–521 (1998). DOI 10.1109/9.664154
294. Tomlin, C.J., Lygeros, J., Sastry, S.: A game theoretic approach to controller design for hybrid systems. Proc. IEEE **88**(7), 949–970 (2000). DOI 10.1109/5.871303
295. Umeno, S., Lynch, N.A.: Proving safety properties of an aircraft landing protocol using I/O automata and the PVS theorem prover: A case study. In: J. Misra, T. Nipkow, E. Sekerinski (eds.) FM, *LNCS*, vol. 4085, pp. 64–80. Springer (2006). DOI 10.1007/11813040_5
296. Umeno, S., Lynch, N.A.: Safety verification of an aircraft landing protocol: A refinement approach. In: Bemporad et al. [41], pp. 557–572. DOI 10.1007/978-3-540-71493-4_43
297. Walter, W.: Ordinary Differential Equations. Springer (1998)
298. Weihrauch, K.: Computable Analysis. Springer (2005)
299. Weispfenning, V.: The complexity of linear problems in fields. J. Symb. Comput. **5**(1/2), 3–27 (1988). DOI 10.1016/S0747-7171(88)80003-8
300. Weispfenning, V.: Quantifier elimination for real algebra — the cubic case. In: ISSAC, pp. 258–263 (1994). DOI 10.1145/190347.190425
301. Weispfenning, V.: Quantifier elimination for real algebra — the quadratic case and beyond. Appl. Algebra Eng. Commun. Comput. **8**(2), 85–101 (1997). DOI 10.1007/s002000050055
302. Witsenhausen, H.S.: A class of hybrid-state continuous-time dynamic systems. IEEE Trans. Automat. Contr. **11**, 161–167 (1966)
303. Wolfram Research, Inc., Champaign, IL: Mathematica, version 5.2 edn. (2005). `http://www.wolfram.com/`
304. Zeidler, E. (ed.): Teubner-Taschenbuch der Mathematik I. Teubner (2003)
305. Zhou, C., Hansen, M.R.: Duration Calculus: A Formal Approach to Real-Time Systems. Monographs in Theoretical Computer Science. Springer (2004)
306. Zhou, C., Ravn, A.P., Hansen, M.R.: An extended duration calculus for hybrid real-time systems. In: Grossman et al. [144], pp. 36–59

Index

Symbols

; **41**, 79, **137**, 164, 218
$[\alpha]$ **47**, 52, **139**, 145, **207**, 213
$\langle\alpha\rangle$ **47**, 52, **139**, 145, **207**
$\wedge$ **40**, **47**, 52, 79, **139**, 145, 164, **207**, 213, 218, **347**, 349, 352
$\leftrightarrow$ **40**, **139**, **207**, **347**
$\exists$ **40**, **47**, 52, 79, 164, **207**, 213, 218, **347**, 350, 352
$\forall^\alpha$ **76**
$\forall$ **40**, **47**, 52, 79, 164, **207**, 213, 218, **347**, 350, 352
$\rightarrow$ **40**, **47**, 52, 79, **139**, 145, 164, **207**, 213, 218, **347**, 349, 352
$\neg$ **40**, **47**, 52, 79, **139**, 145, 164, **207**, 213, 218, **347**, 349, 352
$\vee$ **40**, **47**, 52, 79, **139**, 145, 164, **207**, 213, 218, **347**, 349, 352
$\cup$ **41**, 55, **136**, 145, 211
* **41**, 55, **137**, 145, 211
? **41**, 55, 211
$\Box$ **207**
$\Diamond$ **207**
$\mathsf{d}\mathscr{L}$ **35**
$(\nu,\omega) \models \mathscr{J}$ 142
Σ **37**, **346**
$\vDash$ 52
$\vdash_{\mathrm{DAL}}$ **168**
$\vdash_{\mathsf{d}\mathscr{L}}$ **87**
$\vdash_{\mathrm{dTL}}$ **220**
$\vdash_{\mathscr{D}}$ 109
$\exists x\!:\!\mathbb{N}$ 105
$\forall x\!:\!\mathbb{N}$ 105
$\models$ 52, 145, 350
$\not\models$ 52, 350
$\phi^\#$ 108
$\prec$ **387**
ϕ_x^θ **153**
$\varphi \models \mathscr{D}$ 143
$\sigma_i(\zeta)$ 210
$\prec$ 210
$\mathscr{B}$ 236
$D(F)$ **155**
$\mathrm{HP}(\Sigma)$ **136**
$\mathscr{D}$ 109
$\mathscr{F}$ 236
$F'_{\mathscr{D}}$ 259
FOD 104
$\mathrm{Fml}(\Sigma)$ **47**, **139**, **207**
$\mathrm{Fml}_{\mathrm{FOL}}(\Sigma)$ **40**, **347**
$\mathrm{Fml}_T(\Sigma)$ **207**
$\mathscr{F}(\omega)$ 150
$\mathscr{G}(\rho)$ 150
$\mathrm{HP}(\Sigma)$ **41**
$\mathscr{M}$ 236
$M^\alpha(F)$ 190
QE 77
$\mathscr{S}_\alpha(\vec{x},\vec{v})$ 106
$\mathscr{T}_\alpha(\vec{x},\vec{v})$ 225
$\mathrm{Trm}(\Sigma)$ **39**, **132**
$Z_j^{(n)}$ 105
$[x \mapsto d]$ 52, 55, 349
Λ 210
Σ' 134
$\hat{\nu}$ 210
$\rho(\alpha)$ 144
$\rho_{I,\eta}(\alpha)$ 55
$\tau(\alpha)$ 211
$\bar{\varphi}$ 142
DI 164, 425
DI' 188
DI'' 188

DS 164, 425
DV 164, 425
F' 164
$[']$ 79, 424
$[']\Box$ 218, 426
$[:=]$ 79, 164, 424, 425
$[:=]\Box$ 218, 426
$[;]$ 79, 164, 424, 425
$[;]\Diamond$ 229
$[;]\Box$ 218, 426
$[?]$ 79, 424
$[?]\Box$ 218, 426
$[DR']$ 175
$[DR]$ 164, 425
$[D]$ 164, 425
$[J]$ 164, 425
$[\alpha]\Diamond$ 229
$[\cup]$ 79, 164, 424, 425
$[\cup]\Box$ 218, 426
$[\exists]$ 164, 425
$[]gen$ 79, 164, 424, 425
$[^*]\Box$ 218, 426
$[^{*n}]$ 79, 424
$[^{*n}]\Box$ 218, 426
$\exists$l 79, 424
$\exists$r 79, 424
$\forall$l 79, 424
$\forall$r 79, 424
$\wedge$l 79, 424
$\wedge$r 79, 424
$\langle'\rangle$ 79, 424
$\langle'\rangle'$ 191
$\langle'\rangle\Diamond$ 218, 426
$\langle:=\rangle$ 79, 164, 424, 425
$\langle:=\rangle\Diamond$ 218, 426
$\langle;\rangle$ 79, 164, 424, 425
$\langle;\rangle\Diamond$ 218, 426
$\langle?\rangle$ 79, 424
$\langle?\rangle\Diamond$ 218, 426
$\langle DR\rangle$ 164, 425
$\langle D\rangle$ 164, 425
$\langle J\rangle$ 164, 425
$\langle\cup\rangle$ 79, 164, 424, 425
$\langle\cup\rangle\Diamond$ 218, 426
$\langle\exists\rangle$ 164, 425
$\langle\rangle gen$ 79, 164, 424, 425
$\langle^*\rangle\Diamond$ 218, 426
$\langle^{*n}\rangle$ 79, 424
$\langle^{*n}\rangle\Diamond$ 218, 426
$\rightarrow$l 79, 424
$\rightarrow$r 79, 424
$\neg$l 79, 424
$\neg$r 79, 424
$\vee$l 79, 424
$\vee$r 79, 424
ax 79, 424
con 79, 164, 424, 425
con' 86
cut 79, 424
ind 79, 164, 424, 425
ind' 86
$[']'$ 190
A1 389
A2 389
A3 389
A4 389
A5 389
A6 389
A7 395
i$\exists$ 79, 424
i$\forall$ 79, 424
l$\exists$ 164, 425
l$\forall$ 164, 425
r$\exists$ 164, 425
r$\forall$ 164, 425
$\sim F$ 182
$\vdash$ 76, 351
nat 102
ord_x 134
$val_{I,\eta}(\cdot)$ 349
$val_{I,\eta}(\nu,\cdot)$ 50, 52, 349
$val(\nu,\cdot)$ 141, 213
x' 134
$x' = \theta$ 41, 134
$x' = \theta \,\&\, \chi$ 41
$x^{(n)}$ 134
$x := \theta$ 41, 133

A

abort **210**
accessibility relation 49, 141
acessibility relation 51
adjoint **70**, **74**
admissible *see* substitution, admissible, 70
  monomial order 387
alphabet 37
analytic cut *see* cut
and-operator 47
and/or-branching 245
and/or-parallelism *see* and/or-branching
antecedent **76**, **351**
arithmetic formula *see* formula, $\mathbb{R}$ arithmetic
arity **37**
assignment **50**, **349**
atomic formula *see* formula, atomic
automaton
  hybrid 5–12, 46, **370**, 369–375

axiomatisable 103
axiomatise **102**, 104, 223

B

bi-implication 47
binding priorities 47
bound *see* symbol, bound
  variable renaming **65**
box-modality 47
branch 77, 352
  and **245**
  or **245**

C

calculus
  free-variable **354**
  free variable **79**
  ground **354**
  of DAL **164**, 162–168
  of $d\mathcal{L}$ **79**, 64–93
  of dTL **218**, 217–220
  of FOL **352**, 350–356
Cauchy-Lipschitz 58, 144, 364
changed *see* symbol, changed
check *see* test
closed
  formula *see* formula, closed
closure
  universal **76**
coincidence 93
collision-free 331
collision free 279, 294
complete **102**, **356**
  relatively **104**, 193, 224
    DAL 193
    $d\mathcal{L}$ 104
    dTL 224
compositional
  semantics 50, 52, 54, 144, 145, 210, 213
compositional verification 16
conclusion **77**, **162**
cone **394**
conjunction 47
consequence **98**
constant symbol 38
continued
  to border 363
continuous
  Lipschitz *see* Lipschitz, continuous
  system
    switched *see* dynamical systems, switched
  transition 41, 371
continuous-time 64
control
  graph 370
  refinement 287
  switch 370
control-flow *see* dependency, control-flow
controllability **90**, 287
  constraint **288**, 296
  discovery 287
controllable 294
  state **288**
control structure 36
convex 192
counterexample 351
cut **80**, 352
  differential *see* differential strengthening

D

DA- *see* differential-algebraic
DAL *see* differential-algebraic, logic
  calculus 162–168
  formula 139
  proof 168
  semantics 141–147
  syntax 132–141
  term 132
data-flow *see* dependency, data-flow
DATL 207
decidable **357**
decision
  problem **357**
deductive power 194
definable 38, 44, 76, 84, 102, 114, 139
denotational
  semantics 50, 52, 145, 213
dependency
  control-flow **266**
  data-flow **266**
  differential **264**
derivable **168**
derivation **86**
  lemma 156
  monotonicity **190**
  operator **155**
  syntactic **155**
derived rule *see* rule, derived, 175
design space exploration 278
diamond-modality 47
difference
  equation 35
differential
  axiomatisation 150

constant 155, **156**
cut *see* differential strengthening
dependency *see* dependency, differential
elimination 161
equation 35, **41**, 84, 130, 161, **360**
explicit **161**
normalisation 161
field of fractions **155**
field of quotients **155**
homomorphism **156**
indeterminate **156**
indeterminates 155
induction **167**, 170–185, 194
inequality 130, 161
elimination 161
invariant **167**, 170–181, **259**
as fixed points 256–271
monotonicity 191
polynomial algebra **155**
refinement **166**
saturation **262**
state flow **142**
strengthening 165, 166
substitution 158
substitution lemma 156
symbol **134**, 155
total **154**, 164, 170, 172, 173
transformation 158
variant **167**, 181–185
weakening **166**, 175
differential-algebraic
constraint **125**, 130, **134**
equivalent *see* equivalent, DA-constraint
equation **136**
flow 137
logic **125**, **130**
program **125**, **136**, 136–139
temporal logic **207**
DIFP *see* differential invariants, as fixed points
discrete
jump
constraint **125**, 130, **133**, 137
set **41**
program *see* program, discrete
time 46, 64
transition 35, 41, 371
disjunction 47
disturbance 296
d$\mathscr{L}$ 35
calculus 64–93
formula 47
proof 86
semantics 49–60
syntax 35–49
term 39
dTL **206**
calculus 217–220
formula 207
proof 220
semantics 210–214
syntax 206–209
term 207
dynamic
rule 82, 165
dynamical system
continuous 47, 139
discrete 139
discrete-time 46
hybrid *see* hybrid system, 139
switched 46
dynamical systems
switched 139

E

eager 241
equivalent 47, **98**
DA-constraint **159**
ETCS (European Train Control System) 1–3, 61–64, 118–122, 278, **281**, 281–301
existence 363
existential quantifier *see* quantifier, existential
exists 47
expressible **108**, 160, **225**
extension
conservative **147**, **215**

F

field **382**
real **382**
real-closed **382**
first-order
logic *see* logic, first-order
rule 80, 163, 352
flexible *see* symbol, flexible
flow **55**, **142**
point **210**
state *see* differential, state flow
flyable 305
FOD **104**
FOL 40, 346
calculus 350–356
formula 40, 347
proof 351
semantics 348–350

syntax 346–348
term 347
formula
$\mathbb{R}$ arithmetic **88**
DAL **139**
$\mathsf{d}\mathscr{L}$ **47**
dTL **207**
affirmative **133**
atomic 47
closed 77
FOL **40**, **347**
ground 77
negative 133
nontemporal **207**
positive 133
state 207
trace **207**
for all 47
fragile **271**
free-variable
calculus *see* calculus, free-variable
free variable 77
FTRM **310**, 309–334
function symbol *see* symbol, function

G

generalisation
lemma 110
rule 86
global
rule 85, 167
goal **77**
Gröbner basis **388**
reduced **388**
ground
calculus *see* calculus, ground
formula *see* formula, ground
group
Abelian **381**
guard 36, 370

H

Hoare triple 49
homogeneous
constraint **136**
homomorphic continuation 70
HP *see* program, hybrid
hybrid
automaton *see* automaton, hybrid
dynamical system *see* hybrid system, 103, 139
program *see* program, hybrid
system **4**

I

ideal **387**
generator **387**
vanishing **391**
implication 47
implies-operator 47
incomplete
$\mathsf{d}\mathscr{L}$ 102
dTL 223
induction
differential *see* differential induction
discrete *see* invariant, discrete and variant, discrete
schema *see* invariant, inductive
initial value
problem 84, 191, **360**
symbolic 361
instable **10**
integer
arithmetic 102
intermediate value
property 382
interpretation **50**, *see* valuation, **349**
invariant
continuous **259**
differential *see* differential, invariant
discrete **85**, **259**
inductive **85**
loop *see* invariant, discrete
strong **259**
iterative
refinement process 286

J

jump 370
jump-free **133**

K

key **235**
KeY 378
KeYmaera 377–379
Kleene algebra 41
Kripke semantics 49, 141

L

lazy 241
leading term **387**
Leibniz

rule **155**
lifting *see* quantifier elimination, lifting
Lipschitz 164, 184, 187, **364**
local **364**
literal **155**
live 294
liveness 293
check 287
logic
differential-algebraic 130
differential dynamic **35**
temporal **206**
first-order **40**, 40, 346
logical
variable 37
loop
invariant *see* invariant, discrete
saturation 265
variant *see* variant, discrete
Lyapunov function 128

M

Mandelbrot set 370
MA (movement authority) **61**, 208, 281, 289
modality 47
$[\cdot]\Box$-modality 208
$[\cdot]\Diamond$-modality 208
$[]$-modality 47
$\langle\cdot\rangle\Box$-modality 208
$\langle\cdot\rangle\Diamond$-modality 208
$\langle\rangle$-modality 47
mode 370
model **52**
modification *see* semantic modification
monoid 394
monomial **387**
monotonicity derivation *see* derivation, monotonicity

N

natural
number
definable 102
negation 47
negative
formula *see* formula, negative
nilpotent **365**
non-differential *see* symbol, non-differential
nondeterminism
continuous **135**
discrete **133**
nondeterministic
choice 42, 137
repetition 42, 137
nontemporal
rule 218
not-operator 47

O

ODE *see* differential,equation,ordinary
or-operator 47
oracle 104, **109**, 114
order
partial **134**

P

parameter synthesis 90, 286
parametric
candidate 263
Peano 363
Picard-Lindelöf 58, 144, 187, 364
point flow *see* flow, point
polynomial **387**
position
temporal **210**
positive
formula *see* formula, positive
postcondition **49**
power
deductive *see* deductive power
precondition **49**
predicate symbol *see* symbol, predicate, 47
premise **77**, **162**
product
rule **155**
program
differential-algebraic *see* differential-algebraic, program
discrete 46, 139
hybrid **41**, 41–47, 126
program rendition
DA-programs 193
HP 106
trace 225
proof
calculus *see* calculus
DAL 168
$d\mathscr{L}$ 86
dTL 220
first-order 351
rule 78
proportional-integral **278**
propositional
rule 78, 351

provable **87**, **168**, 220
prover
  background **236**
  foreground 235

Q

QE (quantifier elimination) 77
quantifier
  existential 47
  universal 47
quantifier elimination **77**
  lifting 92
quotient
  rule **155**

R

ℝ-Gödel 105
radical **389**
RBC (radio block controller) 281
reachability 370
reactivity 290, 298
  constraint 290, 298
real-closed field *see* field, real-closed
real field *see* field, real
reduction
  polynomial **387**
  symmetry **317**
remainder 387
reset 370
rich test 42
rigid *see* symbol, rigid
ring **381**
robust **271**
roundabout manoeuvre 3–4, 148–152, 197–201, 309–334
rule **77**, **163**
  derived **86**
  dynamic 82, 165
  first-order 80, 163, 352
  foreground **235**
  global **85**, 167
  nontemporal 218
  propositional 78, 351
  schema **77**, **162**, **163**
  temporal 219

S

safe 199, 294, 331
  dynamics 197
safety 48, 60, 61, 208, 291, 298
  convergence 287
satisfaction relation *see* valuation
satisfiable **52**
satisfied **52**
saturation
  differential *see* differential, saturation
  loop *see* loop, saturation
SB (start braking) 62, 283, 290, 298
semantics
  of DAL 141–147
  of $d\mathcal{L}$ 49–60
  of dTL 210–214
  of FOL 348–350
  trace 211
  transition *see* transition, semantics
semantic modification **52**, **55**
semialgebraic **383**
semidecidable **357**
  $d\mathcal{L}$-fragment 114
semigroup
  commutative **381**
sensor polling 285
separation 200
sequent **76**, **351**
  calculus *see* calculus
sequential composition 42, 137
sign 37
signature **37**, **346**
Skolem
  constant *see* Skolem, function
  function 79, 81, 90, 352, 353
  term 39, 91
solution 57, **359**
sound **98**, **185**, **221**
  algorithm 269
  DAL 185
  $d\mathcal{L}$ 98
  dTL 221
  locally **98**, **185**, **222**
stable **10**
state **50**, **141**, **210**
  differentially augmented **142**
  flow *see* differential, state flow
  test *see* test
  variable 38, 41
ST (start talking) 216, 282
subgoal **77**
substitution **65**, 70
  admissible **66**, 91, **162**
  differential *see* differential substitution
  Lemma 70
  property 75, 157
succedent **76**, **351**
superdense **94**
symbol **37**, **132**, **346**

bound **66**, **162**
changed **133, 134**, 162
constant 38
differential *see* differential, symbol
flexible **38**, **132**
free 77
function 37, 346
non-differential **134**
predicate 37, 346
rigid **38**
variable 346
symbolic decomposition 64, 82, 152
synchronise
branch **239**
syntax
of DAL 132–141
of d$\mathscr{L}$ 35–49
of dTL 206–209
of FOL 346–348

T

temporal
rule 219
term
DAL **132**
d$\mathscr{L}$ **39**
dTL **207**
FOL **347**
leading *see* leading term
terminates **210**
test 36, 42
timed automaton 22
trace **210**
semantics 211
transformation
differential *see* differential transformation
transition 371
relation 144
semantics
DA-programs 144
d$\mathscr{L}$ 55, 56
HP 55
truth-value 38, 346

U

undecidable **357**
unification 355
uniqueness 57, 364
universal quantifier *see* quantifier, universal

V

valid **52**, **98**, **213**, **350**
valuation **50**, **51**
of DA-constraints **143**
of DA-programs **144**
of DAL **145**
of DJ-constraints **142**
of d$\mathscr{L}$ **52**
of dTL 213
of FOL **349**
of formulas *see* of d$\mathscr{L}$,DAL,dTL
of programs **55**, **211**
of state formulas **213**
of terms **50**, **141**, **349**
of trace formulas **213**
variable *see* symbol
cluster **264**
variant
differential *see* differential, variant
discrete **85**
loop *see* variant, discrete
variety
algebraic **391**
vocabulary 37

W

weak negation 182
well-formed
formula 37
term 39

Z

Zeno **146**

Table 2.1 Statements and effects of hybrid programs (HPs)

HP Notation	Operation	Effect
$x_1 := \theta_1, \ldots, x_n := \theta_n$	discrete jump set	simultaneously assigns terms θ_i to variables x_i
$x_1' = \theta_1, \ldots, x_n' = \theta_n \,\&\, \chi$	continuous evolution	differential equations for x_i with terms θ_i within first-order constraint χ (evolution domain)
$?\chi$	state test / check	test first-order formula χ at current state
$\alpha; \beta$	seq. composition	HP β starts after HP α finishes
$\alpha \cup \beta$	nondet. choice	choice between alternatives HP α or HP β
α^*	nondet. repetition	repeats HP α n-times for any $n \in \mathbb{N}$

Tables 2.3, 3.4 and 4.1 Operators of differential dynamic logic ($\mathsf{d}\mathcal{L}$), and additional operators of differential-algebraic dynamic logic (DAL) and differential temporal-dynamic logic (dTL)

$\mathsf{d}\mathcal{L}$ Notation	Operator	Meaning
$p(\theta_1, \ldots, \theta_n)$	atomic formula	true iff predicate p holds for $(\theta_1, \ldots, \theta_n)$
$\neg\phi$	negation / not	true if ϕ is false
$\phi \wedge \psi$	conjunction / and	true if both ϕ and ψ are true
$\phi \vee \psi$	disjunction / or	true if ϕ is true or if ψ is true
$\phi \rightarrow \psi$	implication / implies	true if ϕ is false or ψ is true
$\phi \leftrightarrow \psi$	bi-implication / equivalent	true if ϕ and ψ are both true or both false
$\forall x\, \phi$	universal quantifier / for all	true if ϕ is true for all values of variable x
$\exists x\, \phi$	existential quantifier / exists	true if ϕ is true for some values of variable x
$[\alpha]\phi$	$[\cdot]$ modality / box	true if ϕ is true after all runs of HP α
$\langle\alpha\rangle\phi$	$\langle\cdot\rangle$ modality / diamond	true if ϕ is true after at least one run of HP α
$[\alpha]\phi$	$[\cdot]$ modality / box (DAL)	true if ϕ is true after all runs of DA-program α
$\langle\alpha\rangle\phi$	$\langle\cdot\rangle$ modality / diamond (DAL)	true if ϕ is true after some run of DA-program α
$[\alpha]\Box\phi$	$[\cdot]\Box$ modality nesting (dTL)	if ϕ is true always during all traces of HP α
$\langle\alpha\rangle\Diamond\phi$	$\langle\cdot\rangle\Diamond$ modality nesting (dTL)	if ϕ is true sometimes during some trace of HP α
$[\alpha]\Diamond\phi$	$[\cdot]\Diamond$ modality nesting (dTL)	if ϕ is true sometimes during all traces of HP α
$\langle\alpha\rangle\Box\phi$	$\langle\cdot\rangle\Box$ modality nesting (dTL)	if ϕ is true always during some trace of HP α

Table 3.2 Statements and effects of differential-algebraic programs

DA-program	Operation	Effect
$\mathcal{J}$	discrete jump	jump constraint with assignments holds for discrete jump
$\mathcal{D}$	diff.-alg. flow	differential-algebraic constraint holds during continuous flow
$\alpha; \beta$	seq. composition	DA-program β starts after DA-program α finishes
$\alpha \cup \beta$	nondet. choice	choice between alternative DA-programs α or β
α^*	nondet. repetition	repeats DA-program α n-times for any $n \in \mathbb{N}$

$$(\neg r)\ \frac{\phi \vdash}{\vdash \neg\phi} \quad (\vee r)\ \frac{\vdash \phi, \psi}{\vdash \phi \vee \psi} \quad (\wedge r)\ \frac{\vdash \phi \quad \vdash \psi}{\vdash \phi \wedge \psi} \quad (\to r)\ \frac{\phi \vdash \psi}{\vdash \phi \to \psi} \quad (ax)\ \frac{}{\phi \vdash \phi}$$

$$(\neg l)\ \frac{\vdash \phi}{\neg\phi \vdash} \quad (\vee l)\ \frac{\phi \vdash \quad \psi \vdash}{\phi \vee \psi \vdash} \quad (\wedge l)\ \frac{\phi, \psi \vdash}{\phi \wedge \psi \vdash} \quad (\to l)\ \frac{\vdash \phi \quad \psi \vdash}{\phi \to \psi \vdash} \quad (cut)\ \frac{\vdash \phi \quad \phi \vdash}{\vdash}$$

$$(\langle;\rangle)\ \frac{\langle\alpha\rangle\langle\beta\rangle\phi}{\langle\alpha;\beta\rangle\phi} \quad (\langle^{*n}\rangle)\ \frac{\phi \vee \langle\alpha\rangle\langle\alpha^*\rangle\phi}{\langle\alpha^*\rangle\phi} \quad (\langle:=\rangle)\ \frac{\phi_{x_1}^{\theta_1} \cdots {}_{x_n}^{\theta_n}}{\langle x_1 := \theta_1, .., x_n := \theta_n\rangle\phi}$$

$$([;])\ \frac{[\alpha][\beta]\phi}{[\alpha;\beta]\phi} \quad ([^{*n}])\ \frac{\phi \wedge [\alpha][\alpha^*]\phi}{[\alpha^*]\phi} \quad ([:=])\ \frac{\langle x_1 := \theta_1, .., x_n := \theta_n\rangle\phi}{[x_1 := \theta_1, .., x_n := \theta_n]\phi}$$

$$(\langle\cup\rangle)\ \frac{\langle\alpha\rangle\phi \vee \langle\beta\rangle\phi}{\langle\alpha\cup\beta\rangle\phi} \quad (\langle?\rangle)\ \frac{\chi \wedge \psi}{\langle ?\chi\rangle\psi} \quad (\langle'\rangle)\ \frac{\exists t{\geq}0\,\big((\forall 0{\leq}\tilde{t}{\leq}t\,\langle\mathcal{S}_{\tilde{t}}\rangle\chi) \wedge \langle\mathcal{S}_t\rangle\phi\big)}{\langle x_1' = \theta_1, .., x_n' = \theta_n \,\&\, \chi\rangle\phi}\ {}^{1}$$

$$([\cup])\ \frac{[\alpha]\phi \wedge [\beta]\phi}{[\alpha\cup\beta]\phi} \quad ([?])\ \frac{\chi \to \psi}{[?\chi]\psi} \quad ([\,'\,])\ \frac{\forall t{\geq}0\,\big((\forall 0{\leq}\tilde{t}{\leq}t\,\langle\mathcal{S}_{\tilde{t}}\rangle\chi) \to \langle\mathcal{S}_t\rangle\phi\big)}{[x_1' = \theta_1, .., x_n' = \theta_n \,\&\, \chi]\phi}\ {}^{1}$$

$$(\forall r)\ \frac{\vdash \phi(s(X_1, .., X_n))}{\vdash \forall x\,\phi(x)}\ {}^{2} \qquad (\exists r)\ \frac{\vdash \phi(X)}{\vdash \exists x\,\phi(x)}\ {}^{4}$$

$$(\exists l)\ \frac{\phi(s(X_1, .., X_n)) \vdash}{\exists x\,\phi(x) \vdash}\ {}^{2} \qquad (\forall l)\ \frac{\phi(X) \vdash}{\forall x\,\phi(x) \vdash}\ {}^{4}$$

$$(i\forall)\ \frac{\vdash \mathrm{QE}(\forall X\,(\Phi(X) \vdash \Psi(X)))}{\Phi(s(X_1, .., X_n)) \vdash \Psi(s(X_1, .., X_n))}\ {}^{3} \qquad (i\exists)\ \frac{\vdash \mathrm{QE}(\exists X \bigwedge_i (\Phi_i \vdash \Psi_i))}{\Phi_1 \vdash \Psi_1 \quad \ldots \quad \Phi_n \vdash \Psi_n}\ {}^{5}$$

$$([]gen)\ \frac{\vdash \forall^\alpha(\phi \to \psi)}{[\alpha]\phi \vdash [\alpha]\psi} \qquad (\langle\rangle gen)\ \frac{\vdash \forall^\alpha(\phi \to \psi)}{\langle\alpha\rangle\phi \vdash \langle\alpha\rangle\psi}$$

$$(ind)\ \frac{\vdash \forall^\alpha(\phi \to [\alpha]\phi)}{\phi \vdash [\alpha^*]\phi} \qquad (con)\ \frac{\vdash \forall^\alpha \forall v{>}0\,(\varphi(v) \to \langle\alpha\rangle\varphi(v-1))}{\exists v\,\varphi(v) \vdash \langle\alpha^*\rangle\exists v{\leq}0\,\varphi(v)}\ {}^{6}$$

---

[1] t and $\tilde{t}$ are fresh logical variables and $\langle\mathcal{S}_t\rangle$ is the jump set $\langle x_1 := y_1(t), .., x_n := y_n(t)\rangle$ with simultaneous solutions $y_1, .., y_n$ of the respective differential equations with constant symbols x_i as symbolic initial values.

[2] s is a new (Skolem) function symbol and $X_1, .., X_n$ are all free logical variables of $\forall x\,\phi(x)$.

[3] X is a new logical variable. Further, QE needs to be defined for the formula in the premise.

[4] X is a new logical variable.

[5] Among all open branches, free logical variable X only occurs in the branches $\Phi_i \vdash \Psi_i$. Further, QE needs to be defined for the formula in the premise, especially, no Skolem dependencies on X can occur.

[6] Logical variable v does not occur in α.

Fig. 2.11 Proof calculus for differential dynamic logic (d$\mathcal{L}$)

$$(\mathrm{r}\forall)\ \frac{\mathrm{QE}(\forall x \bigwedge_i(\Gamma_i \vdash \Delta_i))}{\Gamma \vdash \Delta, \forall x \phi}\,{}^{1} \qquad (\mathrm{r}\exists)\ \frac{\mathrm{QE}(\exists x \bigwedge_i(\Gamma_i \vdash \Delta_i))}{\Gamma \vdash \Delta, \exists x \phi}\,{}^{1}$$

$$(\mathrm{l}\forall)\ \frac{\mathrm{QE}(\exists x \bigwedge_i(\Gamma_i \vdash \Delta_i))}{\Gamma, \forall x \phi \vdash \Delta}\,{}^{1} \qquad (\mathrm{l}\exists)\ \frac{\mathrm{QE}(\forall x \bigwedge_i(\Gamma_i \vdash \Delta_i))}{\Gamma, \exists x \phi \vdash \Delta}\,{}^{1}$$

$$(\langle;\rangle)\ \frac{\langle\alpha\rangle\langle\beta\rangle\phi}{\langle\alpha;\beta\rangle\phi} \qquad (\langle\exists\rangle)\ \frac{\exists x\langle\mathscr{J}\rangle\phi}{\langle\exists x\,\mathscr{J}\rangle\phi} \qquad (\langle:=\rangle)\ \frac{\chi \wedge \phi_{x_1}^{\theta_1}\cdots_{x_n}^{\theta_n}}{\langle x_1 := \theta_1 \wedge .. \wedge x_n := \theta_n \wedge \chi\rangle\phi}\,{}^{3}$$

$$([;])\ \frac{[\alpha][\beta]\phi}{[\alpha;\beta]\phi} \qquad ([\exists])\ \frac{\forall x[\mathscr{J}]\phi}{[\exists x\,\mathscr{J}]\phi} \qquad ([:=])\ \frac{\chi \to \phi_{x_1}^{\theta_1}\cdots_{x_n}^{\theta_n}}{[x_1 := \theta_1 \wedge .. \wedge x_n := \theta_n \wedge \chi]\phi}\,{}^{3}$$

$$(\langle\cup\rangle)\ \frac{\langle\alpha\rangle\phi \vee \langle\beta\rangle\phi}{\langle\alpha\cup\beta\rangle\phi} \qquad (\langle J\rangle)\ \frac{\langle\mathscr{J}_1\cup\ldots\cup\mathscr{J}_n\rangle\phi}{\langle\mathscr{J}\rangle\phi}\,{}^{2} \qquad (\langle D\rangle)\ \frac{\langle(\mathscr{D}_1\cup\ldots\cup\mathscr{D}_n)^*\rangle\phi}{\langle\mathscr{D}\rangle\phi}\,{}^{4}$$

$$([\cup])\ \frac{[\alpha]\phi \wedge [\beta]\phi}{[\alpha\cup\beta]\phi} \qquad ([J])\ \frac{[\mathscr{J}_1\cup\ldots\cup\mathscr{J}_n]\phi}{[\mathscr{J}]\phi}\,{}^{2} \qquad ([D])\ \frac{[(\mathscr{D}_1\cup\ldots\cup\mathscr{D}_n)^*]\phi}{[\mathscr{D}]\phi}\,{}^{4}$$

$$([DR])\ \frac{\vdash [\mathscr{E}]\phi}{\vdash [\mathscr{D}]\phi}\,{}^{5} \qquad (\langle DR\rangle)\ \frac{\vdash \langle\mathscr{D}\rangle\phi}{\vdash \langle\mathscr{E}\rangle\phi}\,{}^{5} \qquad (DS)\ \frac{\vdash [\mathscr{D}]\chi \quad \vdash [\mathscr{D}\wedge\chi]\phi}{\vdash [\mathscr{D}]\phi}$$

$$([]gen)\ \frac{\vdash \forall^\alpha(\phi \to \psi)}{[\alpha]\phi \vdash [\alpha]\psi} \qquad (\langle\rangle gen)\ \frac{\vdash \forall^\alpha(\phi \to \psi)}{\langle\alpha\rangle\phi \vdash \langle\alpha\rangle\psi}$$

$$(ind)\ \frac{\vdash \forall^\alpha(\phi \to [\alpha]\phi)}{\phi \vdash [\alpha^*]\phi} \qquad (con)\ \frac{\vdash \forall^\alpha\forall v{>}0\,(\varphi(v) \to \langle\alpha\rangle\varphi(v-1))}{\exists v\,\varphi(v) \vdash \langle\alpha^*\rangle\exists v{\le}0\,\varphi(v)}\,{}^{6}$$

$$(DI)\ \frac{\vdash \forall^\alpha\forall y_1..\forall y_k\,(\chi \to F'^{\theta_1}_{x'_1}\cdots^{\theta_n}_{x'_n})}{[\exists y_1..\exists y_k\,\chi]F \vdash [\exists y_1..\exists y_k\,(x'_1 = \theta_1 \wedge .. \wedge x'_n = \theta_n \wedge \chi)]F}\,{}^{7}$$

$$(DV)\ \frac{\vdash \exists\varepsilon{>}0\,\forall^\alpha\forall y_1..y_k\,(\neg F \wedge \chi \to (F' \ge \varepsilon)^{\theta_1}_{x'_1}\cdots^{\theta_n}_{x'_n})}{[\exists y_1..y_k\,(x'_1 = \theta_1 \wedge .. \wedge x'_n = \theta_n \wedge {\sim}F)]\chi \vdash \langle\exists y_1..y_k\,(x'_1 = \theta_1 \wedge .. \wedge x'_n = \theta_n \wedge \chi)\rangle F}\,{}^{8}$$

---

[1] $\Gamma_i \vdash \Delta_i$ are obtained from the subgoals of side deduction ($\star$) in Fig. 3.10, in which x is assumed to occur in first-order formulas only, as QE is then applicable. The side deduction starts from goal $\Gamma \vdash \Delta, \phi$ at the bottom (or $\Gamma, \phi \vdash \Delta$ for l$\forall$ and l$\exists$), where x does not occur in Γ, Δ using renaming.

[2] $\mathscr{J}_1 \vee \cdots \vee \mathscr{J}_n$ is a disjunctive normal form of the DJ-constraint $\mathscr{J}$.

[3] Rule applicable for any reordering of the conjuncts of the DJ-constraint where χ is jump-free.

[4] $\mathscr{D}_1 \vee \cdots \vee \mathscr{D}_n$ is a disjunctive normal form of the DA-constraint $\mathscr{D}$.

[5] $\mathscr{D}$ implies $\mathscr{E}$, i.e., satisfies the assumptions of Lemma 3.3.

[6] Logical variable v does not occur in α.

[7] Applicable for any reordering of the conjuncts where χ is non-differential. F is first-order without negative equalities, and F' abbreviates $D(F)$, with z' replaced with 0 for unchanged variables.

[8] Like DI, but F contains no equalities and the differential equations are Lipschitz continuous.

Fig. 3.9 Proof calculus for differential-algebraic dynamic logic (DAL)

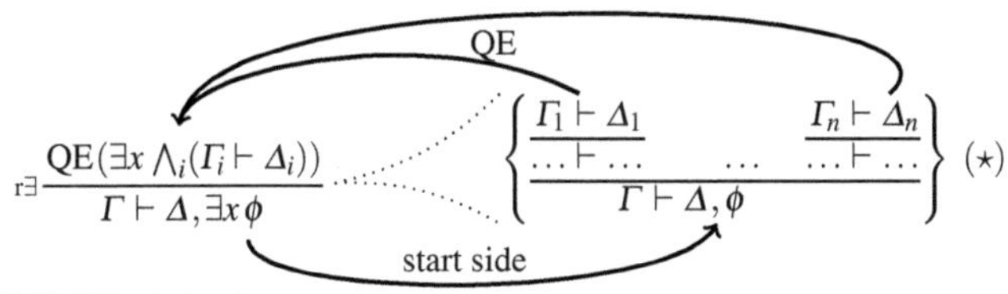

Fig. 3.10 Side deduction

$$([\cup]\Box)\ \frac{[\alpha]\pi \wedge [\beta]\pi}{[\alpha \cup \beta]\pi}\,^{1} \qquad (\langle\cup\rangle\Diamond)\ \frac{\langle\alpha\rangle\pi \vee \langle\beta\rangle\pi}{\langle\alpha \cup \beta\rangle\pi}\,^{1}$$

$$([;]\Box)\ \frac{[\alpha]\Box\phi \wedge [\alpha][\beta]\Box\phi}{[\alpha;\beta]\Box\phi} \qquad (\langle;\rangle\Diamond)\ \frac{\langle\alpha\rangle\Diamond\phi \vee \langle\alpha\rangle\langle\beta\rangle\Diamond\phi}{\langle\alpha;\beta\rangle\Diamond\phi}$$

$$([?]\Box)\ \frac{\phi}{[?\chi]\Box\phi} \qquad (\langle?\rangle\Diamond)\ \frac{\phi}{\langle?\chi\rangle\Diamond\phi}$$

$$([:=]\Box)\ \frac{\phi \wedge [x:=\theta]\phi}{[x:=\theta]\Box\phi} \qquad (\langle:=\rangle\Diamond)\ \frac{\phi \vee \langle x:=\theta\rangle\phi}{\langle x:=\theta\rangle\Diamond\phi}$$

$$([']\Box)\ \frac{[x'=\theta]\phi}{[x'=\theta]\Box\phi} \qquad (\langle'\rangle\Diamond)\ \frac{\langle x'=\theta\rangle\phi}{\langle x'=\theta\rangle\Diamond\phi}$$

$$([^{*n}]\Box)\ \frac{[\alpha;\alpha^*]\Box\phi}{[\alpha^*]\Box\phi} \qquad (\langle^{*n}\rangle\Diamond)\ \frac{\langle\alpha;\alpha^*\rangle\Diamond\phi}{\langle\alpha^*\rangle\Diamond\phi}$$

$$([^{*}]\Box)\ \frac{[\alpha^*][\alpha]\Box\phi}{[\alpha^*]\Box\phi} \qquad (\langle^{*}\rangle\Diamond)\ \frac{\langle\alpha^*\rangle\langle\alpha\rangle\Diamond\phi}{\langle\alpha^*\rangle\Diamond\phi}$$

[1] π is a trace formula and—unlike the state formulas ϕ and ψ—may thus begin with a temporal modality $\Box$ or $\Diamond$.

Fig. 4.3 Proof calculus for differential temporal dynamic logic (dTL)

MIX
Papier aus verantwortungsvollen Quellen
Paper from responsible sources
FSC® C105338

If you have any concerns about our products,
you can contact us on
ProductSafety@springernature.com

In case Publisher is established outside the EU,
the EU authorized representative is:
Springer Nature Customer Service Center GmbH
Europaplatz 3, 69115 Heidelberg, Germany

Printed by Libri Plureos GmbH
in Hamburg, Germany